T0392682

VOLUME ONE HUNDRED AND EIGHTY THREE

Progress in
MOLECULAR BIOLOGY AND TRANSLATIONAL SCIENCE

Dancing Protein Clouds: Intrinsically Disordered Proteins in the Norm and Pathology, Part C

VOLUME ONE HUNDRED AND EIGHTY THREE

Progress in
MOLECULAR BIOLOGY AND TRANSLATIONAL SCIENCE

Dancing Protein Clouds: Intrinsically Disordered Proteins in the Norm and Pathology, Part C

Edited by

VLADIMIR N. UVERSKY

Department of Molecular Medicine, Morsani College of Medicine, University of South Florida, Tampa, Florida, United States;
Laboratory of New Methods in Biology, Institute of Biological Instrumentation, Russian Academy of Sciences, Moscow, Russia

ELSEVIER

ACADEMIC PRESS
An imprint of Elsevier

Academic Press is an imprint of Elsevier
50 Hampshire Street, 5th Floor, Cambridge, MA 02139, United States
525 B Street, Suite 1650, San Diego, CA 92101, United States
The Boulevard, Langford Lane, Kidlington, Oxford OX5 1GB, United Kingdom
125 London Wall, London EC2Y 5AS, United Kingdom

First edition 2021

Notices
Knowledge and best practice in this field are constantly changing. As new research and experience broaden our understanding, changes in research methods, professional practices, or medical treatment may become necessary.

Practitioners and researchers must always rely on their own experience and knowledge in evaluating and using any information, methods, compounds, or experiments described herein. In using such information or methods they should be mindful of their own safety and the safety of others, including parties for whom they have a professional responsibility.

To the fullest extent of the law, neither the Publisher nor the authors, contributors, or editors, assume any liability for any injury and/or damage to persons or property as a matter of products liability, negligence or otherwise, or from any use or operation of any methods, products, instructions, or ideas contained in the material herein.

ISBN: 978-0-323-85299-9
ISSN: 1877-1173

For information on all Academic Press publications
visit our website at https://www.elsevier.com/books-and-journals

Publisher: Zoe Kruze
Acquisitions Editor: Ashlie M. Jackman
Developmental Editor: Jhon Michael Peñano
Production Project Manager: James Selvam
Cover Designer: Matthew Limbert

Typeset by STRAIVE, India

Working together
to grow libraries in
developing countries

www.elsevier.com • www.bookaid.org

Contents

Contributors

Brian J. Aneskievich
Department of Pharmaceutical Sciences, University of Connecticut, Storrs, CT, United States

Elisar Barbar
Department of Biochemistry and Biophysics, Oregon State University, Corvallis, OR, United States

Sveinn Bjarnason
Department of Biochemistry, Science Institute, University of Iceland, Reykjavík, Iceland

Yraima Cordeiro
Faculty of Pharmacy, Federal University of Rio de Janeiro, Rio de Janeiro, Brazil

Orkid Coskuner-Weber
Molecular Biotechnology, Turkish-German University, Sahinkaya Caddesi, Istanbul, Turkey

Mariana J. do Amaral
Faculty of Pharmacy, Federal University of Rio de Janeiro, Rio de Janeiro, Brazil

Zsuzsanna Dosztányi
Department of Biochemistry, ELTE Eötvös Loránd University, Budapest, Hungary

Mateusz Dyla
Department of Molecular Biology and Genetics, Aarhus University; The Danish Research Institute for Translational Neuroscience (DANDRITE), Nordic EMBL Partnership for Molecular Medicine, Aarhus, Denmark

Hebah Fatafta
Institute of Biological Information Processing (IBI-7: Structural Biochemistry), Forschungszentrum Jülich, Jülich, Germany

Heather M. Forsythe
Department of Biochemistry and Biophysics, Oregon State University, Corvallis, OR, United States

Annelies Haegeman
Flanders Research Institute for Agriculture, Fisheries and Food (ILVO), Plant Sciences Unit, Merelbeke, Belgium

Kyou-Hoon Han
Korea Research Institute of Bioscience and Biotechnology, Daejeon, South Korea

Eugénie Hébrard
PHIM, Plant Health Institute, IRD, Cirad, Université de Montpellier, INRAE, Institut Agro, Montpellier, France

Pétur O. Heidarsson
Department of Biochemistry, Science Institute, University of Iceland, Reykjavík, Iceland

Do-Hyoung Kim
Korea Research Institute of Bioscience and Biotechnology, Daejeon, South Korea

Magnus Kjaergaard
Department of Molecular Biology and Genetics, Aarhus University; The Danish Research Institute for Translational Neuroscience (DANDRITE), Nordic EMBL Partnership for Molecular Medicine; The Danish National Research Foundation Center for Proteins in Memory (PROMEMO), Aarhus, Denmark

Sébastien Massart
University of Liège (ULg), Department of Integrated and Urban Phytopathology, Gembloux, Belgium

Jordan McIvor
School of Chemical Science, University of Auckland, Auckland, New Zealand

Davide Mercadante
School of Chemical Science, University of Auckland, Auckland, New Zealand

Olga D. Novikova
G.B. Elyakov Pacific Institute of Bioorganic Chemistry, Far East Branch, Russian Academy of Sciences, Vladivostok, Russia

Mátyás Pajkos
Department of Biochemistry, ELTE Eötvös Loránd University, Budapest, Hungary

Sarah F. Ruidiaz
Department of Biochemistry, Science Institute, University of Iceland, Reykjavík, Iceland

Suman Samantray
Institute of Biological Information Processing (IBI-7: Structural Biochemistry), Forschungszentrum Jülich, Jülich; AICES Graduate School, RWTH Aachen University, Aachen, Germany

Abdallah Sayyed-Ahmad
Department of Physics, Birzeit University, Birzeit, Palestine

Rambon Shamilov
Graduate Program in Pharmacology and Toxicology, Department of Pharmaceutical Sciences, University of Connecticut, Storrs, CT, United States

Birgit Strodel
Institute of Biological Information Processing (IBI-7: Structural Biochemistry), Forschungszentrum Jülich, Jülich; Institute of Theoretical and Computational Chemistry, Heinrich Heine University Düsseldorf, Düsseldorf, Germany

Rachid Tahzima
University of Liège (ULg), Department of Integrated and Urban Phytopathology, Gembloux, Belgium

Vladimir N. Uversky
Department of Molecular Medicine, Morsani College of Medicine, University of South Florida, Tampa, Florida, United States; Laboratory of New Methods in Biology, Institute of Biological Instrumentation, Russian Academy of Sciences, Moscow, Russia

Olga Vinogradova
Department of Pharmaceutical Sciences, University of Connecticut, Storrs, CT, United States

Elena A. Zelepuga
G.B. Elyakov Pacific Institute of Bioorganic Chemistry, Far East Branch, Russian Academy of Sciences, Vladivostok, Russia

Preface: The Patchwork Quilt of Intrinsic Disorder

Protein intrinsic disorder is ubiquitous and inevitable. It cannot be ignored as it penetrates all the nooks of the protein universe and represents a universal lock pick that helps opening seemingly unopenable locks and aids explaining unexplainable. This volume represents a patchwork quilt of intrinsic disorder that covers various aspects of the protein intrinsic disorder phenomenon ranging from the roles of intrinsic disorder in liquid-liquid phase separation (LLPS) to analysis of the evolutionary peculiarities of intrinsically disordered proteins (IDPs) and intrinsically disordered regions (IDRs), to look for the functional roles of intrinsic disorder in nonspecific porins of Gram-negative bacteria and in the integral membrane proteins, to description of various approaches used for the molecular simulations of IDPs, to the analysis of the roles of prestructured motifs (PreSMos), to emphasizing the roles of "dancing duplexes" in biology and diseases, to description of the roles of long IDRs in functionality of A-kinase-anchoring proteins (AKAPs), to the representation of the nucleoprotein landscape of the eukaryotic cell nucleus, and to the analysis of the genomic signatures preserved in the orthologous IDRs of plant viral proteins.

Chapter 1 written by Mariana J. do Amaral and Yraima Cordeiro[1] represents an attempt to use protein intrinsic disorder and liquid-liquid phase separation phenomena to answer some of the existential questions related to prion protein.[1] It seems that the major unsolved mysteries of the exciting prion world are linked to its intrinsically disordered nature. This includes the lack of the high-resolution atomic structures for either full-length endogenous human PrP^C or isolated infectious PrP^{Sc} particles, the lack of a complete understanding of the multifunctionality and binding promiscuity of this protein, mechanisms of its misfolding, strain phenomenon (associated with the heterogeneity of PrP^{Sc} conformations), and peculiar transmissibility of the misfoldedness to the unaffected prion protein molecules, as well as peculiarities of its cellular distribution, the presence of several PrP fragments originating from the conserved proteolytic cleavage, and the existence of a multitude of various posttranslational modifications. Intrinsic disorder nature of prion protein may also define its ability to undergo LLPS, a phenomenon behind the biogenesis of membrane-less organelles.[1]

In Chapter 2, Mátyás Pajkos and Zsuzsanna Dosztányi analyze the evolutionary peculiarities of the intrinsically disordered proteins (IDPs) or intrinsically disordered regions (IDRs).[2] The authors point out that nature uses different rules in the evolution of ordered and disordered proteins. In fact, since the functions of ordered proteins are dependent on their unique structures, the need for the preservation of such unique functional structures serves as the major evolutionary component of such proteins. On the other hand, evolutionary pressure shaping functionality of structureless IDPs/IDPRs is very different, and the authors describe the different types of evolutionary behavior of these proteins and regions and their associated functions.[2]

In Chapter 3, Olga D. Novikova, Vladimir N. Uversky, and Elena A. Zelepuga focus on the various aspects of nonspecific porins of Gram-negative bacteria and the roles of their IDRs in functional properties of these proteins and their abilities to form functional bacterial amyloids.[3] The author use several illustrative examples of the nonspecific porins (such as *Marinomonas primoryensis* MpOmp, *Yersinia pseudotuberculosis* YpOmpF and YpOmC, *Yersinia ruckeri* YrOmpF and YrOmpC, and *Escherichia coli* EcOmpF and EcOmpC) to show that the conformational plasticity of the tertiary structure of porins plays a special role in the manifestation of various types of biological activity by these proteins. In fact, increased conformational plasticity was found in regions, the functional importance of which has long been reliably confirmed as a result of numerous experimental studies. The authors also claim that all these proteins contain significant levels of intrinsic disorder, and their disorder status varies not only within porin types but also within bacteria species.[3] Finally, the authors emphasize that the formation of amyloid-like structures by the nonspecific OM pore-forming proteins of Gram-negative bacteria uncovers a new type of β-structured transmembrane proteins from prokaryotes capable of fibrillation.[3]

In Chapter 4, Brian J. Aneskievich, Rambon Shamilov, and Olga Vinogradova continue with the subject of the prevalence and the overall functional importance of IDR in integral membrane proteins (IMPs).[4] Although many receptor functions of IMPs can be described within the classical "lock-and-key" model, these proteins are known to contain substantial levels of intrinsic disorder. The authors provide an overview of the potential functional advantages of intrinsic disorder in selected single-span IMPs and also consider the possible conformational impacts of disease-associated mutations within IDRs of these proteins. Furthermore, the authors emphasize that building of the more realistic and complete mechanistic models of disorder-containing IMPs requires a complex experimental approach integrating biophysical and computational methods.[4] This more complete view of IMPs is expected to help in improving the conformational flexibility and functionality of these important proteins in both normal and pathological settings.

In Chapter 5, Hebah Fatafta, Suman Samantray, Abdallah Sayyed-Ahmad, Orkid Coskuner-Weber, and Birgit Strodel provide a comprehensive overview of various approaches used for the molecular simulations of IDPs.[5] Based on the premises that conformational flexibility and related entropy-driven motions are crucial for various functions of IDPs, that many IDPs/IDRs can undergo functional disorder-to-order transitions coupled to their binding to specific interaction partners, that the reliable characterization of these proteins requires a methodological utilization of both experimental and theoretical tools, and that the outputs of computational studies depend strongly on the chosen force field parameters and simulation techniques, the authors provide an overview of all-atom force fields developed recently for IDPs, and present molecular dynamics-based simulation methods that are capable of generating IDP conformational ensembles and also provide means for the characterization of functional disorder-to-order transitions.[5]

In Chapter 6, Do-Hyoung Kim and Kyou-Hoon Han focus on the analysis of the roles of prestructured motifs (PreSMos), which are transient secondary structure elements observed in many IDPs/IDRs serving as protein target-binding hot spots.[6] Similar to the binding pockets of globular proteins that preexist before target binding, PreSMos are already present in the target-unbound state of IDPs/IDRs, thereby serving IDP "active sites," being, however, not a spatially assembled binding pocket within the unique structure of a globular protein, but rather a functional secondary structural motif transiently populated within the conformational ensemble of IDPs/IDPRs. The existence of such PreSMos indicates that even highly disordered proteins are not entirely structureless, but might contain some transient structures that (being present prior to the binding to a partner) are conformationally selected by a partner, being then stabilized into the more ordered bound structures as a result of complex formation. Therefore, PreSMos-containing IDPs bind their partners via the combined conformational selection and induced fit mechanism and acts as "structural rationale" for the functionality of unstructured proteins.[6]

In Chapter 7, Heather M. Forsythe and Elisar Barbar emphasize the roles of "dancing duplexes" in biology and diseases.[7] The dancing protein duplex is defined as a flexible homodimer that contains structured domains separated by long IDRs, which is formed via self-association domains or using some other means for strong protein interactions. The major focus of the chapter is on IDP duplexes capable of binding to a dynamic hub protein—dynein light chain 8 (LC8)—that is a highly conserved and functionally diverse protein interacting with IDRs of target proteins containing a highly conserved "TQT" motif. Despite these rather simple founding principles, LC8-binding IDP duplexes are very diverse and can be grouped into several classes, where one can find IDP duplexes containing a self-association domain that is extended by LC8 binding, IDP duplexes without self-association domains but capable of dimerization through binding to several copies of LC8, and multivalent LC8-binding IDP duplexes that also have self-association domains.[7] Since many proteins have domain organization similar to that of IDP duplexes, the authors indicate that such functional units are conserved and can be used as new therapeutic targets.[7]

The subject of Chapter 8 by Mateusz Dyla and Magnus Kjaergaard is A-kinase-anchoring proteins (AKAPs), a diverse class of anchoring proteins regulating protein kinase A (PKA).[8] AKAPs are responsible for targeting PKA to different activators and substrates of this important kinase. The authors show that a characteristic feature of the most AKAPs is the presence of long IDRs that serve as crucial means for the tethered phosphorylation, where substrates are tethered to the kinase, thereby forming grounds for the intracomplex enzymatic reactions, where many phosphorylation events might happen while PKA stays physically connected to its substrates via AKAPs. The authors suggest that the allosteric regulation of PKA signaling is, at least in part, dependent on the structure of the AKAP signaling complex.[8] Interestingly, all the players engaged in the formation of PKA-AKAP signaling complex are shown to contain long IDRs serving as linkers between functional domains. Importantly, the reach zone of the kinase domain of the dynamic AKAP-PKA complexes is shown to be determined by the length of flexible AKAP linkers, with longer linkers defining larger reach zone.[8]

In Chapter 9, Sveinn Bjarnason, Sarah F. Ruidiaz, Jordan McIvor, Davide Mercadante, and Pétur O. Heidarsson looked at the nucleoprotein landscape of the eukaryotic cell nucleus.[9] Many nuclear proteins are intrinsically disordered, and some of them are responsible for the genome compaction at the first fundamental level, the nucleosome. The authors discuss the challenges and recent advances in understanding structural disorder within the context of transcription. They also provide a description of how cellular transcription patterns determining the dynamics, architecture, and epigenetic regulation of the genetic material are controlled by the intrinsic disorder-based interactions with and within nucleosomes. The overall goal of this chapter is to emphasize the prevalence of intrinsic disorder within the nucleus, show that IDPs play key roles in maintaining the genome and regulating its readout, underline the important interplay between IDPs and the nucleosomal landscape that leads to functional outputs, and describe major challenges and developments in the structural characterization of disordered nuclear proteins both experimentally and computationally.[9] The authors conclude the chapter with a very important message that better understanding of the disordered nuclear milieu is greatly dependent on a better synergy between experiments and simulations.[9]

Based on the analysis of the plant viromes, in Chapter 10, Rachid Tahzima, Annelies Haegeman, Sébastien Massart, and Eugénie Hébrard propose an evolutionary quantitative proteomic approach to understand some of the genomic signatures preserved in the amino acid sequences of orthologous IDRs from plant viruses.[10] The final chapter begins with a brief introduction of the proteomes of plant viruses, their evolution, unique modular organization, and introduction of various transmission mechanisms. Then, the prevalence and variability of intrinsic disorder in plant viromes are discussed together with a comprehensive coverage of the functionality of plant viral proteins and emphasis on an important point that in plant viruses, intrinsic disorder represents an important evolutionary functional modulator of adaptability.[10] The authors provide a comprehensive evolutionary analysis of plant viromes and show that IDPs in plant viruses are characterized by multiple specific and phylogenomically preserved features, which are correlated with the genome size and partition or replication strategies, and show that the viral multifunctionality is modulated by different IDPs with similar evolutionary signatures. Based on these and many other important observations, the authors propose a model, where evolutionary proteome-wide intrinsic disorder-associated patterns can be used for the elucidation of the plant virus transmission mechanisms.[10]

VLADIMIR N. UVERSKY
Department of Molecular Medicine and Byrd Alzheimer's Research Institute, Morsani College of Medicine, University of South Florida, Tampa, FL, United States

References

1. do Amaral MJ, Cordeiro Y. Intrinsic disorder and phase transitions: pieces in the puzzling role of the prion protein in health and disease. In: Uversky VN, ed. *Dancing Protein Clouds: Intrinsically Disordered Proteins in the Norm and Pathology. Part C.* Philadelphia, PA, USA: Elsevier Science & Technology Books; 2021. Progress in Molecular Biology and Translational Science. Teplow D, series ed.

2. Pajkos M, Dosztányi Z. Functions of intrinsically disordered proteins through evolution-ary lenses. In: Uversky VN, ed. *Dancing Protein Clouds: Intrinsically Disordered Proteins in the Norm and Pathology. Part C.* Philadelphia, PA, USA: Elsevier Science & Technology Books; 2021. Progress in Molecular Biology and Translational Science. Teplow D, series ed.

3. Novikova OD, Uversky VN, Zelepuga EA. Non-specific porins of Gram-negative bac-teria as proteins containing intrinsically disordered regions with amyloidogenic potential. In: Uversky VN, ed. *Dancing Protein Clouds: Intrinsically Disordered Proteins in the Norm and Pathology. Part C.* Philadelphia, PA, USA: Elsevier Science & Technology Books; 2021. Progress in Molecular Biology and Translational Science. Teplow D, series ed.

4. Aneskievich BJ, Shamilov R, Vinogradova O. Intrinsic disorder in integral membrane proteins. In: Uversky VN, ed. *Dancing Protein Clouds: Intrinsically Disordered Proteins in the Norm and Pathology. Part C.* Philadelphia, PA, USA: Elsevier Science & Technology Books; 2021. Progress in Molecular Biology and Translational Science. Teplow D, series ed.

5. Fatafta H, Samantray S, Sayyed-Ahmad A, Coskuner-Weber O, Strodel B. Molecular simulations of IDPs: from ensemble generation to IDP interactions leading to disorder-to-order transitions. In: Uversky VN, ed. *Dancing Protein Clouds: Intrinsically Disordered Proteins in the Norm and Pathology. Part C.* Philadelphia, PA, USA: Elsevier Science & Technology Books; 2021. Progress in Molecular Biology and Translational Science. Teplow D, series ed.

6. Kim D-H, Han K-H. Target-binding behavior of IDPs via pre-structured motifs. In: Uversky VN, ed. *Dancing Protein Clouds: Intrinsically Disordered Proteins in the Norm and Pathology. Part C.* Philadelphia, PA, USA: Elsevier Science & Technology Books; 2021. Progress in Molecular Biology and Translational Science. Teplow D, series ed.

7. Forsythe HM, Barbar E. The role of dancing duplexes in biology and disease. In: Uversky VN, ed. *Dancing Protein Clouds: Intrinsically Disordered Proteins in the Norm and Pathology. Part C.* Philadelphia, PA, USA: Elsevier Science & Technology Books; 2021. Progress in Molecular Biology and Translational Science. Teplow D, series ed.

8. Dyla M, Kjaergaard M. Intrinsic disorder in protein kinase A anchoring proteins signal-ing complexes. In: Uversky VN, ed. *Dancing Protein Clouds: Intrinsically Disordered Proteins in the Norm and Pathology. Part C.* Philadelphia, PA, USA: Elsevier Science & Technology Books; 2021. Progress in Molecular Biology and Translational Science. Teplow D, series ed.

9. Bjarnason S, Ruidiaz SF, McIvor J, Mercadante D, Heidarsson PO. Protein intrinsic dis-order on a dynamic nucleosomal landscape. In: Uversky VN, ed. *Dancing Protein Clouds: Intrinsically Disordered Proteins in the Norm and Pathology. Part C.* Philadelphia, PA, USA: Elsevier Science & Technology Books; 2021. Progress in Molecular Biology and Translational Science. Teplow D, series ed.

10. Tahzima R, Haegeman A, Massart S, Hébrard E. Flexible spandrels of the global plant virome: proteomic-wide evolutionary patterns of structural intrinsic protein disorder elucidate modulation at the functional virus-host interplay. In: Uversky VN, ed. *Dancing Protein Clouds: Intrinsically Disordered Proteins in the Norm and Pathology. Part C.* Philadelphia, PA, USA: Elsevier Science & Technology Books; 2021. Progress in Molecular Biology and Translational Science. Teplow D, series ed.

CHAPTER ONE

Intrinsic disorder and phase transitions: Pieces in the puzzling role of the prion protein in health and disease

Mariana J. do Amaral* and Yraima Cordeiro*

Faculty of Pharmacy, Federal University of Rio de Janeiro, Rio de Janeiro, Brazil
*Corresponding authors: e-mail address: marianajamaral@yahoo.com.br; yraima@pharma.ufrj.br

Contents

Abstract

After four decades of prion protein research, the pressing questions in the literature remain similar to the common existential dilemmas. *Who am I?* Some structural characteristics of the cellular prion protein (PrP^C) and scrapie PrP (PrP^{Sc}) remain unknown: there are no high-resolution atomic structures for either full-length endogenous human PrP^C or isolated infectious PrP^{Sc} particles. *Why am I here?* It is not known why PrP^C and PrP^{Sc} are found in specific cellular compartments such as the nucleus; while the

Progress in Molecular Biology and Translational Science, Volume 183
ISSN 1877-1173
https://doi.org/10.1016/bs.pmbts.2021.06.001

physiological functions of PrP^C are still being uncovered, the misfolding site remains obscure. *Where am I going?* The subcellular distribution of PrP^C and PrP^{Sc} is wide (reported in 10 different locations in the cell). This complexity is further exacerbated by the eight different PrP fragments yielded from conserved proteolytic cleavages and by reversible post-translational modifications, such as glycosylation, phosphorylation, and ubiquitination. Moreover, about 55 pathological mutations and 16 polymorphisms on the PrP gene (*PRNP*) have been described. Prion diseases also share unique, challenging features: strain phenomenon (associated with the heterogeneity of PrP^{Sc} conformations) and the possible transmissibility between species, factors which contribute to PrP undruggability. However, two recent concepts in biochemistry—intrinsically disordered proteins and phase transitions—may shed light on the molecular basis of PrP's role in physiology and disease.

1. Introduction

1.1 Paradigm shifts in biochemistry are present in prion research: A protein as infectious agent, intrinsic disorder, and phase separation

A half-century has passed since the revolutionary idea emerged of a protein directly triggering a group of central nervous system diseases called transmissible spongiform encephalopathies (TSE).[1,2] This *protein-only* hypothesis challenged the nucleic acid informational flow (DNA➔RNA➔protein), postulated in the central dogma of molecular biology.[a] In 1982, Stanley Prusiner identified the nature of TSEs,[4] supporting the conformational templating proposed by Griffith,[1] according to which the cellular prion protein (PrP),[b] highly expressed in neurons (and other cell types), undergoes an auto-catalyzed misfolding into a self-propagating non–native isoform named scrapie[c] (PrP^{Sc}).[4] This concept shed light on several other proteinopathies, not limited to the central nervous system, which are currently termed prion–like diseases.[5] The so-called protein misfolding diseases include p53 mutant-associated cancers (breast tissue),[6] immunoglobulin light chain amyloidosis (systemic), Alzheimer's disease (amyloid-β and tau aggregates found in the brain), amylin-associated type II diabetes mellitus (endocrine pancreas) and other disorders (reviewed in ref. 7). Paradoxically, even as prion research experienced a boom, especially in the 1980s with the creation of the first transgenic prion mouse model,[8] a key field in understanding prion

---

[a] For a complete historical overview of prion diseases research, see ref. 3.

[b] "***pro***teinaceous ***in***fectious particles" (acronym for prion, coined by Stanley Prusiner).

[c] Scrapie refers to the pathognomonic sign of intense pruritus in sheep TSE. Animals tend to scrape against fences and objects to relieve their itchy skin.

behavior went largely ignored: intrinsically disordered proteins (IDPs) were treated by many biochemists as useless proteins[9,10] with no relevance, as noted by Dobson.[11] Indeed, IDPs were disregarded in theoretical studies on the nature of protein folding and stability.[12,13] Furthermore, high purity recombinant production of IDPs was hampered by their lack of a stable tertiary structure, which facilitates proteolytic cleavage and aggregation, as well as recurrent co-purification with other macromolecules such as nucleic acids. However, since the mid-1990s, investigations of genomic sequences have uncovered the widespread occurrence of IDPs across different species within the three domains of life. Around half of the mammalian proteome has disordered regions with at least 30 residues, as demonstrated in pioneering studies from Dunker's group.[14–17] Additionally, the regulatory role of IDPs in many cellular signaling pathways—especially through nucleic acid-related functions[18,19]—and their enrichment in neurodegenerative diseases[20] have helped speed up their biophysical characterization. Since NMR analysis[d] of full-length recombinant murine PrP (rPrP$^{23-231}$)[21] revealed that half of the protein (spanning from residues 23–120) is disordered, pioneering bioinformatic tools studying the "protein disordered code" have made use of PrP to validate algorithms with different training sets.[15] The disorder in disorders (D^2) concept has found links between several culprits of proteinopathies and their enrichment in intrinsically disordered regions.[20,22] Recently, cell biology has been faced with a novel framework related to nucleic acid-binding IDPs involved in diseases since they have been found to drive liquid-liquid phase separation (LLPS).[23] The formation of protein-rich condensates or granules by LLPS underpins the formation of membraneless organelles (MLOs) *in vivo*.[24,25] PrP has emerged as one of the proteins that form liquid-like condensates,[26] especially in the presence of nucleic acids.[27] The highly dynamic compartments formed by DNA/RNA-PrP may explain various functions of PrP^C and its misfolding. If so, condensates may be part of the PrP^{Sc} pathway and may therefore represent novel targetable structures for therapeutics.

1.2 The prion protein

The human PrP^C (26 kDa; 253 amino acids; UniProt ID P04156) is a soluble, monomeric protein with a long-disordered N-terminal domain

---

[d] Consult other NMR structures for the human PrP C-terminal domain, which has a well-defined three-dimensional structure, at PDB 1HJM (obtained at pH 7.0), PDBs 1QM0, and 5YJ5 (at pH 4.5) PDB 2LSB at pH 5.5.

followed by a globular domain containing three α-helical segments and two short β-strands.[28] The only intramolecular disulfide bridge (Cys179 in α-helix 2 bound to Cys214 in α-helix 3) seems necessary for the conversion event, since the addition of a reducing agent or cysteine mutants result in a failure to generate PrP^{Sc} from PrP.[29,30] Interestingly, the overall three-dimensional conformation of PrP^{C} is maintained across different species of vertebrates (100-residue unfolded N-terminus followed by a globular domain).[21,31–33] Although PrP^{C} has been primarily studied in the central nervous system (CNS), it is expressed ubiquitously in extra-neuronal tissues[34] including skeletal muscle,[35] blood (with the highest expression in lymphocytes,[36]) and endothelial cells[37] to which PrP^{Sc} has been found attached.[38] Despite being regarded as a plasma membrane protein anchored by a glycosylphosphatidylinositol (GPI) unit, about 20% of PrP^{C} is found in the cytosol (the endoplasmic reticulum signal peptide is suggested to be inefficient).[39] In fact, PrP traffics through several cellular compartments, endocytic vesicles, mitochondria, and the nucleus (reviewed in ref. 40).

PrP^{C} does not have a specific physiological function, as its dynamic, intrinsically disordered nature suggests. PrP^{C} is also a multivalent protein, displaying short motifs that bind multiple partners with high affinity.[41,42] Additionally, PrP^{C} can undergo physiological post-translational modifications (PTMs): glycosylation (at asparagine residues 181 and 197; believed to increase folding and modulate interactions),[42,43] phosphorylation (at serine 43; PrP^{Sc} was reported to be phosphorylated in the nucleus),[44] ubiquitination (at lysine 194; for turn-over)[45] and four proteolytic cleavages (β-cleavage; α-cleavage; γ-cleavage and shedding).[46]

PrP^{C} has been associated with several physiological roles,[42] such as copper sensing and trafficking,[47] as well as the unwinding of G-quadruplex structures of its own gene promoter (auto-transcriptional regulation).[48–50] PrP^{C} also influences cellular fates (survival, death, proliferation, differentiation, and myelin maintenance) and systemic effects, such as memory consolidation.[42] Since PrP^{C} engages in various signaling pathways and interacts with different types of ligands, it is regarded as a scaffold protein,[42] acting as a hub to assemble multicomponent molecular complexes. The liquid–liquid phase separation phenomenon thus provides a useful lens for understanding its myriad functions: PrP^{C} might be the "driver" scaffold, with ligands and solutes diffusing in and out of the liquid droplets once specific reactions have taken place. Moreover, PrP condensation is possibly modified by PTMs and has already proven to be finely controlled by ligands *in vitro*, especially by specific nucleic acids, which can bind PrP with micro to nanomolar affinity.

In mammals, the misfolding of PrP^C into the pathological PrP^{Sc} leads to the development of TSEs (or prion diseases), a group of acquired, sporadic, or genetic fatal neurodegenerative disorders with an average survival time of 1 year after diagnosis depending on disease subtype (reviewed in ref. 51). Most TSE cases are sporadic, with Creutzfeldt–Jakob disease (CJD) having the greatest incidence (one case per million people per year).[52] The main histopathological features are spongiform degeneration (vacuoles with a size of 20–1500 μm)[53] and accumulation of PrP^{Sc} plaques. PrP^{Sc}, which is characterized by β-sheet-rich aggregates/multimers, has a self-propagating ability (i.e., PrP^{Sc} catalyzes the conversion of native PrP^C into PrP^{Sc}), and, unlike PrP^C, is insoluble in aqueous solvent, infectious and resistant to proteases.[54]

In addition to the links between PrP^{Sc} and TSEs, recent studies have shown a direct connection between PrP and other age-related disorders such as Alzheimer's (AD) and Parkinson's diseases (PD).[55–59] Some studies have reported that PrP^C acts as a receptor for neurotoxic β-sheet-rich oligomers of α-synuclein, tau, and amyloid-β (Aβ).[60,61] The finding that PrP is involved in pathogenic signaling pathways and that its aggregation or fibrillation might synergize with other misfolded proteins places PrP at the center of neurodegenerative diseases.[62,63]

1.3 The liquid-liquid phase separation phenomenon

Along with membrane-bound compartments, membraneless organelles (MLOs), which are visualized in the cellular *milieu* as micrometer-sized spherical condensates, classically serve to organize biomolecular content. MLOs are numerous in eukaryotes, with about 18 known MLOs in the nucleus, which range from the nucleolus to paraspeckles, along with stress granules and P bodies in the cytoplasm, among others (reviewed in refs. 64, 65). An example of MLO are stress granules (SGs) that form upon environmental or intrinsic cellular stresses, such as oxidative damage, and are rapidly dissolved once the cellular homeostasis is reestablished. These granules sequester specific mRNA and RNA-binding proteins resulting in a pro-survival effect, although the precise functions of SGs remain little understood (reviewed in ref. 66).

The formation and disassembly of MLOs are finely tuned due to the availability of certain NAs and post-translational modifications of the "driver" biomolecules, namely those that form the scaffold. The MLOs are comprised of a scaffold of proteins co-mixed with nucleic acids that form a specialized environment that facilitates a range of biochemical reactions from mRNA translation to synaptic signaling (reviewed in ref. 67). These non-membranous bodies exhibit biophysical properties distinct from lipid vesicles, such as direct contact with their intracellular surroundings and the characteristic fluidity of liquids and are assembled through liquid-liquid phase separation (LLPS) or liquid-liquid demixing.[24,65,68] The main molecular "driver" of this phenomenon consists of proteins containing intrinsically disordered regions (IDRs), oligomerization domains, and NA-binding properties. The LLPS is a physicochemical process in which a binary mixture separates into a dense and a light phase, such as oil droplets dispersed in water. In the case of phase-separating proteins, the dense phase (protein-rich condensates) can reach a concentration 10–300 times higher than the light phase consisting of the same molecules dispersed at a lower concentration.[69] Additionally, RNA can be concentrated in liquid droplets by five orders of magnitude.[70,71]

Recently, the field of protein phase transitions[e] has focused on examining the structure in the dense phase (either liquid-like, hydrogel, or solid-like) at the atomic level[26,72–74] and cracking the "LLPS code." Phase separation was initially envisioned as a property of disordered regions whereby weak homotypic (intra- and/or intermolecular interactions of the "driver" biomolecule) or heterotypic interactions (interactions between proteins and NAs) were responsible for maintaining the phase-separated state. However, the molecular determinants of LLPS are complex, and the dissociation between a solely IDP-driven phenomenon is clear since the condensate-forming protein can have a compact three-dimensional structure when phase-separated.[26,73,75] Collective interactions play an important role in the formation of liquid droplets. This is because the so-called "driver" proteins are usually multivalent (i.e., contain multiple binding points within disordered regions). The formation of liquid droplets can be achieved through the presence of **S**hort **Li**near amino acid **M**otifs (SLiMs) in low-complexity regions.[76] SLiMs are formed by polar amino acid residues interspersed with aromatic or glycine residues in tandem (e.g., RGG and RG). Curiously, liquid droplets can transit to hydrogels, which are reversible upon temperature increment, dissolved by 1,6-hexanediol, and are detergent-sensitive. The key interactions that drive reversible gels (also called reversible amyloid cores, RACs) are kinked β-sheets in the place of

---

[e] Any change in molecular phase, such as from soluble protein to solid aggregates, from liquid condensates to hydrogels, from liquid condensates to a soluble one-phase regimen, etc. Considering the metastable characteristic of liquid condensates, these can transit to gel-like structures and aggregates. Transitions of protein-rich condensates are altered through ligand binding, post-translational modifications, splicing variations, and mutations, for instance.

the highly stable steric zippers found in irreversible fibrils. In these unusual structures, termed low–complexity aromatic–rich kinked segments (LARKS), torsions are evident in the aromatic amino acids and glycine residues that compose the backbone when these interact with other chains, causing the former to serve as weak adhesion points. Notably, PrP is one of the 400 proteins in the human proteome enriched in LARKS,[77] denoted by the five octarepeats (Fig. 1). Of the top eight gene ontology terms related to LARKS-containing proteins, seven are related to NA-related functions such as mRNA transport and stress granule assembly.[77]

The Flory–Huggins theory, formulated for a mixture of homopolymers, has been used to understand the phenomenon applied to proteins or NAs that phase separate. However, the "monotony" of homopolymers is not ideal for understanding biomolecular phase behavior since proteins and

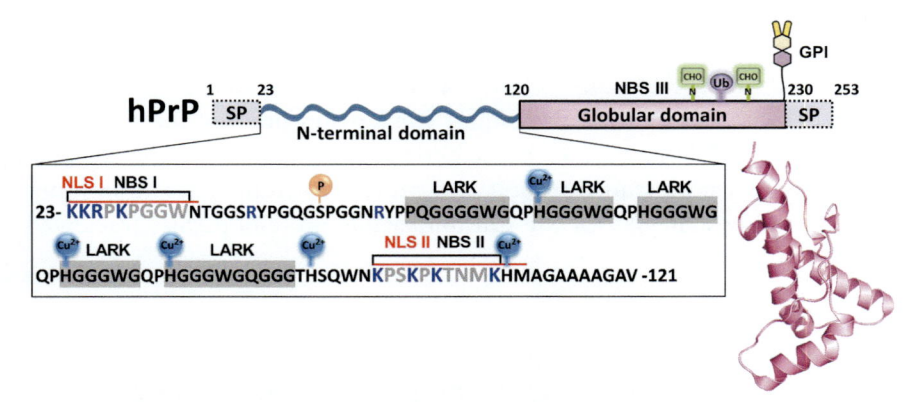

Fig. 1 The prion protein N-terminal primary structure encodes LLPS and possible modulators include post-translational modifications and ligands. Domain organization of human PrP showing the two signal peptides (SP) at the N- and C-extremities; the unfolded N-terminal segment comprising residues 23–120 and the C-terminal globular domain (residues 120 to 230). The N-terminus contains five octarepeats (gray highlight) predicted to be low-complexity aromatic-rich kinked segments (LARKS), likely forming transient β-sheet contacts involved in LLPS. The nucleic acid-binding sites (NBS I and NBS II) are clusters of lysine residues in which the suggested nuclear import signals (NLS) are also located (marked by a red line). The C-terminal domain has been involved in DNA interaction (NBS III). Positive (blue), aromatic amino acids and prolines can potentially interact through cation-pi, pi-pi, hydrophobic and electrostatic contacts, driving LLPS. Sites that coordinate copper ions (Cu^{2+}) are assigned, as well as post-translational modifications (PTMs) such as phosphorylation (P), N-glycosylation sites (CHO), ubiquitination (Ub) and the glycosylphosphatidylinositol (GPI) that anchors to the plasma membrane. The 3D structure of the C-terminal domain (residues 120 to 231) is shown in pink (PDB: 1HJM).

NAs are highly heterogenous, with a combination of different (non)structural elements. Currently, more accurate models have been developed based on associative polymers, which are macromolecules that contain adhesive patches or "stickers" (in the case of the wholly disordered PrP N-terminus, analogous to SLiMs). This model also considers the effect of loops called "spacers" and is, therefore, termed the "stickers-and-spacers" model. See Harmon et al.[76] and Ruff et al.[78] for mechanistic descriptions of homotypic-driven and ligand-based phase transitions, respectively.

However, the same "non-equilibrium" state of protein liquid condensates that provides them with such dynamic properties (rapid assembly/disassembly, e.g., deforming, coalescing, and undergoing fusion/fission events) can promote the transition to irreversible solids upon interaction with certain ligands. The uncontrolled formation of condensates has been linked to the aggregation into highly stable amyloids, which is implicated in cellular dysfunction and disease, as shown for cancer-associated (e.g., p53[79–81] and SPOP[82]) and neurodegeneration-associated proteins (e.g., TDP-43,[83] FUS,[84,85] tau,[86,87] huntingtin,[88] and α-synuclein[89]). There is solid evidence to indicate that liquid condensates precede the formation of aggregates. Therefore, new therapeutic approaches to keep the condensates' liquid nature (functionality) and prevent aberrant phase transitions are promising strategies for the still untreatable proteinopathies (reviewed in ref. 90).

2. Considerations on intrinsic disorder and phase separation ability of the prion protein family

The prion gene family includes three proteins: PrP, Doppel (Dpl, downstream prion protein-like), and Shadoo[f] (Sho, shadow of prion protein). All family members have conserved characteristics: signal peptides at the N- and C-terminus and the potential to be membrane-anchored and undergo N-glycosylation. In the biosynthesis of these proteins, the N-terminal signal peptide targets the protein to the endoplasmic reticulum (ER), while the C-terminal motif drives PrP to the plasma membrane, followed by addition of the glycosylphosphatidylinositol (GPI) anchor.[91] Despite the low primary sequence identity between PrP and their two paralogs (16.5% relative to Dpl and 13.7% relative to Sho based on studies using human sequences aligned using Clustal Omega software[92]), they do share

---

[f] Shadoo comes from the word for shadow in Japanese.

common biochemical and structural features. PrP contains a disordered N-terminus (residues 23–121) with a random-coil conformation, followed by a folded domain (residues 121–230; a "three-helix bundle" conformation plus short regions of antiparallel β-strands).

Several algorithms predict that Sho is fully disordered (Fig. 2, middle panel). This is supported by circular dichroism experiments showing characteristic unordered spectra.[93] Due to its structural plasticity, the atomic resolution structure of Sho is most likely to be hampered. Mature Sho has an arginine-rich basic region covering half of its sequence. Moreover, Sho does not have any disulfide bridges and only two small clusters of residues with opposite charges (Fig. 3, bottom graph).

The physicochemical character of Sho's primary structure resembles the N-terminal region of PrP, while the secondary structure elements and tertiary folding of Dpl are homologous to the PrP C-terminal domain. The main exception is that Dpl has two disulfide bonds, making the structure more compact.[94] Notably, the expression of Dpl is mostly confined to the testis in adults, whereas PrP and Sho are highly expressed in the CNS.[95] Considering the structure-function relationship and the functional redundancy in biology, one would expect that the depletion of PrP^C could be replenished by Sho (mediating the N-terminal functions of PrP) and Dpl (mediating the C-terminal functions). This speculation is based on studies that have shown an interplay between functional and pathological outcomes of the two paralogs. For example, PrP knockout mice have increased Dpl transcripts[96] and Sho has been shown to play a role in neuroprotection.[97] Indeed, PrP and Sho might share important biochemical behavior: nucleic acid-binding, nuclear localization, and potential condensation, as we discuss next.

Among the three proteins, the various disorder algorithms identify only PrP and Sho as having long predicted regions of intrinsic disorder (Fig. 2). It is estimated that approximately 90% of proteins linked to transcriptional regulation have long segments of intrinsic disorder.[98] The most similar region of the primary sequence between PrP and Sho is the alanine-rich hydrophobic region (residues 120–132 in the case of PrP and 68–87 in the case of Sho). This region is a putative dimerization domain[99] and encompasses the neurotoxic peptide 106–126, attributed as the minimal amyloidogenic region of PrP.[100] Note that the Sho N-terminal region contains five repetitions of RGG/RG (arginine/glycine repeats) and other arginine residues that most likely contribute to phase separation mainly mediated by electrostatic interactions (Fig. 3). The arginine 28 resides within a post-translational

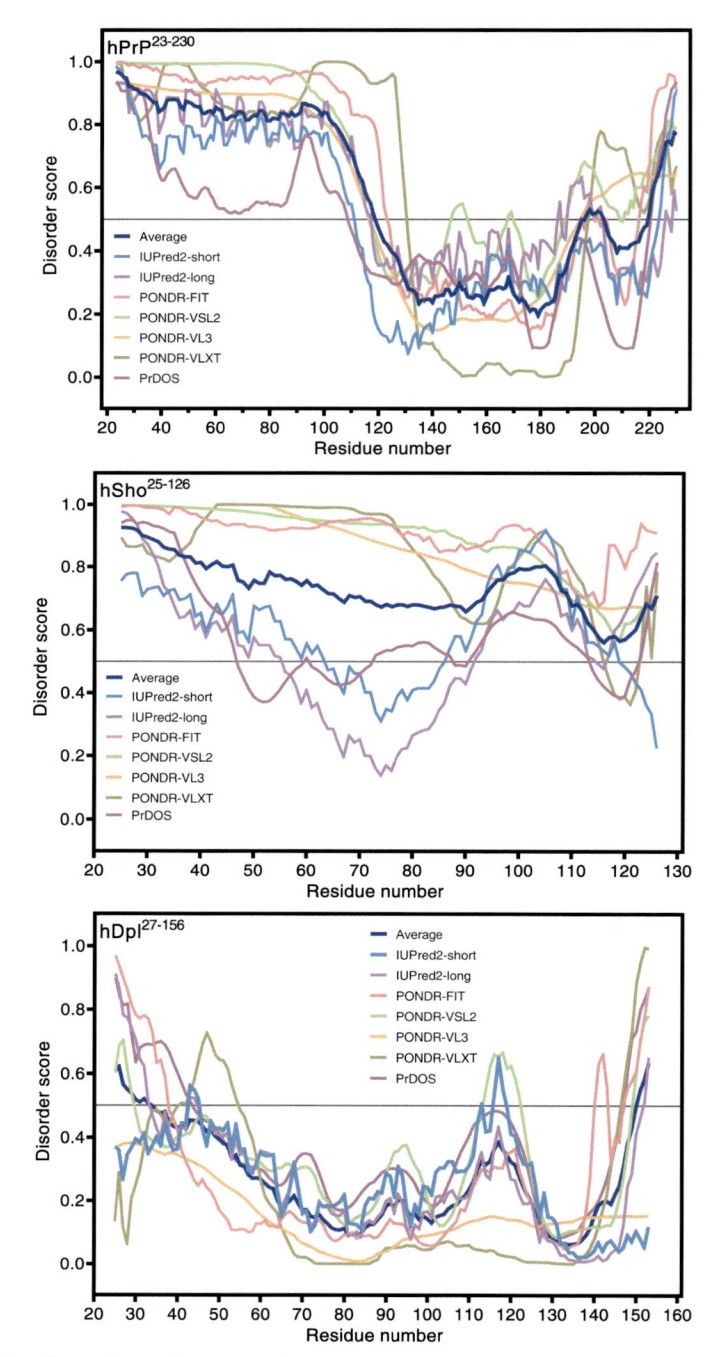

Fig. 2 See figure legend on opposite page.

methylation conserved motif.[93] The protein FUS contains RGG–rich regions crucial to mediate LLPS (when arginine residues were replaced to alanine residues, phase separation is inhibited) and methylated patterns control the material property of condensates.[101] Since cation-pi contacts between Arg and Tyr are responsible to mediate FUS LLPS, methylated Arg tune the strength of these interactions.[101] Considering that post-translational modifications impact on the intra/inter-molecular contacts that sustain the condensates,[102] it is pivotal to assess whether PTMs reported for PrP, such as N-linked glycosylation (Fig. 1) impact on the assembly/disassembly of condensates. It is also unknown whether the addition of methyl groups in Arg residues from Sho occurs *in vivo*. The disordered N–terminal regions of Sho and PrP have low levels of complexity and can also be termed prion–like domains (PrLD) due to their biased composition favoring repeated polar amino acids to the detriment of hydrophobic residues. Glycine residues merit special attention, as they account for about one-third of the PrP N–terminal's composition (34.3%) and full–length Sho (27.5%). For recombinant proteins of FUS members, when glycine residues were replaced by alanine residues in the disordered domain, the fusibility of the droplets was decreased 20-fold, as demonstrated using optical tweezers. Since Gly residues play a key role in the fluidity of liquid droplets, the increased flexibility due to the lack of a side chain likely contributes to this effect.

Electrostatic, cation-pi, and pi-pi planar interactions may mediate heterotypic NA-driven LLPS of PrP or Sho, as suggested by their sparsely arranged aromatic residues and arginine/lysine enriched clusters. In the case of PrP, the LLPS prediction algorithm PScore[103] assigns the extreme N-terminal region (residues 30–90) of PrP as an important segment for planar pi-pi contacts involved in LLPS. No PScore prediction could be

Fig. 2 Prediction of intrinsic disorder in the prion protein family. Analyses of per-residue disorder predisposition of mature isoforms of human PrP (residues 23–230; UniProt ID P04156; top panel); Sho (residues 25–126, UniProt ID Q5BIV9; middle panel), Dpl (27–156, Q9UKY0, bottom panel) by various algorithms as indicated in the legends. Mean disorder predisposition was calculated by averaging all predictor-specific per-residue disorder profiles (dark blue curve). Scores above 0.5 correspond to disordered residues/regions, whereas scores between 0.2 and 0.5 show flexible regions. Access to PONDR® VLXT, PONDR® VL3, and PONDR® VSL2 was provided by http://www.pondr.com/. Access to PONDR® FIT was provided by http://original.disprot.org/pondr-fit.php; access to IUPred was provided by https://iupred2a.elte.hu/plot; access to PrDOS was provided by http://prdos.hgc.jp/cgi-bin/top.cgi.

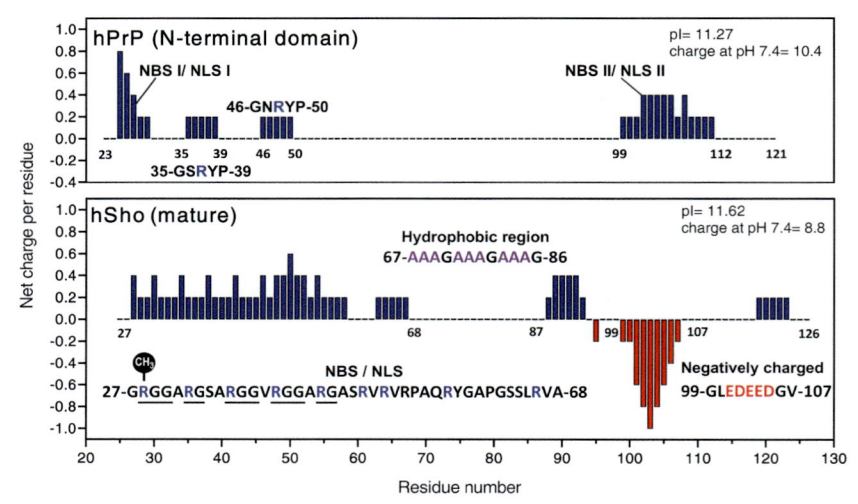

Fig. 3 PrP N-terminal domain and Shadoo have positively charged clusters that can potentially interact with nucleic acids. Analysis using the Classification of Intrinsically Disordered Ensemble Regions (CIDER) webserver (http://pappulab.wustl.edu/CIDER/analysis/) demonstrates the highly basic characteristics of the PrP N-terminal domain (top graph) and Sho (bottom graph), as shown by the plot of net charge per residue distribution (NCPR). Upper panel: PrP has two basic clusters in the N-terminal domain described to be NA-binding sites (NBS I and NBS II) and two other positive motifs (residues 35–39 and 46–50) where arginine side chains could possibly engage on cation-pi stacking with nucleotides. Bottom panel: Sho has RGG/RG short motifs (underscore) predicted to bind NAs (nucleic acid-binding site, NBS) and also configuring a nuclear localization signal (NLS) followed by a hydrophobic segment and negatively charged region. The arginine 28 is a conserved methylation site (CH₃). Positively charged residues are colored in blue and negative ones in red. The isoelectric point and net charge (at pH 7.4) of the PrP N-terminal domain (residues 23 to 121) and mature Sho are shown.

calculated for mature Sho and Dpl due to their small size (shorter than 140 amino acid residues). Despite Dpl's low total score for granule formation, this propensity is high in the region comprising residues 20–40. This segment could therefore mediate LLPS in specific conditions. PrP and Sho are assigned as RNA-binding proteins by the catRAPID algorithm but not Dpl (Fig. 4).

Among the prion family, Sho has the highest overall LLPS score (Fig. 4) predicted by catGranules[104] and has been shown to form amyloid fibrils in physiological conditions.[99] Despite the recently described nuclear localization of Sho,[105] its liquid phase separation behavior has not yet been explored. Indeed, the N-terminal domain of Sho binds to bacterial NAs *in vitro* as evidenced by elution from a chromatographic column immobilized with

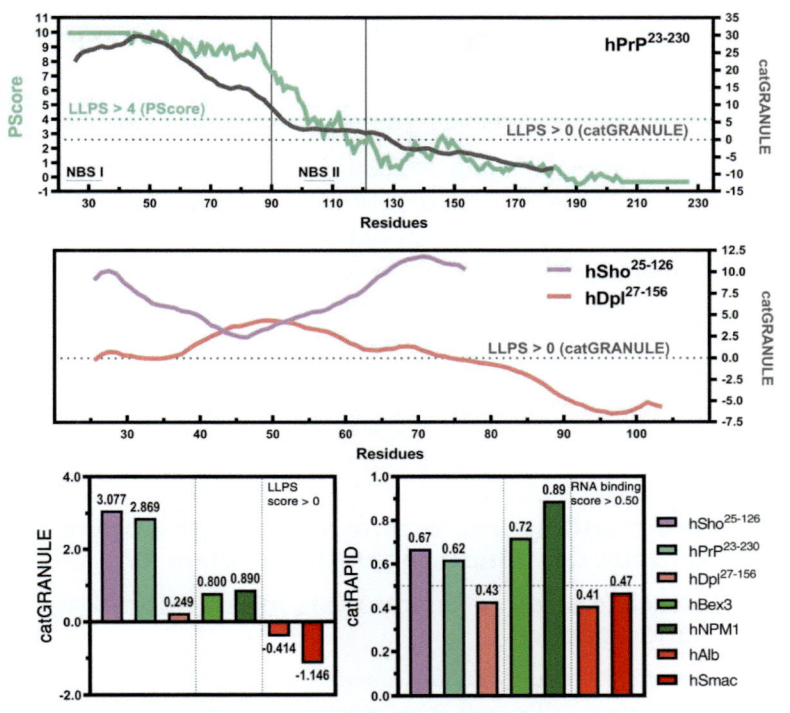

Fig. 4 Liquid-liquid phase separation and RNA-binding propensity of the prion protein gene family predicted using bioinformatic tools. Top and middle: Propensity profiles of liquid-liquid phase separation along the primary structure for mature forms of prion protein members was calculated by catGranule algorithm (http://service.tartaglialab.com/new_submission/catGRANULE) and PScore based in pi-pi planar contacts (http://abragam.med.utoronto.ca/~JFKlab/Software/psp.htm). PScore was only calculated for PrP due to size limitation, primary sequence >140 residues. Bottom: overall scores for liquid-liquid phase separation based on catGranule (score > 0 indicates granule formation; >1.0 strong propensity for LLPS). Scores of hPrP, hSho, and hDpl were compared with controls. For example, controls of proteins that were experimentally proven to undergo LLPS modulated by nucleic acids such as hNPM1 (UniProt ID P06748) and hBEX3 (UniProt ID Q00994) represented by green bars, or negative controls composed by folded proteins such as hAlb (UniProt ID P02768) and hSmac/Diablo (UniProt ID Q9NR28) represented by red bars. Bottom right: catRAPID analysis (http://s.tartaglialab.com/page/catrapid_group) of RNA-binding propensity (overall score above 0.50 denotes interaction with RNA). hSho and hPrP are assigned as RBPs with a similar overall score.

recombinant Sho; this happens through arginine residues since replacing them to histidine residues prevented binding.[106] Sho's RGG–rich region is also assigned as a NLS[107] (Fig. 3). We are currently studying the prion family phase transition behavior *in vitro*. Considering the evolutionary landscape of the prion gene family, with Sho arising earlier than PrP,[108,109] we specifically aim to compare whether the characteristic NA-driven phase separation behavior of PrP is also shared by its ancestor. Moreover, would Dpl interact with NAs and undergo phase separation? This possibility cannot be ruled out, especially considering that DNA interaction still occurs with recombinant PrP Δ32–121 (where the 32–121 region has been deleted).[110] Moreover, rPrP[121–231] in presence of DNA misfolds into amyloid fibrils, suggesting NA interaction with the C-terminus.[111] In agreement, Bera and Biring state that "(…) full length prion protein and its 121–231 fragment bind nucleic acids with comparable affinities (…)" in spite of not showing data.[112] Thus, if the PrP C-terminal globular domain (residues ∼121–231) has DNA binding ability, this property may also be present in Dpl, given that it shares very similar folding.

3. The manifold interactions of PrP with nucleic acid targets: Can phase separation explain it?

3.1 Nucleic acids act as catalysts in PrP misfolding/ aggregation

The concept of PrP[Sc] being the only causative agent of TSEs is challenged by reports of recombinant PrP (rPrP) *in vitro* conversion upon the addition of polyanionic cofactors.[113–115] Glycosaminoglycans, lipids, and nucleic acids (NAs) have been proposed as catalysts for the conversion from soluble rPrP to PrP[Sc]-like aggregated rPrP.[114–117] Indeed, several studies have shown that, over time, NAs can trigger the formation of solid-like β-sheet-rich oligomers. Both wild-type soluble rPrP and PrP[Sc] can interact with DNA and RNA with high affinity in *in vitro* assays, as extensively reviewed in refs. 41, 117. A major caveat in this field is the lack of evidence for PrP-nucleic acid interaction in living cells. Nonetheless, PrP[C] is found in the nuclei of several cell lines and exhibits a *bona fide* nucleic acid-binding behavior, as will be discussed in detail here.

3.2 PrP interacts with different types of nucleic acids

The interaction of PrP with NAs is promiscuous as it binds eukaryotic and prokaryotic NAs regardless of size and sequence, albeit with different binding kinetic and thermodynamic parameters.[41,48,118–120] PrP can interact with polynucleotides (e.g., poly(A), poly(T), poly(G)) and stimulate propagation of PrP[Sc].[115,121] Also, the role of short dsDNAs is controversial; they interact with full-length PrP inducing β-sheet enriched oligomers, but do not stimulate aggregation of a peptide comprising a PrP hydrophobic region (PrP[109–149]).[114] Despite the absence of canonical NA-binding domains, there are three regions of PrP involved in NA-binding (residues 23–31[110,122,123] and 101–110[48,110] in the N-terminus; being the C-terminus also involved[27,110,124]) and two nuclear localization signals (NLS) were suggested.[125,126] PrP generally shows a binding preference towards structured NAs such as hairpins and G-quadruplexes (reviewed in ref. 41). This finding is in agreement with the "structure (of RNA)-driven protein interactivity," observed in a large study involving CLIP (Cross-Linking ImmunoPrecipitation) of protein-RNA networks.[127] It is estimated that 77% of ribonucleoproteins (RBPs) show bias towards binding structure-enriched RNAs. Groot et al. contextualize their results on the high dynamics of single-stranded RNAs that enable fast contacts and poor residence of proteins.[127] On the other hand, RNA folding contributes to the existence of precise binding sites, aiding in the formation of protein-RNA stable complexes.[127,128] If both the RNA and the RBP have multivalency (interleaved binding motifs repeated along sequence), "interconnected polymeric networks" are possibly stimulated as shown in phase separated-states (reviewed in ref. 129).

Curiously, specific NAs containing stable secondary structure are effective in assembling protein-rich condensates even in a known disassembly condition (excess total RNA background).[130,131] Earlier results from our group and others on PrP "aggregation" as a function of NA concentration— we now know that the observed "aggregation" was in fact LLPS at time zero—highlighted the impact of protein:NA stoichiometry on misfolding modulation. Studies on the lifetime (i.e., association and dissociation rates) and stability of different PrP-NA complexes are necessary to understand the effect of specific NAs on PrP misfolding and phase transitions. Moreover, even if PrP displays similar dissociation constant values with different NAs *in vitro* (usually in the nano- to micromolar range), in the cellular millieu small changes in affinity, influenced by ionic currents and competing molecular partners, for example, may drastically influence the assembly/disassembly of PrP-NA complexes. Not only can the soluble rPrP structure be influenced by NA binding—as demonstrated by circular dichroism, fluorescence, and NMR studies—but soluble rPrP also appears to induce NA unfolding (reciprocal structural changes).[27,110,119] Higher-order oligomerization in PrP is remarkable in the presence of guanine-rich oligonucleotides,[120] which may form G-quadruplexes (GQ), tertiary structures formed by guanine-rich sequences. A GQ is a planar structure formed by stacked guanine-tetrads paired by Hoogsteen hydrogen bonds and stabilized by cations (monovalent or divalent, with ideal coordination by potassium ions).[132] Curiously, studies have suggested that soluble rPrP has a "helicase-like" activity *in vitro* in the absence of ATP, given that rPrP can unwind small GQs and stem-loop hairpin DNAs.[48-50,133] In GQs, the unwinding has been reported as occurring within the promoter region (-700 to $+299$ relative to the transcription start site).[50] The *PRNP* promoter contains two putative GQ segments with lengths of 24 and 25 nucleotides. Luciferase reporter assays in human neuroblastoma cells (IMR-32) have shown that PrP^C transcriptional activity is higher in the promoter containing the preserved 25-nt GQ segment.[50] In addition to the predicted GQs within the *PRNP* promoter, the initial portion of PrP mRNA comprising the five octarepeats is predicted to form five GQs.[49] Synthesized RNAs (ranging from 13- to 24-nts) corresponding to four segments were analyzed by biophysical methods, three of them showed GQ conformation in the presence of potassium ions and the other showed stem-loop conformation.[49] Among DNAs and RNAs, structured molecules such as GQs do show a higher affinity towards PrP[27,41,48,49] (reviewed in ref. 41). Taken together, Olsthoorn has proposed that co-translational folding of PrP occurs when recruited by the GQs from its own mRNA and altered interactions would induce the conformational switch (PrP^C→ PrP^{Sc}). In this respect, the cofactor involved in PrP aggregation would be its octarepeats-mRNA.[49] However, a recent study analyzing the conformation of entire octarepeats' mRNA (94–288 in relation to *PRNP* ORF) ruled out the hypothesis of formation of GQs within this region since the conformation was not observed even in presence of stabilizing agents.[134] Instead, the OR-mRNA has a well-defined fold of A-helical stem-loop, being questionable the GQ formation in living cells.[134] Considering the higher affinity of PrP for structured NAs than single-stranded NAs, addressing whether PrP could bind to its own mRNA is needed. In this manner, many intrinsically disordered RBPs (e.g., TDP-43, Hsp70, FMRP, RpS3) exert an auto-regulatory feedback on their expression level achieved by binding to its transcripts.[135,136] For example, when the RBP is overexpressed complexes with its own mRNAs are formed, hence, hampering translation. This is advantageous considering the aggregation-prone nature of these proteins. In many instances a high protein to NA ratio promotes condensates' formation; thus, we speculate whether their own mRNAs can be sequestered. Oppositely, at low protein:NA ratio, condensates' formation is prevented, and further translation might be elicited. The complex interaction of PrP and NAs (extensively reviewed in ref. 41) is becoming better understood in the light of LLPS, whereby the same nucleotide sequence can promote or inhibit condensates' assembly depending on its folding, i.e., mostly single-stranded (heated) vs structured RNA. This was the case for the PGL-3 protein condensation.[137] Despite the vast PrP-NA literature, how/when/where those interactions occur in the cellular context and their functional consequences remains to be dissected. Moreover, the crosstalk between NA and proteins not only argues the possibility that PrP regulates NA fate but the other way around,[138] i.e., NA-binding regulates PrP localization, function and stability. The latter possibility is clearly observed *in vitro*, where different NAs modulate rPrP stability; high-affinity oligonucleotides stabilize PrP^C whereas unspecific NAs could accelerate the conversion of PrP^C to PrP^{Sc} (reviewed in ref. 41). Anyhow, increasing evidence suggest nucleic acids as natural partners of PrP.

3.3 PrP destabilizes and remodels nucleic acids: Is this activity promoted inside condensates?

Since the 2000s, *in vitro* studies have assigned soluble recombinant PrP as a DNA/RNA chaperone with activities similar to those of the highly basic and disordered HIV-1 nucleocapsid protein p7 (NCp7).[133,139–141] Firstly, rPrP has been shown to bind tRNA^{Lys,3} and anneal to the HIV RNA primer-binding site, assisting reverse transcriptase initiation.[141] Secondly, soluble rPrP has been implicated in the catalysis of DNA strand transfer.[140] Thirdly, soluble rPrP has been reported to form RNA dimers as well as DNA and RNA multimers, as shown by electrophoretic mobility shift assays (EMSA).[139] Finally, soluble rPrP has been shown to melt double-stranded oligonucleotides (16- and 28-nts), as demonstrated by fluorescence resonance energy transfer (FRET) and UV spectroscopy data.[133] Whether PrP^C can also exert an annealing activity, given that some helicases display rewinding activity,[142] or stabilize the secondary structure of physiologically relevant NAs, remains a matter for further investigation. More recently, another reported effect of soluble recombinant PrP on DNA has been the compaction of extended DNAs, especially an 840-bp DNA containing only GC repeats, as demonstrated in fluorescence studies. Interestingly, PrP$^{121–231}$ cannot exert this activity, indicating that the N-terminus, the region most prone to LLPS, is required. This observed DNA compaction effect is reasonable in the crowded environment of a condensate.[112]

Concerning the denomination of PrP^C as a NA chaperone, it is very likely that PrP takes advantage of its phase separation behavior in the presence of NAs to modulate various nucleic acid-related processes. Indeed, a single protein with phase separation behavior can lead to the formation of ribonucleoprotein (RBP) complexes with multiple functions depending on the intracellular microenvironment (reviewed in ref. 67). FUS, which is associated with the development of frontotemporal dementia and is among the most studied proteins with phase separation behavior,[84,143–145] has been correlated with the following activities: regulation of the fate of noncoding RNAs and mRNAs as well as modulation of RNA trafficking, splicing, processing, stability, and translation. The factor that dictates the functional outcome of the condensate depends on the clients partitioned in the granule and their biophysical properties, including viscoelasticity. Interestingly, with respect to the activity comparison between PrP and NCp7, it has recently become clear that nucleocapsid proteins initiate encapsidation of genomic material via LLPS.[146,147]

Condensates show a unique inner milieu, with specific charge patterns and mesh size because of interactions (cation-pi, hydrophobic, electrostatic, etc.) that keep the phase-separated state. The activity of rPrP in NA folding could be explained by phase separation processes. The internal crowded environment of PrP liquid droplets would promote conformational changes in NAs, analogous to duplex melting activity of short DNAs promoted inside DEAD-box protein 4 (Ddx4) protein condensates.[148]

3.4 Identification of a small nucleic acid partner of prion: The switch from infectious role to possible therapeutic potential

Some studies have sought to identify a "prion-specific" NA from TSE brain homogenates but have not shown conserved nucleotide sequences.[149,150] Additionally, NAs co-purified with PrP^{Sc} exhibit heterogeneous sizes. Safar et al. found small NAs with about 25 nucleotides co-purified with PrP^{Sc} from the brain of TSE-affected Syrian hamsters.[150] Similarly, recombinant PrP has been found to interact with two small RNA populations purified from scrapie-associated fibrils (~27 and 55 nts).[151] However, other studies have reported nuclease-resistant large NAs ranging from 300 to 5000 nts.[152–155] Interestingly, a monoclonal antibody against genomic DNA (OCD4) was able to specifically detect PrP^{Sc} in human brain homogenates in various forms of prion diseases, supporting the assertion that PrP^{Sc} is bound to DNA.[156] rPrP complexes with total RNA from murine neuroblastoma cells (N2a) resulted in ~80% cell viability loss. However, this cytotoxic effect was not observed when soluble rPrP was complexed with a 17-mer RNA displaying nanomolar affinity for rPrP.[119,157] Aptamers are NAs with high specificity and affinity for a single protein target. Usually, they are small DNAs (20–100 nts) selected by many rounds of a method called SELEX (Systematic Evolution of Ligands by EXponential enrichment),[158] which our group has explored using the globular domain of recombinant PrP (rPrP$^{90-231}$), containing a 31-residue N-terminal flexible region. Among the selected aptamers, the highest binding affinity was observed with a 21-mer harboring secondary structure (named A1). A1 is predicted to have a hairpin consisting of a hydrogen bonding-stabilized stem and an internal loop. While rPrP binding to A1 initiates LLPS, a mutated aptamer with two nucleotide substitutions named A1_mut has been shown to elicit aberrant phase separation of rPrP and to age into amyloid-like aggregates.[27] The latter was the first study to show nucleic acid-driven LLPS with PrP, providing a framework for further exploitation of both the pathophysiological role of PrP-NA phase transitions and the potential therapeutic value of using aptamers to control PrP condensation.

Most studies on the identification of NAs bound to PrP^{Sc} focused on *ex vivo* preparations. Thus, the possibility of non-specific scrapie-NA complexes assembled *in vitro* cannot be ruled out. Since both soluble and scrapie PrPs have been shown to bind NAs, would preferential conformations display an enhanced binding affinity? Future studies on the interaction of aggregated PrP with NAs should account for the amyloid polymorphism (strain phenomenon) and whether NAs form the amyloid core. Such studies should also allow the characterization of the binding of NA to aggregated PrP containing post-translational modifications and PrP mutants. There is also a possibility that PrP forms reversible amyloid fibrils. These possibilities highlight the need for further research addressing the molecular structure of PrP after phase separation. Additionally, PrP exists in solution as monomeric, dimeric, and low-order oligomeric species; thus, assessing which organization is more prone to NA-binding would be mechanistically valuable despite the dynamics between these species. The hypothesis of PrP amyloids serving as a platform to sequester NAs, and impairing their normal functions, has recently been explored by molecular dynamics using small peptides.[159] In the latter study, two PrP C-terminal peptides with steric zip conformations (127-GYMLGS-132 and a polar amyloid surface 170-SNQNNF-175) formed stable complexes with a small RNA (UCCU). RNA molecules displayed a higher affinity for the peptide 170–175, which has exposed side chains, and in many simulations, the RNA showed conformational rearrangements.[159]

Reciprocal conformational changes in rPrP and NAs are reported *in vitro*, with both biomolecules undergoing loss of secondary structure content upon interaction.[27,48,119,120] Physiologically, protein-NA interactions result in bidirectional outcomes, i.e., proteins can regulate NA fate and function and the reverse can also happen (reviewed in ref. 138). This protein-NA crosstalk is likely a consequence of the mutual protein-NA structural changes, tuned by binding specificity and stoichiometry. Reentrant LLPS, as observed for rPrP with addition of nucleic acids,[27,160] should be finely controlled intracellularly, as impaired disassembly of PrP-NA condensates could result in pathological aggregates. Shifting the molar ratio towards excess of NA in relation to protein by antisense oligonucleotides (ASOs, complementary to mRNA targets) or by aptamers at appropriate molar ratio administered directly into the brain[161] might offer promising NA-interventions to control aberrant PS.

3.5 The potential role of non-coding RNA species in PrP pathophysiology

Future directions in the field of PrP-NA studies should include identifying interactions in living cells using specialized approaches, such as RNA-protein co-localization using high-resolution single-molecule fluorescence microscopy, which can also assess the spatiotemporal dynamics of such interactions.[162] Another powerful method is PAR-iCLIP (PhotoActivatable-Ribonucleoside-enhanced individual-nucleotide resolution UV Cross-Linking ImmunoPrecipitation).[163] In this method, protein-RNA complexes are crosslinked by UV light in cells, followed by immunoprecipitation and deep sequencing, enabling the mapping of protein-RNA contacts. The protein tau—linked to the development of frontotemporal dementias—preferentially binds to transfer RNAs (tRNAs) from among a pool of total RNA in living human cells, as shown by PAR-CLIP. Upon tRNA binding, tau undergoes LLPS.[164]

In the case of PrP, tRNAs have shown the most noticeable impact on murine PrP higher-order oligomerization between the different types of RNAs and ribonucleoside triphosphates (rNTPs) analyzed.[165] These RNAs included two different mRNA sequences (domains V, IV, and II from the 23S E. coli rRNA) and a prokaryotic tRNA pool. While rNTPs did not affect recombinant PrP phase separation, mRNAs and rRNAs showed a similar trend, resulting in a ~7-fold increment in PrP light scattering (LS) with different PrP:RNA stoichiometries.[165] Notably, tRNAs resulted in a much more pronounced effect, with ~20-fold higher light scattering. Consistently, upon immediate addition of tRNAs at low molar ratio (at a 20:1 ratio), a homogenous population displaying hydrodynamic radii (R_h) of ~700 nm was observed by DLS.[165] These R_h values were seven times larger than the effect of other RNA types.[165] Indeed, recombinant human PrP binds tRNAs with high affinity (dissociation constant of 1.7 ± 0.7 μM).[166] The tRNAs have noncanonical functions, especially through their derived fragments, a new class named tRNA-derived fragments.[167] Small-derived fragments of tRNAs such as tRFs (14–30 nts) can modulate phase transitions of Brain-Expressed X-linked 3 (BEX3).[75] Short intracellular RNAs such as microRNAs (miRNAs) and tRFs have recently emerged as important players in neurological syndromes. Their expression and chemical modification patterns are altered in diseased brains.[168–170] The tRFs and tRNA halves (tiRNAs) are found across all domains of life and are derived from tRNA upon digestion of ribonucleases.[171] Their vast array of regulatory functions includes gene expression, regulation of mRNA stability (acting as miRNAs), and activation of pro-survival protein translation (in response to cellular stress).[172] The anti-apoptotic role of some tRFs that complex with cytochrome c, ultimately preventing apoptosome formation, has also been described.[173] In another context, some tRFs have been associated with p53-dependent neuronal cell death.[174,175] Further studies dissecting tRNAs and the interaction between their fragments and PrP would be interesting, considering in particular two studies that identified the interaction of PrP with argonaute proteins (AGO1 and AGO2).[176,177] All four argonaute proteins are loaded with tRFs, forming molecular complexes that can post-transcriptionally silence genes.[178,179] AGO proteins are the core of the large RNA-induced silencing complex (RISC), which exerts an endonuclease activity on mRNAs with complementary sequences to specific miRNAs. It is possible that PrP indirectly binds to AGO2 as both can co-bind miRNAs or tRFs. However, PrP contains six glycine-tryptophan (GW/WG) repeats within the octarepeat domain. These motifs can mediate contact with AGO2, similarly to GW182 protein members.[176] The interaction of PrP with AGO2 thus leads to stabilization of the large RISC and ultimately to translational repression.[176] In this case, PrP demixing feature would certainly favor the formation of RNA processing bodies.

Circulating cell-free NAs (cfNAs) are not as scarce and labile as previously thought. Among cfNAs, miRNAs present in cell culture medium have shown outstanding 2-month stability in cell lysates.[180,181] MiRNAs are a type of non-coding RNAs with 19–23 nts that can abrogate mRNA translation by annealing to $3'$-untranslated regions of target mRNAs.[182] It is estimated that around half of the human transcriptome can be post-transcriptionally repressed by miRNAs. Additionally, a single miRNA can regulate approximately 200 mRNAs, a redundant process reviewed in ref. 183. Notably, several miRNAs are differentially expressed following prion infection. In an effort to understand which miRNAs regulate cellular PrP expression, a recent report identified miR-148a-3p and 148b-3p as strong repressors.[184] Studies on miRNAs tackling PrP^C expression may offer therapeutic strategies since PrP is the central factor in prion diseases. This reverse concept of targeting native PrP is based on the fact that *PRNP* knockout mice are resistant to disease[185,186] and that depletion of PrP by the Cre-lox system ameliorates spongiform degeneration.[187]

Since nuclease activity is high in body fluids, most NAs would be expected to be degraded unless compartmentalized or occluded in the RBP binding sites. Indeed, some cfNAs reside inside extracellular vesicles, but prominent studies have shown that most miRNAs are associated with RBP complexes in human blood.[188] Indeed, these miRNAs are associated with AGO2 and nucleophosmin (NPM1).[180,189] NPM1 exhibits polyampholytic properties that mediate self-association and drives its LLPS.[190] Likewise, PrP also has weakly charged clusters in its primary structure, which may be important for electrostatic contacts driving the NA-protein droplets.[27] Turchinovich et al. noted that, despite NPM1 having 33 kDa, the fraction of miRNAs-NPM1 did not flow through 100 kDa filter devices; rather, they stayed in the concentrate.[180] Based on that, it is speculated that NPM1 could be part of a higher-order complex, perhaps forming an extracellular phase-separated condensate, in our view.

Cell-free NAs are released by apoptotic/necrotic cells and extracellular vesicles such as exosomes and small vesicles (30–150 nm) that mediate inter-cellular communication, as reviewed in ref. 191. Exosomes transport different RNA cargos depending on cell type, stimuli, and stress condition. Both PrP^C and PrP^{Sc} can be released into the cell culture medium in association with exosomes.[192,193] Additionally, isolated exosomes containing PrP^{Sc} act as seeds for prion propagation in cells.[193] Could the RNA load in exosomes or cfNAs have a role in the conversion of PrP^C to PrP^{Sc}? In this regard, the encounter of PrP with NAs might occur extracellularly and generate seeds for propagation. Likewise, at physiological state, high amounts of shed PrP (residues 23–228) and the N1 fragment (residues 23–110) can be found in the extracellular space, including CSF and blood (reviewed in ref. 46, 194). Moreover, shedding and α-cleavage products contain the NA-binding regions assigned in the N-terminus. Recently, the largest genome-wide association research involving 5208 sporadic CJD individuals identified two prominent transcripts increased in the brain.[195] These consisted of the mRNAs coding for syntaxin 6 (a member of the SNARE family involved in secretory pathways) and galactose-3-O-sulfotransferase 1 (involved in the synthesis of myelin lipids).[195] Along with their mechanistic role in the disease mechanism, the possibility of direct interaction with PrP should be considered using biophysical approaches. Our group is currently studying the effect of specific mRNAs and miRNAs on PrP phase transitions.

The role of miRNAs in the diagnosis and therapeutics of prion diseases is promising since distinct miRNA molecular fingerprints have been found in prion brains.[184,196,197] Some miRNAs are differentially expressed in the thalamus, and extracellular vesicles are shed in blood serum from sporadic CJD mice. Moreover, the profile of miRNAs in serum at the pre-clinical stage has been altered, potentially playing a role as biomarkers in non-invasive blood collection.[197] In addition to differential expression of miRNAs, their post-transcriptional modification landscape (more than 150 types have been reported, such as pseudouridine[198]) may be altered, providing potential insights into the mechanisms and diagnosis of TSEs. This has been recently observed in post-mortem studies of the brains of individuals with Alzheimer's disease, where unique modifications in small RNA nucleosides were identified by liquid chromatography-mass spectrometry.[170] Additionally, tRFs^{Arg} and tRFs^{Tyr} were significantly reduced.[170] Notably, tau, p53, PrP, and BEX3 are examples of proteins implicated in proteinopathies that bind tRNAs and modulate the former's aggregation.[75,199–201] Importantly, the dysregulation of protein translation is common to many age-related diseases.[202] In the case of neurodegenerative-associated proteins that exhibit phase transition behavior, we query whether specific tRNAs and their derived tRFs could be sequestered in aberrant intracellular condensates.

Given that PrP^C is found in body fluids, it is possible that PrP^C encounters cfNAs and undergoes extracellular phase transitions. In fact, different morphological types of PrP^{Sc} plaques can be found extracellularly.[53] In a broader perspective, would protein-NA phase transitions occur in the CSF and peripheral blood? There is evidence for extracellular phase-separated states, such as hydrogels, which are secreted by marine organisms (e.g., mussels) to provide adhesion to surfaces.[203,204] In addition, RNA-protein granules are found in secreted royal jelly.[205] Proteolytic fragments of PrP encompassing residues 90–121—the binding region for NAs and Aβ, two known ligands that trigger LLPS of PrP—can be expected to exhibit phase separation behavior if the critical concentration and specific protein: ligand stoichiometry is reached. Cell adhesion and intercellular signaling are examples of functions that PrP could mediate extracellularly by LLPS. Further studies should address whether extracellular amyloid deposits composed of PrP^{Sc} and $A\beta$, separately or mixed, could originate from the aging of liquid-like condensates.

4. Phase transitions of PrP in health and disease

Phase separation is a key mechanism that helps explaining the promiscuous interaction with NAs shared by many IDPs and their simultaneous involvement in multiple NA regulatory events. *In vitro* LLPS studies have shown that rPrP is able to condense in the absence of ligands in physiological buffer, an effect promoted by macromolecular crowding (e.g., addition of polyethylene glycol of different sizes). However, in the cellular environment, PrP^C engages in interactions with ions, metabolites, and macromolecules that can modulate its phase transition behavior. Fig. 4 and Table 1 summarize the findings from prominent studies that investigated PrP phase transitions using differential interference contrast (DIC) microscopy. The first study investigating recombinant full-length PrP condensates showed liquid droplet formation in phosphate buffered saline (>30 µM protein, pH 7.4), with acidic pH inhibiting the formation.[26] Addition of $A\beta$ oligomers to PrP solutions triggers gel-like amorphous structures with composition dependent on the $A\beta Os$:PrP stoichiometry. Conformational changes in phase-separated states are different in liquid droplets formed by self-interaction of PrP (unfolding of threonine residues in helices α-2 and α-3) or $A\beta Os$:PrP hydrogels (folding of residues 113–120 and octarepeats to α-helical conformation). In this case, more ordered PrP was found in the higher viscosity state (hydrogel). Based on *in vivo* data, it is proposed that the $A\beta Os$:PrP hydrogels sequester metabotropic glutamate receptor 5 (mGluR5), leading to excessive signaling and neuronal dysfunction.[26]

Intricate mechanisms to avoid the solidification of liquid droplets exist intracellularly, with the stoichiometry between the components that form a condensate being a potential determining factor. For a fixed PrP concentration, a low concentration of DNA triggers numerous small liquid droplets. With increasing DNA concentration, the condensates become larger and dissolve when an excess of aptamers is present. The latter mechanism is described as re-entrant phase transition and is typical of LLPS that is mostly driven by electrostatic interactions.[131,208] Re-entrant LLPS was also observed on recombinant C-terminally truncated PrP (Y145 stop mutation) upon addition of either structured RNAs (tRNA pool) or single-stranded homopolymers (poly-uridine).[160] The phase separation modulation of

Table 1 Studies on the prion protein liquid-liquid phase separation.

Construct	PS-stretches	Modulator	Conclusion	Physiological role[a]	Pathological role[a]	R
Addition of ligands or physicochemical changes						
hPrP[23–231]	23–51 and 95–110[b]	Aβ oligomers	Hydrogels formed at low AβO-PrP ratio, reversible by excess of AβO	N/A, focused on pathological role	PrP:Aβo hydrogels sequester mGluR5, leading to synaptotoxic signaling	26
mPrP[90–231]	90–120 is necessary for NA-driven LLPS	25-mer DNA aptamers	Stoichiometry and structure of NA modulate condensate formation	- Storage - Processing - Transport of specific NAs	Aptamer without secondary structure triggers liquid to solid-like transition	27
mPrP[23–231]	N/S	- 21-mer GC-rich DNA - Cu(II) ions	PrP:DNA droplets formation is minimally influenced by Cu(II)	- Storage - Processing - Transport of specific NAs	After 24 h, a large network of amorphous aggregates is seen in presence of the DNA	118
hPrP[23–231]	23–89	- Kosmotropic anions (*e.g.* $Na_2S_2O_3$) - Macromolecular crowding (concentrations of PEG/Dextran)	Aged PrP gels have proteinase-K resistance and some β-sheet content	N/A, focused on pathological role	Conformational transition from PrP to PrP[Sc] is via LLPS	206

Continued

Table 1 Studies on the prion protein liquid-liquid phase separation.—cont'd

Construct	PS-stretches	Modulator	Conclusion	Physiological role[a]	Pathological role[a]	R
Effect of mutations						
hPrP$^{23-231}$ G127V	N/S	– Protective polymorphism – Mutation of glycine 127 to valine (hydrophobic side chain)	G127V liquid droplets and amyloid formation is reduced	N/A, focused on pathological role	The G127V PrP droplets are more dynamic compare to WT PrP, inhibiting amyloidogenesis and cytotoxicity	207
Y145 (stop codon)[c]	N/S	– Mutant found in GSS – tRNAs, ssRNA	Y145Stop undergoes LLPS, after 5 h-aging solid-like amyloids are formed	N/A, focused on pathological role	Aberrant Y145Stop phase transitions are associated to seeding activity	160

[a]Condensate role suggested in the discussion section.
[b]NA-binding sites overlap Aβ interaction sites. However, Aβ oligomers do not displace aptamer binding to PrP$^{90-231}$ (verified for aptamer named A1).[27]
[c]Species (e.g., mouse, human) not specified in the manuscript.
Only studies providing evidence of PrP micrometer-sized condensates via microscopy (DIC and/or fluorescence images) were included.
As a matter of simplicity, we highlight the main conclusion of each study and do not include results of the extensive physicochemical characterization of LLPS (critical concentration, showing which interactions were involved in LLPS by salt addition, 1,6-hexanediol, pH effect, and so on) nor were included other protein constructs that did not lead to the principal conclusion. In addition, studies were classified based on their novelty, since some fell in more than one category. To the best of our knowledge, up to now, there are no studies directly assessing PrP LLPS in living cells. In all cases, recombinant PrP produced in *E. coli* and purified to homogeneity was used. Studies were organized by publication date within subsections.
Abbreviations: *h*, human; *GSS*, Gerstmann-Sträussler-Scheinker syndrome; *m*, mouse; *NA*, nucleic acids; *N/A*, not applicable; *N/S*, not specified; *R*, reference number.

PrP^{90-231} also depended on the secondary structure of 25-mer aptameric DNAs. In addition, the 90–121 region is necessary for PrP-DNA interaction since the 121–230 PrP construct did not show binding to the rhodamine-conjugated A1 aptamer, as shown by fluorescence anisotropy.[27] Collectively, the first unfolded stretch of PrP^{90-231} (composed of 31 residues) plays a crucial role in NA-driven LLPS.

Since various prion-like proteins exhibit RNA-controlled condensation, a general mechanism to control phase separation (mostly electrostatic-driven) has been suggested: high protein:NA molar ratios—present in the cytoplasm—induce LLPS, while low protein:NA ratios—as in the nucleus—keep the protein in a soluble state.[131] This observation corroborates findings that post-mortem misfolding diseased brains contain no aggregates from prion-like proteins in the neuron's nuclei.[131,209] This mechanism where a low concentration of NA in relation to protein nucleates LLPS with a maximum phase separation point (charge neutrality of protein:NA complexes), and a large excess of NA causes a charge inversion in droplets surface (e.g., negatively charged protein:NA condensates, for example) concomitant to dissolution into an homogenous phase is known as reentrant phase transition.[208] Interestingly, Cordeiro's et al. work on $rPrP^{23-231}$-DNA interaction already pointed out reentrant LLPS 20 years ago: "at high DNA-to-protein ratios, aggregation is inhibited"[114]; followed by Nandi et al. observation: "Due to the experimental difficulty encountered from the appearance of turbidity and coagulation during thermal unfolding studies of mouse recombinant prion protein in nucleic acid solution, we have undertaken studies of structural properties of the isolated C-terminal MoPrP (121⁻231) fragment (…)"[111]—this observation of cloudiness was most likely due to the formation of PrP:NA liquid droplets, as recently shown by our group using microscopy techniques.[114]

A number of neurodegeneration-associated proteins can transition from soluble to less dynamic material states over time, which resemble hydrogels and solid-like amyloids.[67] This aging process can be monitored *in vitro* by the characteristic deep-blue intrinsic fluorescence of amyloids.[27,210] We have shown that mouse PrP^{90-231}, in the presence of a single-stranded aptamer (A1_mut), forms condensates with a darkly refractile appearance that show deep-blue autofluorescence with 1 h of incubation with fibril-like species sticking out of the droplets. Furthermore, the same solution kept for 24 h did not show any droplet formation but rather the appearance of amorphous aggregates and fibrils.[27] Similarly, Agarwal et al. recently characterized the liquid-to-solid transition of recombinant Y145Stop PrP, a stop codon mutant associated with GSS. The study showed an increase in intrinsic blue fluorescence as a function of maturation time for recombinant Y145Stop PrP.[160] Curiously, this N-terminal construction of PrP has been shown to be located in the nucleus in the presence of proteasome inhibitors.[211] These findings have been corroborated by a study of the NA-driven LLPS of Y145Stop, which showed a 15-fold lower critical concentration for LLPS in the presence of tRNAs. tRNAs were also shown to promote solidification of Y145Stop droplets, decreasing protein diffusion inside droplets. Aggregates revealed by thioflavin T fluorescence were observed after 24 h.[160] In addition to NAs, PrP phase separation can be modulated by mutations and protein-ligands[26,27,118,160,207,206] (Table 1 and Fig. 5). In their study about the disease-resistant polymorphism G127V, Huang et al. proposed that the change in the phase behavior observed between G127V and wild-type PrP prevents fibrillation and further cytotoxic effects.[207]

Together with the potential biological importance of NA-driven LLPS for PrP, targeting of PrP condensates to prevent aberrant phase transitions opens a new avenue for therapeutic applications. While we characterized the LLPS of PrP tuned by oligonucleotides *in vitro*, further study is needed to understand whether this process occurs in cells, where a multitude of molecules is present. With the current knowledge, the notion that some NAs induce *in vitro* aggregation whereas others prevent it is clearly an oversimplification, as the biophysics of condensates are affected by the stoichiometry of binding and the interaction time window. Proteins that are scaffolds of LLPS have their condensation regulated by specific cellular responses and PTMs.[67] The fine control of assembly/disassembly prevents the dense phase formed by the scaffolding protein from evolving into a less dynamic solid-like species such as amyloid fibrils.

PrP[Sc] has been reported as being phosphorylated at serine 43 in the nuclei of brain cells from 22A-infected mice.[44] This post-translational modification could play a role in nuclear shuttling, as well as regulating PrP phase transitions as verified for other proteins.[102,212] Additional reports have shown the presence of PrP-immunoreactive material in the nuclei of infected mouse neuroblastoma cells (N2a) and the tight association of cellular PrP with chromatin (only dissociated with 400 mM sodium chloride).[213] Various other studies mention that PrP can be localized in the cell nucleus and that nucleolar localization has been shown by confocal microscopy.[214,213,215] Indeed, full-length PrP has at least three motifs/domains that can bind nucleic acids with high affinity[110,114,119,120,123] (illustrated in

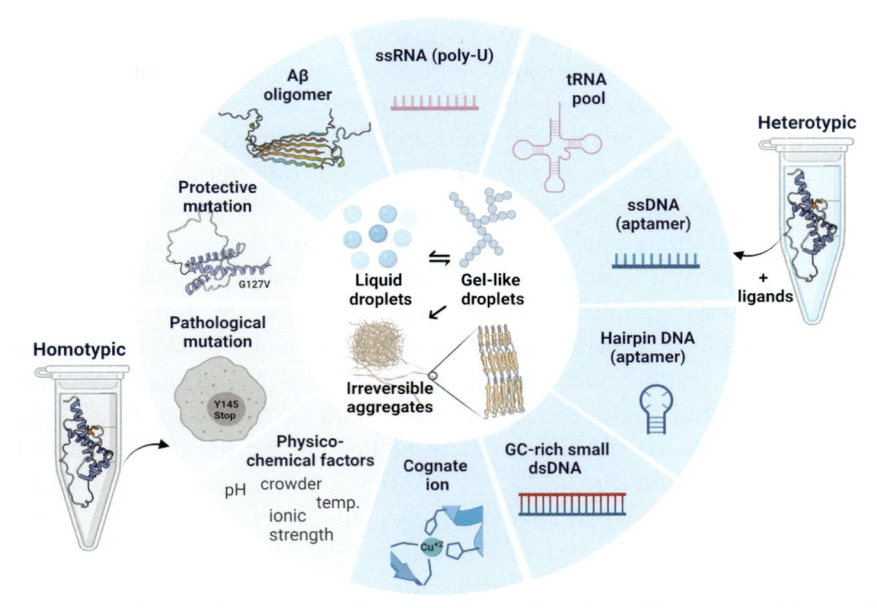

Fig. 5 Regulators of PrP *in vitro* phase transitions. Overview of known modulators of PrP phase transitions as evidenced in studies (published to date) using Differential Interference Contrast (DIC) microscopy. Heterotypic interactions between rPrP (PDB: 5YJ5) and ligands or "clients" (blue shading) modulate condensate formation and their transition to more ordered material states. Homotypic interactions between monomers of PrP that maintain phase separated states are influenced by mutations and physico-chemical factors (gray shading). Heterotypic: Amyloid-β oligomers (AβOs, PDB: 6RHY) complex with PrP forming a reversible hydrogel (ref. 26). Nucleic acids such as a single-stranded (ss) RNA composed of a homopolymer of uridine (poly-U) and crude transfer RNA (tRNA) in presence of rPrP mutant Y145Stop elicit reentrant phase behavior (ref. 160). A selected aptamer for $rPrP^{90-231}$ also showed the characteristic concentration-dependent control of LLPS. A ssDNA aptamer (mutation to disrupt base pairing) triggers few liquid droplets with gel-like characteristic and with long incubation, aggregates with amyloid features are observed (ref. 27). A small dsDNA (22-mer, named D67) triggers full-length $rPrP^{23-231}$ phase separation (10 µM protein: 10 µM D67) and addition of copper ions (60 µM $CuCl_2$) are unable to modify D67-driven LLPS (ref. 118). After 24 h, D67-PrP complexes form solid-like branched aggregates. PrP coordinates Cu(II) *in vivo* and most of its functions are attributed to copper-binding ability. rPrP does not show droplet formation in presence of 200 mM $CuSO_4$; instead, amorphous aggregates are seen after 24 h. Homotypic: macromolecular crowding agents, temperature (T), pH variation, addition of salts and 1,6-hexanediol (HD) influence rPrP LLPS (refs. 26,27,206). The Y145Stop mutant has increased permanence in the nucleus, forming liquid droplets prone to aging in β-sheet-rich solid-like structures (ref. 160). The disease-resistant mutation G127V (PDB: 5YJ4) shows reduced LLPS and fibrillation (ref. 207). Created with BioRender.com.

Fig. 1). Moreover, PrP has two suggested nuclear localization signals (NLS),[125,126] which are curiously embedded on the NA-binding regions from the N-terminus. However, little attention has been devoted to determining the triggers of nucleocytoplasmic shuttling and how it correlates with its phase transitions in cells. In this regard, one study reports PrP in the nuclei of neurons in response to DNA damage induced by an alkylating agent. This localization was verified in both mice and human neuroblastoma cell lines, and, as a result, PrP was shown to activate DNA repair, which coincided with granular immunoreactive staining in the nucleus.[216] Taken together, PrP may form liquid-like condensates inside the nucleus that serve as scaffolds for "clients" involved in DNA damage response (DDR) since LLPS has been associated with DDR (reviewed in ref. 217).

Furthermore, the intrinsic activity of PrP on the remodeling of NA structures verified *in vitro*[50,112,133,139–141] might be important for the repair activity. Another way that condensate-forming proteins modify NA structure is through the biophysics of condensates. Condensates are selective, and not all NAs can diffuse into their interior. This partitioning is achieved by the biophysical properties of condensates such as mesh size and its surface charge, roughly analogous to a "2D gel electrophoresis." Moreover, the interior of condensates favors biochemical reactions without the need of ATP. For instance, Ddx4 is an intrinsically disordered protein that forms condensates *in vitro* in association with DNA or RNA, which are regarded as "passive helicases" since double-stranded nucleic acids are melted inside while single-stranded structures are stabilized.[148] We suggest that the PrP condensates could also function as "biomolecular filters" involved in NA processing.

We cannot rule out the possibility that the molecular structures within the PrP condensates are reversible amyloid fibrils (RACs) with a more labile cross-β conformation, as observed in the X-ray structure of the RNA-binding proteins FUS and hnRNPA.[77,218,219] Is it possible that RACs possess a self-propagating ability similar to irreversible fibrils? According to a recent study, PrP liquid droplets matured for 72 h into hydrogels displayed proteinase-K resistance, with Fourier transform infrared (FTIR) analysis showing a β-sheet structure.[206] Although mechanistically insightful, the study analyzes PrP gels produced in unbuffered aqueous solution under macromolecular crowding and kosmotropic salt addition ($Na_2S_2O_3$), an artificial condition.[206] Additionally, the authors state that Congo red was able to stain PrP aged gels, though no characteristic birefringence was observed under polarized light.[206] In fact, we recently found that monomeric recombinant PrP binds

Congo red, as demonstrated in spectrophotometric assays.[118] Furthermore, thioflavin T and Congo red are capable of modulating phase transitions.[220] The extrinsic dye partition in the crowded condensates could also happen nonspecifically. Results on the use of amyloid dyes to characterize PrP phase-separated states should therefore be evaluated with caution.

Regarding a possible function of PrP liquid droplets, one study showed that recombinant PrP in the presence of a DNA plasmid is endocytosed by mammalian cells.[123] Subsequently, the complex can act as a "transfection" complex since the expression of the protein encoded by the plasmid was achieved.[123] Extrapolating to the cellular *milieu*, liquid droplets of PrP-NA could serve as a means of intracellular storage and transport. When stimulated, neuronal protein-RNA granules migrate to microenvironments within the cell interior to promote local protein synthesis. These RNA granules have various advantages in terms of energy economy and fine spatiotemporal control of cellular processes and can migrate to and from the nucleus and cytosol to remote sites like synaptic terminals (reviewed in ref. 221).

Once liquid droplets are formed, their transition to a more solid material is uncontrolled in the solution unless modulators are added. As time passes, the intermolecular contacts involved in the dense phase are favored, forming ordered structures with tight interactions (perhaps β-solenoid-like folds in the case of PrP) (reviewed in ref. 222). As the biophysics of condensates continuously change with aging, dehydration can occur[223] and stronger interactions lead to increased stability. How does aging take place? Jawerth et al. addressed the rheology of evolving FUS and PGL-3 condensates (up to ~46 h) and found a glass transition (disordered with viscoelastic properties) instead of spontaneous gelation (more ordered, highly crosslinked structures with elastic properties).[224] This intermediate stage between a liquid and a gel displays constant elasticity but higher viscosity with long incubation. Considering the ability of a range of phase-separated systems to be tuned depending on cellular conditions expands the already intricate roles of disordered proteins in signaling pathways. Also, the different material states formed by PrP certainly co-exist in a single cell at a particular moment, as do FUS condensates, which correspond to 1% of total nuclear FUS in living cells.[131] Furthermore, we speculate that some polymeric states of PrP could have functional implications, as shown with functional amyloids (e.g., amyloids of CPEB mediate memory consolidation; reviewed in ref. 225). Considering phase separation being the "new phase in cell biology,"[25] it is rather imprecise to describe microscopy patterns as "inclusion," "foci," and "puncta" formed by misfolding-prone proteins in cells. Moreover, these terms should be dissociated from the irreversible pathological aggregates since liquid condensates can be morphologically identical. To differentiate aggregates from liquid droplets, fluorescence recovery after photobleaching (FRAP) experiments are valuable, as they ascertain the mobility of proteins in the granules.[226] Liquid-like droplets contain macromolecules with a higher diffusion coefficient, meaning that recovering from the photobleached spot is rapid. Instead, solid-like aggregates display poor molecular kinetics (slower recovery).[227] Biomolecular condensates might not always be the intermediate between the soluble phase-separated and solid-like aggregates, though some evidence does exist for NA-binding prion-like proteins (reviewed in ref. 67). Interestingly, data on early oligomeric species have shown liquid-like behavior in mammalian cells.[228]

Aptamers would be an advantageous approach to tackle the aging of PrP biomolecular condensates that likely preclude *scrapie* formation. Since they are highly specific, they can compete with other NAs or proteins that drive aberrant transition. High concentrations of aptamers, however, can dissolve PrP condensates. With the potential effect of LLPS on TSEs and the development of prion-like diseases, new therapeutic strategies based on condensate formation and dissolution are starting to be discussed.[90] How drugs would partition into them represents an even greater challenge to the already complex scenario of IDP drug targeting. We conclude that the perspective embraced by Silva and Cordeiro on the "Jekyll and Hyde" actions of NAs on prion conformation[117] should be extended to the phase transitions phenomenon, which involves soft matter physics. Depending on how NAs modulate the material properties of PrP biomolecular condensates, they can behave as "physiological" Mr. Jekyll (dynamic liquid droplets) or "aberrant" Dr. Hyde (ordered solid-like condensates).

Acknowledgments

We thank Lucas Machado Ascari for careful revision of the text. This study was financed in part by the Fundação de Amparo à Pesquisa do Estado do Rio de Janeiro (FAPERJ) and by the Conselho Nacional de Desenvolvimento Científico e Tecnológico (CNPq).

References

1. Griffith JS. Nature of the scrapie agent: self-replication and scrapie. *Nature*. 1967;215 (5105):1043–1044.
2. Pattison IH, Jones KM. The possible nature of the transmissible agent of scrapie. *Vet Rec*. 1967;80(1):2–9.
3. Aguzzi A, De Cecco E. Shifts and drifts in prion science. *Science*. 2020;370(6512): 32–34.
4. Prusiner SB. Novel proteinaceous infectious particles cause scrapie. *Science*. 1982;216 (4542):136–144.

5. Prusiner SB. Cell biology. A unifying role for prions in neurodegenerative diseases. *Science*. 2012;336(6088):1511–1513.

6. Ano Bom AP, Rangel LP, Costa DC, et al. Mutant p53 aggregates into prion-like amyloid oligomers and fibrils: implications for cancer. *J Biol Chem*. 2012;287(33): 28152–28162.

7. Chiti F, Dobson C. Protein misfolding, amyloid formation, and human disease: a summary of progress over the last decade. *Annu Rev Biochem*. 2017;86:27–68.

8. Scott M, Foster D, Mirenda C, et al. Transgenic mice expressing hamster prion protein produce species-specific scrapie infectivity and amyloid plaques. *Cell*. 1989;59(5): 847–857.

9. Mompean M, Laurents DV. Intrinsically disordered domains, amyloids and protein liquid phases: evolving concepts and open questions. *Protein Pept Lett*. 2017;24 (4):281–293.

10. Kornberg A. *For the Love of Enzymes: The Odyssey of a Biochemist*. Harvard University Press; 1991.

11. Dobson C. Unfolded proteins, compact states and molten globules. *Curr Opin Struct Biol*. 1992;2(1):6–12.

12. Shakhnovich EI, Gutin AM. Engineering of stable and fast-folding sequences of model proteins. *Proc Natl Acad Sci U S A*. 1993;90(15):7195–7199.

13. Uversky VN, Dunker A. The case for intrinsically disordered proteins playing contributory roles in molecular recognition without a stable 3D structure. *F1000 Biol Rep*. 2013;5:1. https://doi.org/10.3410/B5-1.

14. Dunker AK, Obradovic Z, Romero P, Garner EC, Brown CJ. Intrinsic protein disorder in complete genomes. *Genome Inform Ser Workshop Genome Inform*. 2000;11: 161–171.

15. Dunker AK, Lawson JD, Brown CJ, et al. Intrinsically disordered protein. *J Mol Graph Model*. 2001;19(1):26–59.

16. Romero P, Obradovic Z, Kissinger C, Villafranca J, Dunker AK. Identifying disordered regions in proteins from amino acid sequence. In: *Paper presented at: Proceedings of International Conference on Neural Networks (ICNN'97)*; 1997.

17. Oldfield CJ, Cheng Y, Cortese MS, Brown CJ, Uversky VN, Dunker AK. Comparing and combining predictors of mostly disordered proteins. *Biochemistry*. 2005;44(6): 1989–2000.

18. Varadi M, Zsolyomi F, Guharoy M, Tompa P. Functional advantages of conserved intrinsic disorder in RNA-binding proteins. *PLoS One*. 2015;10(10), e0139731.

19. Xie H, Vucetic S, Iakoucheva LM, et al. Functional anthology of intrinsic disorder. 1. Biological processes and functions of proteins with long disordered regions. *J Proteome Res*. 2007;6(5):1882–1898. https://doi.org/10.1021/pr060392u.

20. Uversky V. Targeting intrinsically disordered proteins in neurodegenerative and protein dysfunction diseases: another illustration of the D2 concept. *Expert Rev Proteomics*. 2010;7(4):543–564.

21. Riek R, Hornemann S, Wider G, Glockshuber R, Wüthrich K. NMR characterization of the full-length recombinant murine prion protein, mPrP(23−231). *FEBS Lett*. 1997;413(2):282–288.

22. Uversky V, Oldfield C, Dunker A. Intrinsically disordered proteins in human diseases: introducing the D2 concept. *Annu Rev Biophys*. 2008;37:215–246.

23. Elbaum-Garfinkle S. Matter over mind: liquid phase separation and neurodegeneration. *J Biol Chem*. 2019;294(18):7160–7168.

24. Brangwynne CP, Eckmann CR, Courson DS, et al. Germline P granules are liquid droplets that localize by controlled dissolution/condensation. *Science*. 2009;324(5935): 1729–1732.

25. Boeynaems S, Alberti S, Fawzi NL, et al. Protein phase separation: a new phase in cell biology. *Trends Cell Biol*. 2018;28(6):420–435.

26. Kostylev MA, Tuttle MD, Lee S, et al. Liquid and hydrogel phases of PrP(C) linked to conformation shifts and triggered by Alzheimer's amyloid-beta oligomers. *Mol Cell.* 2018;72(3):426–443.e412.

27. Matos CO, Passos YM, do Amaral MJ, et al. Liquid-liquid phase separation and fibrillation of the prion protein modulated by a high-affinity DNA aptamer. *FASEB J.* 2020;34(1):365–385.

28. Zahn R, Liu A, Luhrs T, et al. NMR solution structure of the human prion protein. *Proc Natl Acad Sci U S A.* 2000;97(1):145–150.

29. Herrmann LM, Caughey B. The importance of the disulfide bond in prion protein conversion. *Neuroreport.* 1998;9(11):2457–2461.

30. Tompa P, Tusnády GE, Friedrich P, Simon I. The role of dimerization in prion replication. *Biophys J.* 2002;82(4):1711–1718.

31. Hornemann S, Schorn C, Wuthrich K. NMR structure of the bovine prion protein isolated from healthy calf brains. *EMBO Rep.* 2004;5(12):1159–1164.

32. Calzolai L, Lysek DA, Perez DR, Guntert P, Wuthrich K. Prion protein NMR structures of chickens, turtles, and frogs. *Proc Natl Acad Sci U S A.* 2005;102(3):651–655.

33. Lysek DA, Schorn C, Nivon LG, et al. Prion protein NMR structures of cats, dogs, pigs, and sheep. *Proc Natl Acad Sci U S A.* 2005;102(3):640–645.

34. Bendheim PE, Brown HR, Rudelli RD, et al. Nearly ubiquitous tissue distribution of the scrapie agent precursor protein. *Neurology.* 1992;42(1):149–156.

35. Bosque PJ, Ryou C, Telling G, et al. Prions in skeletal muscle. *Proc Natl Acad Sci U S A.* 2002;99(6):3812–3817.

36. Cashman NR, Loertscher R, Nalbantoglu J, et al. Cellular isoform of the scrapie agent protein participates in lymphocyte activation. *Cell.* 1990;61(1):185–192.

37. Starke R, Drummond O, MacGregor I, et al. The expression of prion protein by endothelial cells: a source of the plasma form of prion protein? *Br J Haematol.* 2002;119 (3):863–873.

38. Koperek O, Kovacs GG, Ritchie D, Ironside JW, Budka H, Wick G. Disease-associated prion protein in vessel walls. *Am J Pathol.* 2002;161(6):1979–1984.

39. Rane NS, Yonkovich JL, Hegde RS. Protection from cytosolic prion protein toxicity by modulation of protein translocation. *EMBO J.* 2004;23(23):4550–4559.

40. Campana V, Sarnataro D, Zurzolo C. The highways and byways of prion protein trafficking. *Trends Cell Biol.* 2005;15(2):102–111.

41. Macedo B, Cordeiro Y. Unraveling prion protein interactions with aptamers and other PrP-binding nucleic acids. *Int J Mol Sci.* 2017;18(5):1023.

42. Linden R. The biological function of the prion protein: a cell surface scaffold of signaling modules. *Front Mol Neurosci.* 2017;10:77.

43. Ermonval M, Mouillet-Richard S, Codogno P, Kellermann O, Botti J. Evolving views in prion glycosylation: functional and pathological implications. *Biochimie.* 2003;85 (1–2):33–45.

44. Giannopoulos PN, Robertson C, Jodoin J, Paudel H, Booth SA, LeBlanc AC. Phosphorylation of prion protein at serine 43 induces prion protein conformational change. *J Neurosci.* 2009;29(27):8743–8751.

45. Yedidia Y, Horonchik L, Tzaban S, Yanai A, Taraboulos A. Proteasomes and ubiquitin are involved in the turnover of the wild-type prion protein. *EMBO J.* 2001;20 (19):5383–5391.

46. Linsenmeier L, Altmeppen HC, Wetzel S, Mohammadi B, Saftig P, Glatzel M. Diverse functions of the prion protein—does proteolytic processing hold the key? *Biochim Biophys Acta Mol Cell Res.* 2017;1864(11 pt B):2128–2137.

47. Brown DR, Qin K, Herms JW, et al. The cellular prion protein binds copper in vivo. *Nature.* 1997;390(6661):684–687.

48. Cavaliere P, Pagano B, Granata V, et al. Cross-talk between prion protein and quadruplex-forming nucleic acids: a dynamic complex formation. *Nucleic Acids Res.* 2013;41(1):327–339.
49. Olsthoorn RC. G-quadruplexes within prion mRNA: the missing link in prion disease? *Nucleic Acids Res.* 2014;42(14):9327–9333.
50. Pradhan P, Srivastava A, Singh J, et al. Prion protein transcription is auto-regulated through dynamic interactions with G-quadruplex motifs in its own promoter. *Biochim Biophys Acta Gene Regul Mech.* 1863;2020(3):194479.
51. Ascari LM, Rocha SC, Gonçalves PB, TCRG V, Cordeiro Y. Challenges and advances in antemortem diagnosis of human transmissible spongiform encephalopathies. *Front Bioeng Biotechnol.* 2020;8(1228):585896.
52. Uttley L, Carroll C, Wong R, Hilton DA, Stevenson M. Creutzfeldt-Jakob disease: a systematic review of global incidence, prevalence, infectivity, and incubation. *Lancet Infect Dis.* 2020;20(1):e2–e10.
53. Jankovska N, Olejar T, Matej R. Extracellular amyloid deposits in Alzheimer's and Creutzfeldt–Jakob disease: similar behavior of different proteins? *Int J Mol Sci.* 2021; 22(1):7.
54. Prusiner SB. Prions. *Proc Natl Acad Sci U S A.* 1998;95(23):13363–13383.
55. Aulic S, Masperone L, Narkiewicz J, et al. Alpha-synuclein amyloids hijack prion protein to gain cell entry, facilitate cell-to-cell spreading and block prion replication. *Sci Rep.* 2017;7(1):10050.
56. Bistaffa E, Rossi M, De Luca CMG, et al. Prion efficiently replicates in alpha-synuclein knockout mice. *Mol Neurobiol.* 2019;56(11):7448–7457.
57. Kostylev MA, Kaufman AC, Nygaard HB, et al. Prion-protein-interacting amyloid-beta oligomers of high molecular weight are tightly correlated with memory impairment in multiple Alzheimer mouse models. *J Biol Chem.* 2015;290(28): 17415–17438.
58. Lauren J, Gimbel DA, Nygaard HB, Gilbert JW, Strittmatter SM. Cellular prion protein mediates impairment of synaptic plasticity by amyloid-beta oligomers. *Nature.* 2009;457(7233):1128–1132.
59. Resenberger UK, Harmeier A, Woerner AC, et al. The cellular prion protein mediates neurotoxic signalling of beta-sheet-rich conformers independent of prion replication. *EMBO J.* 2011;30(10):2057–2070.
60. Corbett GT, Wang Z, Hong W, et al. PrP is a central player in toxicity mediated by soluble aggregates of neurodegeneration-causing proteins. *Acta Neuropathol.* 2020;139 (3):503–526.
61. Ferreira DG, Temido-Ferreira M, Vicente Miranda H, et al. Alpha-synuclein interacts with PrP(C) to induce cognitive impairment through mGluR5 and NMDAR2B. *Nat Neurosci.* 2017;20(11):1569–1579.
62. Katorcha E, Makarava N, Lee YJ, et al. Cross-seeding of prions by aggregated alpha-synuclein leads to transmissible spongiform encephalopathy. *PLoS Pathog.* 2017; 13(8):e1006563.
63. Masliah E, Rockenstein E, Inglis C, et al. Prion infection promotes extensive accumulation of alpha-synuclein in aged human alpha-synuclein transgenic mice. *Prion.* 2012;6 (2):184–190.
64. Uversky VN. Recent developments in the field of intrinsically disordered proteins: intrinsic disorder–based emergence in cellular biology in light of the physiological and pathological liquid–liquid phase transitions. *Annu Rev Biophys.* 2021;50: 135–156.
65. Shin Y, Brangwynne CP. Liquid phase condensation in cell physiology and disease. *Science.* 2017;357(6357):eaaf4382. https://doi.org/10.1126/science.aaf4382.

66. Guzikowski AR, Chen YS, Zid BM. Stress-induced mRNP granules: form and function of processing bodies and stress granules. *Wiley Interdiscip Rev RNA.* 2019;10(3): e1524.

67. Alberti S, Hyman AA. Biomolecular condensates at the nexus of cellular stress, protein aggregation disease and ageing. *Nat Rev Mol Cell Biol.* 2021;22(3):196–213.

68. Nott TJ, Petsalaki E, Farber P, et al. Phase transition of a disordered nuage protein generates environmentally responsive membraneless organelles. *Mol Cell.* 2015;57 (5):936–947.

69. Fonin AV, Darling AL, Kuznetsova IM, Turoverov KK, Uversky VN. Intrinsically disordered proteins in crowded milieu: when chaos prevails within the cellular gumbo. *Cell Mol Life Sci.* 2018;75(21):3907–3929.

70. Frankel EA, Bevilacqua PC, Keating CD. Polyamine/nucleotide coacervates provide strong compartmentalization of Mg2+, nucleotides, and RNA. *Langmuir.* 2016;32 (8):2041–2049.

71. Nakashima KK, Vibhute MA, Spruijt E. Biomolecular chemistry in liquid phase separated compartments. *Front Mol Biosci.* 2019;6:21.

72. Berkeley RF, Kashefi M, Debelouchina GT. Real-time observation of structure and dynamics during the liquid-to-solid transition of FUS LC. *Biophys J.* 2021;120: 1276–1287.

73. Emmanouilidis L, Esteban-Hofer L, Damberger FF, et al. NMR and EPR reveal a compaction of the RNA-binding protein FUS upon droplet formation. *Nat Chem Biol.* 2021;1–7.

74. Brady JP, Farber PJ, Sekhar A, et al. Structural and hydrodynamic properties of an intrinsically disordered region of a germ cell-specific protein on phase separation. *Proc Natl Acad Sci U S A.* 2017;114(39):E8194–E8203.

75. do Amaral MJ, Araujo TS, Díaz NC, et al. Phase separation and disorder-to-order transition of human brain expressed X-linked 3 (hBEX3) in the presence of small fragments of tRNA. *J Mol Biol.* 2020;432(7):2319–2348.

76. Harmon TS, Holehouse AS, Rosen MK, Pappu RV. Intrinsically disordered linkers determine the interplay between phase separation and gelation in multivalent proteins. *Elife.* 2017;6, e30294.

77. Hughes MP, Sawaya MR, Boyer DR, et al. Atomic structures of low-complexity protein segments reveal kinked β sheets that assemble networks. *Science.* 2018;359(6376): 698–701.

78. Ruff KM, Dar F, Pappu RV. Ligand effects on phase separation of multivalent macromolecules. *Proc Natl Acad Sci U S A.* 2021;118(10), e2017184118.

79. Kamagata K, Kanbayashi S, Honda M, et al. Liquid-like droplet formation by tumor suppressor p53 induced by multivalent electrostatic interactions between two disordered domains. *Sci Rep.* 2020;10(1):580.

80. Safari MS, Wang Z, Tailor K, Kolomeisky AB, Conrad JC, Vekilov PG. Anomalous dense liquid condensates host the nucleation of tumor suppressor p53 fibrils. *iScience.* 2019;12:342–355.

81. Petronilho EC, Pedrote MM, Marques MA, et al. Phase separation of p53 precedes aggregation and is affected by oncogenic mutations and ligands. *Chem Sci.* 2021;12 (21):7334–7349.

82. Bouchard JJ, Otero JH, Scott DC, et al. Cancer mutations of the tumor suppressor SPOP disrupt the formation of active, phase-separated compartments. *Mol Cell.* 2018;72(1):19–36.e18.

83. Conicella AE, Dignon GL, Zerze GH, et al. TDP-43 alpha-helical structure tunes liquid-liquid phase separation and function. *Proc Natl Acad Sci U S A.* 2020;117(11): 5883–5894.

84. Patel A, Lee HO, Jawerth L, et al. A liquid-to-solid phase transition of the ALS protein FUS accelerated by disease mutation. *Cell*. 2015;162(5):1066–1077.

85. Qamar S, Wang G, Randle SJ, et al. FUS phase separation is modulated by a molecular chaperone and methylation of arginine cation-pi interactions. *Cell*. 2018;173(3): 720–734.e715.

86. Ambadipudi S, Biernat J, Riedel D, Mandelkow E, Zweckstetter M. Liquid-liquid phase separation of the microtubule-binding repeats of the Alzheimer-related protein tau. *Nat Commun*. 2017;8(1):275.

87. Wegmann S, Eftekharzadeh B, Tepper K, et al. Tau protein liquid-liquid phase separation can initiate tau aggregation. *EMBO J*. 2018;37(7):e98049.

88. Peskett TR, Rau F, O'Driscoll J, Patani R, Lowe AR, Saibil HR. A liquid to solid phase transition underlying pathological huntingtin Exon1 aggregation. *Mol Cell*. 2018;70(4):588–601.e586.

89. Ray S, Singh N, Kumar R, et al. Alpha-synuclein aggregation nucleates through liquid-liquid phase separation. *Nat Chem*. 2020;12:705–716.

90. Wheeler RJ. Therapeutics-how to treat phase separation-associated diseases. *Emerg Top Life Sci*. 2020;4(3):331–342.

91. Prusiner S. Prions. *Proc Natl Acad Sci U S A*. 1998;95(23):13363–13383.

92. Soding J. Protein homology detection by HMM-HMM comparison. *Bioinformatics*. 2005;21(7):951–960.

93. Ciric D, Rezaei H. Biochemical insight into the prion protein family. *Front Cell Dev Boil*. 2015;3:5.

94. Lührs T, Riek R, Güntert P, Wüthrich K. NMR structure of the human doppel protein. *J Mol Biol*. 2003;326(5):1549–1557.

95. Watts JC, Westaway D. The prion protein family: diversity, rivalry, and dysfunction. *Biochim Biophys Acta Mol Basis Dis*. 2007;1772(6):654–672.

96. Moore RC, Lee IY, Silverman GL, et al. Ataxia in prion protein (PrP)-deficient mice is associated with upregulation of the novel PrP-like protein doppel. *J Mol Biol*. 1999;292 (4):797–817.

97. Watts JC, Drisaldi B, Ng V, et al. The CNS glycoprotein Shadoo has PrP(C)-like protective properties and displays reduced levels in prion infections. *EMBO J*. 2007;26 (17):4038–4050.

98. Liu J, Perumal NB, Oldfield CJ, Su EW, Uversky VN, Dunker AK. Intrinsic disorder in transcription factors. *Biochemistry*. 2006;45(22):6873–6888.

99. Daude N, Ng V, Watts JC, et al. Wild-type Shadoo proteins convert to amyloid-like forms under native conditions. *J Neurochem*. 2010;113(1):92–104.

100. Forloni G, Chiesa R, Bugiani O, Salmona M, Tagliavini F. PrP 106-126—25 years after. *Neuropathol Appl Neurobiol*. 2019;45(5):430–440.

101. Qamar S, Wang G, Randle SJ, et al. FUS phase separation is modulated by a molecular chaperone and methylation of arginine cation-π interactions. *Cell*. 2018;173 (3):720–734. e715.

102. Hofweber M, Dormann D. Friend or foe—post-translational modifications as regulators of phase separation and RNP granule dynamics. *J Biol Chem*. 2019;294(18):7137–7150.

103. Vernon RM, Chong PA, Tsang B, et al. Pi-Pi contacts are an overlooked protein feature relevant to phase separation. *Elife*. 2018;7:e31486.

104. Bolognesi B, Gotor NL, Dhar R, et al. A concentration-dependent liquid phase separation can cause toxicity upon increased protein expression. *Cell Rep*. 2016;16 (1):222–231.

105. Kang S-G, Mays CE, Daude N, Yang J, Kar S, Westaway D. Proteasomal inhibition redirects the PrP-like Shadoo protein to the nucleus. *Mol Neurobiol*. 2019;56(11): 7888–7904.

106. Lau A, Mays CE, Genovesi S, Westaway D. RGG repeats of PrP-like Shadoo protein bind nucleic acids. *Biochemistry*. 2012;51(45):9029–9031.
107. Tóth E, Kulcsár P, Fodor E, et al. The highly conserved, N-terminal (RXXX) 8 motif of mouse Shadoo mediates nuclear accumulation. *Biochim Biophys Acta Mol Cell Res*. 2013;1833(5):1199–1211.
108. Premzl M, Gready JE, Jermiin LS, Simonic T, Marshall Graves JA. Evolution of vertebrate genes related to prion and Shadoo proteins—clues from comparative genomic analysis. *Mol Biol Evol*. 2004;21(12):2210–2231.
109. Castle AR, Gill AC. Physiological functions of the cellular prion protein. *Front Mol Biosci*. 2017;4:19. https://doi.org/10.3389/fmolb.2017.00019.
110. Lima LM, Cordeiro Y, Tinoco LW, et al. Structural insights into the interaction between prion protein and nucleic acid. *Biochemistry*. 2006;45(30):9180–9187.
111. Nandi P, Leclerc E, Nicole J-C, Takahashi M. DNA-induced partial unfolding of prion protein leads to its polymerisation to amyloid. *J Mol Biol*. 2002;322(1):153–161.
112. Bera A, Biring S. A sequence-dependent DNA condensation induced by prion protein. *J Nucleic Acids*. 2018;2018.
113. Abid K, Morales R, Soto C. Cellular factors implicated in prion replication. *FEBS Lett*. 2010;584(11):2409–2414.
114. Cordeiro Y, Machado F, Juliano L, et al. DNA converts cellular prion protein into the beta-sheet conformation and inhibits prion peptide aggregation. *J Biol Chem*. 2001;276 (52):49400–49409.
115. Deleault NR, Harris BT, Rees JR, Supattapone S. Formation of native prions from minimal components in vitro. *Proc Natl Acad Sci U S A*. 2007;104(23):9741–9746.
116. Deleault NR, Lucassen RW, Supattapone S. RNA molecules stimulate prion protein conversion. *Nature*. 2003;425(6959):717–720.
117. Silva JL, Cordeiro Y. The "Jekyll and Hyde" actions of nucleic acids on the prion-like aggregation of proteins. *J Biol Chem*. 2016;291(30):15482–15490.
118. Passos YM, do Amaral MJ, Ferreira NC, et al. The interplay between a GC-rich oligonucleotide and copper ions on prion protein conformational and phase transitions. *Int J Biol Macromol*. 2021;173:34–43.
119. Gomes MP, Millen TA, Ferreira PS, et al. Prion protein complexed to N2a cellular RNAs through its N-terminal domain forms aggregates and is toxic to murine neuroblastoma cells. *J Biol Chem*. 2008;283(28):19616–19625.
120. Macedo B, Millen TA, Braga CA, et al. Nonspecific prion protein-nucleic acid interactions lead to different aggregates and cytotoxic species. *Biochemistry*. 2012;51(27): 5402–5413.
121. Saá P, Sferrazza GF, Ottenberg G, Oelschlegel AM, Dorsey K, Lasmézas CI. Strain-specific role of RNAs in prion replication. *J Virol*. 2012;86(19):10494–10504.
122. Mercey R, Lantier I, Maurel M-C, Grosclaude J, Lantier F, Marc D. Fast, reversible interaction of prion protein with RNA aptamers containing specific sequence patterns. *Arch Virol*. 2006;151(11):2197–2214.
123. Yin S, Fan X, Yu S, Li C, Sy M-S. Binding of recombinant but not endogenous prion protein to DNA causes DNA internalization and expression in mammalian cells. *J Biol Chem*. 2008;283(37):25446–25454.
124. Rhie A, Kirby L, Sayer N, et al. Characterization of 2′-fluoro-RNA aptamers that bind preferentially to disease-associated conformations of prion protein and inhibit conversion. *J Biol Chem*. 2003;278(41):39697–39705.
125. Gu Y, Hinnerwisch J, Fredricks R, Kalepu S, Mishra RS, Singh N. Identification of cryptic nuclear localization signals in the prion protein. *Neurobiol Dis*. 2003;12 (2):133–149.
126. Jaegly A, Mouthon F, Peyrin J-M, Camugli B, Deslys J-P, Dormont D. Search for a nuclear localization signal in the prion protein. *Mol Cell Neurosci*. 1998;11(3):127–133.

127. De Groot NS, Armaos A, Graña-Montes R, et al. RNA structure drives interaction with proteins. *Nat Commun.* 2019;10(1):1–13.
128. Hackermüller J, Meisner N-C, Auer M, Jaritz M, Stadler PF. The effect of RNA secondary structures on RNA-ligand binding and the modifier RNA mechanism: a quantitative model. *Gene.* 2005;345(1):3–12.
129. Langdon EM, Gladfelter AS. A new lens for RNA localization: liquid-liquid phase separation. *Annu Rev Microbiol.* 2018;72:255–271.
130. Langdon EM, Qiu Y, Niaki AG, et al. mRNA structure determines specificity of a polyQ-driven phase separation. *Sci Adv.* 2018;360(6391):922–927.
131. Maharana S, Wang J, Papadopoulos DK, et al. RNA buffers the phase separation behavior of prion-like RNA binding proteins. *Science.* 2018;360(6391):918–921.
132. Parkinson G. Fundamentals of quadruplex structures. In: *Quadruplex Nucleic Acids.* UK: Royal Society of Chemistry; 2006:1–30.
133. Bera A, Roche A-C, Nandi P. Bending and unwinding of nucleic acid by prion protein. *Biochemistry.* 2007;46(5):1320–1328.
134. Czech A, Konarev PV, Goebel I, Svergun DI, Wills PR, Ignatova Z. Octa-repeat domain of the mammalian prion protein mRNA forms stable A-helical hairpin structure rather than G-quadruplexes. *Sci Rep.* 2019;9(1):1–10.
135. Kim HD, Kim TS, Joo YJ, et al. RpS3 translation is repressed by interaction with its own mRNA. *J Cell Biochem.* 2010;110(2):294–303.
136. Armaos A, Zacco E, Sanchez de Groot N, Tartaglia GG. RNA-protein interactions: central players in coordination of regulatory networks. *BioEssays.* 2021;43(2):2000118.
137. Saha S, Weber CA, Nousch M, et al. Polar positioning of phase-separated liquid compartments in cells regulated by an mRNA competition mechanism. *Cell.* 2016;166(6):1572–1584. e1516.
138. Hentze M, Castello A, Schwarzl T, Preiss T. A brave new world of RNA-binding proteins. *Nat Rev Mol Cell Biol.* 2018;19(5):327.
139. Derrington E, Gabus C, Leblanc P, et al. PrPC has nucleic acid chaperoning properties similar to the nucleocapsid protein of HIV-1. *C R Biol.* 2002;325(1):17–23.
140. Gabus C, Auxilien S, Pechoux C, et al. The prion protein has DNA strand transfer properties similar to retroviral nucleocapsid protein. *J Mol Biol.* 2001;307(4):1011–1021.
141. Gabus C, Derrington E, Leblanc P, et al. The prion protein has RNA binding and chaperoning properties characteristic of nucleocapsid protein NCP7 of HIV-1. *J Biol Chem.* 2001;276(22):19301–19309.
142. Wu Y. Unwinding and rewinding: double faces of helicase? *J Nucleic Acids.* 2012; 2012:140601.
143. Wang J, Choi J-M, Holehouse AS, et al. A molecular grammar governing the driving forces for phase separation of prion-like RNA binding proteins. *Cell.* 2018;174(3):688–699. e616.
144. Burke KA, Janke AM, Rhine CL, Fawzi NL. Residue-by-residue view of in vitro FUS granules that bind the C-terminal domain of RNA polymerase II. *Mol Cell.* 2015;60(2):231–241.
145. Murakami T, Qamar S, Lin JQ, et al. ALS/FTD mutation-induced phase transition of FUS liquid droplets and reversible hydrogels into irreversible hydrogels impairs RNP granule function. *Neuron.* 2015;88(4):678–690.
146. Guseva S, Milles S, Jensen MR, et al. Measles virus nucleo-and phosphoproteins form liquid-like phase-separated compartments that promote nucleocapsid assembly. *Sci Adv.* 2020;6(14):eaaz7095.
147. Lu S, Ye Q, Singh D, et al. The SARS-CoV-2 nucleocapsid phosphoprotein forms mutually exclusive condensates with RNA and the membrane-associated M protein. *Nat Commun.* 2021;12(1):1–15.

148. Nott TJ, Craggs TD, Baldwin AJ. Membraneless organelles can melt nucleic acid duplexes and act as biomolecular filters. *Nat Chem.* 2016;8(6):569–575.

149. Kellings K, Meyer N, Mirenda C, Prusiner SB, Riesner D. Further analysis of nucleic acids in purified scrapie prion preparations by improved return refocusing gel electrophoresis. *J Gen Virol.* 1992;73(Pt 4):1025–1029.

150. Safar JG, Kellings K, Serban A, et al. Search for a prion-specific nucleic acid. *J Virol.* 2005;79(16):10796–10806.

151. Simoneau S, Thomzig A, Ruchoux MM, et al. Synthetic scrapie infectivity: interaction between recombinant PrP and scrapie brain-derived RNA. *Virulence.* 2015;6(2): 132–144.

152. Aiken J, Williamson J, Borchardt L, Marsh R. Presence of mitochondrial D-loop DNA in scrapie-infected brain preparations enriched for the prion protein. *J Virol.* 1990;64 (7):3265–3268.

153. Akowitz A, Sklaviadis T, Manuelidis L. Endogenous viral complexes with long RNA cosediment with the agent of Creutzfeldt-Jakob disease. *Nucleic Acids Res.* 1994;22 (6):1101–1107.

154. Manuelidis L. Transmissible encephalopathies: speculations and realities. *Viral Immunol.* 2003;16(2):123–139.

155. Oesch B, Groth D, Prusiner S, Weissmann C. Search for a scrapie-specific nucleic acid: a progress report. In: *Paper presented at: Ciba Found Symp;* 1988.

156. Zou W-Q, Zheng J, Gray DM, Gambetti P, Chen SG. Antibody to DNA detects scrapie but not normal prion protein. *Proc Natl Acad Sci U S A.* 2004;101(5):1380–1385.

157. Sayer NM, Cubin M, Rhie A, Bullock M, Tahiri-Alaoui A, James W. Structural determinants of conformationally selective, prion-binding aptamers. *J Biol Chem.* 2004;279 (13):13102–13109.

158. Tuerk C, Gold L. Systematic evolution of ligands by exponential enrichment: RNA ligands to bacteriophage T4 DNA polymerase. *Science.* 1990;249(4968):505–510.

159. Meli M, Gasset M, Colombo G. Are amyloid fibrils RNA-traps? A molecular dynamics perspective. *Front Mol Biosci.* 2018;5:53.

160. Agarwal A, Rai SK, Avni A, Mukhopadhyay S. An intrinsically disordered pathological variant of the prion protein Y145Stop transforms into self-templating amyloids via liquid-liquid phase separation. *bioRxiv.* 2021. https://doi.org/10.1101/2021.01.09. 426049.

161. Hammond SM, Aartsma-Rus A, Alves S, et al. Delivery of oligonucleotide-based therapeutics: challenges and opportunities. *EMBO Mol Med.* 2021;13, e13243.

162. Das S, Singer RH, Yoon YJ. The travels of mRNAs in neurons: do they know where they are going? *Curr Opin Neurobiol.* 2019;57:110–116.

163. Hafner M, Landthaler M, Burger L, et al. Transcriptome-wide identification of RNA-binding protein and microRNA target sites by PAR-CLIP. *Cell.* 2010;141 (1):129–141.

164. Zhang X, Lin Y, Eschmann NA, et al. RNA stores tau reversibly in complex coacervates. *PLoS Biol.* 2017;15(7):e2002183.

165. Kovachev PS, Gomes MPB, Cordeiro Y, et al. RNA modulates aggregation of the recombinant mammalian prion protein by direct interaction. *Sci Rep.* 2019;9(1):12406.

166. Bera A, Biring S. A quantitative characterization of interaction between prion protein with nucleic acids. *Biochem Biophys Rep.* 2018;14:114–124.

167. Su Z, Wilson B, Kumar P, Dutta A. Noncanonical roles of tRNAs: tRNA fragments and beyond. *Annu Rev Genet.* 2020;54:47–69.

168. Prehn JHM, Jirstrom E. Angiogenin and tRNA fragments in Parkinson's disease and neurodegeneration. *Acta Pharmacol Sin.* 2020;41(4):442–446.

169. Qin C, Xu PP, Zhang X, et al. Pathological significance of tRNA-derived small RNAs in neurological disorders. *Neural Regen Res.* 2020;15(2):212–221.

170. Zhang X, Trebak F, Souza LA, et al. Small RNA modifications in Alzheimer's disease. *Neurobiol Dis*. 2020;145:105058.
171. Kumar P, Anaya J, Mudunuri SB, Dutta A. Meta-analysis of tRNA derived RNA fragments reveals that they are evolutionarily conserved and associate with AGO proteins to recognize specific RNA targets. *BMC Biol*. 2014;12:78.
172. Li S, Xu Z, Sheng J. tRNA-derived small RNA: a novel regulatory small non-coding RNA. *Genes (Basel)*. 2018;9(5):246. https://doi.org/10.3390/genes9050246.
173. Saikia M, Jobava R, Parisien M, et al. Angiogenin-cleaved tRNA halves interact with cytochrome c, protecting cells from apoptosis during osmotic stress. *Mol Cell Biol*. 2014;34(13):2450–2463.
174. Hanada T, Weitzer S, Mair B, et al. CLP1 links tRNA metabolism to progressive motor-neuron loss. *Nature*. 2013;495(7442):474–480.
175. Inoue M, Hada K, Shiraishi H, et al. Tyrosine pre-transfer RNA fragments are linked to p53-dependent neuronal cell death via PKM2. *Biochem Biophys Res Commun*. 2020;525 (3):726–732.
176. Gibbings D, Leblanc P, Jay F, et al. Human prion protein binds Argonaute and promotes accumulation of microRNA effector complexes. *Nat Struct Mol Biol*. 2012;19 (5):517–524 [S511].
177. Satoh J, Obayashi S, Misawa T, Sumiyoshi K, Oosumi K, Tabunoki H. Protein microarray analysis identifies human cellular prion protein interactors. *Neuropathol Appl Neurobiol*. 2009;35(1):16–35.
178. Kuscu C, Kumar P, Kiran M, Su Z, Malik A, Dutta A. tRNA fragments (tRFs) guide ago to regulate gene expression post-transcriptionally in a dicer-independent manner. *RNA*. 2018;24(8):1093–1105.
179. Haussecker D, Huang Y, Lau A, Parameswaran P, Fire AZ, Kay MA. Human tRNA-derived small RNAs in the global regulation of RNA silencing. *RNA*. 2010;16(4):673–695.
180. Turchinovich A, Weiz L, Langheinz A, Burwinkel B. Characterization of extracellular circulating microRNA. *Nucleic Acids Res*. 2011;39(16):7223–7233.
181. Tosar J. Die hard: resilient RNAs in the blood. *Nat Rev Mol Cell Biol*. 2021;22(6):373.
182. Bartel DP. MicroRNAs: genomics, biogenesis, mechanism, and function. *Cell*. 2004;116(2):281–297.
183. Goodall EF, Heath PR, Bandmann O, Kirby J, Shaw PJ. Neuronal dark matter: the emerging role of microRNAs in neurodegeneration. *Front Cell Neurosci*. 2013; 7:178.
184. Pease D, Scheckel C, Schaper E, et al. Genome-wide identification of microRNAs regulating the human prion protein. *Brain Pathol*. 2019;29(2):232–244.
185. Bueler H, Aguzzi A, Sailer A, et al. Mice devoid of PrP are resistant to scrapie. *Cell*. 1993;73(7):1339–1347.
186. Sailer A, Büeler H, Fischer M, Aguzzi A, Weissmann C. No propagation of prions in mice devoid of PrP. *Cell*. 1994;77(7):967–968.
187. Mallucci G, Dickinson A, Linehan J, Klöhn P-C, Brandner S, Collinge J. Depleting neuronal PrP in prion infection prevents disease and reverses spongiosis. *Science*. 2003;302(5646):871–874.
188. Arroyo JD, Chevillet JR, Kroh EM, et al. Argonaute2 complexes carry a population of circulating microRNAs independent of vesicles in human plasma. *Proc Natl Acad Sci U S A*. 2011;108(12):5003–5008.
189. Wang K, Zhang S, Weber J, Baxter D, Galas DJ. Export of microRNAs and microRNA-protective protein by mammalian cells. *Nucleic Acids Res*. 2010;38(20): 7248–7259.
190. Mitrea DM, Cika JA, Stanley CB, et al. Self-interaction of NPM1 modulates multiple mechanisms of liquid-liquid phase separation. *Nat Commun*. 2018;9(1):842.

191. Bruno D, Donatti A, Martin M, et al. Circulating nucleic acids in the plasma and serum as potential biomarkers in neurological disorders. *Braz J Med Biol Res.* 2020;53(10): e9881.

192. Fevrier B, Vilette D, Archer F, et al. Cells release prions in association with exosomes. *Proc Natl Acad Sci U S A.* 2004;101(26):9683–9688.

193. Vella L, Sharples R, Lawson V, et al. Packaging of prions into exosomes is associated with a novel pathway of PrP processing. *J Pathol.* 2007;211(5):582–590.

194. Llorens F, Villar-Piqué A, Schmitz M, et al. Plasma total prion protein as a potential biomarker for neurodegenerative dementia: diagnostic accuracy in the spectrum of prion diseases. *Neuropathol Appl Neurobiol.* 2020;46(3):240–254.

195. Jones E, Hummerich H, Viré E, et al. Identification of novel risk loci and causal insights for sporadic Creutzfeldt-Jakob disease: a genome-wide association study. *Lancet Neurol.* 2020;19(10):840–848.

196. Bellingham SA, Coleman BM, Hill AF. Small RNA deep sequencing reveals a distinct miRNA signature released in exosomes from prion-infected neuronal cells. *Nucleic Acids Res.* 2012;40(21):10937–10949.

197. Cheng L, Quek C, Li X, et al. Distribution of microRNA profiles in pre-clinical and clinical forms of murine and human prion disease. *Commun Biol.* 2021;4(1):1–12.

198. Boccaletto P, Machnicka MA, Purta E, et al. MODOMICS: a database of RNA modification pathways. 2017 update. *Nucleic Acids Res.* 2018;46(D1):D303–D307.

199. Koren S, Galvis-Escobar S, Abisambra JF. Tau-mediated dysregulation of RNA: evidence for a common molecular mechanism of toxicity in frontotemporal dementia and other tauopathies. *Neurobiol Dis.* 2020;141:104939.

200. Cordeiro Y, Vieira T, Kovachev PS, Sanyal S, Silva JL. Modulation of p53 and prion protein aggregation by RNA. *Biochim Biophys Acta Proteins Proteomics.* 2019;1867 (10):933–940.

201. Navas-Pérez E, Vicente-García C, Mirra S, et al. Characterization of an eutherian gene cluster generated after transposon domestication identifies Bex3 as relevant for advanced neurological functions. *Genome Biol.* 2020;21(1):1–27.

202. Tuorto F, Parlato R. rRNA and tRNA bridges to neuronal homeostasis in health and disease. *J Mol Biol.* 2019;431(9):1763–1779.

203. Astoricchio E, Alfano C, Rajendran L, Temussi PA, Pastore A. The wide world of coacervates: from the sea to neurodegeneration. *Trends Biochem Sci.* 2020;45(8): 706–717.

204. Cui M, Wang X, An B, et al. Exploiting mammalian low-complexity domains for liquid-liquid phase separation-driven underwater adhesive coatings. *Sci Adv.* 2019; 5(8):eaax3155.

205. Maori E, Navarro IC, Boncristiani H, et al. A secreted RNA binding protein forms RNA-stabilizing granules in the honeybee royal jelly. *Mol Cell.* 2019;74(3): 598–608.e596.

206. Tange H, Ishibashi D, Nakagaki T, et al. Liquid–liquid phase separation of full-length prion protein initiates conformational conversion in vitro. *J Biol Chem.* 2021;296: 100367.

207. Huang JJ, Li XN, Liu WL, et al. Neutralizing mutations significantly inhibit amyloid formation by human prion protein and decrease its cytotoxicity. *J Mol Biol.* 2020; 432(4):828–844.

208. Banerjee PR, Milin AN, Moosa MM, Onuchic PL, Deniz AA. Reentrant phase transition drives dynamic substructure formation in ribonucleoprotein droplets. *Angew Chem.* 2017;129(38):11512–11517.

209. Ederle H, Dormann D. TDP-43 and FUS en route from the nucleus to the cytoplasm. *FEBS Lett.* 2017;591(11):1489–1507.

210. Chan FT, Schierle GSK, Kumita JR, Bertoncini CW, Dobson CM, Kaminski CF. Protein amyloids develop an intrinsic fluorescence signature during aggregation. *Analyst.* 2013;138(7):2156–2162.

211. Lorenz H, Windl O, Kretzschmar H. Cellular phenotyping of secretory and nuclear prion proteins associated with inherited prion diseases. *J Biol Chem.* 2002;277(10): 8508–8516.

212. Pinho R, Paiva I, Jerčić KG, et al. Nuclear localization and phosphorylation modulate pathological effects of alpha-synuclein. *Hum Mol Genet.* 2019;28(1):31–50.

213. Mange A, Crozet C, Lehmann S, Beranger F. Scrapie-like prion protein is translocated to the nuclei of infected cells independently of proteasome inhibition and interacts with chromatin. *J Cell Sci.* 2004;117(pt 11):2411–2416.

214. Hosokawa T, Tsuchiya K, Sato I, et al. A monoclonal antibody (1D12) defines novel distribution patterns of prion protein (PrP) as granules in nucleus. *Biochem Biophys Res Commun.* 2008;366(3):657–663.

215. Strom A, Wang GS, Picketts DJ, Reimer R, Stuke AW, Scott FW. Cellular prion protein localizes to the nucleus of endocrine and neuronal cells and interacts with structural chromatin components. *Eur J Cell Biol.* 2011;90(5):414–419.

216. Bravard A, Auvre F, Fantini D, et al. The prion protein is critical for DNA repair and cell survival after genotoxic stress. *Nucleic Acids Res.* 2015;43(2):904–916.

217. Pessina F, Gioia U, Brandi O, et al. DNA damage triggers a new phase in neurodegeneration. *Trends Genet.* 2020;37:337–354.

218. Gui X, Luo F, Li Y, et al. Structural basis for reversible amyloids of hnRNPA1 elucidates their role in stress granule assembly. *Nat Commun.* 2019;10(1):1–12.

219. Luo F, Gui X, Zhou H, et al. Atomic structures of FUS LC domain segments reveal bases for reversible amyloid fibril formation. *Nat Struct Mol Biol.* 2018;25(4):341–346.

220. Babinchak WM, Dumm BK, Venus S, et al. Small molecules as potent biphasic modulators of protein liquid-liquid phase separation. *Nat Commun.* 2020;11(1):1–15.

221. Fernandopulle MS, Lippincott-Schwartz J, Ward ME. RNA transport and local translation in neurodevelopmental and neurodegenerative disease. *Nat Neurosci.* 2021;1–11.

222. Wille H, Requena JR. The structure of PrPSc prions. *Pathogens.* 2018;7(1):20.

223. Linsenmeier M, Hondele M, Grigolato F, Secchi E, Weis K, Arosio P. Dynamic arrest and aging of biomolecular condensates are regulated by low-complexity domains, RNA and biochemical activity. *bioRxiv.* 2021. https://doi.org/10.1101/2021.02.26. 433003.

224. Jawerth L, Fischer-Friedrich E, Saha S, et al. Protein condensates as aging Maxwell fluids. *Science.* 2020;370(6522):1317–1323.

225. Roberts RG. Good amyloid, bad amyloid—what's the difference? *PLoS Biol.* 2016; 14(1):e1002362.

226. Taylor NO, Wei M-T, Stone HA, Brangwynne CP. Quantifying dynamics in phase-separated condensates using fluorescence recovery after photobleaching. *Biophys J.* 2019;117(7):1285–1300.

227. Soranno A. The trap in the FRAP: a cautionary tale about transport measurements in biomolecular condensates. *Biophys J.* 2019;117(11):2041.

228. Narayanan A, Meriin A, Andrews JO, Spille J-H, Sherman MY, Cisse II. A first order phase transition mechanism underlies protein aggregation in mammalian cells. *Elife.* 2019;8:e39695.

CHAPTER TWO

Functions of intrinsically disordered proteins through evolutionary lenses

Mátyás Pajkos and Zsuzsanna Dosztányi*
Department of Biochemistry, ELTE Eötvös Loránd University, Budapest, Hungary
*Corresponding author: e-mail address: dosztanyi@caesar.elte.hu

Contents

Abstract

Protein sequences are the result of an evolutionary process that involves the balancing act of experimenting with novel mutations and selecting out those that have an undesirable functional outcome. In the case of globular proteins, the function relies on a well-defined conformation, therefore, there is a strong evolutionary pressure to preserve the structure. However, different evolutionary rules might apply for the group of intrinsically disordered regions and proteins (IDR/IDPs) that exist as an ensemble of fluctuating conformations. The function of IDRs can directly originate from their disordered state or arise through different types of molecular recognition processes. There is an amazing variety of ways IDRs can carry out their functions, and this is also reflected in their evolutionary properties. In this chapter we give an overview of the different types of evolutionary behavior of disordered proteins and associated functions in normal and disease settings.

Progress in Molecular Biology and Translational Science, Volume 183
ISSN 1877-1173
https://doi.org/10.1016/bs.pmbts.2021.06.017

1. Introduction

Intrinsically disordered regions and proteins cannot be characterized by a single well-defined structure, instead they exist as an ensemble of rapidly fluctuating conformations.[1] At a closer look, protein disorder is a heterogeneous phenomenon encompassing a continuum of conformational states. These include random coil-like conformational ensembles or more compact structural states with varying tendencies to form transient secondary and tertiary structure elements.[2–4] Protein disorder can span entire proteins or can correspond to short regions located within ordered globular domains. IDRs defy the traditional structure-function paradigm that stated that a stable 3D-structure is required for functionality.[5] Instead, IDRs rely on their inherent flexibility and plasticity for their function. Disordered regions are integral components of many proteins, in particular those that are involved in cell signaling and regulation.[6–8] Further underlying their importance, IDPs are also associated with a wide range of diseases including various types of cancers or neurodegenerative diseases and can also be direct sites of disease mutations.[6,9–12]

The characteristic molecular features of IDRs, such as their increased molecular dimension, the lack of transition to a denatured state, or their conformational heterogeneity, can be captured by a range of experimental methods.[2,13] Structures deposited into the PDB database can also provide indirect evidence by the presence of disorder in terms of missing electron density in the case of X-ray structures or high structural variations in the case of NMR structures. The most detailed information about the conformational preferences of IDRs can be obtained by NMR.[14,15] However, in order to obtain a complete view of the structural ensemble of IDRs, multiple techniques are needed, including various types of NMR measurements, SAXS, or FRET, in combination with computational calculations.[16] The DisProt database was established to systematically collect experimentally verified disordered regions. It currently contains over 3000 regions IDRs in more than 1700 proteins.[17]

The sequence analysis of experimentally verified disordered proteins revealed that IDRs have a distinct composition compared to that of globular proteins. In general, IDRs are enriched in polar and charged amino acids and are depleted in hydrophobic residues.[2,3] These characteristic differences enable the prediction of disordered and ordered regions from the amino acid sequence. Several dozen prediction methods have been developed, exploiting different principles ranging from simple amino acid scales, biophysical models or machine learning methods, increasingly relying on deep-learning techniques.[18–20] The state-of-art of the predictions methods was recently evaluated in a community-wide effort[21] highlighting a trade-off between speed and accuracy. Nevertheless, sequence based prediction methods make it possible to gain insights into the prevalence of disordered proteins at a genome scale.

Known genome analyses showed that IDPs are present across the tree of life.[22] By measuring the occurrence of predicted long (>30 residue) disordered regions showed that disorder content correlates with the complexity of organisms.[23] While the content of disordered residues in archaeal and eubacterial proteins is 2% and 4.2%, respectively, in eukaryotic proteins this number reaches 33%.[24] This result was confirmed more specifically by analyzing and comparing the whole proteome of *E. coli* and *S. cerevisiae* in terms of 30 residue long disordered regions. Results showed that the yeast proteome contains significantly more disorder than *E. coli*, with ~50% and 15% of proteins containing disordered segments, respectively.[25] At a more detailed level, however, disorder content shows great variability of IDR type and frequency even between closely related taxa and is strongly influenced by environment and pathogenic lifestyle, in both prokaryotes and eukaryotes.[22,26] Interestingly, viral proteins, in particular RNA virus proteins are also enriched in disordered proteins.[27] For example, in the recently emerged SARS-CoV-2 virus, disordered regions play important roles in viral genome packaging.[28] Overall, protein disorder seems to be an important invention of evolution[26] that provides advantages in certain types of functions over globular domains.[29]

2. Functional categories of IDPs

The growing number of IDPs reveal large functional diversity in this class of proteins.[30] The first functional classification schemes for IDPs was first introduced by A. Keith Dunker and Peter Tompa in 2002,[31,32] identifying 28 IDP functions that could be classified into five main classes. These functional categories provide the basis of the IDPO ontology in the DisProt database.[17]

One of the most distinct functions of IDPs corresponds to entropic chains.[1,32] This category includes linkers, which allow movement of domains relative to each other; spacers, which are responsible for regulating the distance between domains; and entropic bristles which can provide barriers for certain molecules.[1] Another main functional group of IDPs is based on molecular recognition.[32] In this group, IDPs can be assemblers that bind multiple partners to create functional complexes or can be effectors, such as inhibitors or activators, that regulate the activity of the partners. The functional class of display sites is also based on molecular recognition and includes IDPs whose regions are targets of post-translational modifications (PTM), such as phosphorylation, methylation, glycosylation or acetylation. In addition, IDPs can function as chaperones, which play a role in the folding mechanism of proteins and RNAs.[33] IDPs that are involved in the formation of biological condensates correspond to a distinct functional category, which was created recently. These IDPs can drive the formation of condensates, such as liquid droplets, through phase separation.[34]

IDPs often exert their function through macromolecular interaction.[32] Their partners can be proteins, RNA, DNA but also small molecules such as metal ions. Upon interactions, IDRs often undergo a disorder-to-order transition through a coupled folding and binding process.[35] As a result, IDRs can adopt a well-defined conformation which is amenable for traditional structure determination methods. The binding regions are often referred to as Molecular Recognition Features.[36] The length of these regions range from a few residues to over 70 amino acids. In the bound form, IDRs can adopt α-helices, β-strands, and irregular (but rigid) secondary structure elements and the unbound states can be biased toward the structural elements adopted in the complex.[1,37] It was suggested that IDRs draw a functional advantage from these preformed structural elements, which can facilitate the binding process.[37] The structural complexes formed between disordered regions with ordered domains are collected in the DIBS database.[38] The focus of a similar database, the MFIB database is to collect complexes formed exclusively between two or more disordered proteins.[39] Both databases require the disordered status to be experimentally verified while the disordered-to-order transition is substantiated by the known structure of the complex. Some IDRs maintain significant conformational heterogeneity even in their bound states.[40] Examples of these so-called fuzzy complexes are collected in the FuzDB database.[41] Although in most cases the complex formation induces structural transition into more ordered forms, a recent example revealed two disordered proteins which remain completely disordered even in their bound state.[42]

IDRs often contain compact functional modules, called short linear motifs (SLiMs), which mediate interactions to specific binding domains.[43] SLiMs usually consist of 3–10 contiguous residues. Typically, only a few amino acids contribute directly to the interaction and are shared among the interaction partners of the binding domain. The binding motif is usually characterized by regular expressions, highlighting fixed positions corresponding to the key residues mediating the interactions, while the other positions of the motif can show larger variations.[43] SLiMs can mediate their interactions independently from the other parts of the protein sequences, however the flanking regions of the motifs can modulate the binding. SLiMs often undergo a disorder-to-order transition upon binding and overlap with disordered binding regions.[44] Validated SLiMs are collected and annotated in the ELM database which currently stores 3542 instances in 235 motif classes.[45]

In the ELM database, SLiMs are classified into two main categories, which can be further classified into 3–3 groups.[1,43] The first category consists of modification motifs. The three groups of this family are (i) the proteolytic cleavage sites, such as motifs that are recognized by the apoptosis and inflammation promoting; (ii) PTM removal/addition, which are recognized by many PTM catalyzing enzymes; (iii) motifs that are involved in structural modifications. The second category comprises ligand motifs and its three subgroups are the following. (i) Complex promoting motifs, which often function in protein scaffolding and can increase the avidity of interactions by their multivalency; (ii) Docking motifs, which increase the specificity and efficiency of PTM events by providing a distinct, additional binding region in the target protein of the PTM; (iii) Targeting motifs are responsible for the localization of proteins in the cell.[1]

Functional regions of IDPs can also correspond to larger segments which show evolutionary conservation, often referred to as **Intrinsically Disordered Domains** (IDDs).[46,47] These regions share two important characteristics of globular domains: they can be structurally and functionally independent from the remainder of the protein and have specific biological function. However, they are fundamentally different from a structural point of view, as they do not correspond to a folding unit. The PFAM database,[48] which collects evolutionary conserved sequence families, contains several examples that are already classified as disordered based on experimental evidence. In general 14% of all PFAM families are predicted as mostly disordered.[47] IDDs are involved in a variety of functions, but are most commonly involved in DNA, RNA, and protein binding.[1]

Another important concept strongly associated with IDRs corresponds to low complexity regions (LCRs) or compositionally biased regions (CBRs). These features are defined based purely on sequence properties using various computational approaches.[49] Certain LCR/CBRs can correspond to fibrillar or repeat regions which are structured, however, there is a large overlap between disordered regions and LCRs/CBRs.[50] For several decades, the function LCRs/CBRs have remained a puzzle. Recently, however, pieces have started to fall into places by recognizing that these regions are often involved in driving or regulating the formation of various biological condensates through the formation of weak, multivalent interactions with RNA, DNA or other proteins.[51]

IDPs have an amazing ability to form interaction often with many partners, thereby participating in the protein-protein interaction networks as hub proteins.[52] They can be static hubs, when a large number of partners are simultaneously bound via distinct interaction regions.[52] However, IDPs often correspond to dynamic hubs, interacting with partners that compete for the same binding site.[53] Overlapping functional modules formed by PTMs, SLiMs and MoRFs can be mutually exclusive or enhance functional outcome through avidity. This can create highly complex molecular switches that are critical elements of the decision making of dynamic signaling networks.[54] Consequently, structural disorder is enriched in proteins involved in signaling and regulation, and is mostly associated with functional categories such as transcription and its regulation, cell cycle, mRNA processing, apoptosis or development.[7,55]

Protein disorder is a heterogeneous phenomenon encompassing a continuum of conformational states. However, it is also important for generating different functioning proteoforms: different molecular forms in which the protein product of a single gene can be found. Basic mechanisms for generating various proteoforms include alternative splicing or PTM events, which often involve IDRs[56]. In a more direct way, proteoforms can also correspond to different conformational states of IDRs or to their promiscuous binding forms adopting different conformations depending on their interacting partner. All these considerations indicate that IDPs/IDRs represent a very rich source of proteoforms. The proteoform concept highlights that the actual "one-gene-one-protein-one-function" model is oversimplified, the real gene-protein relationship is much complex, which can be described by a more general "protein structure-function continuum" model. This model states that a given protein exists as a dynamic conformational ensemble containing multiple proteoforms characterized by a broad spectrum and structural features and possessing different functional forms.[56] These wide range of different functions and the associated molecular features shape the specific evolutionary behavior of disordered proteins.

2.1 Extending the basic toolkit of sequence alignments for IDPs

Evolutionary conservation analyses are widely used to identify and characterize functional residues within protein regions. One of the most important steps of molecular evolutionary analyses is the generation of sequence alignments which has three key elements: (1) substitution matrices, which are used for scoring the matching position pairs of amino acid position between sequences, (2) the gap penalty system, which controls the introduction of indels into the alignment, (3) and the algorithm, typically a dynamic programming method, which aims to find the optimal solution by maximizing the similarity of sequences given the substitution matrix and the penalty system.[57] From pairwise comparisons multiple sequence alignments (MSAs) can be built. Most MSA algorithms are based on progressive approach, first introduced in the Clustalw program.[58] For improved quality, MSA algorithms now largely use an iterative process to refine multiple sequence alignments.[59] Globular domains generally show high evolutionary conservation in their amino acid sequences. This stems from purifying selection to preserve critical functional residues but are also due to structural constraints.[60]

Orthology and paralogy predictions are another important element of evolutionary conservation analyses that rely on MSAs.[61] Identification of orthologs and paralogs provide the basis of inter- and intra-species comparisons, respectively. In principle, there are two main approaches for orthology predictions: reciprocal best hit approaches, such as GOPHER,[62] and phylogenetic approaches that use gene and species tree reconciliation. PhylomeDB[63] and ENSEMBL-Compara[64] are two examples of this latter approach. Sequence conservation is usually calculated for orthologous sequences as paralog sequences often show larger functional divergence.[61]

The development of MSA algorithms has largely focused on globular domains with well defined 3D structures. However, globular domains and IDRs are different from the aspect of sequence homology. Amino acid composition of IDRs differs from the ordered ones, and amino acid substitutions are likely to follow different logic. Further, it has been shown that disorder commonly overlaps with both insertion/deletion events which can lead to larger variations in the length of IDRs.[65] In general, IDRs and in particular, low complexity regions can be challenging to align. Similarly, SLiMs often have weak conservation signals due to their short and degenerate nature, which makes their alignment very difficult.

The choice of substitution matrices is an important element of sequence alignments and can influence the ability to detect and discriminate related disordered proteins.[66] Novel substitution matrices have been derived specifically from IDRs.[66,67] To further improve on alignment quality and to capture conserved features in IDPs, a Knowledge-based MSA (KMAD) was introduced for IDPs. KMAD produces insertion-free alignments, when insertions from homologous sequences (compared to query sequence) are removed. The concept of the insertion-free alignment was introduced for transfer information from the whole alignment to one sequence of interest.[68] In KMAD, elements of the alignment matrix are augmented with feature scores which incorporate metadata, such as SLiM, domain and PTM annotations.[69]

Details of the algorithms can also largely influence the quality of sequence alignments. The MAFFT program is one of the new-generation alignment algorithms that incorporates several adjustments that improves the performance for IDRs. Generally, MAFFT uses an iterative refinement method.[70] For the rapid detection of homologous segments, it uses Fourier Transformation which can drastically reduce execution time without compromising alignment accuracy.[70] An important technical element of MAFFT—that is also relevant for IDPs—is the rescaling of the substitution matrix according to the sequence similarity. Thus a different scoring system is applied for each sequence pair in the alignment procedure, which controls the over-alignment in the MSA.[71] MAFFT also uses generalized affine gap cost, which was first introduced by Altschul.[72] This penalty system allows non-conserved regions to be effectively ignored in the alignments, which can increase the quality of MSAs for sequences containing IDRs. Since short conserved segments of IDRs, such as core regions of SLiMs, are frequently separated by non-conserved regions, this gap penalty system promotes the correct alignment of adjacent functional sites located within non-conserved disordered regions.

The BaliBASE benchmark suite introduced novel test sets in order to specifically evaluate the efficiency of MSA methods to correctly align SLiMs.[73,74] For the novel test set, a specific pipeline with manual validation was applied to generate high-quality MSAs of related sequences that included functional SLiMs. The evaluation of the different alignment methods to be able to correctly align SLiM positions was focused on two important aspects, (i) in all sequences containing a given SLiM should be correctly aligned and (ii) sequences that do not contain these SLiMs should not be aligned in the motif region.[74] Applying this new reference dataset, the most popular alignment algorithms were evaluated. One of the best performing methods was the popular MAFFT algorithm.[74]

Despite these improvements, the alignment of distantly related IDRs remains challenging and further methods are needed. Novel machine learning approaches could also bring improvements into this field.[75] Nevertheless, many important insights have been gained into the evolutionary properties of IDRs using existing methods.

2.2 Evolutionary rate and selection

Estimating the evolutionary rate is a central subject in molecular evolution and is used to characterize the dynamics of evolutionary changes in sequences. Commonly used method for estimating the evolutionary rate is based on the ratio of nonsynonymous (dN) and synonymous (dS) substitutions in nucleotide sequences. Evolutionary rate can be calculated in a intra- or inter-species manner, and can be focused on whole proteins, sequence regions or only individual sites. There are multiple factors that influence evolutionary rates. In general, slowly evolving proteins tend to be regulated by more transcription factors, involved in characteristic biological functions (such as translation), and are generally more abundant, more essential, and enriched for interaction partners.[76] The expression level of protein is also a determinant of the evolutionary rate.[77] However, the presence of structural constraints is a one of the most dominant factors and in general, globular domains typically have a slow evolutionary rate among orthologs.

The first systematic investigation of rate heterogeneity that focused on IDRs was carried out by Brown et al. in 2002.[78] They collected 26 homologous protein families having at least one member with experimentally characterized at least 30 residue long disordered regions (based on X-ray crystallography or NMR) and compared the pairwise genetic distances between the ordered and the disordered regions.[78] The results showed that disordered proteins evolved more rapidly than ordered ones in the majority of cases with 19 out of 26 families exhibiting this evolutionary behavior. For five families, no significant difference between disordered and ordered regions was observed. Interestingly, a slower rate of evolution was observed for two IDRs. The first such region is the 34 amino acid long flexible loop of the adenovirus ssDNA binding protein. The region is part of the DNA binding domain and is essential for high-affinity binding to ssDNA.[79] The second case with slower evolution corresponded to the disordered segment in flagellin that forms the flagellar filament by becoming ordered upon polymerization.[78]

IDRs have an elevated rate of evolution, in general.[78,80] It was suggested that positive selection might have an important role in their rapid evolution.

Detection of molecular selection on coding genes is an important element of the discovery of gene evolution, which contributes to the understanding of the emergence of novel gene functions at the molecular level. The molecular selection prediction is based on the effect of the genetic substitutions. Evolutionary pressure within genes can be calculated from the ratio of synonymous (dS) and nonsynonymous (dN) substitutions. If this ratio is nearly 1, it refers to neutral evolution, when $0 < dN/dS < 1$, negative selection pressure is assumed and when >1, the sequence evolves under positive selection. The CODEML algorithm of the PAML program package[81] is the most widely used tool based on the ratio of synonymous and nonsynonymous substitutions, and it provides several models to calculate molecular selection both individual residue and whole gene levels.[82]

In a recent study, IDRs of human proteins were analyzed to investigate the evolutionary pressures that acted on these regions using mammalian homologous genes.[83] This study confirmed the elevated evolutionary rates in IDRs compared to ordered protein regions. Moreover, it also suggested that positive selection is one of the main contributors for the elevated rates and found evidence for positive selection in 377 IDR.[83] This work presented the first examples for positive selection detected within disordered regions in mammals. The analysis was based on a site-specific test of positive selection that specifically reveals amino acids in proteins that have evolved under positive selection. Overall, examples of site-specific positive selection are difficult to identify for IDRs. This could originate from the higher evolutionary variability of these regions but could also be influenced by technical challenges.

The evolutionary rate based approach of molecular selection strongly depends on the data, especially if positive selection has affected only one evolutionary branch. For example, in order to an accurate calculation, sequences in the MSA must not be too closely related, or too distant, they must be sufficiently divergent.[84,85] Another important factor is the alignment length, which has to be long enough (minimum 30–40 residues) to avoid loss of evolutionary information.[84] Therefore, amino acid substitutions on a given branch do not necessarily present strong evidence for positive selection. However, detection of positive selection can be still useful in protein evolutionary studies even if the individual affected sites cannot be identified reliably.[86]

Species-specific positive selection may be detected more effectively by applying McDonald and Kreitman (MK) test. The MK test compares divergence to polymorphism data in closely related species, such as human and chimp. In a recent study this approach was applied to identify human-specific positive selection and 198 of 9785 human genes were found to show positive selection.[87] Although this study was not focused on IDPs, some proteins of the 198 hits with human-specific selection have IDRs. One of them is the estrogen receptor 1 (ESR1), which is a member of the nuclear hormone receptor family. For the position 300 and 306 of ESR1 human-specific substitutions were observed. These amino acids are required to create a phosphorylation site for protein kinase A in human, therefore it was suggested that these changes are a consequence of human-specific positive selection.[88]

3. Distinct evolutionary scenarios of IDRs

While IDRs generally evolve at a faster evolutionary rate, deviations from this general trend also occur commonly. This was also observed in a comprehensive study that analyzed protein domains containing conserved long (>20 residues) disordered segments.[46] These domains containing conserved disordered segments were found in proteins from all kingdoms of life, however, domains from eukaryotic and viral proteins had more long conserved disordered regions compared to domains from archaea and bacteria. Analyzing the sequence conservation of these disordered segments, it was found that the average value is slightly lower than in the case of ordered regions. This strong conservation of long disordered regions occurred within domains indicates their functional importance. The most common function across all kingdoms was the DNA/RNA binding. In addition, several ribosomal proteins with conserved disordered segments containing domain were from both archaea and eukaryota. The function of binding cytoskeletal components was observed in bacteria and eukaryota.[46] These results highlight that conserved disordered regions are present to a greater extent than first thought.[29]

In many cases it was observed that while there was no sequence conservation detected, the disorder property was still conserved. Bellay et al. systematically analyzed the evolutionary behavior of disorder in the context of sequence conservation for the yeast clade.[89] For this analysis, the conservation of predicted disorder and the conservation of sequence over MSAs of orthologous sequences was calculated for each residue over sequences for 23 yeast species. The disorder tendency of a residue was considered to be conserved if it was conserved in at least half of the species. Similarly, the amino acid type of a residue was considered conserved if it was present in

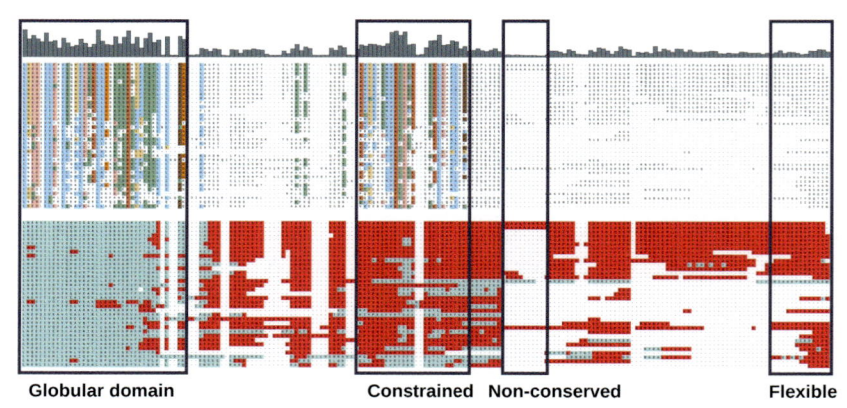

Globular domain   Constrained   Non-conserved   Flexible

Fig. 1 Schematic depiction of the three disorder conservation categories. The upper part of the figure represents multiple sequence alignment of orthologs where colors indicate conserved columns. The bars at the top of the alignment show sequence conservation values highlighting conserved residues of the alignment. At the bottom part, predicted disordered and ordered residues of each sequence are indicated by red and blue colors, respectively, indicating the conservation of structural state among orthologs. Black rectangles show globular domain conservation and examples for the three IDP conservation types.

at least half of the species. Based on this, they established three categories of IDRs exhibiting distinct conservation properties. (i) **constrained disorder**, IDRs, where both disorder and the amino acid sequence are conserved; (ii) **flexible disorder**, where the structural property of disorder itself is conserved rather than the amino acid sequences of IDRs; (iii) **non-conserved disorder**, IDRs that are not conserved across orthologs[89] (Fig. 1). IDRs of these three classes are different not only in their evolutionary conservation, but were also associated with different functional properties. These categories can guide us to introduce some basic scenarios for IDP conservation and their associated functions.

3.1 Constrained disorder

Constrained disorder encompasses cases when the conservation of disorder is accompanied by sequence conservation. This represented approximately the third of disordered residues in yeast. Based on functional enrichment analysis, different functions of proteins with constrained disorder were identified, such as RNA binding of ribosomal proteins and chaperone activity. In general, proteins with constrained disordered were enriched for multi-interface hubs.[89]

Many protein regions with constrained disorder tend to adopt a fixed conformation undergoing disorder to order transition upon binding to their targets. The structural constraints associated with this transition might impose evolutionary constraints on the sequence.[29] Therefore, it was suggested that the ability to undergo structural transition can be a hallmark of constrained disorder. An example of this behavior is provided by the C-terminal domain of the ribosomal protein Rpl5.[89] This disordered region was shown to undergo structural transition in the C-terminal domain upon binding 5S rRNA in Xenopus.[90] It showed high sequence conservation among yeasts.[89] However, sequence conservation of the disordered region is present beyond yeasts and the corresponding segments can be identified in human too. The sequence alignment together with the predicted disorder using the IUPred3[91] disorder prediction tool is shown on Fig. 2.

IDDs naturally represent examples for constrained disorder. VHL, the Von Hippel-Lindau disease tumor suppressor protein has E3 ligase activity and plays a key role in cellular oxygen sensing by targeting hypoxia-inducible factors for ubiquitylation and proteasomal degradation. VHL has two IDDs which are also represented in the Pfam database. The α-domain (VHL-box) and a larger β-domain, which are responsible for binding elongin C and HIF1α, respectively. However, VHL exists in a compact disordered state in isolation[92] and only adopts a well-defined structure in complex. In humans, both domains of VHL are evolutionarily conserved in all vertebrates, while the β-domain is also conserved in the VHLL paralog.[88]

The Wiskott–Aldrich syndrome protein (WASP) homology domain 2 (WH2) of actin-binding proteins corresponds to the WH2 Pfam family (PF02205). The WH2 domain family is about 30 amino acid long conserved sequence module that is almost always embedded in a Pro-rich protein segment. This domain was shown experimentally to be fully disordered in isolation.[47] Similarly to many globular domains, the WH2 domain can also occur in different sequence contexts in multiple, unrelated proteins (Fig. 3). In this particular case, actin binding is coupled to different context-dependent outcomes through the disordered WH2 domain.

In general, the category of constrained disorder is associated with macromolecular binding involving a disorder-to-order transition. Therefore, the sequence conservation is the consequence of the high degree of local structural constraints. However, constrained disorder is also observed in chaperones without a transition to a fixed conformation. For example, the HSP90 heat shock protein is conserved in the bacterial kingdom.

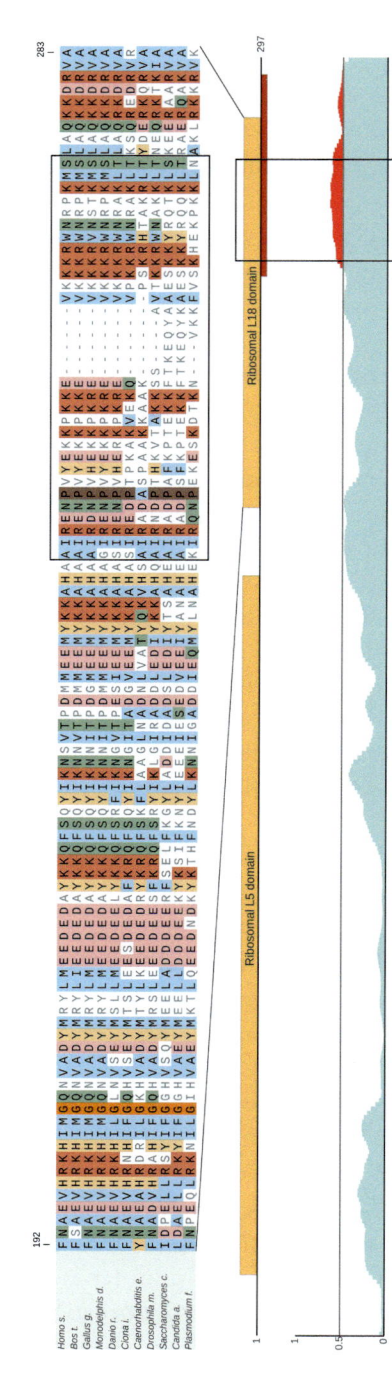

Fig. 2 Representation of evolutionary conservation of the Ribosomal L18 C-terminal domain from human RPL5. The upper part of the figure shows the multiple sequence alignment of the domain including the conserved disordered region, which is highlighted by a black rectangle in the alignment. Indicated positions at the top of the alignment correspond to the human Uniprot sequence. The bottom part shows the domain organization and disorder profile generated by IUPred3 of protein.[91] Yellow and red boxes depict the domains and the disordered region, respectively. On the profile, the disordered region is highlighted by red.

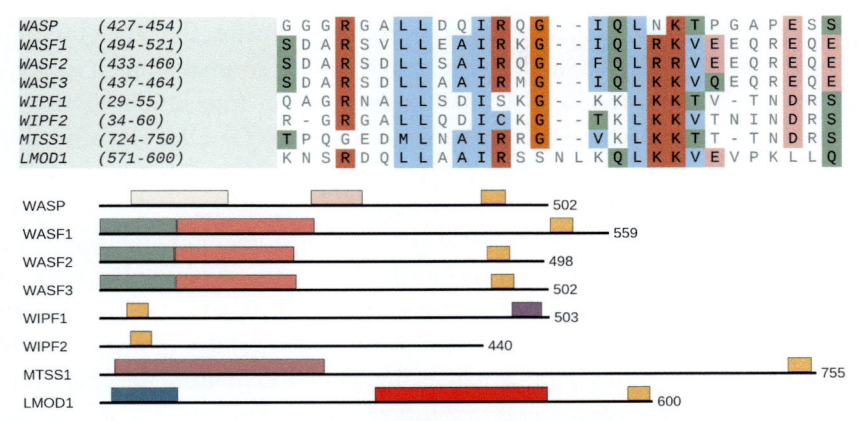

Fig. 3 Conservation representation of WH2 domain among human proteins. At the upper part of the figure the multiple sequence alignment of the WH2 domain is shown. The positions of the aligned domain regions of the proteins are shown in parentheses. The bottom part represents the domain organization of the WH2 domain containing proteins, which was retrieved by applying the interproscan tool.[93] The WH2 domains are shown in yellow boxes in all cases. The WH1 and the PDB domain of WASP are depicted with warmwhite and light pink, respectively. In the WASF family green and orange boxes show unknown Gene3D families. Purple box marks the C-terminal PH-like domain of WIPF1. The IMD domain of MTSS1 is depicted by brown, and in LMOD1 the Tropomodulin domain and Leu-rich repeat are shown by blue and red boxes, respectively.

Within this protein family, a constrained disorder region is localized at the inner surface of the barrel–shaped protein and it may play a role in entropy transfer and the refolding of clients, only interacting with partners in a transient fashion.[89] Conserved disorder can also be associated with entropic chain function. The projection domain corresponds to the IPR013588 interpro family and it was described as an IDD in microtubule–associated protein 2 (MAP2).[47] In the cytoskeleton, MAP2 serves as a spacer by repealing molecules that approach microtubules, therefore it has an entropic chain function.[1]

3.2 Flexible disorder

IDRs with flexible disorder conservation type show high sequence divergence while the property of disorder is conserved. One common function associated with flexible disorder is the entropic chain function. An illustrative example is the 70 kDa subunit of replication protein A (RPA70). RPA70 is a eukaryotic protein that plays a role in replication, recombination and DNA repair. RPA70 contains a linker region between residues 100 and

200, approximately. This connects two globular domains, the DNA binding domain (from 200 to 400 in the sequence) to an N-terminal domain responsible for single stranded DNA binding and protein interactions (DBD F, ~1–100 residues). The tethering provides mobility for DBD F to be able to interact with ssDNA, other proteins, or the other subunits of RPA heterotrimer. Exploiting the conservation of the adjacent domains, the linker sequence of RPA70 was identified in 19 orthologs ranging from yeast to mammals. In order to confirm disorderedness, the dynamic behavior of the linker was verified in five cases by solution NMR analysis.[94] Although the disorder linkers of orthologs could be identified using a sequence-based pipeline, they showed a high sequence divergence, exhibiting the characteristics of flexible disorder conservation type[94] (Fig. 4).

In general, PTM sites, in particular phosphorylated Ser and The residues tend to evolve more slowly compared to non-phosphorylated ones.[95] However, phosphorylation based protein function is not strongly dependent on individual sites, often only the density and the accessibility of PTM sites is relevant. In these cases, clusters of phosphorylation sites within a disordered region can act in groups, representing a single functional unit. This can be observed in the case of inhibitors of cyclin dependent kinases (CDKs). These proteins show conserved regulation in animals and fungi but diverge in the position and number of clusters of phosphorylation sites. For example, Ste5 inhibition by CDK is proportional to the charge in the disordered region surrounding the PM domain of Ste5. Emergence of CDK consensus phosphorylation sites in this region will increase, while loss of these sites will decrease the strength of inhibition. As long as the total strength of inhibition is within an optimal range, the exact number and location of sites are not crucial. Accordingly, these regulatory sites of Ste5 in related yeasts show a diversity in the location of phosphorylation sites.[96]

The preservation of groups of phosphorylation sites within IDRs over long evolution time provided the basis of the **stabilizing selection model**. In this model the individual amino acid sites are under relatively weak functional constraints and mutations at these positions are tolerated as long as there are enough phosphorylation sites to retain function. In other words, a given group of phosphorylation sites can be considered as a quantitative trait where selection does not act on a particular site but rather on the number or density of sites.[96]

Stabilizing selection can also have an effect at the proteome level. Flexible disorder regions with no detectable similarity in their amino acid sequences can be associated with specific biological functions among

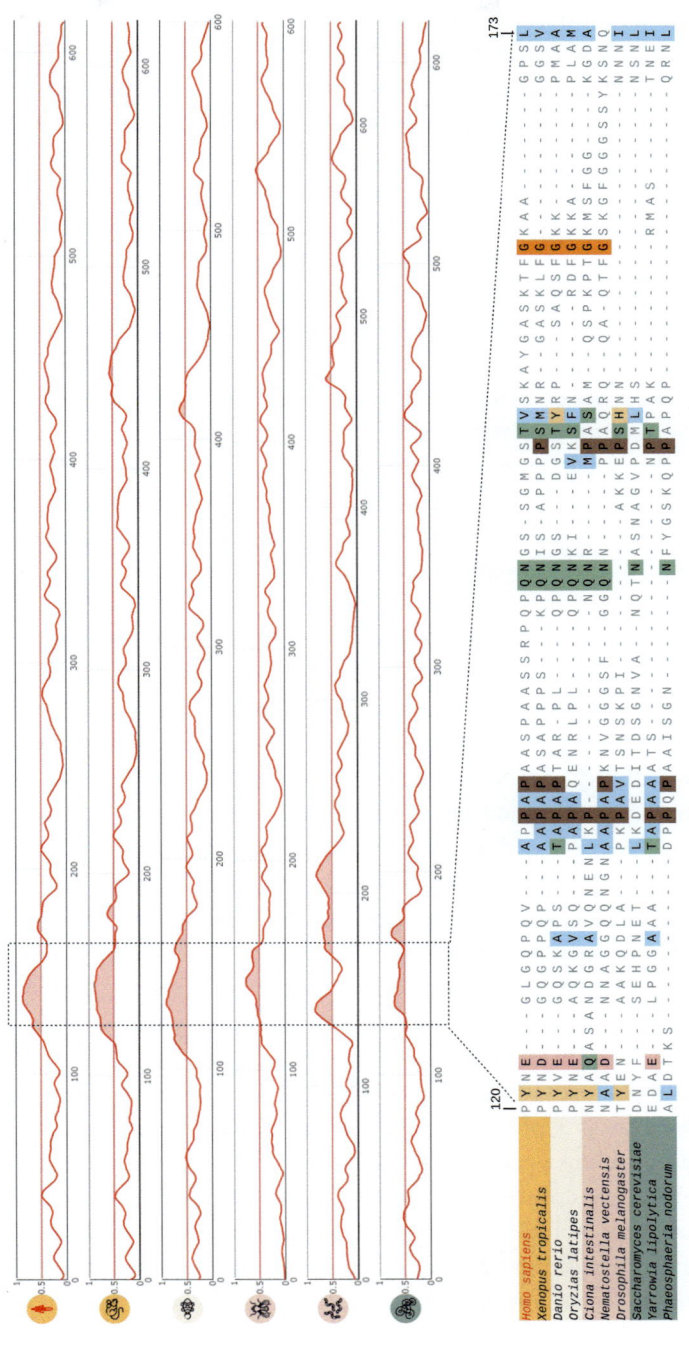

Fig. 4 The top of the figure represents the disorder profiles of the human RPA70 and its orthologs from five generally known model organisms (*Mus musculus, Danio rerio, Drosophila melanogaster, Caenorhabditis elegans, Saccharomyces cerevisiae*). Disorder profiles were generated by IUPred3 disorder prediction tool.[9] At the bottom of the figure, MSA of the linker regions of orthologs is depicted using an extended set of model organisms. Indicated positions of the alignment correspond to the human Uniprot sequence.

non-homologous proteins in a given species. For example, in yeast more than 150 IDRs are believed to represent mitochondrial N-terminal targeting signals that share similar constraints on their molecular features, such as the abundance of positively charged residues. These signals have been preserved independently over evolution as mitochondrial genes were transferred individually to the nuclear genome. Swapping the N-terminal targeting signals that have no detectable similarity in their primary amino acid sequences in mitochondrial proteins, such as Atm1 and Cox15, has no effect on localization indicating that protein localization requires only the appropriate molecular feature of signal peptides with limited constraints on their sequences.[97]

Several other types of molecular features were identified to be evolutionary preserved within flexible disordered regions. These features include charge properties, asparagine or other residue content, physicochemical features such as isoelectric point, etc.[97] For example, the ~100 residues long IDR of yeast DNA repair Srs2 protein showed a pronounced separation of positive and negatively charged residues along the sequence and this feature was conserved in yeast species.[97]

Another important stabilizing feature is associated with low complexity regions. An illustrative example is the budding yeast Sup35. This protein is a translation termination factor that is a key element of protein synthesis and has the ability to form reversible biomolecular condensates.[98] Sup35 can be divided into a disordered N-terminal region and an essential C-terminal GTPase domain which catalyzes termination of protein synthesis. The N-terminal region has ionizable groups, specifically glutamic acid residues, and the density of these residues was observed to be conserved.[97] This charged N-terminal region can contact the C-terminal domain and basically acts as a pH sensor. This region also provides the interactions that drive liquid phase separation as a response to environmental stress caused by pH changes. This liquid phase separation based mechanism is capable of rescuing the GTPase domain from irreversible aggregation during environmental condition changes, providing the functionality of Sup35. The ability of Sup35 to function as a pH sensor through biomolecular condensation is shared among budding yeast and fission yeast.[97,98] Moreover, this feature of Sup35 seems to be conserved not only in opisthokonts, but also in other eukaryota orthologs including mammalian sequences as well, which suggests that condensate formation of the translational termination factor is a highly conserved and ancestral function.

In the case of the low complexity disordered region of Ccr4, the polyglutamine repeats represent an evolutionary conserved molecular feature in yeasts that are associated with the so-called "prionogenicity."[97] The conservation of critical amino acid responsible for the liquid forming potential was also observed to be conserved in RNA-binding protein Fused in Sarcoma (FUS). While the prion-like domain evolves three times as rapidly as the remainder of FUS, the tyrosine residues that are crucial for phase separation were conserved over the 160 My of mammalian evolution.[99]

While flexible disordered regions provide many amazing examples of functionality through disordered regions, their evolutionary studies remain highly challenging and need novel approaches that can trace back the evolutionary history of the specific molecular features associated with function.[100]

3.3 Non-conserved disorder and the limitations of the concept

In the case of non-conserved IDRs not even the property of being disordered is conserved in orthologous sequences. Nonconserved disorder accounts for around 17% of disordered residues in yeast.[89] Neither SLiMs or phosphosites showed enrichment within IDRs that exhibit non-conserved disorder. This category was not associated with any specific function. Although specific functional roles were not identified for non-conserved disorder regions, their functionality cannot be ruled out.

The three conservation categories were established based on yeast species using cutoff values of 50% of sequences for both the conservation of disorder and sequence. However, different cutoff values were applied in a similar analysis of metazoa species (30% for disorder conservation and 90% for sequence conservation).[101] These differences highlight a potential problem with the proposed classification scheme of disorder conservation. Namely, results are largely dependent on the applied cutoff values, the evolutionary range of sequences or even selected sequences.

4. Evolutionary conservation of SLIMs

In terms of evolutionary conservation SLiMs fall into a special case showing varied types of conservation.

Similarly to the non-conserved disordered cases in yeast, SLiMs commonly show species or clade specific evolutionary conservation. Moreover, SLiMs can occur within a single paralog among homologous proteins from the same species, which is a special case of intra-species non-conserved disorder. These conservation features are the direct consequences of the specific features of SLiMs. They are defined by only a few fixed amino acids and this makes them likely to arise or disappear by tiny changes in the protein sequence, such as a given substitution or a few residues long indel. Examples for the phenomenon when nonfunctional disorder protein regions change into functional motifs were collected and studied by Davey et al. and based on these case studies they formulated the ex-nihilo motif birth model.[102] In this model, SLiM acquisition is driven by random mutations leading to the appearance of the motif "ex nihilo." Once the motif becomes fixed, purifying selection will retain it over evolution during which further fine tuning mutations can occur.[102] An example of ex-nihilo motif birth is the Ser2Gly substitution in the N-terminus of the human leucine-rich repeat protein SHOC-2, that was observed in multiple Noonan-like syndrome patients. This mutation results in the birth of an N-myristoylation motif that was the first example of an acquired N-terminal lipid modification of a protein causing a human disease.[103] Another example for the ex-nihilo motif birth is a clade specific motif emergence of the RxL Cyclin binding motif of E3 ubiquitin-protein ligase Mdm2 in rodents, which is a result of a four amino acid long deletion.[104] Because SLiMs can arise and be lost very simply over evolution, they have a high propensity to convergently evolve. One example for convergent evolution is the retinoblastoma-binding motif, which has more than 20 convergently evolved instances.[105]

Other SLiM examples occurring within IDRs are representative of flexible disorder. In these cases, the core SLiM positions can show high evolutionary conservation among homologs. One notable example is the YxPPxxR pattern of the human Eukaryotic translation initiation factor 2A (eIF2A), which corresponds to a currently unverified consensus translation initiation factor (eIF4E) binding motif (YxPPxΦR). The pattern of eIF2A motif shows conservation across many diverse species from unicellular eukaryotic organisms to vertebrates.[106] Another example is the PCNA-binding PIP box motif in the Flap endonuclease 1 (FEN1) protein. The functionally important positions of the motif are almost invariable across Eukaryotes and Archaea. Comparing these motif positions to the surrounding residues in MSA of homologous sequences, they are relatively more conserved, and show a specific, island-like conservation pattern.[107] This island-like conservation of SLiMs within disordered regions can be used as a discriminatory technique for motif discovery. SLiMPrints, a de novo motif discovery tool, is based on this approach identifying relatively conserved proximal residues within disordered protein regions.[106] Similarly, exploiting that SLiMs are preferentially conserved such that substitutions and insertions or deletions occur more frequently adjacent to the motif than within it, a phylo–HMM-based computational framework was developed and used to systematically detect SLIMs in 13 related yeast species.[47]

In many cases several adjacent and/or nested SLiMs can form larger functional modules with strong evolutionary conservation corresponding to constrained disorder. The disordered C-terminal of the retinoblastoma-associated protein (RB1) is one such example. RB1 is a key regulator of entry into cell division. It promotes G0-G1 transition when phosphorylated by CDK3/cyclin-C and acts as a transcription repressor of E2F1 target genes.[108] SLiMs occurring in the C-terminal region of RB1 are highly conserved at the whole range of vertebrate species. This disordered region is approximately 30 residues long and only a few changes can be observed among the orthologous sequences. In agreement with these conserved sequence features, this region is part of the Rb C-terminal PFAM domain.

Interestingly, binding partners of a single linear motif binding domain can exhibit all three types of motif conservation behavior. The dynein light chain LC8 is a eukaryotic hub protein, whose binding motifs were identified in a wide variety of proteins. The LC8 protein is a highly conserved protein from human to yeast, however, its binding motif instances show varied conservation as it was shown in systematic evolutionary analyses of the known human binding partners.[109] Many of the LC8 binding motifs showed an ancient, island-like conservation pattern, like the LDVSSQTD motif (from 278) of WWC1/KIBRA. In this case, highlighting a very strong conservation, residues of the core motif are identical from human to invertebrates, while the adjacent positions are significantly less conserved in the MSA of orthologs (Fig. 5A). The region involving the LC8 binding motif in Cytoplasmic dynein 1 intermediate chain (DC1|1) DC1|1 and its paralog, DC1|2, are examples of constrained disorder. This region overlaps with the Dynein_IC2 (PF11540) PFAM sequence family, which includes two adjacent binding motifs recognized by the light chains DYNLT1/Tctex1 and DYNLL1/LC8. The 30 residues long Dynein_IC2 binding region show high sequence conservation from human to fruit fly (Fig. 5B).[109,110] In contrast, the LC8 binding motif of MAP4 is mammalian specific, but motifs specific to slightly larger taxonomy ranges were also identified, for example in the AMBRA1 or MTCL1 proteins, which are conserved only in vertebrates.

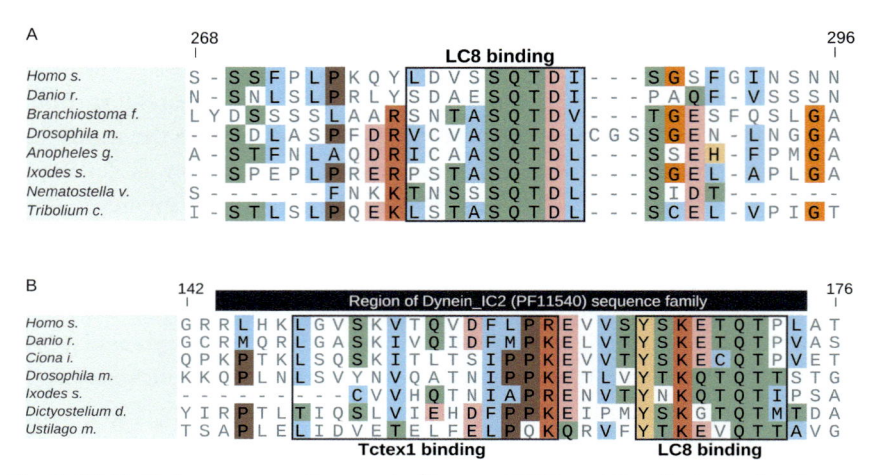

Fig. 5 (A) Multiple sequence alignment of human KIBRA and its orthologs with the representation of the LC8 binding LDVSSQTD instance. Indicated sequence positions belong to the human Uniprot sequence. (B) Multiple sequence alignment of the Dynein_IC2 PFAM family from orthologous sequences of the cytoplasmic dynein 1 intermediate chain. Black rectangles at the bottom of the alignment indicate the separate Tctex1 and LC8 binding regions. At the top of the alignment, the positions of the human Uniprot sequence are indicated.

5. Evolutionary origin IDRs mutated in cancer

The determination of gene origin is a common aim of evolutionary conservation analyses. Gene origin prediction can contribute to the discovery of evolutionary histories of genes and can also provide insights about the function of genes. Gene origin prediction is commonly based on orthologous and/or homologous relationships, of which orthology is a more commonly used criterion. Using orthology, the origin of a single gene can be defined as the deepest speciation node of all descendents of the corresponding orthologous group.[111] This approach gives information about single gene histories, however the evolution at gene family level is not inferred. An alternative method is based on the so-called founder gene approach. Founder genes represent evolutionary novelties in protein sequence space that emerged not simply duplications of existing genes or genes with re-shuffled functional domains.[112] Domazet-Loso et al. developed the phylostratigraphy method to identify the origin of founder genes.[113] The technique relies on BLAST sequence similarity searches to gather descendants of founder genes which will represent a given clade (e.g., Mammalia, Eumetazoa, Arthropoda, etc.) whose common ancestor first carried the ancient founder gene, and thus defining its origin. This BLAST based approach relies on the conservation of globular domains (e.g., the kinase catalytic domain) and can produce clades which do not share an evolutionary history beyond a common domain.

Domazet-Loso et al. analyzed the origin of genes that are associated with cancer.[112] The results revealed that the emergence of cancer genes largely corresponds to the evolutionary transition when multicellularity evolved. These cancer genes typically have a gatekeeper role and promote tumor progression directly by changing cell differentiation, growth and death rates. Gatekeeper genes that control cell division are among the best known cancer-associated oncogenes and tumor suppressors.[114] However, many cancer genes have a more ancient origin and can be traced back to the unicellular life.[112] These genes were suggested to play a caretaker role and contribute to tumorigenesis by increasing mutation rates and genome instability. The emergence of disease genes, as we as novel genes in general followed a similar trend.[112,115] However, these results are based on the founder gene approach, which is heavily biased toward globular domains.

IDRs are important components of many cancer-associated proteins and can also be direct targets of cancer mutations.[116] In general, cancer samples contain a large number of mutations, but only a small number of the observed variations, the so-called driver mutations, are responsible for actively driving tumorigenesis. A recent study collected disordered regions that were enriched in cancer mutations and were likely to play driver roles in tumorigenesis.[12] The detailed analysis of these disordered drivers revealed their functional diversity. The mutated disordered regions included not only SLiMs, but also PTM sites, linker and autoregulatory regions, disordered domains involved in binding to other proteins, DNA or RNA. It was shown that these disordered drivers had increased interaction capacity and were enriched in specific biological processes such as transcription, gene expression regulation and protein degradation.[12] An interesting question is what is the evolutionary origin of these disordered regions which are specifically targeted in cancer.

In our recent work we explored the evolutionary origin of cancer drivers that were primarily mutated in their disordered regions.[88] In this analysis, the origin of cancer regions was determined at gene family level that incorporated paralogs as well. This was necessary to trace back the origin of regions which were lost in one paralog after gene duplication events, which can occur more frequently for IDRs. This way the first emergence of regions can be traced back even when gene duplication events occurred.

For this, the orthology and paralogy relationships of homologous sequences provided by the ENSEMBL-Compara[64] project were used.

Surprisingly, our results indicated that most of these disordered cancer genes had an ancient evolutionary origin.[88] Out of the 36 cases collected, the majority, 23 IDRs have emerged at the eumetazoa evolutionary stage, while a more recent, vertebrate origin was identified in only eight cases. Interestingly, five IDRs mutated in human cancer were very ancient and their evolutionary origin was traced back to unicellular organisms.[88] An interesting example of ancient IDR origin is the linker region of the mismatch repair protein MLH1. This region connects the DNA mismatch repair and C-terminal domains and can regulate both DNA interactions and enzymatic activities of neighboring structured domains.[117] Cancer mutations targeted an approximately 20 amino acid long segment within the linker region.[118] The linker region showed relatively low sequence conservation in general, showing flexible disorder-like evolutionary behavior. However, the mutated positions were highly conserved from human to yeast, suggesting that region was already functional in the common ancestor of human and yeast. Although the exact function of this 20 residues long region is not known, the strong evolutionary conservation indicates a highly important function.[88]

6. Conclusion

Protein disorder is a diverse and complex phenomenon. The function of IDRs can directly originate from their disordered state or arise through a molecular recognition process. In general, IDRs evolve faster than structured domains due to the lack of structural constraints. However, disorder tendency can also be highly conserved with or without sequence conservation. Based on the relationship between the conservation of sequence and disorder property, three basic scenarios were outlined corresponding to conserved disorder, flexible disorder and non-conserved disorder. These categories were originally suggested to correlate with specific functional categories. However, specific examples paint a more nuanced picture. Conserved disordered regions are not restricted to DNA and RNA binding regions which become structured upon binding, but can also correspond to longer loops within domains, or regulatory modules incorporating a combination of SLiMs. There are also many different types of functional modules that exhibit the characteristics of flexible disorder. While the evolutionary constraints on the sequence are more relaxed, there are often specific molecular features that are preserved besides the disorder tendency. Overall, the evolutionary constraints can be as strong for IDRs as for functional residues for globular domains, supporting the fundamental biological importance of these residues. This is further underlied by the strong evolutionary conservation of cancer-associated disordered proteins. While currently the evolutionary studies largely rely on approaches developed for globular domains, novel methods and approaches would enable to gain further insights on how evolutionary processes shaped the versatile group of intrinsically disordered proteins.

References

1. van der Lee R, Buljan M, Lang B, et al. Classification of intrinsically disordered regions and proteins. *Chem Rev*. 2014;114(13):6589–6631.
2. Dunker AK, Lawson JD, Brown CJ, et al. Intrinsically disordered protein. *J Mol Graph Model*. 2001;19(1):26–59.
3. Uversky VN. Natively unfolded proteins: a point where biology waits for physics. *Protein Sci*. 2002;11(4):739–756.
4. Habchi J, Tompa P, Longhi S, Uversky VN. Introducing protein intrinsic disorder. *Chem Rev*. 2014;114(13):6561–6588.
5. Wright PE, Dyson HJ. Intrinsically unstructured proteins: re-assessing the protein structure-function paradigm. *J Mol Biol*. 1999;293(2):321–331.
6. Iakoucheva LM, Brown CJ, Lawson JD, Obradović Z, Dunker AK. Intrinsic disorder in cell-signaling and cancer-associated proteins. *J Mol Biol*. 2002;323(3):573–584.
7. Wright PE, Dyson HJ. Intrinsically disordered proteins in cellular signalling and regulation. *Nat Rev Mol Cell Biol*. 2015;16(1):18–29.
8. Tompa P. Intrinsically disordered proteins: a 10-year recap. *Trends Biochem Sci*. 2012;37(12):509–516.
9. Uversky VN. Intrinsic disorder in proteins associated with neurodegenerative diseases. *Front Biosci*. 2009;14:5188–5238.
10. Deng C-X. BRCA1: cell cycle checkpoint, genetic instability, DNA damage response and cancer evolution. *Nucleic Acids Res*. 2006;34(5):1416–1426.
11. Joerger AC, Fersht AR. Structural biology of the tumor suppressor p53. *Annu Rev Biochem*. 2008;77:557–582.
12. Mészáros B, Hajdu-Soltész B, Zeke A, Dosztányi Z. Mutations of intrinsically disordered protein regions can drive cancer but lack therapeutic strategies. *Biomolecules*. 2021;11(3). https://doi.org/10.3390/biom11030381.
13. Piovesan D, Tabaro F, Mičetić I, et al. DisProt 7.0: a major update of the database of disordered proteins. *Nucleic Acids Res*. 2017;45(D1):D219–D227.
14. Dyson HJ, Wright PE. Perspective: the essential role of NMR in the discovery and characterization of intrinsically disordered proteins. *J Biomol NMR*. 2019;73 (12):651–659.
15. Sormanni P, Piovesan D, Heller GT, et al. Simultaneous quantification of protein order and disorder. *Nat Chem Biol*. 2017;13(4):339–342.
16. Lazar T, Martínez-Pérez E, Quaglia F, et al. PED in 2021: a major update of the protein ensemble database for intrinsically disordered proteins. *Nucleic Acids Res*. 2021;49(D1): D404–D411.
17. Hatos A, Hajdu-Soltész B, Monzon AM, et al. DisProt: intrinsic protein disorder annotation in 2020. *Nucleic Acids Res*. 2020;48(D1):D269–D276.

18. He B, Wang K, Liu Y, Xue B, Uversky VN, Dunker AK. Predicting intrinsic disorder in proteins: an overview. *Cell Res*. 2009;19(8):929–949.

19. Meng F, Uversky VN, Kurgan L. Comprehensive review of methods for prediction of intrinsic disorder and its molecular functions. *Cell Mol Life Sci*. 2017;74(17):3069–3090.

20. Hanson J, Paliwal KK, Litfin T, Zhou Y. SPOT-Disorder2: improved protein intrinsic disorder prediction by ensembled deep learning. *Genomics Proteomics Bioinformatics*. 2019;17(6):645–656.

21. Necci M, Piovesan D, CAID Predictors, DisProt Curators, SCE T. Critical assessment of protein intrinsic disorder prediction. *Nat Methods*. April 2021;18:472–481. https://doi.org/10.1038/s41592-021-01117-3.

22. Kastano K, Erdős G, Mier P, et al. Evolutionary study of disorder in protein sequences. *Biomolecules*. 2020;10(10):1413. https://doi.org/10.3390/biom10101413.

23. Romero P, Obradovic Z, Dunker AK. Natively disordered proteins: functions and predictions. *Appl Bioinformatics*. 2004;3(2-3):105–113.

24. Ward JJ, Sodhi JS, McGuffin LJ, Buxton BF, Jones DT. Prediction and functional analysis of native disorder in proteins from the three kingdoms of life. *J Mol Biol*. 2004;337(3):635–645.

25. Tompa P, Dosztanyi Z, Simon I. Prevalent structural disorder in E. coli and S. cerevisiae proteomes. *J Proteome Res*. 2006;5(8):1996–2000.

26. Schlessinger A, Schaefer C, Vicedo E, Schmidberger M, Punta M, Rost B. Protein disorder—a breakthrough invention of evolution? *Curr Opin Struct Biol*. 2011;21(3):412–418.

27. Mishra PM, Verma NC, Rao C, Uversky VN, Nandi CK. Intrinsically disordered proteins of viruses: involvement in the mechanism of cell regulation and pathogenesis. *Prog Mol Biol Transl Sci*. 2020;174:1–78.

28. Cubuk J, Alston JJ, Incicco JJ, et al. The SARS-CoV-2 nucleocapsid protein is dynamic, disordered, and phase separates with RNA. *Nat Commun*. 2021;12(1):1936.

29. Ahrens JB, Nunez-Castilla J, Siltberg-Liberles J. Evolution of intrinsic disorder in eukaryotic proteins. *Cell Mol Life Sci*. 2017;74(17):3163–3174.

30. Dyson HJ, Wright PE. Intrinsically unstructured proteins and their functions. *Nat Rev Mol Cell Biol*. 2005;6(3):197–208.

31. Dunker AK, Brown CJ, Lawson JD, Iakoucheva LM, Obradović Z. Intrinsic disorder and protein function. *Biochemistry*. 2002;41(21):6573–6582.

32. Tompa P. The interplay between structure and function in intrinsically unstructured proteins. *FEBS Lett*. 2005;579(15):3346–3354.

33. Tompa P, Csermely P. The role of structural disorder in the function of RNA and protein chaperones. *FASEB J*. 2004;18(11):1169–1175.

34. Boeynaems S, Alberti S, Fawzi NL, et al. Protein phase separation: a new phase in cell biology. *Trends Cell Biol*. 2018;28(6):420–435.

35. Dyson HJ, Wright PE. Coupling of folding and binding for unstructured proteins. *Curr Opin Struct Biol*. 2002;12(1):54–60.

36. Yang J, Gao M, Xiong J, Su Z, Huang Y. Features of molecular recognition of intrinsically disordered proteins via coupled folding and binding. *Protein Sci*. 2019;28(11):1952–1965.

37. Fuxreiter M, Simon I, Friedrich P, Tompa P. Preformed structural elements feature in partner recognition by intrinsically unstructured proteins. *J Mol Biol*. 2004;338(5):1015–1026.

38. Schad E, Fichó E, Pancsa R, Simon I, Dosztányi Z, Mészáros B. DIBS: a repository of disordered binding sites mediating interactions with ordered proteins. *Bioinformatics*. 2018;34(3):535–537.

39. Fichó E, Reményi I, Simon I, Mészáros B. MFIB: a repository of protein complexes with mutual folding induced by binding. *Bioinformatics*. 2017;33(22):3682–3684.

40. Fuxreiter M, Tompa P. Fuzzy complexes: a more stochastic view of protein function. *Adv Exp Med Biol.* 2012;725:1–14.
41. Miskei M, Antal C, Fuxreiter M. FuzDB: database of fuzzy complexes, a tool to develop stochastic structure-function relationships for protein complexes and higher-order assemblies. *Nucleic Acids Res.* 2017;45(D1):D228–D235.
42. Borgia A, Borgia MB, Bugge K, et al. Extreme disorder in an ultrahigh-affinity protein complex. *Nature.* 2018;555(7694):61–66.
43. Van Roey K, Uyar B, Weatheritt RJ, et al. Short linear motifs: ubiquitous and functionally diverse protein interaction modules directing cell regulation. *Chem Rev.* 2014;114(13):6733–6778.
44. Mészáros B, Dosztányi Z, Simon I. Disordered binding regions and linear motifs—bridging the gap between two models of molecular recognition. *PLoS One.* 2012;7(10):e46829.
45. Kumar M, Gouw M, Michael S, et al. ELM-the eukaryotic linear motif resource in 2020. *Nucleic Acids Res.* 2020;48(D1):D296–D306.
46. Chen JW, Romero P, Uversky VN, Dunker AK. Conservation of intrinsic disorder in protein domains and families: II. functions of conserved disorder. *J Proteome Res.* 2006;5(4):888–898.
47. Tompa P, Fuxreiter M, Oldfield CJ, Simon I, Dunker AK, Uversky VN. Close encounters of the third kind: disordered domains and the interactions of proteins. *Bioessays.* 2009;31(3):328–335.
48. Punta M, Coggill PC, Eberhardt RY, et al. The Pfam protein families database. *Nucleic Acids Res.* 2012;40(Database issue):D290–D301.
49. Mier P, Paladin L, Tamana S, et al. Disentangling the complexity of low complexity proteins. *Brief Bioinform.* 2020;21(2):458–472.
50. Romero P, Obradovic Z, Li X, Garner EC, Brown CJ, Dunker AK. Sequence complexity of disordered protein. *Proteins.* 2001;42(1):38–48.
51. Ruff KM, Pappu RV, Holehouse AS. Conformational preferences and phase behavior of intrinsically disordered low complexity sequences: insights from multiscale simulations. *Curr Opin Struct Biol.* 2019;56:1–10.
52. Dosztányi Z, Chen J, Dunker AK, Simon I, Tompa P. Disorder and sequence repeats in hub proteins and their implications for network evolution. *J Proteome Res.* 2006;5(11):2985–2995.
53. Mohan A, Oldfield CJ, Radivojac P, et al. Analysis of molecular recognition features (MoRFs). *J Mol Biol.* 2006;362(5):1043–1059.
54. Gibson TJ. Cell regulation: determined to signal discrete cooperation. *Trends Biochem Sci.* 2009;34(10):471–482.
55. Galea CA, Pagala VR, Obenauer JC, Park C-G, Slaughter CA, Kriwacki RW. Proteomic studies of the intrinsically unstructured mammalian proteome. *J Proteome Res.* 2006;5(10):2839–2848.
56. Uversky VN. Protein intrinsic disorder and structure-function continuum. *Prog Mol Biol Transl Sci.* 2019;166:1–17.
57. Apostolico A, Giancarlo R. Sequence alignment in molecular biology. *J Comput Biol.* 1998;5(2):173–196.
58. Thompson JD, Higgins DG, Gibson TJ. CLUSTAL W: improving the sensitivity of progressive multiple sequence alignment through sequence weighting, position-specific gap penalties and weight matrix choice. *Nucleic Acids Res.* 1994;22(22):4673–4680.
59. Feng DF, Doolittle RF. Progressive sequence alignment as a prerequisite to correct phylogenetic trees. *J Mol Evol.* 1987;25(4):351–360.
60. Goldman N, Thorne JL, Jones DT. Assessing the impact of secondary structure and solvent accessibility on protein evolution. *Genetics.* 1998;149(1):445–458.

61. Gabaldón T, Koonin EV. Functional and evolutionary implications of gene orthology. *Nat Rev Genet.* 2013;14(5):360–366.
62. Davey NE, Edwards RJ, Shields DC. The SLiMDisc server: short, linear motif discovery in proteins. *Nucleic Acids Res.* 2007;35(Web Server issue):W455–W459.
63. Huerta-Cepas J, Bueno A, Dopazo J, Gabaldón T. PhylomeDB: a database for genome-wide collections of gene phylogenies. *Nucleic Acids Res.* 2008;36(Database issue):D491–D496.
64. Herrero J, Muffato M, Beal K, et al. Ensembl comparative genomics resources. *Database.* 2016;2016:baw053. https://doi.org/10.1093/database/bav096.
65. Light S, Sagit R, Sachenkova O, Ekman D, Elofsson A. Protein expansion is primarily due to indels in intrinsically disordered regions. *Mol Biol Evol.* 2013;30(12):2645–2653.
66. Radivojac P, Obradovic Z, Brown CJ, Dunker AK. Improving sequence alignments for intrinsically disordered proteins. *Pac Symp Biocomput.* 2002;589–600.
67. Trivedi R, Nagarajaram HA. Amino acid substitution scoring matrices specific to intrinsically disordered regions in proteins. *Sci Rep.* 2019;9(1):16380.
68. Sander C, Schneider R. Database of homology-derived protein structures and the structural meaning of sequence alignment. *Proteins.* 1991;9(1):56–68.
69. Lange J, Wyrwicz LS, Vriend G. KMAD: knowledge-based multiple sequence alignment for intrinsically disordered proteins. *Bioinformatics.* 2016;32(6):932–936.
70. Katoh K, Misawa K, Kuma K-I, Miyata T. MAFFT: a novel method for rapid multiple sequence alignment based on fast Fourier transform. *Nucleic Acids Res.* 2002;30(14):3059–3066.
71. Katoh K, Standley DM. A simple method to control over-alignment in the MAFFT multiple sequence alignment program. *Bioinformatics.* 2016;32(13):1933–1942.
72. Altschul SF. Generalized affine gap costs for protein sequence alignment. *Proteins.* 1998;32(1):88–96.
73. Thompson JD, Koehl P, Ripp R, Poch O. BAliBASE 3.0: latest developments of the multiple sequence alignment benchmark. *Proteins.* 2005;61(1):127–136.
74. Perrodou E, Chica C, Poch O, Gibson TJ, Thompson JD. A new protein linear motif benchmark for multiple sequence alignment software. *BMC Bioinformatics.* 2008;9:213.
75. Bileschi ML, Belanger D, Bryant D, et al. Using deep learning to annotate the protein universe. *bioRxiv.* 2019;626507. https://doi.org/10.1101/626507.
76. Xia Y, Franzosa EA, Gerstein MB. Integrated assessment of genomic correlates of protein evolutionary rate. *PLoS Comput Biol.* 2009;5(6):e1000413.
77. Zhang J, Yang J-R. Determinants of the rate of protein sequence evolution. *Nat Rev Genet.* 2015;16(7):409–420.
78. Brown CJ, Takayama S, Campen AM, et al. Evolutionary rate heterogeneity in proteins with long disordered regions. *J Mol Evol.* 2002;55(1):104–110.
79. Dekker J, Kanellopoulos PN, van Oosterhout JA, Stier G, Tucker PA, van der Vliet PC. ATP-independent DNA unwinding by the adenovirus single-stranded DNA binding protein requires a flexible DNA binding loop. *J Mol Biol.* 1998;277(4):825–838.
80. Brown CJ, Johnson AK, Dunker AK, Daughdrill GW. Evolution and disorder. *Curr Opin Struct Biol.* 2011;21(3):441–446.
81. Yang Z. PAML 4: phylogenetic analysis by maximum likelihood. *Mol Biol Evol.* 2007;24(8):1586–1591.
82. Yang Z, Nielsen R. Codon-substitution models for detecting molecular adaptation at individual sites along specific lineages. *Mol Biol Evol.* 2002;19(6):908–917.
83. Afanasyeva A, Bockwoldt M, Cooney CR, Heiland I, Gossmann TI. Human long intrinsically disordered protein regions are frequent targets of positive selection. *Genome Res.* 2018;28(7):975–982.
84. Anisimova M, Bielawski JP, Yang Z. Accuracy and power of the likelihood ratio test in detecting adaptive molecular evolution. *Mol Biol Evol.* 2001;18(8):1585–1592.

85. Jeffares DC, Tomiczek B, Sojo V, dos Reis M. A beginners guide to estimating the non-synonymous to synonymous rate ratio of all protein-coding genes in a genome. *Methods Mol Biol*. 2015;1201:65–90.

86. Zhang J, Nielsen R, Yang Z. Evaluation of an improved branch-site likelihood method for detecting positive selection at the molecular level. *Mol Biol Evol*. 2005;22 (12):2472–2479.

87. Gayà-Vidal M, Albà MM. Uncovering adaptive evolution in the human lineage. *BMC Genomics*. 2014;15:599.

88. Pajkos M, Zeke A, Dosztányi Z. Ancient evolutionary origin of intrinsically disordered cancer risk regions. *Biomolecules*. 2020;10(8):1115. https://doi.org/10.3390/biom10081115.

89. Bellay J, Han S, Michaut M, et al. Bringing order to protein disorder through comparative genomics and genetic interactions. *Genome Biol*. 2011;12(2):R14.

90. DiNitto JP, Huber PW. Mutual induced fit binding of Xenopus ribosomal protein L5 to 5S rRNA. *J Mol Biol*. 2003;330(5):979–992.

91. Erdős G, Pajkos M, Dosztányi Z. IUPred3—prediction of protein disorder enhanced with unambiguous experimental annotation and visualization of evolutionary conservation. *Nucleic Acids Res*. 2021;49:W297–W303.

92. Sutovsky H, Gazit E. The von Hippel-Lindau tumor suppressor protein is a molten globule under native conditions: implications for its physiological activities. *J Biol Chem*. 2004;279(17):17190–17196.

93. Jones P, Binns D, Chang H-Y, et al. InterProScan 5: genome-scale protein function classification. *Bioinformatics*. 2014;30(9):1236–1240.

94. Daughdrill GW, Narayanaswami P, Gilmore SH, Belczyk A, Brown CJ. Dynamic behavior of an intrinsically unstructured linker domain is conserved in the face of negligible amino acid sequence conservation. *J Mol Evol*. 2007;65(3):277–288.

95. Chen SC-C, Chen F-C, Li W-H. Phosphorylated and nonphosphorylated serine and threonine residues evolve at different rates in mammals. *Mol Biol Evol*. 2010;27 (11):2548–2554.

96. Landry CR, Freschi L, Zarin T, Moses AM. Turnover of protein phosphorylation evolving under stabilizing selection. *Front Genet*. 2014;5:245.

97. Zarin T, Strome B, Nguyen Ba AN, Alberti S, Forman-Kay JD, Moses AM. Proteome-wide signatures of function in highly diverged intrinsically disordered regions. *Elife*. 2019;8:46883. https://doi.org/10.7554/eLife.46883.

98. Franzmann TM, Jahnel M, Pozniakovsky A, et al. Phase separation of a yeast prion protein promotes cellular fitness. *Science*. 2018;359(6371):5654. https://doi.org/10.1126/science.aao5654.

99. Dasmeh P, Wagner A. Natural selection on the phase-separation properties of FUS during 160 my of mammalian evolution. *Mol Biol Evol*. 2021;38(3):940–951.

100. Zarin T, Strome B, Peng G, Pritišanac I, Forman-Kay JD, Moses AM. Identifying molecular features that are associated with biological function of intrinsically disordered protein regions. *Elife*. 2021;10:60220. https://doi.org/10.7554/eLife.60220.

101. Khan T, Douglas GM, Patel P, Nguyen Ba AN, Moses AM. Polymorphism analysis reveals reduced negative selection and elevated rate of insertions and deletions in intrinsically disordered protein regions. *Genome Biol Evol*. 2015;7(6):1815–1826.

102. Davey NE, Cyert MS, Moses AM. Short linear motifs—ex nihilo evolution of protein regulation. *Cell Commun Signal*. 2015;13:43.

103. Cordeddu V, Di Schiavi E, Pennacchio LA, et al. Mutation of SHOC2 promotes aberrant protein N-myristoylation and causes Noonan-like syndrome with loose anagen hair. *Nat Genet*. 2009;41(9):1022–1026.

104. Zhang T, Prives C. Cyclin a-CDK phosphorylation regulates MDM2 protein interactions. *J Biol Chem*. 2001;276(32):29702–29710.

105. Davey NE, Van Roey K, Weatheritt RJ, et al. Attributes of short linear motifs. *Mol Biosyst.* 2012;8(1):268–281.
106. Davey NE, Cowan JL, Shields DC, Gibson TJ, Coldwell MJ, Edwards RJ. SLiMPrints: conservation-based discovery of functional motif fingerprints in intrinsically disordered protein regions. *Nucleic Acids Res.* 2012;40(21):10628–10641.
107. Davey NE, Shields DC, Edwards RJ. Masking residues using context-specific evolutionary conservation significantly improves short linear motif discovery. *Bioinformatics.* 2009;25(4):443–450.
108. Hirschi A, Cecchini M, Steinhardt RC, Schamber MR, Dick FA, Rubin SM. An overlapping kinase and phosphatase docking site regulates activity of the retinoblastoma protein. *Nat Struct Mol Biol.* 2010;17(9):1051–1057.
109. Erdős G, Szaniszló T, Pajkos M, et al. Novel linear motif filtering protocol reveals the role of the LC8 dynein light chain in the hippo pathway. *PLoS Comput Biol.* 2017;13(12):e1005885.
110. Merino-Gracia J, Zamora-Carreras H, Bruix M, Rodríguez-Crespo I. Molecular basis for the protein recognition specificity of the dynein light chain DYNLT1/Tctex1: characterization of the interaction with activin receptor IIB. *J Biol Chem.* 2016;291(40):20962–20975.
111. Liebeskind BJ, McWhite CD, Marcotte EM. Towards consensus gene ages. *Genome Biol Evol.* 2016;8(6):1812–1823.
112. Domazet-Loso T, Tautz D. Phylostratigraphic tracking of cancer genes suggests a link to the emergence of multicellularity in metazoa. *BMC Biol.* 2010;8:66.
113. Domazet-Loso T, Brajković J, Tautz D. A phylostratigraphy approach to uncover the genomic history of major adaptations in metazoan lineages. *Trends Genet.* 2007;23(11):533–539.
114. Kinzler KW, Vogelstein B. Cancer-susceptibility genes Gatekeepers and caretakers. *Nature.* 1997;386(6627):761. 763.
115. Domazet-Loso T, Tautz D. An ancient evolutionary origin of genes associated with human genetic diseases. *Mol Biol Evol.* 2008;25(12):2699–2707.
116. Pajkos M, Mészáros B, Simon I, Dosztányi Z. Is there a biological cost of protein disorder? Analysis of cancer-associated mutations. *Mol Biosyst.* 2012;8(1):296–307.
117. Kim Y, Furman CM, Manhart CM, Alani E, Finkelstein IJ. Intrinsically disordered regions regulate both catalytic and non-catalytic activities of the MutLα mismatch repair complex. *Nucleic Acids Res.* 2019;47(4):1823–1835.
118. Shcherbakova PV, Kunkel TA. Mutator phenotypes conferred by MLH1 overexpression and by heterozygosity for mlh1 mutations. *Mol Cell Biol.* 1999;19(4):3177–3183.

CHAPTER THREE

Non-specific porins of Gram-negative bacteria as proteins containing intrinsically disordered regions with amyloidogenic potential

Olga D. Novikova[a,*], Vladimir N. Uversky[b,c], and Elena A. Zelepuga[a]

[a]G.B. Elyakov Pacific Institute of Bioorganic Chemistry, Far East Branch, Russian Academy of Sciences, Vladivostok, Russia
[b]Department of Molecular Medicine, Morsani College of Medicine, University of South Florida, Tampa, Florida, United States
[c]Laboratory of New Methods in Biology, Institute of Biological Instrumentation, Russian Academy of Sciences, Moscow, Russia
*Corresponding author: e-mail address: novolga_05@mail.ru

Contents

Abstract

Features of the structure and functional activity of bacterial outer membrane porins, coupled with their dynamic "behavior," suggests that intrinsically disordered regions (IDPRs) are contained in their structure. Using bioinformatic analysis, the quantitative content of amyloidogenic regions in the amino acid sequence of non-specific porins inhabiting various natural niches was determined: from terrestrial bacteria of the genus *Yersinia* (OmpF and OmpC proteins of *Y. pseudotuberculosis* and *Y. ruckeri*) and from the marine bacterium *Marinomonas primoryensis* (MpOmp). It was found that OmpF and OmpC porins can be classified as moderately disordered proteins, while MpOmp can be classified as highly disordered protein. Mapping of IDPRs, performed using 3D structures of monomers of the proteins, showed that the regions of increased conformational plasticity fall on the regions, the functional importance of which has been reliably confirmed as a result of numerous experimental studies. The revealed correlation made it possible to explain the differences in the physicochemical characteristics and properties of not only porins from terrestrial and marine bacteria, but also non-specific porins of different types, OmpF and OmpC proteins. First of all, this concerns the flexible outer loops that form the pore vestibule, as well as regions of the barrel with an increased "ability" for aggregation, the so-called "hot spots" of aggregation.

75

The abnormally high content of IDPRs in the MpOmp structure made it possible to suggest that the high adaptive potential of bacteria may correlate with an increase in the number of IDPRs and/or regions with increased conformational variability.

Protein molecules, due to their astonishing structural and functional versatility, play a central role in the distinctive properties of eukaryotic and prokaryotic cells. The functions of proteins are extremely diverse and are of great importance for maintaining the vital activity of both an individual cell and the entire organism as a whole. The most important of these functions are structural, energy metabolism, transport, motor, catalytic, signaling, and regulatory. Presently, the researchers in the field of protein structure and function have no doubts that many functional proteins do not have well-folded structures. On the contrary, so-called intrinsically disordered proteins (IDPs) and proteins with intrinsically disordered protein regions (IDPRs), which play a number of crucial roles in a living cell, are widespread in nature.[1] Under certain conditions, some IDPs are able to form amyloid-like structures.

Amyloids are protein fibrillar aggregates with an ordered β-sheet secondary structure (known as cross-β-structure). While one part of amyloids is associated with the development of about 50 incurable diseases in humans and animals, the other performs various important physiological functions. The greatest variety of functions was found in amyloids of prokaryotic species. Therefore, bacteria represent the simplest cellular model for studying the process of functional amyloidosis. Bacterial amyloids form layers of extracellular protein, acting as adhesins, regulating the activity of toxins and virulence factors, and being components of biofilms as a matrix.[2] It is also known that various species of pathogenic bacteria widely use the amyloid form of the protein to interact with eukaryotic organisms.[3] Therefore, the repertoire of disease-associated amyloids includes not only dozens of pathological mammalian amyloids, formation of which is associated with various amyloidoses, but also numerous microbial ones utilized by the bacteria for more efficient host infection. Although the ability of symbiotic microorganisms to produce amyloids has recently been demonstrated, their functional role in host-symbiont interactions, as well as in interactions between species within prokaryotic communities, remains poorly understood. In this regard, the formation of protein assemblies with amyloid properties is currently the subject of intensive research.[4]

All functional amyloids of microorganisms known to date have been identified using targeted approaches that did not allow assessing the entire set of amyloids in the cell. However, as a result of proteomic screening of candidates for the role of new amyloid-forming proteins using the PSIA (Proteomic Screening and Identification of Amyloids) method developed by Nizhnikov et al.[5] 61 proteins present in the fractions resistant to ionic detergents were identified in *Escherichia coli*, one of the most important model and biotechnological subjects. The proportion of proteins containing potentially amyloidogenic regions predicted by bioinformatics algorithms among the identified proteins was 3–5 times higher than the proportion of such proteins in the entire proteome of *E. coli*.[5]

The discovery and characterization of IDPs and IDPRs disrupted the established structure-function paradigm. Disordered proteins exist as highly dynamic conformational ensembles of different states that interconvert at different time scales. Structured, ordered proteins can fold into unique stable structure that correspond to a deep minimum in the energy landscape, which is achieved via the process of spontaneous folding. In contrast, the IDP energy landscape is characterized by multiple, shallow/local minima that correspond to a number of different states, and this structural plasticity is an integral part of their function.[1,6,7] However, it is interesting and important to note that protein amyloids are the most stable, conceptually simple, and universal macromolecular structures that have ever been encountered in nature, and therefore, the study of their structure and properties is fraught with a huge potential of new knowledge useful for problems of protein engineering.[8]

OmpA is an important structural protein of *E. coli*, which plays a vital role in the stability of the outer membrane (OM) of this microorganism, presumably due to the presence of the long hydrophilic extracellular loops, which represent a kind of "molecular rivet."[9] With the exception of OmpA, the porins of the outer bacterial membrane have not yet been characterized and described in literature as prokaryotic proteins containing IDPRs and amyloidogenic regions in their amino acid sequences.[10] OmpA is one of a few transmembrane proteins, which folding and stability has been studied in detail. It was found that only half of the mass of OmpA encodes its transmembrane β-barrel, and the remaining sequence constitutes a soluble domain located on the periplasmic side of OM. To understand how the periplasmic domain of OmpA contributes to the stability and folding of the full-length OmpA protein, the periplasmic domain of the protein was cloned, expressed, purified, and explored independently of the transmembrane β-barrel region.[11] Experiments have shown that the periplasmic domain of OmpA exists as an independently folded unit, which, according to the circular dichroism (CD) spectroscopy, has a mixed α/β secondary structure. *In vitro* folding analysis has shown that self-association competes with the β-barrel folding if the folding process occurs outside the membrane, and the periplasmic domain increases the folding efficiency of the full-length protein by decreasing its self-association. It was also found that the periplasmic domain of OmpA reduces the tendency for self-association of the unfolded barrel-shaped domain, but only in the cis-position; i.e., being covalently attached to the protein body. These results define a novel chaperone function for the OmpA periplasmic domain that may be relevant to the *in vivo* folding. In addition, the study of self-assembly of OmpA showed that the transmembrane region of the protein must form a critical core consisting of three molecules before further oligomerization with the formation of new molecular form of the protein with a large molecular weight.[11]

The formation of amyloid fibril-like structure by the transmembrane protein OmpA is a very remarkable example of bacterial functional amyloids, which has important consequences. First of all, it opens up for the researchers another class of bacterial proteins that can be used to study the process of fibrillation and to determine the properties of the amino acid sequence that contribute to the formation of amyloid. On the other hand, it raises a very important question related to the folding process of OM proteins, namely, the role of periplasmic chaperone proteins that can prevent the formation of amyloid fibrils by β-barrel proteins.

Porins are OM proteins of Gram-negative bacteria. They belong to the class of integral membrane β-structured proteins. Using the example of non-specific *E. coli* porins (OmpF and OmpC types), defined as classical in the modern literature,[12] the main characteristics and properties of these channel-forming proteins have been determined. In quantitative terms, they are the major proteins of bacterial OM, accounting for about 2% of the total protein content in the cell. Being structural proteins that have an exit to the surface of a bacterial cell, they can act as receptors in many cellular processes. However, their major function, which defines the name of these proteins, is their ability to form a network of water-filled pores in the OM of microorganisms, which are intended for the non-specific diffusion of soluble nutrients and metabolites through the membrane.[13] Many external factors influencing the expression of porins of one type or another have been determined, the main of which are osmolarity, temperature, and pH of the medium.[14] The significance of this complex regulation of pore-forming proteins in the vital activity of the cell has not yet been fully elucidated. Nevertheless, it is assumed that these two types of non-specific porins

(OmpF and OmpC) are involved in the balanced supply of nutrients through the OM and protect the bacterial cell from toxins.[12,13]

As membrane proteins, porins are unique, since, unlike most other membrane proteins with α-helical structure, they consist of β-cylinders (barrels).[15] In addition, a characteristic feature of their sequence is that it is predominantly polar and does not contain long hydrophobic segments similar to those found in the membrane α-helices of bacteriorhodopsin and the bacterial photosynthetic reaction center.[16] This is due to the noticeable differences in the geometry of β-strands and α-helices. In fact, one turn of α-helix that includes 3.6 residues provides a pitch (advance per turn) of 5.4 Å (i.e., 1.5 Å rise per residue). On the other hand, in a β-strand, the distance between C_i^α and C_{i+2}^α ranges from 7.6 Å in a fully extended peptide to 6 Å in "pleated" β-strands. This means that with the average thickness of a membrane of 40 Å, it takes a transmembrane α-helix (which is typically 40 Å long) 7–8 turns (or 25–29 residues) to cross the membrane. On the other hand, the length of the transmembrane β-strands typically ranges from 7 to 12 residues. Furthermore, due to the alternation of charged and hydrophobic residues, the β-barrel surface shows noticeable alternating polarity, which in turn, allows porins to interact both with a non-polar lipid membrane and with hydrophilic substances diffusing through the water-filled channel. In native membranes, porins exist as homotrimers. Their monomer units have a very stable barrel-shaped (β-barrel) structure (e.g., see Fig. 1A and B), which is stabilized by a whole network of interactions of various nature, such as hydrophobic interactions, as well as ionic and hydrogen bonds.[17]

Undoubtedly, the presence in proteins such as porins of intrinsically disordered regions, which have some advantages over folded domains including the capability of formation of various bonds,[18] including polyvalent electrostatic interactions that occur on the surface of lipid bilayers and inside the porin channel, can explain manifestations of various porin properties. Therefore, in the last decade, an increasing amount of attention has been paid to studying the features of the conformational and functional activity of porins, coupled with the analysis of their dynamic "behavior" under the influence of various factors, including the environment. Nevertheless, despite the clear progress in studying the structural and functional features of the pore-forming proteins, the factors determining their multifunctionality, as well as the mechanisms underlying their conformational variability (plasticity), are still insufficiently understood.

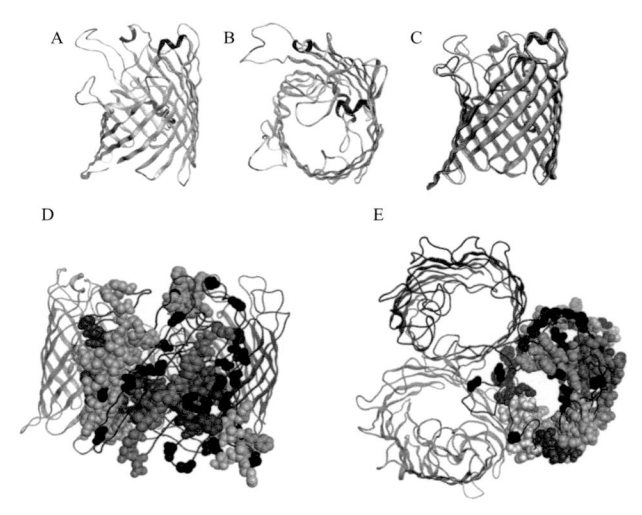

F
>OmpC_m31758.
AEIYNK**DGNKLDLYGK**IDGLH**Y**F**SDNKN**LDGDQS**YMR**FGL**KGETQ**ITDQLTGYG**QWE**YQVNLNKAENEDGNHDSFT
RVGFAGLKFADYGSLDYGRNYG**VV**YDVTS**WT**DVL**PEFGG**DTYGSDNFMQQRGNGFATYRNTDFF**GLVDGLNFALQY
QGKNGSSSETNNG**RG**VADQNGDGYG**MSLS**YDYLG**WG**VSASAAMASSLRTTAQNDLQYG**QG**KRANAYT**GG**LKYDANN
VYLAANYTQTYNLTRF**GDF**SNRSSDAAF**GF**ADKAHNIEVVAQYQFDF**GL**RPSVAYLQSKGKDI**GI**YGDQDLL**KY**VD
IGATYFF**NKNMSTY**VDYK**I**NLLDDNQF**TN**AR**INTDD**IVAGMVYQF

Fig. 1 Theoretical model of the spatial structure of the OmpC porin from
Y. pseudotuberculosis (strain 3260). The structure of the OmpC porin is shown as ribbon
diagram (side view—A, top view—B). Superposition of Cα atoms of 3D structures of
porins OmpC *Y. pseudotuberculosis* and osmoporin *K. pneumoniae* (PDB code 1OSM.
A) (C). The value of the standard deviation for all Cα atoms is 1.1 Å. The superposition
and the figure were obtained using the MOE™ 9.10 program. The structure of the
OmpC porin trimer from *Y. pseudotuberculosis* is shown as ribbon diagram (side
view—D, top view—E). In one of the monomers, AK, residues are represented as
spheres corresponding to Van der Waals radii; glycine residues are marked in black.
Amino acid sequence YpOmpC with "hot spots" of aggregation marked as bold font (F).

In this regard, bioinformatics analysis of the porin sequence would
undoubtedly be a welcome approach to solving this problem. The relationship between the amino acid sequences of structural elements, such as
β–hairpin, in porins and amyloidogenic motifs that form the β–arcade of typical amyloid proteins is of particular interest. This because, like amyloid
fibrils, most β–hairpin repeat (BHR) structures have a so-called crossover
fold β–forms. Furthermore, the β–hairpin is considered a likely intermediate
structure in amyloidogenesis. Bioinformatics analysis shows that the occurrence of predicted β–arcade motifs in BHR regions varies greatly depending

on the BHR structural fold, cell localization, and phylogeny. One of the most striking observations is the high level of sequence similarity between BHR membrane porins and β-arcade motifs.[19] This sequence similarity is the further evidence that the membrane porins and the ring-shaped amyloid oligomers may have structural similarities. Furthermore, these results explain how some amyloidogenic sequence can fold into either ring-shaped oligomers or elongated amyloid fibrils. It was also found that potentially lethal amyloidogenic β-arcade motifs are absent in the elongated BHR structures of intracellular eukaryotic proteins. In this case, the selective evolutionary pressure is suggested to act against aggregation.[19]

In this study, we made an attempt to comparatively analyze the quantitative content of amyloidogenic regions in the amino acid sequence of non-specific porins from terrestrial bacteria of the genus *Yersinia* (OmpF and OmpC proteins from *Y. pseudotuberculosis* and *Y. ruckeri*) and from the marine bacterium *Marinomonas primoryensis*, as well as using theoretical models of the proteins considered to determine the localization of IDPRs in their structure, and, if possible, to assess how these regions coincide with the functionally important regions of porins.

For many years, we have been studying the structure and function of *Yersinia* porins as typical representatives of β-barrel integral membrane proteins that form ion channels in the bacterial membrane and serve as immunodominant antigens of these microorganisms. *Yersinia* is a genus of Gram-negative bacteria with unique adaptive capabilities: they live and multiply both in the external environment and in a warm-blooded organism (including humans), combining saprophytic and parasitic lifestyles.[20] As a result of studying the structural transformations of *Yersinia* porins under the influence of denaturing factors, as well as their functional and biological properties, we came to the unambiguous conclusion that these porins possess all the characteristic properties of proteins containing IDPRs in their sequence. The most important and indicative of these features are conformational plasticity and multifunctionality of the *Yersinia* porins.

Porins are extremely stable proteins that are resistant to denaturation in the presence of 5 M guanidine hydrochloride or 2% SDS at elevated temperatures. Studies have shown that the L2 OmpF fixing loop directed toward the adjacent monomer contributes significantly to this exceptionally high conformational stability. In this respect, *Yersinia* porins are no exception among the other transmembrane β-barrel proteins. At the same time, they easily form various conformational intermediates, the spatial structure of which is determined by the nature of the detergent used to solubilize these proteins. The conformation of porins under various conditions changes not only at the level of the tertiary structure, but also at the level of secondary structure. In solutions of ionic amphiphilic (zwitterionic) and hydrophobic detergents, the ratio of the elements of the regular secondary structure of porins differs significantly. In the presence of SDS, the content of α-helical structure increases markedly, and this trend is even more pronounced when the temperature is raised.[21] On the contrary, in the solutions of n-octyl-β-D-glucopyranoside (octylglucoside) and Zwittergent 3–14 (Zw), the amount of α-helices was minimal, no more than 5%,[22] which, according to Garavito and Rosenbusch, corresponds to the native structure of the protein in lipid environment (in the bacterial membrane).[23] Interestingly, an increase in the temperature did not result in an increase in the α-helical content in the structure of porin from *Y. ruckeri* (YrOmpF) solubilized in a Zw solution. Instead, a significant increase in the amount of disordered structure was observed.[22]

As known, an important feature of intrinsically unstructured proteins is their ability to undergo a disorder–order transition during or before their biological function. Functionally, IDPs/IDPRs are described as the "protein quartet" model, where the protein function is determined by four specific conformations (ordered forms, molten globules, pre-molten globules and random coils) and transitions between these states.[10] Interestingly, it was shown that upon pH titration, the *Yersinia* porins a capable of formation of a number of partially folded intermediates that differ from each other in the degree of their unfolding.[23] The OmpF porin from *Y. pseudotuberculosis* (YpOmpF) at pH 4.5 forms a specific intermediate with the minimal content of the regular β-structure and the largest amount of β-turns, which is known to be a characteristic feature of the structural instability of the protein molecule. At pH 3.0, YpOmpF is characterized by a significant decrease in the degree of ordering of the environment of the side chains of the protein, and, on the other hand, by a maximum increase in the affinity to the hydrophobic quenchers of protein fluorescence, which indicates the exposure of the hydrophobic regions of the molecule. With a further decrease in the pH of the medium to 2.0, a new intermediate is formed, which, in terms of the degree of its tertiary structure ordering is practically similar to porin at pH 3.0, but differs from it in the content of the regular secondary structure. On the basis of these data, it was concluded that at pH 3.0, the molecular organization of YpOmpF satisfies the criteria of the molten globule conformational intermediate, whereas at pH 2.0, porin adopts the pre-molten globule-like structure.[24]

One of the main internal characteristics that determines the nature of a particular protein and, as a consequence, its biological properties is the ability to aggregate and/or to be engaged in protein-protein interactions. Using various bioinformatics approaches,[25] it is possible to accurately determine the localization of regions with an increased predisposition for aggregation, based on the primary structure of a protein. The prediction of such aggregation-prone segments, the so-called "hot spots" of aggregation, in the OmpC amino acid sequence of the porin from *Y. pseudotuberculosis* (YpOmpC) was carried out using the AGGRESCAN webserver.[26] As our studies have shown, some of these zones are located in the region of the inter-monomer contacts of the porin (Fig. 1E). However, most of them are located on the outer surface of the β-barrel.[27] Fig. 1D and E show the localization of these "hot spots" on the spatial model of the YpOmpC porin. In addition, as can be seen from Fig. 1D, these regions either include glycine residues or are limited to the β-strands containing so-called glycine clusters. The high content of glycine residues (about 14%) and the frequency of their occurrence (more than 13%) are characteristic of pore-forming proteins of OM bacteria. It turned out that porin OmpC from *Y. pseudotuberculosis* is no exception. As shown by our calculations, 78% of glycine residues in this protein are conservative and are located mainly in β-strands.[28] There is an opinion that the glycine residues form a kind of glycine backbone of the β-barrel. On the other hand, according to the views of a number of authors,[29] an increased content of glycine residues imparts conformational plasticity to the protein structure. The Gly residue is unique among the amino acids in that its entire side chain is just a hydrogen atom. As a result, its conformation has greater freedom than other amino acids and furthermore can provide flexibility for adjacent residues. Possibly, this compensates to a certain extent for the increased stability of the β-barrel and ensures the mobility of the tertiary structure of porins, which is manifested in their ability to form various conformational intermediates.

Another characteristic feature of IDPs is their multifunctionality. It is well known that the structure and location on the bacterial surface endows pore-forming proteins with a variety of functions. However, it should be noted that the acquisition of the status of functionally important components of OM bacteria by porins in various manifestations of their interaction with eukaryotic cells of the host organism has a long history. In early studies, porins were considered exclusively as a neutral carrier of lipopolysaccharides (LPSs). At present, researchers have no doubts about the fundamental roles of porins not only in the life support of a bacterial cell, but also as a target for the innate immune system of the host organism. Porins play a number of important roles as effectors of pathogenesis: they promote adhesion (and possibly invasion) of bacteria, affect the nature of the immune response, providing resistance to phagocytosis. In addition, they interact with the complement binding system and inhibit the production of lymphocytes.[30] The adhesion sites on the surface of Gram-negative bacteria are external loops of porins, which are involved in the binding of bactericidal compounds and in interaction with other cells. Changes in the loop structure as a mechanism of "escape from the influence of the immune response" as well as modulation of porin expression in response to the presence of antibiotics are survival strategies developed by some pathogenic bacteria.

In connection with the aforementioned considerations, in addition to structural and functional studies of *Yersinia* porins, we were interested in their role in the pathogenesis of pseudotuberculosis. We have shown that these proteins are highly immunogenic antigens. Antibodies to them are found both after vaccination of laboratory animals, and during the natural development of infection. They are targets for the innate immunity system of the host organism, and also act as effectors of pathogenesis, suppressing individual stages of the host's immune defense, and thereby ensure the survival of the pathogen in the host.[31] It was found that *Y. ruckeri* OmpF (YrOmpF) induces the arrest of the S-phase of the cell cycle both in normal chum salmon heart cells (CHH-1) and in THP-1 tumor cells. In cancer cells, the observed effect was most pronounced. A significant cytotoxic effect of YrOmpF on primary mouse peritoneal macrophages was also found. YrOmpF stimulated the activity of bactericidal systems of phagocytes, especially the oxygen-independent subsystem. Antibodies against YrOmpF decreased the release of myeloperoxidase and the synthesis of cationic peptides by peritoneal macrophages, and increased their viability.[31]

The study of the biological activity of the recombinant OmpF porin of *Y. pseudotuberculosis in vivo* showed that the *Yersinia* porins play an important role in the formation of the cellular and humoral immune response, namely, they act as inducers of some mediators of the immune response. It was possible to reveal the actual effects of porin on the synthesis and dynamics of accumulation of cytokines on the first day of the development of a bacterial infection. The data obtained suggested that the presence of LPS in the protein sample can compensate for both suppressive (against IL-12, IL-1β and IL-17A) and pronounced stimulating (against IL-6, MIP-1α, MIP-1β) effects of porin on some mediators. Taken together these data allowed us to conclude that during the development of a bacterial infection *in vivo*, porins and LPS act on the host cells not only as separate components of OM, but also as an endotoxin-protein complex. Obviously, such a combined effect on the specific receptors of the host immune cells contributes to the development of a balanced immune response.[32]

The significance of the conformational plasticity of the outer loops of porin in the manifestation of the immunogenic properties of OmpF *Y. pseudotuberculosis* (YpOmpF) was convincingly demonstrated when studying the antigenic and immunogenic activity of mutant OmpF porins with deletions of the outer loops,[33] as well as when elucidating the causes of cross-serological reactivity between OmpF porin and thyroid-stimulating hormone receptor (hTSHR).[34] As a result of immunochemical analysis, a significant decrease in the immunochemical activity of mutant porins was shown when interacting with immune serum to the native trimer YpOmpF. Obviously, not all outer loops are equally involved in the formation of antigenic determinants of porin, or the conformational determinants of YpOmpF formed at the level of the tertiary and quaternary structure of the protein undergo significant changes (i.e., might represent neotopes[35]), which seems to us the most probable.[33]

In experiments to identify the immunochemical relationship between YpOmpF and hTSHR, it was shown that the monoclonal antibodies to the receptor (mAb) interact equally effectively with both antigens in thyroid tissue extracts and with YpOmpF. However, according to immunoblotting data, the trimeric form of porin was not detected by these antibodies. To explain the obtained experimental data, theoretical models of the spatial structures of monomeric and trimeric complexes of YpOmpF with antibodies to hTSHR were constructed. According to the results of molecular modeling, YpOmpF, being in monomeric form, can, like hTSHR, freely interact with mAbs. But when a porin trimer is formed, the hydrophobic region, which is located in the interaction zone of the porin with the antibody, is closed. This feature, as well as other spatial rearrangements of amino acid residues, which determine the efficiency of binding, interfere with the interaction between the porin trimer and the monoclonal antibodies to the receptor.[34] These results, obtained *in vitro* and *in silico*, confirmed the existence of the phenomenon of molecular mimicry. Thus, autoimmune thyroid disease (Graves' disease), which sometimes occurs in patients suffering from pseudotuberculosis, may be due to structural and antigenic similarities between the immunodominant OM protein, porin, and the human thyroid-stimulating hormone receptor. In addition to the very fact of the experimentally and theoretically proven existence of the phenomenon of molecular mimicry, according to the results of this study, one more interesting conclusion can be drawn. Since the amino acid sequences of porin YpOmpF and the hTSHR receptor are not found to have sufficiently long regions with high sequence similarity, the reason for the antigenic relationship is, most likely, the similarity of the conformation of individual regions of their spatial structure.[34]

The aforementioned examples clearly indicate that the conformational plasticity of the tertiary structure of porins plays a special role in the manifestation of various types of biological activity by these proteins. Next, we conducted comparative bioinformatics analysis of the *Yersinia* porins (*Y. pseudotuberculosis* and *Y. ruckeri*) from bacteria inhabiting noticeably different natural niches (environmental conditions). In earlier studies, these porins demonstrated significant differences in the thermal stability of their trimeric structure and in the electrophysiological characteristics of the channels they form. In order to identify conformationally flexible regions in the amino acid sequences of these proteins, as well as to determine their overall intrinsic disorder predispositions, we involved several web resources, such as D^2P^2 (http://d2p2.pro/)[36] and ODiNPred (https://stprotein.chem.au.dk/odinpred).[37] We assumed that it is these intrinsically disordered regions that may turn out to be functionally important structural elements of the porin molecules under consideration, and, possibly, a comparison of their quantitative content and location in the polypeptide chain would explain the differences in the properties of these proteins. The results of these analyses are summarized in the Table 1 and Fig. 2 and show a good agreement between the prediction methods we used for IDPRs. For comparison, the Table 1 also shows data for the classical porin of *E. coli* and marine porin from *M. primoryensis*, the trimeric structure of which demonstrated extremely low resistance to temperature.[40]

Table 1 Percent of predicted intrinsically disordered residues (PPIDR) in the porins of bacteria in marine, terrestrial, and freshwater habitats, %.

The source of porins	PPIDR in OmpF	PPIDR in OmpC
Marinomonas primoryensis (MpOmp)	53.8	
Yersinia pseudotuberculosis (YpOmpF/C)	12.4	29.1
Yersinia ruckeri (YrOmpF/C)	25.9	24.8
Escherichia coli (EcOmpF/C)	14.1	25.1

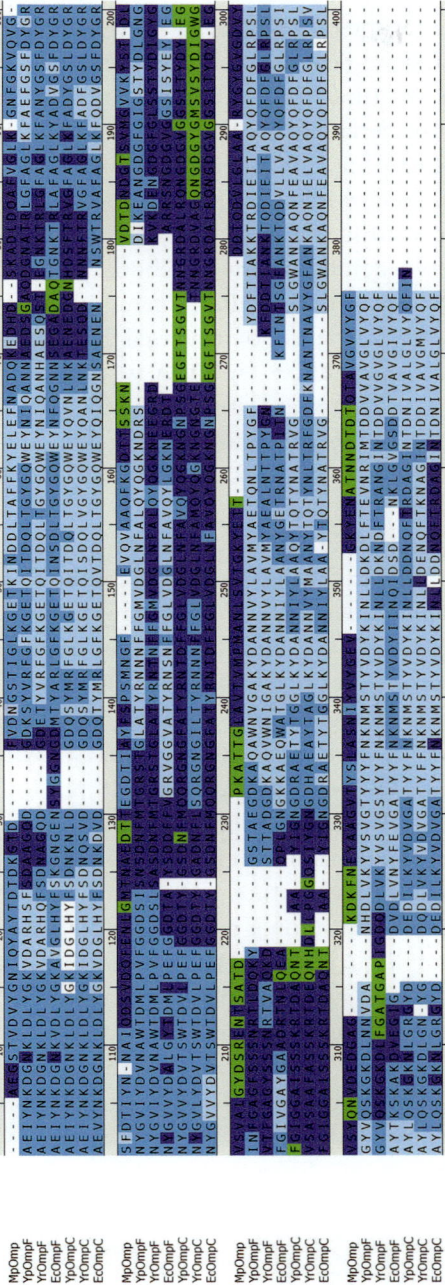

Fig. 2 Multiple sequence alignment of OmpF and OmpC porins from *Y. pseudotuberculosis, Y. ruckeri, E. coli* as well as from *M. primoryensis* (MpOmp) colored according to Cumulative Distributions of ODiNPred Z-score for individual residues indicating their distribution between ordered and disordered conformation. Residues with Z-scores <3.0 (colored green) can be considered fully disordered (DRs),[38,39] whereas 3.0 < Z < 8.0 (colored violet) corresponds to regions with fractional formation of local, ordered structure. Conversely, residues with Z > 11.0 (colored blue) correspond to segments of regular secondary structure or structured rigid loops, whereas 8.0 < Z < 11.0 (colored light blue) corresponds to flexible loops between ordered segments.

The data obtained indicate that the percent of predicted intrinsically disordered residues (PPIDR) for the proteins varies from 12.4% for YpOmpF to 53.8% for MpOmp, which indicates essential differences in the ordering and rigidity of the protein structural organization not only of different porin types, but also of different bacteria species. Based on their content of disordered residues, all considered OmpF and OmpC porins can be classified as moderately disordered proteins, while MpOmp can be classified as highly disordered protein.[41] Noteworthy is the fact that OmpC proteins, in contrast to OmpF, do not show a significant scatter in both the PPIDR content (25.1–29.1%) (Table 1), as well as in the DR-containing domains localization (Figs. 2 and 3). However, the degree of structural flexibility (conformational plasticity) in these areas may differ, especially when comparing E. coli and Yersinia OmpC porins. A common feature of OmpC porins is a fairly stable structural organization of the C-terminal part of the sequence, while its N-terminal domain and, in particular, the central part can exhibit sufficient structural plasticity.

In the case of bacterial OmpF porins from different habitats, Z-score indicates a large variability both in the distribution of conformationally flexible regions along the protein sequence and in the degree of disorder of these regions. In YrOmpF, 25.9% of the residues are predicted as intrinsically disordered, while YpOmpF contains only 12.4% of such residues (Table 1). This correlates with our earlier data on a significant difference in the temperature values of heat denaturation for these porins, at which the protein trimer dissociates: 76 °C for YpOmpF and 88 °C for YrOmpF. Possibly, the presence of a larger number of flexible elements in the Y. ruckeri porin defines a greater variability of its secondary structure without destruction of the porin quaternary structure until the critical temperature of the irreversible conformational transition is reached.[22] Alternatively, greater content of IDPRs in YrOmpF can be linked to the capability of this protein to format highly intertwined and conformationally stable complexes originating as a result of the folding at binding events.

Using the 3D structure of the E. coli porin monomer and theoretical models of the Yersinia porin monomers, IDPRs and areas of increased conformational flexibility were mapped (Fig. 3), which allows us to visualize the spatial localization of these regions. Even with a cursory glance, it can be seen that the areas with increased conformational plasticity are located in regions, the functional importance of which has long been reliably confirmed as a result of numerous experimental studies. This is the area of the L2 loop, which is responsible for the stability of the trimeric structure

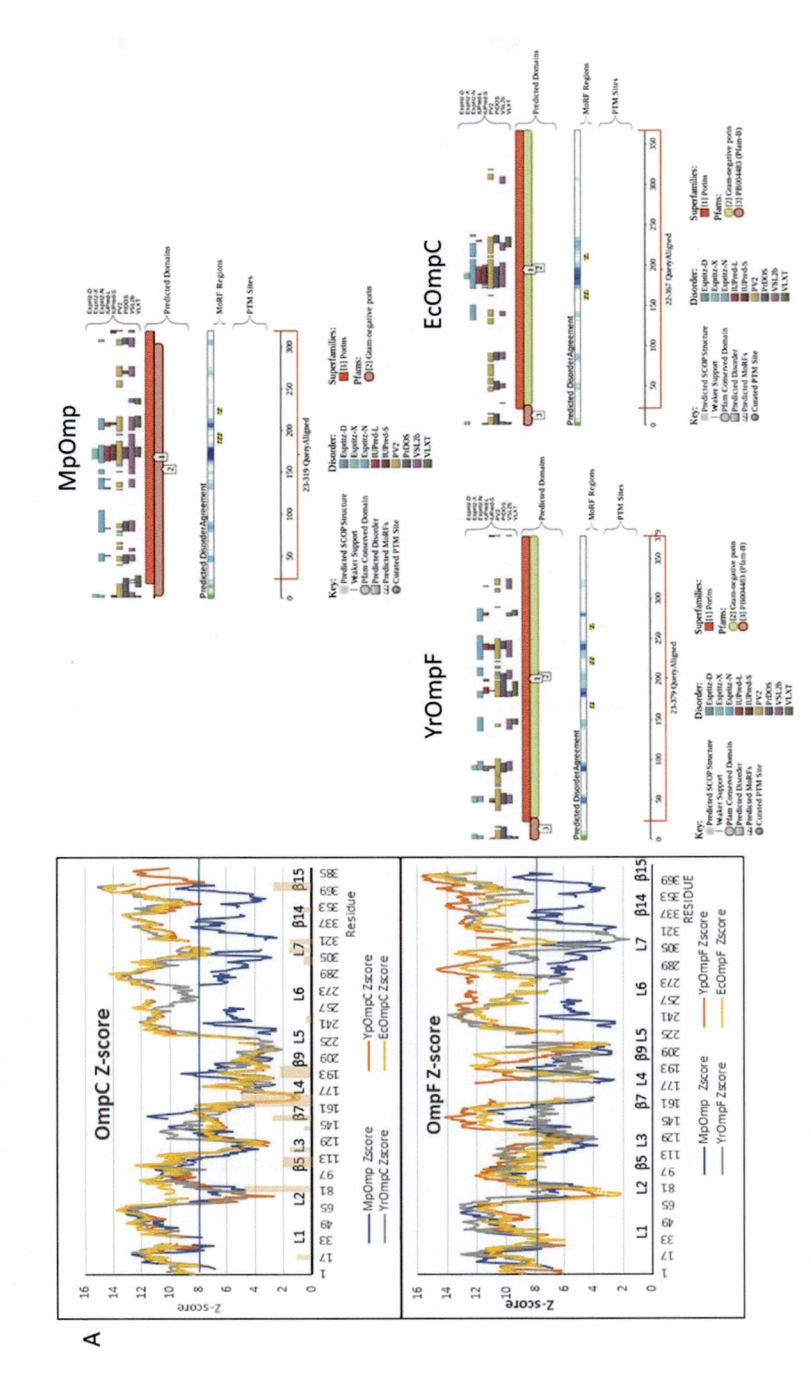

Fig. 3 See figure legend on next page.

of porins, a quite expected area that is part of the aforementioned aggregation "hot spots," which includes the L6 loop and β11 strand and, finally, the outer loops (L4, L5, and L7) that form the pore vestibule with adjacent barrel sections: β5, β8 and β9.

It should be noted that it was somewhat unexpected that the L3 loop, which is directly involved in the regulation of channel conduction, was identified, based its sequence, as just conformationally flexible region, but not IDPR. However, current conception of the mechanism of pore functioning are quite correlated with the result obtained. As known, the OmpF channel remains in an open state most of the time, providing the flow of ions and hydrophilic molecules into the cell. The longest loop L3 is buried in the pore cavity up to the middle, thereby limiting the size of the pore and forming a narrowing zone of constriction, or "pore eyelet." The β-barrel wall is formed mainly by positively charged residues, while the L3 loop, on the contrary, contains a large number of acidic residues. The spatial configuration of charged residues is such that an electrostatic field is formed inside the channel, which determines the selectivity of the channel with respect to the charges of penetrating ions and hydrophilic compounds.[42] However, most porins have the ability to pass into a stable closed state, for example, with an increase in the medium acidity or under the action of an applied external potential (voltage-dependent closure).[43–45] There are various hypotheses about the mechanism of such channel sensitivity. Based on the molecular dynamics (MD) simulation data, a model of a movable loop L3, whose displacement leads to the channel blockage was proposed as a possible gating mechanism.[46] However, since L3 forms multiple interactions with the barrel wall (salt bridges, hydrogen bond network), this idea seems unlikely. In addition, no significant changes in the loop position were shown to accompany the channel closure. For example, no noticeable differences in this

---

Fig. 3 Disorder predictions (A) Cumulative Distributions of ODiNPred Z-score for individual residues indicating their division between ordered and disordered conformation of OmpF and OmpC porins from *Y. pseudotuberculosis*, *Y. ruckeri*, *E. coli* as well as marine porin from *M. primoryensis* (MpOmp) (ODiNPred; https://stprotein.chem.au.dk/odinpred). Residues with Z-scores <3.0 can be considered fully disordered,[38,39] whereas 3.0 < Z < 8.0 corresponds to regions with fractional formation of local, ordered structure. Conversely, residues with Z > 11.0 correspond to segments of regular secondary structure or structured rigid loops, whereas 8.0 < Z < 11.0 corresponds to flexible loops between ordered segments. Predicted areas of aggregation "hot spots" are indicated as a slightly painted vertical square. (B) Evaluation of the functional intrinsic disorder propensity of marine porin from *M. primoryensis*, OmpF porin from *Y. ruckeri* and OmpC from *E. coli* conducted by D^2P^2 (http://d2p2.pro/).

property were found for EcOmpF with the L3 loop modified with disulfide bridges.[47] This fact indicates the some local changes in the tertiary structure of the definite parts of the L3 loop become a possible cause of channel closure. MD studies of perturbations have suggested that at least part of the L3 loop of porin from *R. capsulatus* is flexible.[48]

According to our predictions, the L2 loop of YpOmpF and EcOmpF is a partially structured region with elements of IDPRs, while in porin YrOmpF, this region is a flexible loop with a regular structure. Probably, this may indicate a higher heat stability of the *Y. ruckeri* porin trimer in comparison with those of *Y. pseudotuberculosis* and *E. coli* porins. On the other hand, the possible reason for this distinction derives from the special position of *Y. ruckeri* among *Yersinia*, which is confirmed by the formation of a separate cluster by it on the phylogenetic tree.[49]

Regarding the regions containing β-strands with an increased glycine clusters content (β5–β9), according to the results of our analysis, most of them do not contain IDPR, but differ significantly in the degree of their conformational flexibility. Therefore, the OmpF porins of *E. coli* and *Y. pseudotuberculosis* in the β6-β7 region with Z-Score value of 11.0–14.0 have a regular secondary structure, while the structure of the OmpF porin of *Y. ruckeri* and other OmpC proteins considered here is characterized by a higher conformational mobility in this region (Z-score of 6.0–8.5). A similar pattern is observed in the region of β8-β9 strands, where the OmpF proteins of *E. coli* and *Y. pseudotuberculosis* also have a more rigid structure in comparison with the similar region of OmpF of *Y. ruckeri* and OmpC porins (Z-score of 3.0–6.0). Nevertheless, it should be emphasized that the presence of IDPRs in the region of the β7 strand and the L4 loop for EcOmpC and YpOmpC was unambiguously predicted both using D^2P^2 (http://d2p2.pro/) and using ODiNPred (https://stproteinchem.au.dk/odinpred).

The L5, L7, and L8 loops, along with the L3 loop, are known to make a significant contribution to the efficiency of pore conductivity.[50] In this regard, the prediction of the most of structure variability in the region of the loops forming the vestibule of the porin channel (L4, L5, L7) for all studied porins was quite an expected result. In our opinion, the conformational flexibility of these regions that distinguishes two types of these proteins is the evidence of the peculiarities of porin functioning, associated primarily with the adaptation of bacteria to environmental conditions. Loops of OmpF porins have a quite regular structure, with the exception of small regions with increased conformational flexibility and rare inclusions of IDPRs in L4 and L5 loops. At the same time, in the case of OmpC porins, IDPRs are presented as rather extended areas (in EcOmpC) or as regions even covering the entire loop sequence (in YrOmpC).

The differences we found may be due to macro- and microevolutionary processes that contribute to the *ompC* gene divergence. It is well known that the reservoir of *Yersinia* in natural ecosystems are the soil and soil protozoa. *Yersinia* species are resistant to low temperatures, therefore the adaptation to the conditions of a warm-blooded organism is a strategy for their survival in various stressful conditions, including such as a new habitat niche. Recently, based on the phylogenetic analysis of all currently known bacterial species of the genus *Yersinia*, the mechanism of the molecular evolution of OmpC porins was established.[49] It was shown that positive selection observed in the outer loops and homologous recombination provide the exchange of new *ompC* variants in *Yersinia* populations.[49]

In this regard, the appearance of IDPRs in the L4 and L5 loops in the OmpC proteins of *Yersinia*, in contrast to the OmpF proteins, obviously indicates that in this way, the *Yersinia* realize their unique adaptive potential. On the other hand, the habitat conditions of *Y. ruckeri* dictate to a lesser extent the need to capture a new niche, since the aquatic environment and cold-blooded organisms are natural for it (it is a fish pathogen). Therefore, the OmpC porins of *Y. ruckeri*, from our point of view, cannot serve as a model for studying the mechanisms of adaptive evolution. In contrast, YrOmpF demonstrates greater diversity in the distribution of IDPRs and conformationally flexible regions, including in the region of the outer loops. This may be a consequence of the aforementioned separation of this microorganism at an earlier stage of evolution and/or the need to adapt to frequently changing conditions of the aquatic environment. Therefore, YrOmpF has an IDPR within the L7 loop in contrast to other porins, the C-terminus of which is structured. As aforementioned, the L7 loop refers to the areas that form the pore vestibule.[51] Therefore, the appearance of such a site in L7 can be considered as additional benefit to support the adaptive potential of this *Yersinia* specie.

Recently we have isolated the novel pore-forming protein from the marine psychrophilic bacterium *Marinomonas primoryensis* (MpOmp).[40] *M. primoryensis* KMM 3633 T, a marine bacterium living at extreme conditions, was isolated from a sample of coastal sea ice in the Amursky Bay near Vladivostok, Russia. The amino acid composition of MpOmp is distinguished by a high content of acidic and low content of sulfur-containing amino acids, but there are no tryptophan residues in its molecule. The porin of *M. primoryensis* KMM 3633 T was classified as Porin_4 according to the search for conserved domains.[52] We were the first to construct a spatial structure model of this porin by homologous modeling technique, using several prototypes of porins, such as EcOmpF, EcOmpC, Omp32 from *Delftia acidovorans* and OmpK36 from *Klebsiella pneumoniae* with the following loop refinement, optimization, and energy minimization in Amber14:EHT forcefield with MOE2020.09 CCG software (unpublished data). The generated MpOmp model allowed the localization of the conformationally flexible regions and IDPRs to be determined (Fig. 4C and D). One can see from the data summarized in Table 1, Figs. 3A and 4C, MpOmp has 53.8% PPIDR, which indicates the porin belongs to a group of highly disordered proteins, while almost all outer loops (except for L1 and L2) are IDPRs with a probability of 85–98%. In addition, in contrast to all considered porins, both OmpF and OmpC types, the C-terminal fragment of the β-barrel of MpOmp consists of conformationally flexible regions. It is noteworthy that the length of the L3 loop is not only shorter than that of its analogs of other porins of both types, but also consists exclusively of conformationally flexible regions and IDPRs that were not found in other proteins (Fig. 4C). In addition, this functionally important loop forms a network of non-covalent intramolecular interactions within the pore with flexible regions of the β-barrel (Fig. 4D), which is also not observed in the case of other porins.

The appearance of so unprecedented flexible β-barrel structure in the case of porin from marine bacterium may be associated with its marine environmental conditions, in comparison with freshwater ones, which are distinguished by an even larger set of changing (stress) factors (osmolarity, temperature, pressure) contributing to the development of greater adaptive capabilities in this bacterium (the acquisition of a greater adaptive potential). The results of bioinformatics analysis of the porin sequence, in our opinion, confirmed our rather bold assumption that the need for broad adaptive capabilities can correlate with an increase in the number of IDPRs and/or sites with increased conformational flexibility.

In conclusion, as an experimental confirmation of the results of bioinformatics analyses, we present electron microscopic photographs of an amyloid-like structures formed by the OmpF porin from *Y. pseudotuberculosis*, obtained using a LIBRA-120 electron microscope (Carl Zeiss, Germany) (Fig. 5). This conformational species is formed under the rather harsh conditions for obtaining a lipid A—protein complex according to Alaupović; i.e., during treatment of the endotoxin, or LPS—protein complex (LPPC) from *Y. pseudotuberculosis*,[53] with 1% acetic acid (AcOH) at 100 °C for 5 h.[54] Under these conditions, the O-specific chains of LPS are cleaved off, and,

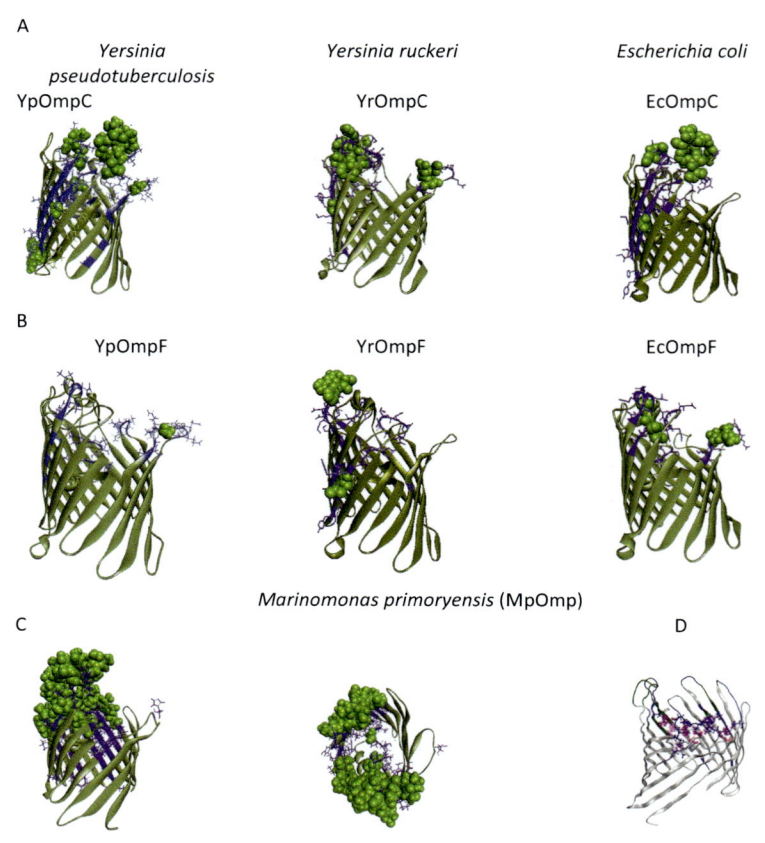

Fig. 4 IDPRs and conformationally flexible regions mapped to porin 3D-structure models/structures. Ribbon diagrams of 3D-structures of porins OmpC (A), OmpF (B) from *Y. pseudotuberculosis*, *Y. ruckeri*, *E. coli* as well as marine porin *M. primoryensis* (C, D). Cumulative Distributions of ODiNPred Z-score and ODiNPred Disorder Probability (ODiNPred; https://stprotein.chem.au.dk/odinpred) for individual residues indicating their distribution between ordered and disordered conformation of OmpC and OmpF (A, B) porins as well as MpOmp (C): ordered conformation is presented as light green ribbons, conformationally flexible—as violet ribbon and sticks, disordered—as green ribbon and balls. Visualization was conducted with *Biovia* Discovery Studio 2020 software (Dassault Systèmes BIOVIA, Discovery Studio Modeling Environment, Release 2020, San Diego: Dassault Systèmes, 2021). (C) Non-covalent intramolecular interactions of conformationally flexible regions and IDPRs (green and violet, respectively) in MpOmp β-barrel with L3 loop. Residues involved in non-covalent intramolecular interactions of the internal loop (L3) are presented as violet ball and sticks. Hydrogen bonds and distance (hydrophobic, WDW) interactions are indicated by magenta clouds (the stronger, the darker). Non-covalent intramolecular interactions analysis is performed with MOE 2020.09 CCG (Chemical Computing Group ULC, 1010 Sherbrooke St. West, Suite #910, Montreal, QC, Canada, H3A 2R7, 2020).

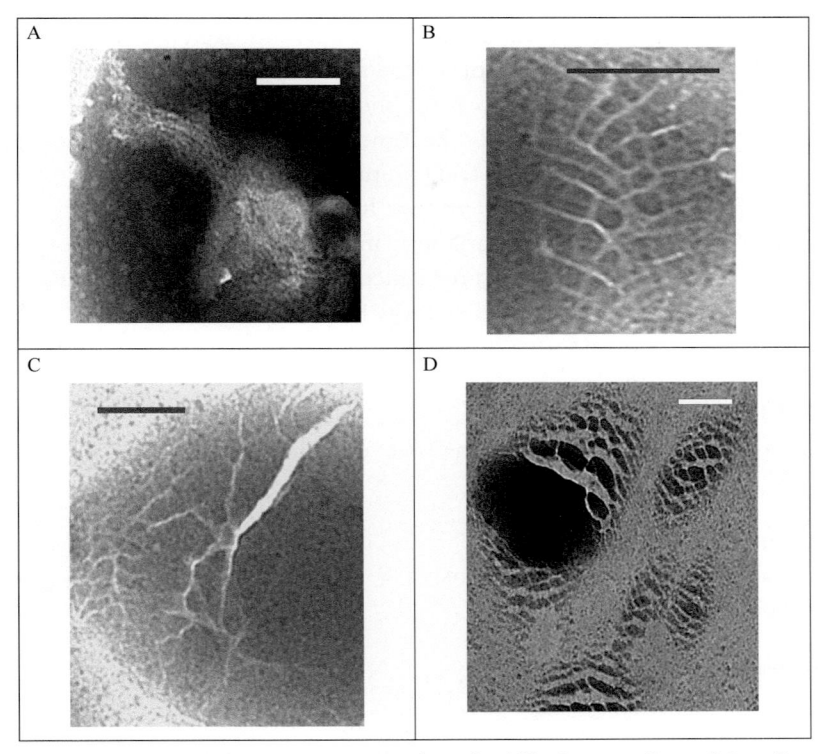

Fig. 5 Transmission electron microscopy of amyloid-like intermediate of OmpF porin from *Y. pseudotuberculosis*. For electron microscopy observations, the protein solutions were placed on grids coated with formvar film stabilized with carbon. The samples were contrasted for 3 min with 2% phosphotungstic acid (PTA), pH 7.0. Then, the samples were carefully dried using a piece of filter paper and examined with a LIBRA-200 electron microscope (Carl Zeiss, Germany). (A) The diameter of the filamental structures of about 5.6 Nm, (B) the diameter of the threaded structures 7.8–8.7 nm, (C) the diameter of the thread structures is about 7.1 nm, (D) the diameter of the filamentary structures 4.8–8 nm. PTA-stained protein preparations. Bar = 200 nm.

therefore, the relative content of LPS in the protein preparation decreases. Based on these considerations, we assumed that the treatment of protein sample with acid would further purify YpOmpF from LPS.

As can be seen from the data in Fig. 5, the aggregation of YpOmpF yields the formation of morphologically diverse variants: amorphous (Fig. 5A) or ordered typical amyloid fibrils, long and rigid (Fig. 5B–D).

The formation of amyloid-like structures by the non-specific OM pore-forming proteins of Gram-negative bacteria uncovers a new type of β-structured transmembrane proteins from prokaryotes, which represent a novel model for the in-depth elucidation of amyloidogenesis *in vitro*, primarily to the search for the inhibitors of the formation of pathological amyloid forms. Due to their ability to form morphologically diverse aggregated species, they can undoubtedly become an effective tool for studying amyloidogenesis, an intriguing and complex phenomenon process inherent in both pro- and eukaryotic organisms. It is possible that the existence of a certain correlation between variations in the manifestation of their main channel-forming function and prevalence of IDPRs, as well as peculiarities of disorder distribution within their sequences, will be useful to highlight the molecular basis of the formation of the adaptive potential in bacteria.

Funding

This work was supported by the Russian Foundation for Basic Research grant № 19-04-00318.

Acknowledgment

We thank Dr. Sci. A.V. Reunov and PhD L.A. Lapshina for the conduction of electron microscopy studies.

References

1. Uversky VN. Unusual biophysics of intrinsically disordered proteins. *Biochim Biophys Acta*. 2013;1834:932–951. https://doi.org/10.1016/j.bbapap.2012.12.008.
2. Marcoleta A, Wien F, Arluison V, Lagos R, Giraldo R. Bacterial amyloids. In: *eLS*; 2019:1–9. https://doi.org/10.1002/9780470015902.a0028401.
3. Molina-Garcia L, Gasset-Rosa F, Alamo MM, de la Espina SM, Giraldo R. Addressing intracellular amyloidosis in bacteria with RepA-WH1, a prion-like protein. *Methods Mol Biol*. 2018;1779:289–312. https://doi.org/10.1007/978-1-4939-7816-8_18.
4. Kosolapova AO, Antonets KS, Belousov MV, Nizhnikov AA. Biological functions of prokaryotic amyloids in interspecies interactions: facts and assumptions. *Int J Mol Sci*. 2020;21:7240. https://doi.org/10.3390/ijms21197240.
5. Antonets KS, Volkov KV, Maltseva AL, Arshakian LM, Galkin AP, Nizhnikov AA. Proteomic analysis of Escherichia coli protein fractions resistant to solubilization by ionic detergents. *Biochemistry (Mosc)*. 2016;81:34–46. https://doi.org/10.1134/S00062979160 10041.
6. Oldfield CJ, Dunker AK. Intrinsically disordered proteins and intrinsically disordered protein regions. *Annu Rev Biochem*. 2014;83:553–584. https://doi.org/10.1146/annurev-biochem-072711-164947.
7. Granata D, Baftizadeh F, Habchi J, et al. The inverted free energy landscape of an intrinsically disordered peptide by simulations and experiments. *Sci Rep*. 2015;5:15449. https://doi.org/10.1038/srep15449.
8. Giraldo R. Amyloid assemblies: protein legos at a crossroads in bottom-up synthetic biology. *Chembiochem*. 2010;11:2347–2357. https://doi.org/10.1002/cbic.201000412.
9. Wang Y. The function of OmpA in Escherichia coli. *Biochem Biophys Res Commun*. 2002;292:396–401. https://doi.org/10.1006/bbrc.2002.6657.
10. Uversky VN. Natively unfolded proteins: a point where biology waits for physics. *Protein Sci*. 2002;11:739–756. https://doi.org/10.1110/ps.4210102.

11. Danoff EJ, Fleming KG. Aqueous, unfolded OmpA forms amyloid-like fibrils upon self-association. *PLoS One*. 2015;10: e0132301. https://doi.org/10.1371/journal.pone. 0132301.

12. Nikaido H. Molecular basis of bacterial outer membrane permeability revisited. *Microbiol Mol Biol Rev*. 2003;67:593–656. https://doi.org/10.1128/mmbr.67.4.593-656.2003.

13. Delcour AH. Solute uptake through general porins. *Front Biosci*. 2003;8: d1055–d1071. https://doi.org/10.2741/1132.

14. Alcaraz A, Nestorovich EM, Aguilella-Arzo M, Aguilella VM, Bezrukov SM. Salting out the ionic selectivity of a wide channel: the asymmetry of OmpF. *Biophys J*. 2004; 87:943–957. https://doi.org/10.1529/biophysj.104/043414.

15. Schulz GE. Porins: general to specific, native to engineered passive pores. *Curr Opin Struct Biol*. 1996;6:485–490. https://doi.org/10.1016/s0959-440x(96)80113-8.

16. Wimley WC. The versatile beta-barrel membrane protein. *Curr Opin Struct Biol*. 2003;13:404–411. https://doi.org/10.1016/s0959-440x(03)00099-x.

17. Haltia T, Freire E. Forces and factors that contribute to the structural stability of membrane proteins. *Biochim Biophys Acta*. 1995;1228:1–27. https://doi.org/10.1016/0005-2728(94)00161-w.

18. Cornish J, Chamberlain SG, Owen D, Mott HR. Intrinsically disordered proteins and membranes: a marriage of convenience for cell signalling? *Biochem Soc Trans*. 2020; 48:2669–2689. https://doi.org/10.1042/BST20200467.

19. Villain E, Nikekhin AA, Kajava AV. Porins and amyloids are coded by similar sequence motifs. *Proteomics*. 2019;19: e1800075. https://doi.org/10.1002/pmic.201800075.

20. Smirnov IV. The causative agent of yersiniosis and related microorganisms. *Clin Microbiol Antimicrob Chemother*. 2004;6:1–21.

21. Khomenko VA, Portniagina O, Novikova OD, et al. Isolation and characterization of recombinant OmpF-like porin from the Yersinia pseudotuberculosis outer membrane. *Bioorg Khim*. 2008;34:177–184. https://doi.org/10.1134/s1068162008020040.

22. Novikova OD, Chistyulin DK, Khomenko VA, et al. Peculiarities of thermal denaturation of OmpF porin from Yersinia ruckeri. *Mol Biosyst*. 2017;13:1854–1862. https://doi.org/10.1039/c7mb00239d.

23. Garavito RM, Rosenbusch JP. Isolation and crystallization of bacterial porin. *Methods Enzymol*. 1986;125:309–328. https://doi.org/10.1016/s0076-6879(86)25027-2.

24. Novikova OD, Kim NY, Luk'yanov PA, Likhatskaya GN, Emel'yanenko VI, Solov'eva TF. Effects of pH on structural and functional properties of porin from the outer membrane of Yersinia pseudotuberculosis. II. Characterization of pH-induced conformational intermediates of Yersinin. *Biochem (Mosc) Suppl Ser A Membr Cell Biol*. 2007;1: 154–162.

25. Tartaglia GG, Vendruscolo M. The Zyggregator method for predicting protein aggregation propensities. *Chem Soc Rev*. 2008;37:1395–1401. https://doi.org/10.1039/b706784b.

26. Conchillo-Sole O, de Groot NS, Aviles FX, Vendrell J, Daura X, Ventura S. AGGRESCAN: a server for the prediction and evaluation of "hot spots" of aggregation in polypeptides. *BMC Bioinf*. 2007;8:65. https://doi.org/10.1186/1471-2105-8-65.

27. Novikova OD, Khomenko VA, Emelyanenko VI, et al. OmpC-like porin from Yersinia pseudotuberculosis: molecular characteristics, physico-chemical and functional properties. *Biochem (Mosc) Suppl Ser A Membr Cell Biol*. 2011;5:263–277.

28. Jeanteur D, Lakey JH, Pattus F. The bacterial porin superfamily: sequence alignment and structure prediction. *Mol Microbiol*. 1991;5:2153–2164. https://doi.org/10.1111/j.1365-2958.1991.tb02145.x.

29. Branden C, Tooze J. *Introduction to Protein Structure*. New York: Garland Science; 1999.

30. Achouak W, Heulin T, Pages JM. Multiple facets of bacterial porins. *FEMS Microbiol Lett*. 2001;199:1–7. https://doi.org/10.1111/j.1574-6968.2001.tb10642.x.

31. Portnyagina O, Chistyulin D, Dyshlovoy S, et al. OmpF porin from Yersinia ruckeri as pathogenic factor: surface antigenic sites and biological properties. *Microb Pathog.* 2021;150:104694. https://doi.org/10.1016/j.micpath.2020.104694.

32. Portnyagina OY, Kuzmitch AS, Khomenko VA, Isaeva MP, Solov'eva TF, Novikova OD. Induction of the synthesis of interleukins and chemotaxis activators under the action of OmpF porin from Yersinia pseudotuberculosis in an in vivo model. *Cytokines Inflamm.* 2019;1–2:36–42.

33. Sidorova OV, Khomenko VA, Portnyagina OY, et al. Mutant OmpF porins of Yersinia pseudotuberculosis with deletions of external loops: structure-functional and immuno-chemical properties. *Biochem Biophys Res Commun.* 2014;445:428–432. https://doi.org/10.1016/j.bbrc.2014.02.018.

34. Portnyagina O, Zelepuga E, Khomenko V, Solov'eva E, Solov'eva T, Novikova O. In silico and in vitro analysis of cross-reactivity between Yersinia pseudotuberculosis OmpF porin and thyroid-stimulating hormone receptor. *Int J Biol Macromol.* 2018;107:2484–2491. https://doi.org/10.1016/j.ijbiomac.2017.10.133.

35. Van Regenmortel MH. Plant virus serology. *Adv Virus Res.* 1966;12:207–271. https://doi.org/10.1016/s0065-3527(08)60850-7.

36. Oates ME, Romero P, Ishida T, et al. D(2)P(2): database of disordered protein predictions. *Nucleic Acids Res.* 2013;41:D508–D516. https://doi.org/10.1093/nar/gks1226.

37. Dass R, Mulder FAA, Nielsen JT. ODiNPred: comprehensive prediction of protein order and disorder. *Sci Rep.* 2020;10:14780. https://doi.org/10.1038/s41598-020-71716-1.

38. Nielsen JT, Mulder FAA. Quality and bias of protein disorder predictors. *Sci Rep.* 2019;9:5137. https://doi.org/10.1038/s41598-019-41644-w.

39. Nielsen JT, Mulder FA. There is diversity in disorder-"in all Chaos there is a Cosmos, in all disorder a secret order". *Front Mol Biosci.* 2016;3:4. https://doi.org/10.3389/fmolb.2016.00004.

40. Novikova OD, Khomenko VA, Kim NY, et al. Porin from marine bacterium Marinomonas primoryensis KMM 3633(T): isolation, physico-chemical properties, and functional activity. *Molecules.* 2020;25:3131. https://doi.org/10.3390/molecules25143131.

41. Rajagopalan K, Mooney SM, Parekh N, Getzenberg RH, Kulkarni P. A majority of the cancer/testis antigens are intrinsically disordered proteins. *J Cell Biochem.* 2011;112:3256–3267. https://doi.org/10.1002/jcb.23252.

42. Cowan SW, Schirmer T, Rummel G, et al. Crystal structures explain functional properties of two *E. coli* porins. *Nature.* 1992;358:727–733. https://doi.org/10.1038/358727a0.

43. Nestorovich EM, Rostovtseva TK, Bezrukov SM. Residue ionization and ion transport through OmpF channels. *Biophys J.* 2003;85:3718–3729. https://doi.org/10.1016/S0006-3495(03)74788-2.

44. Todt JC, McGroarty EJ. Acid pH decreases OmpF and OmpC channel size in vivo. *Biochem Biophys Res Commun.* 1992;189:1498–1502. https://doi.org/10.1016/0006-291x(92)90244-f.

45. Lakey JH, Pattus F. The voltage-dependent activity of Escherichia coli porins in different planar bilayer reconstitutions. *Eur J Biochem.* 1989;186:303–308. https://doi.org/10.1111/j.1432-1033.1989.tb15209.x.

46. Watanabe M, Rosenbusch J, Schirmer T, Karplus M. Computer simulations of the OmpF porin from the outer membrane of Escherichia coli. *Biophys J.* 1997;72:2094–2102. https://doi.org/10.1016/S0006-3495(97)78852-0.

47. Phale PS, Schirmer T, Prilipov A, Lou KL, Hardmeyer A, Rosenbusch JP. Voltage gating of Escherichia coli porin channels: Role of the constriction loop. *Proc Natl Acad Sci U S A.* 1997;94:6741–6745. https://doi.org/10.1073/pnas.94.13.6741.

48. Soares CM, Bjorksten J, Tapia O. L3 loop-mediated mechanisms of pore closing in porin: a molecular dynamics perturbation approach. *Protein Eng.* 1995;8:5–12. https://doi.org/10.1093/protein/8.1.5.

49. Stenkova AM, Bystritskaya EP, Guzev KV, Rakin AV, Isaeva MP. Molecular evolution of the Yersinia major outer membrane protein C (OmpC). *Evol Bioinformatics Online.* 2016;12:185–191. https://doi.org/10.4137/EBO.S40346.

50. Altschul SF, Gish W, Miller W, Myers EW, Lipman DJ. Basic local alignment search tool. *J Mol Biol.* 1990;215:403–410. https://doi.org/10.1016/S0022-2836(05)80360-2.

51. Likhatskaya GN, Solov'eva TF, Novikova OD, et al. Homology models of the Yersinia pseudotuberculosis and Yersinia pestis general porins and comparative analysis of their functional and antigenic regions. *J Biomol Struct Dyn.* 2005;23:163–174. https://doi.org/10.1080/07391102.2005.10507056.

52. El-Gebali S, Mistry J, Bateman A, et al. The Pfam protein families database in 2019. *Nucleic Acids Res.* 2019;47:D427–D432. https://doi.org/10.1093/nar/gky995.

53. Ovodov YS, Solovjeva TF, Khomenko VA, et al. Porin as a component of Yersinia pseudotuberculosis endotoxin. *Adv Exp Med Biol.* 1990;256:185–187. https://doi.org/10.1007/978-1-4757-5140-6_15.

54. Wober W, Alaupovic P. Studies on the protein moiety of endotoxin from gram-negative bacteria. Characterization of the protein moiety isolated by acetic acid hydrolysis of endotoxin from Serratia marcescens 08. *Eur J Biochem.* 1971;19:357–367. https://doi.org/10.1111/j.1432-1033.1971.tb01324.x.

CHAPTER FOUR

Intrinsic disorder in integral membrane proteins

Brian J. Aneskievich[a], Rambon Shamilov[b], and Olga Vinogradova[a],*

[a]Department of Pharmaceutical Sciences, University of Connecticut, Storrs, CT, United States
[b]Graduate Program in Pharmacology and Toxicology, Department of Pharmaceutical Sciences, University of Connecticut, Storrs, CT, United States
*Corresponding author: e-mail address: olga.vinogradova@uconn.edu

Contents

Abstract

The well-defined roles and specific protein-protein interactions of many integral membrane proteins (IMPs), such as those functioning as receptors for extracellular matrix proteins and soluble growth factors, easily align with considering IMP structure as a classical "lock-and-key" concept. Nevertheless, continued advances in understanding protein conformation, such as those which established the widespread existence of intrinsically disordered proteins (IDPs) and especially intrinsically disordered regions (IDRs) in otherwise three-dimensional organized proteins, call for ongoing reevaluation of transmembrane proteins. Here, we present basic traits of IDPs and IDRs, and, for some select single-span IMPs, consider the potential functional advantages intrinsic disorder might provide and the possible conformational impact of disease-associated mutations. For transmembrane proteins in general, we highlight several investigational approaches, such as biophysical and computational methods, stressing the importance of integrating them to produce a more-complete mechanistic model of disorder-containing IMPs.

Progress in Molecular Biology and Translational Science, Volume 183
ISSN 1877-1173
https://doi.org/10.1016/bs.pmbts.2021.06.002

These procedures, when synergized with in-cell assessments, will likely be key in translating *in silico* and *in vitro* results to improved understanding of IMP conformational flexibility in normal cell physiology as well as disease, and will help to extend their potential as therapeutic targets.

1. Distinction between intrinsically disordered and structured proteins

Since the very beginning of scientific attempts to understand how life is organized at a molecular level and with the comprehension that proteins play key roles in virtually all processes within living cells, the "one structure–one function" paradigm remained unchallenged for almost a century. The versatility of proteins is indeed based upon the variability of their structures and the specificity of their actions that arises from the uniqueness of a particular fold(s). The transition from a static "lock–and–key" model[1,2] of the early 20th century to a more modern induced-fit (glove-and-hand) concept[3] in protein–ligand interactions reflects the advances in modern understanding of biopolymers, acknowledging that proteins are not rock–solid and often require remarkable flexibility to perform their functions. The further idea of conformational selection,[4] postulating that a protein can accommodate an ensemble of different conformations, often characterized by similar energy levels, with a certain fraction of these states being preferable for a particular binding mode, appears necessary to explain a vast majority of accumulating experimental data in signal transduction and cooperative binding.[5] The appreciation of the complexity and continuity of the protein universe has further evolved to the recognition of an additional class, with the members being either intrinsically disordered proteins (IDPs) or hybrid, meaning that they contain uniquely folded domains intercalated with intrinsically disordered regions (IDRs).[6–9] In addition, more than one unique polypeptide chain can be generated from a single gene, due to allelic variations, alternative splicing, other pre-translational mechanisms or post-translational modifications (PTMs),[10] and these various isoforms can adopt distinctive conformations responsible for different functions.[11,12] Altogether, this molecular diversity and continuum of conformational states is defined as the proteoform of one protein and provides an important foundation for protein multifunctionality.

The concept of the proteoform helps to alleviate protein diversity restriction and likely the parallel functional limits otherwise imposed by a classical one gene – one protein model. This spectrum of building in variation on protein form then loops back to intrinsic disorder by recalling[13] that IDR often arise *via* alternative splicing and that once IDR are present in proteins they are often the site of PTM. Thus, as much as alternative splicing and PTM may add to the entirety of a particular proteoform, the conformational ensemble derived from intrinsic disorder may well be on equal footing for adding different molecular forms of a protein (see Ref. 13 for review). Nevertheless, it is not always easy, straightforward or even necessary to quantitatively distinguish highly dynamic proteins/regions from the ones classified as disordered. Important, however, is the notion that without structural heterogeneity and conformational flexibility, life, as we know it, would not be possible.

When an IDP is mentioned during discussion, it is a soluble protein that most often first comes to mind. This is likely due to the fact that structure-function relationships in soluble proteins have been studied much longer and more extensively than in integral membrane proteins (IMPs) and, also, because of the better developed understanding of how amino acid sequence of soluble IDPs is defined by the polar properties of water. Water molecules form hydrogen bonds with exposed polar groups and, thus, force burial of hydrophobic groups into the interior core of the protein as hydration of non-polar groups is entropically very costly. Shielding the hydrophobic groups from the polar environment of aqueous solution provides the major driving force in soluble protein folding. Therefore, low hydrophobicity and high net charge of soluble IDPs[8] can explain the lack of thermodynamic impetus for their folding. But how do the same thermodynamic principles apply to proteins that are fully or partially membrane incorporated and should we expect to see IDPs/IDRs among IMPs?

Before starting this discussion it is important to keep in mind that efforts to understand structure-function relationship in IMPs were halted for a long period of time due to an additional layer of complexity: for structural studies, IMPs have to be extracted from the native lipid bilayer environment into a carefully chosen membrane mimetic. Also, IMPs are notorious for the difficulties associated with their recombinant expression and purification as well as for their tendency to aggregate. Nevertheless, in the recent years a lot have been accomplished in the field with an explosion of technological advances in high resolution structure determination techniques such as X-ray crystallography, Nuclear Magnetic Resonance (NMR) and, particularly, cryo-Electron Microscopy (cryoEM). However, even before these technological advances, it was predicted from thermodynamic principles[14] and later experimentally confirmed, that polar amide protons and carbonyl oxygens of the polypeptide backbone have to form intra–molecular hydrogen bonds to be shielded from the hydrophobic environment of the inner lipid bilayer core composed of non-polar aliphatic chains. Therefore, in almost all known 3D structures of IMPs, the transmembrane (TM) domain is composed of either one/several α–helice(s) or β–barrel(s) (the outliers represent auto-transporters composed of β-barrels with α-helical bundles partially inserted into their pores). This means that at least ordered secondary structure elements are always present within the membrane spanning part, explaining the notion that the term "IDP" was not initially associated with membrane proteins. However, if we take into account the often disordered cytoplasmic or extracellular domains and/or flexible loops interconnecting TM helices or β-strands, we envision a totally different story, where IMPs represent a set of hybrid proteins containing both folded domains and IDRs (Fig. 1). On a separate note, Kjaergaard[15] has argued that a putative poly-topic membrane embedded protein, adopting a set of conformations analogous to a soluble molten globule state, with the secondary structure elements formed but no defined tertiary structure, should also be regarded as an IDP (Fig. 1). As a matter of fact, freely diffusing TM-helices have been proposed as general intermediates in the process of folding for any helical IMP[16] and may represent a state that most membrane proteins have to pass.

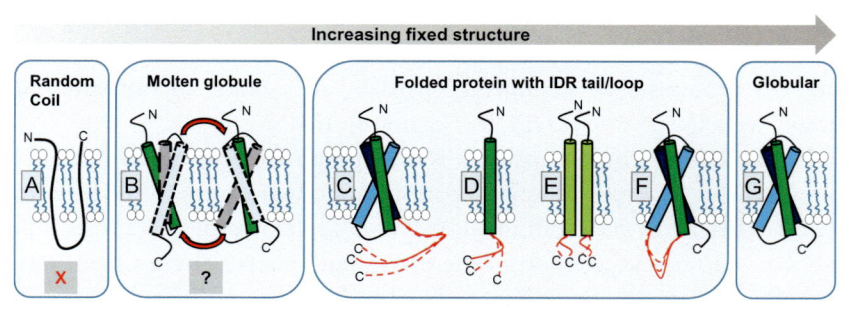

Fig. 1 Examples of transmembrane proteins featuring varying degrees of intrinsic disorder. Embedded receptors representing (left to right) (A) Unlikely random coil (red X) embedded featuring no secondary or tertiary structure. (B) Theoretical molten globule-like receptor featuring fixed secondary structure and dynamic tertiary structure promoting conformational switching (question mark) between embedded loops. Mostly globular (C) multi-pass, (D) single-pass and (E) dimeric receptors which feature C-terminus IDR tails. (F) Globular membrane receptors featuring dynamic cytosolic loops. (G) Totally globular membrane bound protein featuring minimal dynamic conformation in inactive form. Dotted lines used to show conformational switching of disordered protein regions.

However, to-date, we are not aware of any experimental data confirming that this structural state is not transient.

Over the years it became apparent that a certain degree of flexibility is essential for the functionality of all proteins, although dynamics may vary significantly in rates and amplitudes. But is there a phenomenological quantitative difference between a disordered and merely dynamic protein? It seems impractical to define a single rigid threshold in a rather smooth transition from fully unstructured to folded states in the continuum of structural propensities.[17] The distinction can be made, though qualitatively, when we consider this problem from the perspective of energy landscapes (Fig. 2). The dynamic IMPs are characterized by a few local minima with steep walls, while a truly disordered protein exists in a wide energy well with many local

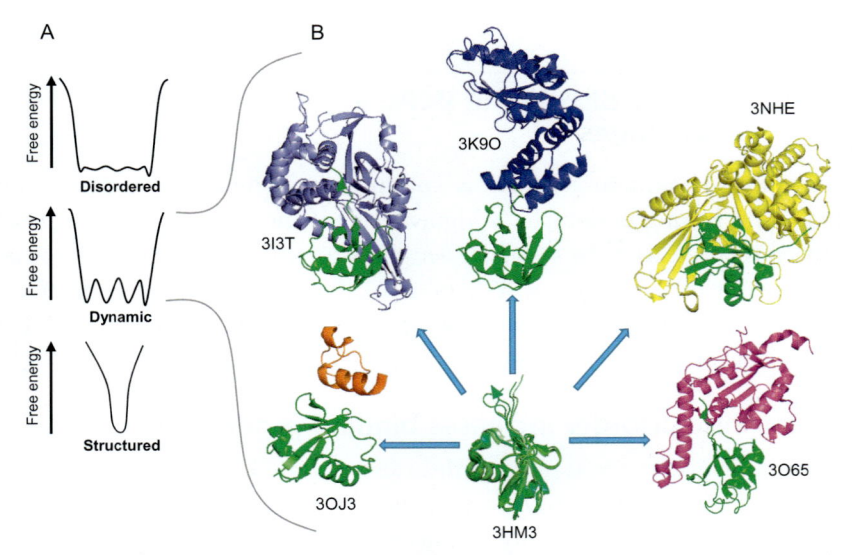

Fig. 2 Ubiquitin as a representative dynamic protein which exist across many conformations. (A) *top to bottom*, representative energy well diagrams with a disordered protein represented as existing with a large ensemble of conformations with low energy barriers separating them. Dynamic proteins feature greater than one conformation but are more discreetly separated energy-wise with individual energy minima for each conformation. Lastly, globular or structured proteins are typically thought to have one energy minimum. (B) Ubiquitin (shown in green in all diagrams) as an example of a dynamic protein featuring many distinct conformations associated with protein-protein interactions with different binding partner proteins, each with its own theoretical associated free energy. Structures prepared using Pymol.[18] PDBIDs of shown structures cited. Image for ubiquitin (bottom center) is alignment of ubiquitin only model (3HM3) with ubiquitin molecules from other shown images.

minima and low energy barriers. Transmembrane regions of IMPs are often dynamic, but they spend most of their time within discrete well-defined states, and thus cannot be described as truly disordered. Disordered regions lack stable conformations, rather occupying a continuous ensemble, and are predominately found within cytoplasmic and extracellular domains of IMPs.[19,20]

In this chapter, we are going to focus on two major topics related to disorder in IMPs: (i) we are going to speculate about functional advantages that IDRs might provide, followed by a brief discussion of few examples (for a more comprehensive set of examples good reviews are available, e.g.,[21]); and (ii) we are going to highlight experimental and computational methods, that we believe need to be integrated through a divide-and-conquer approach, to produce a complete mechanistic model of these unique proteins containing both folded domains and disordered regions.

2. Intrinsic disorder in IMPs: Potential functional advantages

Intrinsic protein disorder is common in complete genomes, with eukaryotes having potentially a higher percentage of native disorder than archaea or bacteria.[22] In sequence length, proteins with intrinsic disorder can account for up to about 33% of a eukaryotic proteome.[23] Therefore, important reasons must exist allowing it to prevail rather than be eliminated by millions of years of evolution.

2.1 Intrinsic disorder increases binding promiscuity

Among the most obvious intrinsic disorder conferred advantage is the binding promiscuity of IDPs/IDRs, i.e., the ability to interact with multiple partners and to act as highly connected nodes, or hubs, often within protein–protein interaction (PPI) networks.[24] Among IMPs, cell surface receptors, which transduce extra-cellular signals across the membrane to the cytoplasm, are extensively involved in PPI networks. According to the computational analysis by Burgi and co-workers,[25] membrane proteins containing IDRs more often play a role in ion binding and signal transduction whereas fully folded membrane proteins are usually involved in enzymatic functions and GPCR signaling. In this study, an IMP was considered to be an IDP if it contained at least one IDR with the length of 30 or more amino acids. Herein, we will follow the same nomenclature for simplicity, although it should be kept in mind that in precise classification an IMP with

IDR(s) is a hybrid protein containing both folded domain(s) and disordered region(s). In the same study, it was also predicted that IDRs can cover large proportions, on average of 60%, of the cytosolic domain of single spanning membrane proteins and that IDPs tend to localize to specific cellular sub-domains, such as cell projections, dendrites, and presynaptic membranes. As those structures are specific for high order multicellular organisms, the findings could indicate that the IDR functions tend to evolve continuously. Ordered transmembrane proteins were also more abundant in peroxisomes and in the inner mitochondria membrane, but not in the outer mitochon-drial membrane. In contrast, the lack of disordered proteins in the inner mitochondria is consistent with their prokaryotic origin, since archea and bacteria were predicted to have far less IDPs than eukaryotic cells. Another earlier genome-wide study predicted that about 40% of human membrane proteins are IDPs, with intracellular domains carrying most of the predicted disorder.[19]

Integrin $\alpha_{IIb}\beta_3$: A perfect example of this remarkable structural dexterity can be found in cytoplasmic tail (CT) of the major platelet integrin $\alpha_{IIb}\beta_3$, which belongs to probably the most extensively studied cell surface receptor family. Integrins constitute a major class of heterodimeric (α/β) transmem-brane receptors that connect cells to the extracellular matrix (ECM). Integrin-mediated adhesion is controlled in a remarkable bi-directional manner,[27] termed "inside-out" (or activation) and "outside-in" signaling.[28] Through a cross-talk with other cell surface receptors, including G-protein coupled receptors (GPCRs) and/or growth factor receptors,[29] a series of intricate intracellular events is initiated, which induce an inside-out signal that increases the ligand binding affinity/avidity of the extracellular domains of integrins. Subsequently, ligand binding triggers conformational changes and clustering of integrin receptors, an outside-in signaling event, ultimately leading to the assembly of large intracellular protein complexes linked to the cytoskeleton. This physical linkage between extracellular and intracellular compartments allows dynamic regulation of many cellular processes and is of fundamental interest in signal transduction investigations.

Notably for this chapter, integrin $\alpha_{IIb}\beta_3$, a major platelet non-covalent heterodimer IMP, would not even be considered as an IDP by conventional computational analysis (bioinf.cs.ucl.ac.uk), as shown in Fig. 3A,[30] due to the short length of its cytoplasmic domains (predicted IDRs are less than 30 amino acids long). Numerous biophysical and biochemical studies[31–37] have confirmed, however, that its cytoplasmic tails have the potential for forming different secondary structural elements despite the high

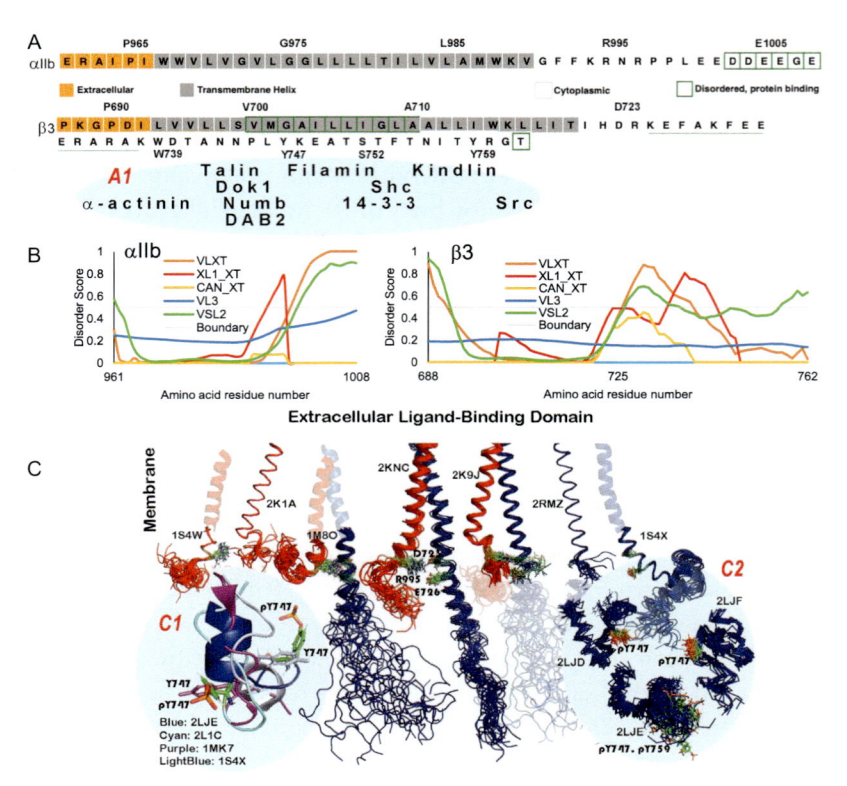

Fig. 3 Analysis summary for the TM domain and cytoplasmic tail (CT) of major platelet integrin $\alpha_{IIb}\beta_3$ (E^{961}-E^{1008}; P^{688}-T^{762}): (A) Computational predictions of TM helices and disordered regions of the hetero-dimer by PSIPRED, MEMSAT-SVM, DISOPRED3 (bioinf.cs.ucl.ac.uk); the most disordered segment of β_3 CT determined by PONDR[26] is underlined in green; *A1* inset presents coarsely aligned binding sites for β_3 interactions with its several cytoplasmic targets; (B) Disorder prediction by PONDR; (C) Structural data from the PDB (prepared using Pymol,[18] PDBIDs of presented ensembles are cited): faded regions represent parts of TMCT missing in a particular construct, side-chains of R^{995}, D^{723} and E^{726}, forming a salt-bridge, are shown by sticks and generally reside in cytosol; *C1* inset shows the overlay of β_3 CT stretch W^{739}-E^{749} from different structures illuminating the presence of diverse secondary structural elements within same region (helix turn, β-strand and loop); *C2* inset shows different phosphorylation states of β_3 CT and the effect of phosphorylation on its interaction with membrane; phosphorylated side-chains of Y^{747} and Y^{759} are represented by sticks.

PONDR[26] scores presented in Fig. 3B. These include membrane proximal helices in both subunits, distant membrane associated helices in β_3, various turns in NPxY and NxxY sorting motifs of β_3, and extended conformations in the regions adjacent to these motifs. It is also important to stress at this point that studying the TM and soluble domains of an individual protein at the same time is a notoriously difficult task. To circumvent the problem people use the "divide and conquer" approach, utilizing constructs of various composition investigated with numerous membrane mimicking conditions, including micelles, bicelles, and, occasionally, organic solvents (more details can be found in our recent review Puthenveetil et al.[38]), in addition to studying separate soluble parts in aqueous solution and in the complexes with their soluble binding targets. Fig. 3C highlights just a fraction of such studies of integrin $\alpha_{IIb}\beta_3$ TM domain and CT that resulted in structural characterization. In general, the ability of the juxta-membrane cytoplasmic IDRs to shift between different conformations in response to changes in their environment can serve as a functional advantage in cross-membrane signal transduction. Integrin cytoplasmic tails are capable of accommodating significantly different structural features (Fig. 3C), in several cases even shifting between different secondary elements within the same region. Shown as an example in Fig. 3C, inset C1, the β_3-CT stretch of W^{739}-E^{749} can form various helical motifs (PDBID: 1S4X, colored light-blue, in DPC; 2LJE, colored blue, in DPC with pY^{747}), turns/loops (PDBID: 2L1C, colored cyan, bound to Shc pTB domain, pY^{747}) and extended strands (PDBID: 1MK7, purple, bound to talin F3 domain with Y^{747} in the non-phosphorylated form) depending upon binding partners, phosphorylation states, and potential interactions with lipids of the membrane. This remarkable dexterity may be the underlying foundation for the crucial bidirectional flow of information through integrins.[31]

L1-CAM: Another example of binding promiscuity based upon intrinsic disorder is L1, a highly conserved transmembrane glycoprotein member of the immunoglobulin superfamily of cell adhesion molecules (CAMs). It mediates many developmental processes in the nervous system including cell migration, axon extension, fasciculation, guidance, and myelination.[39] L1 is essential for dynamically regulated axon morphogenesis[40] and functions through numerous interactions with extracellular ligands, adjacent molecules on the same membrane, and intracellular second messenger systems. We have studied its interactions with two adaptor proteins.[41] Through an employment of various techniques, including NMR, analytical ultracentrifugation, size exclusion chromatography and dynamic light scattering, we have confirmed the intrinsically disordered nature of its CT and mapped its binding interfaces with ezrin FERM domain and AP2-μ2. L1-CAM is present in the DisProt database (https://www.disprot.org) which contains experimentally studied and manually annotated IDPs/IDRs. Fig. 4A shows

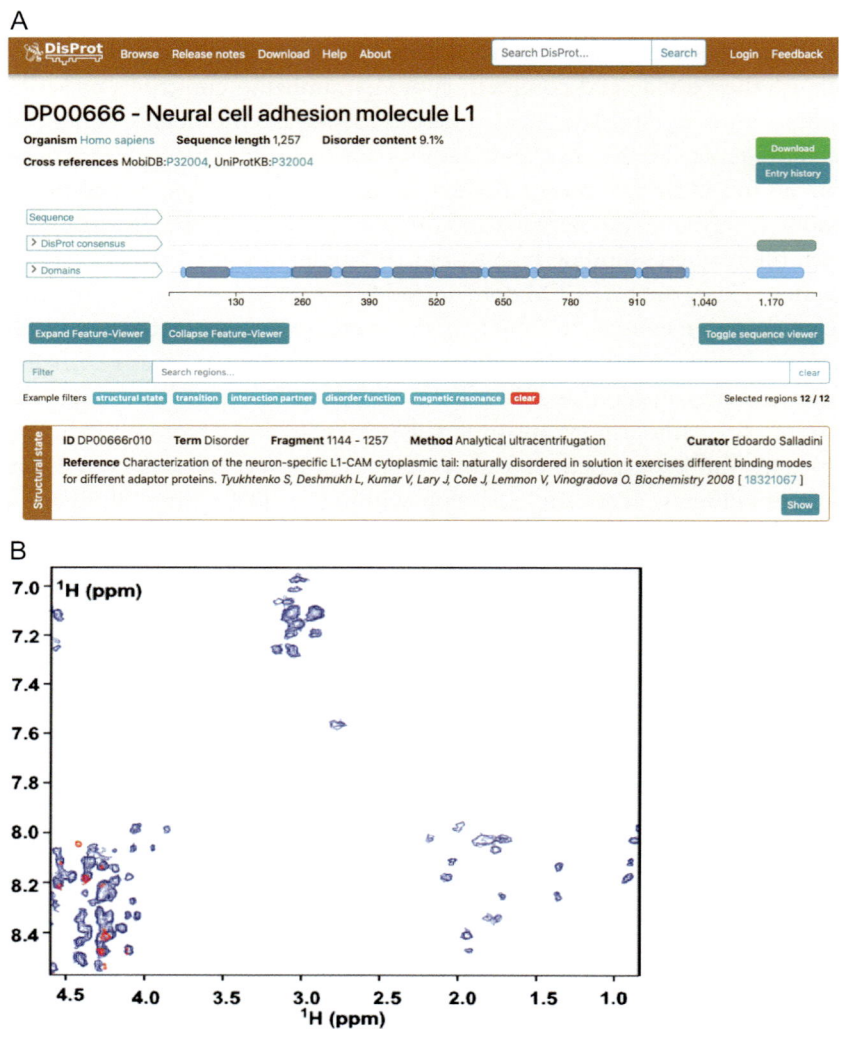

Fig. 4 Neuronal cell adhesion molecule L1: (A) Top of DisProt screenshot with the information available for L1-CAM. (B) ^{1}H-NOESY spectra of unlabeled L1-CT in absence (red) and in presence (blue, 10:1 ratio) of AP2-μ2 (1 mM, 295 K, pH 6.8, 50 mM KPi, 300 mM NaCl, mix = 300 ms) showing additional trNOEs (in blue) of the complex.

a screenshot of its DisProt record (database identifier DP00666) highlighting analysis (e.g., domains) and data (publications) available at the site. Fig. 4B further illustrates the type of NMR data, which allows monitoring the transition from disordered state of CT in aqueous solution into partially folded conformation upon interaction with its binding target, AP2-μ2 in the shown case, to be discussed later in the chapter.

BP180: BP180 can serve as another interesting though less known and less structurally characterized example for disorder in an IMP. It is also referred to as the 180 kDa bullous pemphigoid antigen 2 (BPAG2) or the collagen alpha-1(XVII) chain. Bullous pemphigoid (BP) is a blistering disease of stratified epithelia such as the skin and oral cavity where cells of the lowest layer lose their attachment to the underlying extracellular matrix proteins. It is encoded by COL17A1 but is atypical of collagens with its function as a transmembrane component of hemidesmosomes (HM).[42] It essentially serves as a membrane spanning hub within hemidesmosomes connecting cytoplasmic cytoskeletal elements to the extracellular matrix.[43] From its comparatively large N-terminus in the cytoplasm, BP180 passes through the cell membrane with its C-terminus in the extracellular space. It is the repeating amino acid collagenous GLY-X-Y motif in its extracellular C-terminus[44] that earned BP180 the collagen name designation and likely contributes to its homo-trimeric structure as a transmembrane protein (Fig. 5 for BP180 domains, panel A). Among other proteins in addition to BP180, the skin keratinocyte hemidesmosome protein complex contains another IMP, the integrin dimer $\alpha_6\beta_4$, which is proposed to interact with BP180 through both, intracellular and extracellular domains as described below.

Although physical associations responsible for them are as yet undefined, BP180 has been implicated in multiple cell functions beyond attachment including stem cell maintenance,[45] cancer cell invasion,[46] and several intracellular signaling pathways such as p38 MAPK and those downstream of TGFβ binding (see Ref. 44 for review). Thus, for BP180, its structural involvement in epithelial adhesion, its genetic and immune connection to severe tissue pathologies, and its emerging functional connection to several cell regulatory pathways make it an attractive candidate for new angles of protein conformational investigation. The ongoing challenge, as with many other proteins to which disorder is attributed from recombinantly-expressed or *in silico*-assessed protein, will be to experimentally detect BP180 conformations *in situ* within the cell and see directly whether and how the proposed disorder affects BP180 interaction with partner proteins and hemidesmosome organization and function. Here of course, cell-matrix interactions both under instances of tissue homeostasis and more malleable conditions such as wound repair where migration requires cycling cell-matrix attachment/release will need to be considered. BP180 ability to progress through dynamic conformations may be advantageous in either situation.

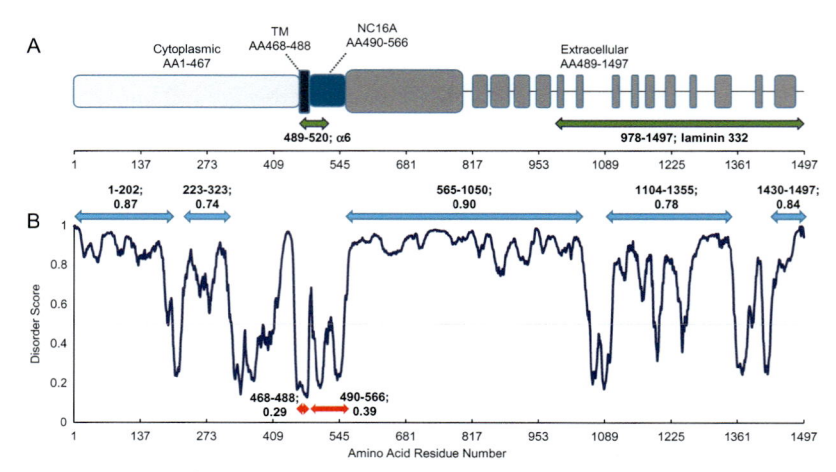

Fig. 5 Alignment of BP180 subdomains with predicted regional disorder. The BP180 regions (A) are relatively to scale with typical, major subdomains labeled. TM, transmembrane; NC, non-collagenous. Cytoplasmic, TM, and extracellular domains are as per UniProt Q9UMD9. By convention, NC domain numbering starts at the end of the extracellular domain, at the protein's C-terminus. This places NC16A internal to the full-length protein. Other lengths below BP180 stick figure are called out by amino acid positions, e.g., 978–1497, for interaction with other proteins. These are the regions used experimentally; it is not necessarily equivalent that all of the particular length was required for interaction. Numbering and subdomains in (A) align with amino acid residue numbers on the PONDR-FIT x-axis (B). Two cytoplasmic and three extended subregions of the N- and C-termini, respectively, are marked by double-headed arrows for local high disorder (average ≥ 0.50) prediction above plot peaks. They are labeled as to amino acid positions with the average PONDR-FIT score within that subregion (e.g., 1–200; 0.87). The short TM and NC16A regions called out below PONDR-FIT plot valleys (e.g., 468–488; 0.29) have relatively low predicted disorder.

With a PONDR-FIT assessment of the full-length BP180 (Fig. 5B), we confirm the IDRs noted by Liu et al.[47] for the N-terminal intracellular domain, amino acids (AA)1–467, (≥0.5 disorder score: 1–202; 223–323; 412–452). We here report extremely high ID scoring for three extensive regions of the sizeable extracellular domain, AA489–1497: AA565–1050 (average score 0.90); the more-carboxyl AA1104–1355 (average score 0.78) and ultimately along the AA1430–1497C-terminus (average score 0.84). In contrast and as might be expected, there is a precipitous ID drop within the transmembrane region (AA468–488; average score 0.29) and an extended length containing it and immediately flanking it on the cytoplasmic and extracellular sides (AA453–554; average score 0.31). The cytoplasmic IDRs of BP180, its singe TM span, its participation in a discrete membrane assembly, and its being involved in signal transduction completely fits the earlier computational analysis expectations mentioned above from Burgi and colleagues.[25]

Disorder is predicted *in silico* for the intracellular domain of BP180 and physically detected for it in aqueous solution. Nevertheless, this subdomain can accommodate secondary structural elements upon interaction with the negatively charged inner leaflet of the lipid bilayer. This region has been shown by circular dichroism to gain significant α–helical content in the presence of an anionic membrane mimetics.[47] Thus, BP180 is another example of the ability of the juxta-membrane cytoplasmic IDRs to shift between different conformations in response to changes in their environment.

While several proteins are localized to the HD, it is only the $\alpha_6\beta_4$ dimer and BP180 that are TM proteins.[48] From fluorescence co-localization studies,[49] it has been suggested BP180 may interact with either or both partners of the integrin dimer. The extracellular N-terminus of α_6 has been broadly referred to as interacting with the extracellular domain of BP180.[50] Particularly, it is a non-collagenous, extracellular BP180 region (Fig. 5A) proximal to the membrane (AA489–520), when expressed in the context of the BP180 N-terminus (AA1–520), that is sufficient for physical protein association with the extracellular domain of α_6.[49,51] Notably, from the first residue of the BP180 extracellular domain, through this α_6-interaction region, up to and including those residues associated with epitopes for blistering diseases (altogether AA489–566), there is an average PONDR-FIT score for this region of 0.39 (Fig. 5B) followed by an extensive region of ~500 amino acids (AA565–1050) where disorder is predicted to increase dramatically to an average score of 0.90. Whatever conformational flexibility this may bring to BP180 in the hemidesmosome, it does not seem to negatively impact its association with α_6.

For β_4, it is again fluorescence co-localization studies that suggest association with BP180.[52] The β_4 integrin, with a cytoplasmic C-terminus of ~1000 amino acids, is distinct from the much shorter cytoplasmic lengths of other β integrins and thus may have more opportunity for protein-protein interactions. Using yeast two hybrid assays,[53] the BP180 AA1–520 region was ranked for its relative interaction with specific regions of β_4. β_4 AA1,320–1668 was sufficient for performing among the highest ranked pairings of BP180 and β_4. Unfortunately, with no dissection of the BP180 cytoplasmic domain, we cannot conclude what affect its sizeable IDRs (AA1–202 0.87; AA223–323, 0.74) may have on interaction with even this relatively limited ~340 amino acid stretch of β_4. Assessment and consequences of intrinsic disorder are much better appreciated since these studies of BP180 interaction and colocalization with the integrin $\alpha_6\beta_4$ dimer were reported. BP180 intracellular intrinsic disorder as reported by Tussa et al.[47] and the assessment we provided for the extracellular domain suggest timing is right to experimentally investigate effects of BP180 intrinsic disorder in hemidesmosome assembly and function.

Although not a partner protein within the hemidesmosome, the cruciate trimer laminin 322 (three protein chains, alpha3, beta3, and gamma2)[54] is an extracellular matrix protein binding target of BP180. The carboxyl-terminus of BP180 AA978–1497 is necessary for that protein binding[55] suggesting the predicted disorder (Fig. 5 for BP180 domains, panel B) in this extended region is at least compatible with, if not a requirement for, interaction.

2.2 Intrinsic disorder optimizes utilization of limited space within cell

Based upon the observation that cellular contents such as nucleic acids and proteins are profoundly crowded, it has been proposed that a IDP might have an evolutionary advantage. The unstructured nature of an IDP can provide a large binding interface, which would require a globular protein to be 2–3 times larger to present the same surface, potentially increasing the overall size of the cell.[56] Vast experimental data from our research group and numerous other investigators presented above for β_3 integrin direct interactions with its cytoplasmic binding partners supports this hypothesis: as illustrated in Fig. 3B, inset A1, the C-terminus of β_3 CT binds to so many cytoplasmic targets through exceedingly overlapping regions that it would require a much longer sequence for the globular fold to accommodate all the potential conformations for providing a necessary degree of specificity. Our data related to neuronal L1-CAM, another disordered IMP discussed earlier and much less characterized than integrin, also supports this hypothesis as we have mapped significant and overlapping binding interfaces of its interactions with ezrin FERM domain and AP2-μ2.[41]

Liquid–liquid phase separation (LLPS) of bio-macromolecules, driven by weak interactions between multivalent molecules, has been proposed as a phenomenological basis for the formation of membraneless compartments within the cell. Mostly studied in soluble proteins, LLPS contributes to formation of condensates such as stress granules and the nucleolus. These meso–scale structures, termed biomolecular condensates, concentrate specific collections of proteins and nucleic acids without a surrounding membrane.[57] In general, many proteins that can undergo LLPS contain IDRs and are quite dynamic in solution. Due to this observation, this ability became almost a hallmark for IDPs. It should not be forgotten, however, that well-folded proteins can be induced to undergo LLPS *in vitro* too when placed under the right conditions and/or in presence of crowding agents.[58] Interestingly for this chapter, LLPS has been recently shown to promote assembly of IMPs with their cytoplasmic binding partners into clusters. Examples include nephrin, a TM adhesion molecule required to form the glomerular filtration barrier in kidneys,[59] and linker for the activation of T cells (LAT), a TM protein required for T cell activation.[60,61] Receptor tyrosine kinase[59] and even integrin[62] clustering has been proposed to be coupled to phase separation in multivalent cytoplasmic proteins.[63] However, it is important to note that oligomerization and phase separation are distinct physical processes that can be experimentally separated; therefore, phase separation should not be automatically expected whenever oligomerization occurs.[64] Even more important, the causal relationship and functional consequences of LLPS *in vivo* are not well defined and it is necessary to perform quantitative measurements on proteins in their endogenous state and physiological abundance to come up with more reliable conclusions about the biological significance of LLPS in general and its relationship to IMPs.[65]

2.3 Intrinsic disorder facilitates higher susceptibility for post-translational modifications

The conformational flexibility within IDPs is expected to provide increased access for protein-processing enzymes increasing their frequency of post-translational modifications in comparison to folded proteins. Interestingly, the number phosphorylation sites in IMPs containing IDRs is predicted to be more than two times higher on an average than for soluble protein with IDRs, although not all of these sites belong to disordered regions.[25,66] PTM for TM proteins occurs on both intra- and extracellular domains, such as with phosphorylation at tyrosine, serine, and/or threonine and glycosylation at asparagine, serine, and/or threonine, respectively, with effects on cell-cell and cell-matrix interactions.[67,68] The potential consequences of other PTM, e.g., methylation, acetylation and ubiquitination, for all IDP in general have also been broadly considered.[69] Of all of these, phosphorylation is among the most intensively investigated and the modification we will focus on here with particular IMPs we introduced above, integrins, L1-CAM, and BP180. PTMs have recently been reviewed[70] for other diverse IMPs such as cystic fibrosis transmembrane conductance regulator (CFTR) and the Na+/H+ exchanger (NHE-1).

In general, phosphorylation is utilized as a common switch regulating cell surface receptors.[71] In regards to integrin $\alpha_{IIb}\beta_3$, discussed as an example above, phosphorylation of its CT could constitute one of the spatiotemporal mechanisms for imparting selective recognition of its proximal effectors. Integrin β_3 CT is laden with various phosphorylation sites, including two tyrosines, one serine, and multiple threonines. The tyrosines phosphorylation was found to be specific for the outside-in signaling[72] and Shc (in particular its p52 isoform) was identified as the primary signaling partner.[73] We have performed detailed NMR analysis of the effects of tyrosine(s) phosphorylation on integrin β_3 CT under both aqueous and membrane-mimetic conditions.[32] We have shown that the phosphorylation causes significant conformational rearrangement in β_3 CT under solution conditions where the pY^{747} containing segment folds back and interacts with the membrane-proximal region (Fig. 3C, inset C2, PDBID: 2LJF), making the overall CT more compact and rigid. This arrangement prevents the β_3 CT from binding to α_{IIb} CT, thus likely dictating an unclasped state of the receptor necessary to mediate integrin outside-in signaling. Moreover, tyrosine(s) phosphorylation under membrane-mimetic conditions (in DPC micelles) modifies β_3 CT interaction with the membrane and perturbs its overall fold (Fig. 3C, inset C2, PDBIDs: non-phosphorylated - 1S4X, pY^{747} - 2LJD, pY^{747} & pY^{759} - 2LJE). By preventing the phosphorylated tyrosines containing regions from being inserted into or associated with the lipid bilayer, phosphorylation shifts the equilibrium of integrin interactions with different cytoplasmic adaptor proteins, adjacent receptors, and/or the cytoskeleton. These findings shed light upon molecular details of how phosphorylation may play multiple roles in regulating different states and dynamics of cell surface receptors, suggesting a more complex paradigm than a simple two state (active/inactive) model.

The disordered cytoplasmic tail of our second example, L1-CAM, is also rich in potential phosphorylation sites. These include S^{1152}, phosphorylated by p90rsk and shown to be important for neurite outgrowth,[74] and Y^{1176}, potentially phosphorylated by p60src kinase and involved in cytoskeleton binding through ezrin FERM domain[41,75] and endocytosis regulation which is discussed later. Mutation of Y^{1176} to F to mimic an unphosphorylated tyrosine side group prevents ezrin and AP2-μ2 binding.[75] Y^{1176} is within the membrane-proximal disordered segment, AA 1144–1205 (average score 0.77), and substitution with F does not change the disorder score. This leaves open the interpretation of its interaction with other proteins to being more dependent on local charge alone, or charge-affected conformation, when phosphorylated.

The subcellular localization of BP180 does appear to be sensitive to its phosphorylation status,[76] at least in response to cell culture manipulations. Although *in vitro* substrates for cultured keratinocytes do not recapitulate complete hemidesmosome formation,[77] human keratinocytes can be successfully grown in culture and greatly enriched for relatively less- or more-differentiated cells, respectively, under low *versus* standard calcium concentration media conditions (typically 0.07 mM vs 1.8 mM). Upon shift from low to high calcium conditions, mimicking calcium gradient signals for the onset of differentiation and preparation for release from underlying substrates, there is redistribution of BP180 coincident with activation of cytoplasmic protein kinase C (PKC). Separately, experimental activation of PKC by 12-O-tetradecanoylphorbol-13-acetate induces similar reorganization and increased serine phosphorylation. The relocalization has been interpreted as disintegration of hemidesmosomes with BP180 release from such discrete foci.[78] Our cursory overview of potential phosphorylation sites with GasPhos[79] for possible PKC targets returns over 10 hits for SER throughout the BP180 cytoplasmic N-terminus. Specific phospho-residues involved have yet to be experimentally determined. It is reasonable to expect, however, that BP180 cytoplasmic tail conformational flexibility from intrinsic disorder facilitates access of phosphorylation-modifying enzymes and consequently BP180 interaction with other hemidesmosome-localized proteins.

2.4 Intrinsic disorder affects turn-over of IMPs

Endocytosis of L1 provides an interesting example for how an IDR can play an important role in regulation of CAMs mobility during internalization and resorting within the cell. Although the L1-CT is not required for cell adhesion, it was found to be very important for the dynamic regulation of this process.[40] Encoded with inclusion of exon 27, neuronal L1-CT contains four extra amino acids (RSLE) in contrast to L1-CT found in non-neuronal cells such as Schwann cells or a variety of human tumor cell lines including neuroblastomas, melanomas and lung carcinomas.[80,81] Together, Y^{1176} and the immediately adjacent RSLE stretch compose a Y-based sorting motif (YRSL–YxxΦ, where Φ is a residue with a bulky hydrophobic side chain) that is required for endocytosis of L1 *via* clathrin-coated pits.[82]

Like Y-based sorting signals found in many proteins,[83] the YRSL sequence of L1 serves as a docking site for the µ2 chain of clathrin-associated AP2-complex *in vitro*. Mutating Y^{1176} or removing the RSLE exon disrupts the L1-AP2-µ2 interaction and consequently prevents clathrin-mediated endocytosis.[82] The neuronal form of L1, containing the RSLE sequence, is significantly less adhesive than the non-neuronal form due to its more rapid internalization and shorter dwell time on the cell surface,[84] with potential regulatory mechanisms depending on the phosphorylation state. Phosphorylation of Y^{1176}, possibly by p60src kinase, prevents L1 from binding to AP-µ2 *in vitro*[85] and disrupts endocytosis. Therefore, dephosphorylation of Y^{1176} may be critical for L1 mediated cell adhesion and signaling.

With respect to our third example, Liu and colleagues[45] recently reported BP180 as the "most unstable component of hemidesmosomes." BP180 instability in cells is consistent with the above-described multiple and extensive IDR within its recombinantly-expressed intracellular domain.[47] Intrinsic disorder derived conformational flexibility, possibly facilitating BP180 protein-protein interactions with its hemidesmosome partners, would be consistent with the collateral expense of its protein stability although overall IDP protease sensitivity *in vivo* is reported to be less so than that detected during *in vitro* assays.[86] As with the internalization of L1, turnover of other IMP offers ready remodeling for cellular adaptations such as those needed for migration or changes in cell-cell associations.

3. Mutations in IMP: Consideration of IDR

Mutations in many IMPs are causative of several human pathologies. However, these amino acid sequence alterations have as yet little or no direct biophysical and biological assessment for their possible impact on conformation in terms of order/disorder. To begin to address this deficit, we present initial bioinformatic consideration of some IMP discussed above in terms of disease-linked mutations and possible consequences on protein subregion disorder.

All of the mutations identified in patients with Glanzmann's thrombasthenia (GT), an inherited blood-clotting disorder, result in functional deficiency of major platelet integrin $\alpha_{IIb}\beta_3$,[87] with the absence of agonist-induced platelet aggregation being a hallmark of this disease.[88] Most of these mutations reside in the extracellular ligand-binding domain of the receptor and directly reduce the binding affinity of integrin in its activated state. However, functional deficiency can also result from two β_3 CT mutations, a $R^{724}X$ nonsense mutation resulting in a truncated protein with the deletion of the C-terminal 39 residues and a $S^{752}P$ missense mutation. Both CT mutations are proposed to disrupt the activated state of the receptor, as resting platelets from both patients express significant levels of stable $\alpha_{IIb}\beta_3$ complexes that are unresponsive to agonists (though do respond to conformational activators). Functional analyses show normal adhesion to immobilized fibrinogen but abnormal cell spreading. The $S^{752}P$ mutation shows reduced focal adhesion plaque formation, and the $R^{724}X$ mutation results in undetectable tyrosine phosphorylation of focal adhesion kinase pp125^{FAK}. The S^{752} does not belong to the most disordered segment of β_3 CT (AA 725–738, average score 0.69) according to PONDER-FIT assessment; its mutation to P does not affect the predicted degree of disorder at all. Instead, it lies within a unique region of β_3 CT covering two Y-based sorting motifs (NPxY^{747} and NxxY^{759}). The segment between two tyrosines is shown to accommodate different secondary structural elements. For example, it forms an α-helix in DPC micelles[36] and an extended β-strand when bound to the groove of filamin Ig21 domain.[89] The exact reason why the $S^{752}P$ mutation is inhibitory remains to be found.

Rare gain-of-function (or activating) mutations within $\alpha_{IIb}\beta_3$ CT have been also shown to cause autosomal dominant macrothrombocytopenia (MTP). An $R^{995}W$ in α_{IIb} and/or a $D^{723}H$ in β_3 mutation[90] disrupt the R^{995}-D^{723} salt bridge linking the juxta–membrane helical portions of the two subunits of the receptor in a resting state.[37] For all affected patients, Favier and colleagues have found that the bleeding syndrome and MTP were mild to moderate; platelet aggregation tended to be reduced but not absent, platelets were on average larger than normal, and some contained abnormally large α-granules showing signs of fusion. Analysis of the maturation and development of megakaryocytes revealed no defect in their early maturation but abnormal proplatelet formation was observed with increased size of the tips.

Mutations in L1 are responsible for a wide variety of neurological abnormalities and mental retardation (such as MASA, Mental Retardation, Aphasia, Shuffling Gait, Adducted Thumbs, or X-linked hydrocephalus). However, mutations in L1 CT are often less severe than other mutations; they lead to defects consistent with failures of axon guidance.[91,92] In contrast, mutations that would lead to L1 expression loss or expression of a markedly truncated form are the most severe and lead to death during the first year of life.[93–95]

Proteolytic processing of the BP180 extracellular domain yields two large fragments, a 120 kDa protein and then from it a 97 kDa fragment. The 120 kDa fragment is released from the cell surface by a disintegrin and metalloprotease (ADAM) *via* digestion at different amino acids within a non-collagenous (NC) spacer known as the NC16A domain (Fig. 5A). Importantly, these liberated fragments of the BP180 extracellular domain can, for some individuals, become an auto-antigen leading to antibody deposition and initiation of the skin auto-immune blistering disease referred to as linear IgA dermatosis (LAD) . Junctional epidermolysis bullosa (JEB) is a genetic skin blistering disease, caused by recessive mutations in BP180 (*COL17A1* gene), integrin α_6 (*ITGA6* gene), or integrin β_4 (*ITGB4* gene) and presumably loss of function in HD assembly. The structural failure of any one of these proteins leads to loss of contact between basal keratinocytes and the lamina lucida of the underlying basement membrane. Intriguingly, autoantibody reactivity in bullous pemphigoid (BP), a skin blistering disease distinct from LAD,[42] is almost exclusively directed to the non-collagenous juxta-membrane NC16A region (AA490–566; average score 0.39) (Fig. 5A and B) which in the backbone of the full-length protein appears to have relatively minimal disorder as the antigenic region. Thus, disorder across the extracellular BP180 C-terminus may allow for TM protein cleavage and presentation of novel auto-antigenic epitopes (LAD) as well as allow access of antibodies, once generated, to key structured regions (BP).

Intrinsic disorder is prominently associated with protein misfolding degenerative diseases where the entire length of a protein such as α-synuclein and β-amyloid may be involved in aggregation and pathological deposits (see Ref. 96 for review). Conclusively linking TM protein mutation or other dysfunction (improper folding, insufficient amount, incorrect PTM) that definitively affect disorder/order conformations and in turn are responsible for a distinct pathogenic effect lies ahead of us but is a worthwhile exploration. As just one example of IMP deficient function with likely some instances linked to protein conformation, over 60 mutations associated with JEB have been mapped across the BP180 N-terminal cytoplasmic tail and C-terminal extracellular collagenous regions.[97] These cases have variable presentations; their severity does not necessarily correlate with particular residues and none are as clinically debilitating as the protein null situation. Perhaps most interesting in the context of protein conformation are changes reported for a small number of patients with recessive JEB where glycine was mutated in a GLY-X-Y collagenous repeat in the juxta-membrane extracellular region (Fig. 5). In each of four clinical cases, mutation of one glycine (G609, 621, 627, or 633) in that region led to decreased *in vitro* thermal stability and increased sensitivity to limited trypsin proteolysis.[98] The spectrum of JEB mutations has diverse but yet significant human health impact. Nevertheless, there is therapeutic promise from observations that even a relatively small amount of partially functional BP180 protein can lessen or delay clinical severity (see Ref. 99 for review). Small molecule or other synthetic chaperoning to improve stability or function of insufficient wildtype or dys/non-functioning of mutated TM proteins could have widespread clinical benefit for diverse IMP where disorder has been associated with pathology as suggested for proteins in other cellular compartments.[100]

Another interesting example more definitively highlighting the role of disorder-order transition associated with disease causing mutations can be found in epidermal growth factor receptor (EGFR), a representative of the family of asymmetric tyrosine kinases which undergo ligand-induced dimerization. A regulatory feature of this process is a shift in conformation of the αC-helix of the kinase domain from the "off" to the "on" position.[101] This kinase region of the EGFR features a high proportion of all mutations found in EGFR-associated cancers.[102,103] This includes the most common EGFR kinase domain-associated oncogenic mutation $L^{834}R$ (sometimes referred to as $L^{858}R$).[104] This mutation is exceedingly important in clinical settings. Although associated with overactivation of EGFR, it is susceptible to binding the tyrosine kinase inhibitor gefitinib providing for significantly increased responsiveness with this kind of EGFR-targeting treatment.[105] It is of interest to this chapter as it is associated with an IDR region within EGFR. This IDR region was first proposed and later confirmed in the αC-helix within the kinase domain of EGFR through crystallographic evidence followed by hydrogen/deuterium exchange[103,106] and *in silico* studies revealing several short IDR or flexible regions with this domain.[107] Shan and colleagues proposed that $L^{834}R$ mutation promotes overactive EGFR through disruption of the intrinsic disorder in this kinase domain. This mutation was shown to promote conformational rigidity by introducing a positively charged R^{834} that interacts favorably with localized negatively charged residues (E^{734}, D^{831}, and D^{813}) in the αC-helix, and, by doing so, protects the placement of the αC helix in the way that the catalytically important *KE salt bridge* between K^{721} and E^{738} is maintained.[101,103] Ultimately, because the αC-helix is located near the dimerization interface, the stabilized helix and reduced local structural flexibility have been shown to promote dimerization and subsequent EGFR activity. Reported free energy surface analysis is consistent with this notion demonstrating the presence of active conformations in the $L^{834}R$ mutant ensemble. This is in contrast to the wild type ensemble, which lacks active conformations due to disorder in the αC-helix.[101] As to increased susceptibility of EGFR carrying the $L^{834}R$ mutation to inhibition, it has been proposed that this mutation stabilizes the EGFR dimer compensating for slow binding kinetics of inhibitors, which bind the active form of the receptor, and is thus responsible for inhibitor efficacy.[103] Often when discussing the $L^{834}R$ mutation in EGFR, it becomes important to mention the $T^{790}M$ mutation which is found very commonly and is associated with acquired resistance to inhibitors and increased survival of cancer cells.[108] Alone, this mutation has been both reported to increase overall intrinsic disorder while stabilizing the active form of EGFR.[107,108] However, both these reports agree that one consequence of the $T^{790}M$ mutation is an increased propensity to form the active conformation of EGFR. Understanding how these mutations function to impact local disorder is likely to enhance pharmaceutical efforts, particularly in overcoming the acquired resistance with the $T^{790}M$ mutation. Resolving how conformational dynamics impact receptor activity, such as possible interplay of charge and local secondary structure[109]), could dictate new strategies for small molecule-induced conformational switching.

4. IMP intrinsic disorder: Investigational approaches

In general, the most appropriate experimental method for structure analysis depends upon dynamic properties of a protein in question. X-ray crystallography works really well for folded domains, but, since disordered regions do not crystallize, it cannot be used to determine their structures. CryoEM, a rising star of structural biology, does not require crystallization. However, to produce a high-resolution structure an assembly of class averages needs to be performed. This can be done for IMPs showing moderate degrees of flexibility,[110] but is unlikely to be useful for truly disordered regions. Small-angle X-ray scattering (SAXS) has been valuable for describing sizes and distribution of the disordered protein ensembles, but resolution is low.[111] NMR spectroscopy is our favorite technique to explore IDRs, as it is currently the only technique that can provide atomic resolution.[112] Intrinsically disordered IMPs contain both folded domains and disordered regions. Therefore, experimental methods optimal for studying well-folded, rigid as well as disordered, flexible biomolecules have to be integrated, potentially through a divide-and-conquer approach, to produce a complete mechanistic model of these unique targets. Interestingly, this strategy of combining different complementary techniques has become an emerging trend within the structural biology community. The key question, however, remains of how well IDRs studied alone represent IDRs in the context of full-length proteins in the presence of membrane mimetics, with the even more dazzling question of potential IDR conformations *in situ* within live cells.

4.1 Computational approach adjusted for IMPs

A starting, and low cost, point to investigate disorder in IMPs is to utilize various computational *in silico* approaches. However, as most of the programs predicting disorder from protein's primary sequence were originally developed using information derived from soluble proteins, there is a clear distinction between programs that accurately predict disordered regions in membrane proteins and programs that perform poorly. A meta-predictor PONDR-FIT, with artificial neural network which could be trained on specific data subsets to predict the disorder propensity, was developed to circumvent this problem.[113] PONDR-FIT and numerous other disorder predictors, with some reporting potential transmembrane regions, have been recently reviewed.[114] Still, as we have shown in our integrin example, remarkable structural dexterity could be found within the regions with reasonably low disorder scores as well as secondary structural elements may exist in the segments with high disorder scores. Therefore, direct experimental methods are necessary to explore and define disorder and its functionality in IMPs.

4.2 Thermostable enrichment

The amino acid compositional bias of IDP or proteins with extensive or multiple IDRs confers on them atypical biophysical characteristics, among them an *in vitro* tendency of thermostability (TS). TS is a relative resistance to denaturation and precipitation following heat treatment, e.g., 98 °C for >15 min followed by rapid chilling.[115,116] Unlike globular or structured proteins, IDPs are more likely to remain soluble after such extreme treatment. Importantly though, this is more preferential enrichment of IDPs and reduction of globular proteins rather than an absolute separation of them.[116] Nevertheless, the approach yields protein fractions suitable for large scale, proteome level analysis (an IDP-ome[115]). IDPs and protein with IDRs retrieved from TS fractions of cell lysates cover a very wide range of biological processes and cellular compartments,[115] including BP180 discussed above (Shamilov and Aneskievich, unpublished). Kriwacki and colleagues also recovered TM proteins from a mouse fibroblast TS preparation.[115] They noted TM proteins were however a minor component of the overall >1300 TS proteins isolated. Sixty-five proteins with TM domains were identified by computational assessment from the entire TS population. While just over half (55%) of the TM cohort were evaluated to have one or two TM passes, Kriwacki et al. reported a range of up to 16 TM helices in the remaining proteins. Interestingly, among the TS-retrieved proteins were those TM spanning the nuclear and golgi membranes as well as the cell membrane. Along with other criteria, such as charge hydropathy, all TS proteins including these TM proteins were relatively scored as a disordered protein (average disorder score > 0.50) or a protein of IDR mixed with more-ordered regions (average disorder >0.32). TM proteins were an enrichment for and not an absolute purification of proteins with only unstructured domains. Like other classes of proteins, numerous TM proteins in the TS cohort were predicted to be mostly ordered. Nevertheless, TS provides for a relatively easy and quantifiable approach to IDP throughout the cell.

4.3 Solution NMR technology

Among all the methods presently at the researcher's disposal, solution NMR, in our opinion, is the best-suited technique for studying conformational heterogeneity of IDRs in IMPs, both in an aqueous solution and membrane mimetics. Chemical shifts (or rather their deviations from random coil values) provide direct reflection of secondary and tertiary structural elements, while changes in chemicals shifts allow to examine protein-protein interactions[117] at the atomic level and at near physiological conditions. For those IDRs that acquire stable fold upon binding to their targets a classical method known as transferred Nuclear Overhauser Effect (trNOE) measurement[118] can be used to define the bound state. The major idea of this approach is that cross relaxation between protons of the IDR in the bound state, which is governed by large correlation time of the complex, is transferred to the protons of IDR in free state, tumbling rapidly in solution, through chemical exchange. This results in an appearance of additional cross-peaks in its NOESY (Nuclear Overhauser Effect Spectroscopy) spectrum, as exemplified in Fig. 4B, from which restraints can be derived for structural characterization of the bound conformation.

More recent developments allow generation of IDP/IDR ensembles, through the ENSEMBLE algorithm,[119] consistent with an expanded set of experimental NMR restraints, which can include residual dipolar couplings (RDCs), paramagnetic relaxation enhancements (PREs), O_2-induced ^{13}C paramagnetic shifts, hydrogen-exchange protection factors, $^{15}N\ R_2$ relaxation data, NOEs, J-couplings, chemical shifts, hydrodynamic radius, solvent accessibility restraints, plus SAXS measurements. It is important to remember that for IDPs, the mentioned NMR observables represent a weighted average over the conformational ensemble. The ENSEMBLE algorithm assigns equal populations to all conformers for computational simplicity and to reduce the problem of overfitting. In this approach, conformers that can be thought of as containing an "implicit" entropy, i.e., largely extended conformations with few structural contacts, have a high entropy and represent a relatively large "sub-ensemble." Distinct compact conformers, containing many contacts, represent fewer but enthalpically more favorable conformations. Overall, the calculated ensembles may differ in many ways from random coil ensembles and contain significant transient structures.[120]

The LLPS phenomenon can be studied very efficiently by solution NMR as well, with the detailed review of the method found elsewhere.[121] Briefly, signals from disperse and condensed phases can be distinguished. It can be done either through physical separation, by studying the protein below the concentration required for LLPS or by creating a macroscopic phase that fills the coil volume of the NMR spectrometer, or through the application of different filters during signal acquisition in bi-phasic samples: with an R_2 relaxation rate filter selecting for signals arising from the dispersed phase or with a pulsed-field gradient diffusion rate filter selecting for signals arising from the condensed phase. Based upon NMR data collected so far, we can start to speculate about common features of condensed phases, such as the maintenance of protein disorder, restricted motions due to high viscosities and protein concentrations, and transient, "fuzzy" interactions. However, a disconnect between *in vitro* and *in vivo* studies as we discussed earlier[65] still remains an issue, which, in the future, might be tackled by *in-cell* NMR methods as they develop.[122]

4.4 Other techniques

Although being a low resolution technique even with the most recent improvements in data quality and reconstitution algorithms, SAXS can directly measure flexibility of biomolecules in solution.[123] Detecting flexibility earlier in the process is essential, as it steers modeling efforts from the rigid-body toward ensemble approaches when generating molecular models.[124] In the presence of conformational flexibility traditional *ab initio* shape reconstruction programs based upon optimal placement of spherical beads within a fixed volume, such as DAMMIN[125] and GASBOR,[126] often fail. The recently developed *ab initio* shape reconstruction program, DENSS (DENsity from Solution Scattering),[127] however, is more robust, providing low-resolution insight into macromolecular architecture through the iterative retrieval of structure factors directly from experimental scattering data. Its advantage is based upon capturing non-uniform biomolecular volumes (e.g., particle cavities), which is beneficial for modeling of flexible and disordered systems, and detecting differences in electron density among different biomolecular phases (e.g., protein vs lipid, as might be encountered for IMP under membrane mimicking conditions).

Another popular method that is well suited for quantifying solvent accessibility and conformational dynamics, as well as ligand-induced perturbations of IDRs, is hydrogen–deuterium exchange combined with mass spectrometry (HDX-MS). It is similar to measurements of hydrogen-exchange protection factors by solution NMR, with the MS being a detector in this case, but it is not limited by the size of macromolecule under investigation. It can be applied to both, soluble IDPs and IDRs within IMPs. The example of HDX-MS utility is presented by quantification of the ligand-induced conformational perturbations of intrinsically disordered CT of β2-adrenergic receptor (β2-AR).[128] This study showed that conformational flexibility of β2-AR is enhanced on binding to agonists, while a stabilizing effect was observed upon binding to inverse agonists.

Fluorescence depolarization methodology using time-correlated single-photon counting (TCSPC) can be utilized to distinguish between collapsed globules and expanded coils of IDRs. A detailed description of the method can be found elsewhere.[129] Briefly, fluorescence depolarization kinetics of an intrinsic Trp or a fluorescent dye, site-specifically attached to a protein, provides the picosecond time-resolved decay of fluorescence anisotropy from the fundamental (time-zero) anisotropy. The slow rotational correlation time, ϕ_{slow}, was shown to be a unique and reliable indicator of the conformational preference of an IDR. For a collapsed IDR, ϕ_{slow} is expected to represent the global tumbling of the compact globule, whereas, for an expanded IDP, ϕ_{slow} corresponds to the intrinsic backbone segmental mobility that is independent of the global tumbling.[130] This methodology has confirmed that κ-casein forms a collapsed globule while α-synuclein is an expanded polypeptide chain.

SmFRET, single molecule (sm) fluorescence (or Förster) resonance energy transfer (FRET), is the method that can resolve signals from individual molecules. In this application of FRET, a pair of donor and acceptor fluorophores are excited and detected on a single molecule level. Therefore, it provides information about the conformational distribution within ensembles rather than just average properties of the entire ensemble. It has been increasingly applied to studies of disordered IMPs *in vivo*,[131,132] expanding the scope of *in vitro* biophysical approaches.

An accurate description of potential IMP conformational ensembles crucially depends on the amount and quality of the experimental data, as well as the way it has been integrated. An integrative modeling approach has been recently proposed and tested to understand how conformational restraints imposed by the most common structural techniques used for IDPs investigation, such as NMR, SAXS, and smFRET, reach concordance on structural ensembles for non-phosphorylated and phosphorylated Sic1.[133]

5. Conclusions

IMP disorder has been formally hypothesized[15] now for over 5 years with many earlier reports of membrane protein dynamics[16] open to reinterpretation in such a context. With the premise and existence of disorder within extracellular and cytoplasmic domains of TM proteins further conceptually developed,[21] the scene was clearly set for new two-way evaluations of disorder and mechanisms TM proteins utilize to perform their function, including clustering, trafficking, and the inter-relationship of PTM and protein conformation. The single-pass IMPs emphasized here in the context of TM protein disorder highlight the probable impact of regional flexibility in normal physiological function of IMP as well as challenges and possible insight into pharmacological control of regional disorder to modulate IMP function in various disease states. Incorporation of disordered regions into essential cell surface receptors likely enhances formation of functional networks necessary for adaptable and efficient cross-membrane signal transduction. Nevertheless, there is much yet to be deciphered as to the consequences of intrinsic disorder for the conformation and in turn function of IMP with unstructured domains. This includes the specific assignment of presumed increased functionality to regions of disorder as well as the physiological effects of the order-disorder (and *vice versa*) transitions possibly occurring due to interaction with natural ligands or to pathological mutations. As highlighted in this chapter for a few TM protein examples, some separate facets of this have been done for individual proteins. It is clear that an integrative approach of bioinformatics, biophysical assessments, *in vitro* assembly models, and targeted mutagenesis will be the vital key for future successes in the field.

References

1. Fischer E. Einfluss der Configuration auf die Wirkung der Enzyme. *Berichte der deutschen chemischen Gesellschaft*. 1894;27(3):2985–2993.
2. Lindorff-Larsen K, Best RB, Depristo MA, Dobson CM, Vendruscolo M. Simultaneous determination of protein structure and dynamics. *Nature*. 2005;433(7022):128–132.
3. Koshland DE. Application of a theory of enzyme specificity to protein synthesis. *Proc Natl Acad Sci U S A*. 1958;44(2):98–104.
4. Rubin MM, Changeux JP. On the nature of allosteric transitions: Implications of non-exclusive ligand binding. *J Mol Biol*. 1966;21(2):265–274.
5. Changeux JP, Edelstein S. Conformational selection or induced fit? 50 years of debate resolved. *F1000 Biol Rep*. 2011;3:19.
6. Dunker AK, Lawson JD, Brown CJ, et al. Intrinsically disordered protein. *J Mol Graph Model*. 2001;19(1):26–59.
7. Tompa P. Intrinsically unstructured proteins. *Trends Biochem Sci*. 2002;27(10):527–533.
8. Uversky VN, Gillespie JR, Fink AL. Why are "natively unfolded" proteins unstructured under physiologic conditions? *Proteins*. 2000;41(3):415–427.
9. Wright PE, Dyson HJ. Intrinsically unstructured proteins: Re-assessing the protein structure-function paradigm. *J Mol Biol*. 1999;293(2):321–331.
10. Smith LM, Kelleher NL, Consortium for Top Down Proteomics. Proteoform: A single term describing protein complexity. *Nat Methods*. 2013;10(3):186–187.
11. Malaney P, Uversky VN, Dave V. PTEN proteoforms in biology and disease. *Cell Mol Life Sci*. 2017;74(15):2783–2794.
12. Uversky VN. p53 Proteoforms and intrinsic disorder: An illustration of the protein structure-function continuum concept. *Int J Mol Sci*. 2016;17(11):1–37.
13. Uversky VN. Protein intrinsic disorder and structure-function continuum. *Prog Mol Biol Transl Sci*. 2019;166:1–17.
14. White SH, Wimley WC. Membrane protein folding and stability: Physical principles. *Annu Rev Biophys Biomol Struct*. 1999;28:319–365.
15. Kjaergaard M. Can proteins be intrinsically disordered inside a membrane? *Intrinsically Disord Proteins*. 2015;3(1): e984570.
16. Popot JL, Engelman DM. Membrane protein folding and oligomerization: The two-stage model. *Biochemistry*. 1990;29(17):4031–4037.
17. Dyson HJ, Wright PE. Intrinsically unstructured proteins and their functions. *Nat Rev Mol Cell Biol*. 2005;6(3):197–208.
18. DeLano WL. *The PyMOL Molecular Graphics System*. CA, USA, DeLano Scientific LLC: Palo Alto; 2008.
19. Minezaki Y, Homma K, Nishikawa K. Intrinsically disordered regions of human plasma membrane proteins preferentially occur in the cytoplasmic segment. *J Mol Biol*. 2007;368(3):902–913.
20. Tovo-Rodrigues L, Roux A, Hutz MH, Rohde LA, Woods AS. Functional characterization of G-protein-coupled receptors: A bioinformatics approach. *Neuroscience*. 2014;277:764–779.

21. Kjaergaard M, Kragelund BB. Functions of intrinsic disorder in transmembrane proteins. *Cell Mol Life Sci.* 2017;74(17):3205–3224.
22. Dunker AK, Obradovic Z, Romero P, Garner EC, Brown CJ. Intrinsic protein disorder in complete genomes. *Genome Inform Ser Workshop Genome Inform.* 2000; 11:161–171.
23. Ward JJ, Sodhi JS, McGuffin LJ, Buxton BF, Jones DT. Prediction and functional analysis of native disorder in proteins from the three kingdoms of life. *J Mol Biol.* 2004;337 (3):635–645.
24. Patil A, Kinoshita K, Nakamura H. Hub promiscuity in protein-protein interaction networks. *Int J Mol Sci.* 2010;11(4):1930–1943.
25. Burgi J, Xue B, Uversky VN, van der Goot FG. Intrinsic disorder in transmembrane proteins: Roles in signaling and topology prediction. *PLoS One.* 2016;11(7): e0158594.
26. Xue B, Dunbrack RL, Williams RW, Dunker AK, Uversky VN. PONDR-FIT: A meta-predictor of intrinsically disordered amino acids. *Biochim Biophys Acta.* 2010;1804(4):996–1010.
27. Hynes RO. Integrins: Bidirectional allosteric signaling machines. *Cell.* 2002;110(6): 673–687.
28. Qin J, Vinogradova O, Plow EF. Integrin bidirectional signaling: A molecular view. *PLoS Biol.* 2004;2(6):726–729.
29. Byzova TV, Goldman CK, Pampori N, et al. A mechanism for modulation of cellular responses to VEGF: Activation of the integrins. *Mol Cell.* 2000;6(4):851–860.
30. Jones DT, Cozzetto D. DISOPRED3: Precise disordered region predictions with annotated protein-binding activity. *Bioinformatics.* 2015;31(6):857–863.
31. Deshmukh L, Gorbatyuk V, Vinogradova O. Integrin beta3 phosphorylation dictates its complex with Shc PTB domain. *J Biol Chem.* 2010;285:34875–34884.
32. Deshmukh L, Meller N, Alder N, Byzova T, Vinogradova O. Tyrosine phosphorylation as a conformational switch: A case study of integrin beta3 cytoplasmic tail. *J Biol Chem.* 2011;286(47):40943–40953.
33. Garcia-Alvarez B, de Pereda JM, Calderwood DA, et al. Structural determinants of integrin recognition by Talin. *Mol Cell.* 2003;11(1):49–58.
34. Katyal P, Puthenveetil R, Vinogradova O. Structural insights into the recognition of β3 integrin cytoplasmic tail by SH3 domain of Src kinase. *Protein Sci.* 2013;22: 1358–1365.
35. Vinogradova O, Haas T, Plow EF, Qin J. A structural basis for integrin activation by the cytoplasmic tail of the alpha IIb-subunit. *Proc Natl Acad Sci U S A.* 2000;97(4): 1450–1455.
36. Vinogradova O, Vaynberg J, Kong X, Haas TA, Plow EF, Qin J. Membrane-mediated structural transitions at the cytoplasmic face during integrin activation. *Proc Natl Acad Sci U S A.* 2004;101(12):4094–4099.
37. Vinogradova O, Velyvis A, Velyviene A, et al. A structural mechanism of integrin alpha(IIb)beta(3) "inside-out" activation as regulated by its cytoplasmic face. *Cell.* 2002;110(5):587–597.
38. Puthenveetil R, Vinogradova O. Solution NMR: A powerful tool for structural and functional studies of membrane proteins in reconstituted environments. *J Biol Chem.* 2019;294(44):15914–15931.
39. Hortsch M. The L1 family of neural cell adhesion molecules: Old proteins performing new tricks. *Neuron.* 1996;17(4):587–593.
40. Long KE, Lemmon V. Dynamic regulation of cell adhesion molecules during axon outgrowth. *J Neurobiol.* 2000;44(2):230–245.
41. Tyukhtenko S, Deshmukh L, Kumar V, et al. Characterization of the neuron-specific L1-CAM cytoplasmic tail: Naturally disordered in solution it exercises different binding modes for different adaptor proteins. *Biochemistry.* 2008;47(13):4160–4168.

42. Hammers CM, Stanley JR. Mechanisms of disease: Pemphigus and bullous pemphigoid. *Annu Rev Pathol.* 2016;11:175–197.
43. Van den Bergh F, Eliason SL, Giudice GJ. Type XVII collagen (BP180) can function as a cell-matrix adhesion molecule via binding to laminin 332. *Matrix Biol.* 2011;30(2):100–108.
44. Natsuga K, Watanabe M, Nishie W, Shimizu H. Life before and beyond blistering: The role of collagen XVII in epidermal physiology. *Exp Dermatol.* 2019;28(10):1135–1141.
45. Liu N, Matsumura H, Kato T, et al. Stem cell competition orchestrates skin homeostasis and ageing. *Nature.* 2019;568(7752):344–350.
46. Jones VA, Patel PM, Gibson FT, Cordova A, Amber KT. The role of collagen XVII in cancer: Squamous cell carcinoma and beyond. *Front Oncol.* 2020;10:352.
47. Tuusa J, Koski MK, Ruskamo S, Tasanen K. The intracellular domain of BP180/collagen XVII is intrinsically disordered and partially folds in an anionic membrane lipid-mimicking environment. *Amino Acids.* 2020;52(4):619–627.
48. Walko G, Castanon MJ, Wiche G. Molecular architecture and function of the hemidesmosome. *Cell Tissue Res.* 2015;360(3):529–544.
49. Hopkinson SB, Baker SE, Jones JC. Molecular genetic studies of a human epidermal autoantigen (the 180-kD bullous pemphigoid antigen/BP180): Identification of functionally important sequences within the BP180 molecule and evidence for an interaction between BP180 and alpha 6 integrin. *J Cell Biol.* 1995;130(1):117–125.
50. Powell AM, Sakuma-Oyama Y, Oyama N, Black MM. Collagen XVII/BP180: A collagenous transmembrane protein and component of the dermoepidermal anchoring complex. *Clin Exp Dermatol.* 2005;30(6):682–687.
51. Hopkinson SB, Findlay K, deHart GW, Jones JC. Interaction of BP180 (type XVII collagen) and alpha6 integrin is necessary for stabilization of hemidesmosome structure. *J Invest Dermatol.* 1998;111(6):1015–1022.
52. Borradori L, Koch PJ, Niessen CM, Erkeland S, van Leusden MR, Sonnenberg A. The localization of bullous pemphigoid antigen 180 (BP180) in hemidesmosomes is mediated by its cytoplasmic domain and seems to be regulated by the beta4 integrin subunit. *J Cell Biol.* 1997;136(6):1333–1347.
53. Schaapveld RQ, Borradori L, Geerts D, et al. Hemidesmosome formation is initiated by the beta4 integrin subunit, requires complex formation of beta4 and HD1/plectin, and involves a direct interaction between beta4 and the bullous pemphigoid antigen 180. *J Cell Biol.* 1998;142(1):271–284.
54. Kiritsi D, Has C, Bruckner-Tuderman L. Laminin 332 in junctional epidermolysis bullosa. *Cell Adh Migr.* 2013;7(1):135–141.
55. Nishie W, Kiritsi D, Nystrom A, Hofmann SC, Bruckner-Tuderman L. Dynamic interactions of epidermal collagen XVII with the extracellular matrix: Laminin 332 as a major binding partner. *Am J Pathol.* 2011;179(2):829–837.
56. Gunasekaran K, Tsai CJ, Kumar S, Zanuy D, Nussinov R. Extended disordered proteins: Targeting function with less scaffold. *Trends Biochem Sci.* 2003;28(2):81–85.
57. Banani SF, Lee HO, Hyman AA, Rosen MK. Biomolecular condensates: Organizers of cellular biochemistry. *Nat Rev Mol Cell Biol.* 2017;18(5):285–298.
58. Asherie N. Protein crystallization and phase diagrams. *Methods.* 2004;34(3):266–272.
59. Banjade S, Rosen MK. Phase transitions of multivalent proteins can promote clustering of membrane receptors. *Elife.* 2014;3:e04123. https://doi.org/10.7554/eLife.04123.
60. Su X, Ditlev JA, Hui E, et al. Phase separation of signaling molecules promotes T cell receptor signal transduction. *Science.* 2016;352(6285):595–599.
61. Zeng M, Chen X, Guan D, et al. Reconstituted postsynaptic density as a molecular platform for understanding synapse formation and plasticity. *Cell.* 2018;174(5):1172–118. [e1116].

62. Cluzel C, Saltel F, Lussi J, Paulhe F, Imhof BA, Wehrle-Haller B. The mechanisms and dynamics of (alpha)v(beta)3 integrin clustering in living cells. *J Cell Biol*. 2005;171 (2):383–392.

63. Case LB, Ditlev JA, Rosen MK. Regulation of transmembrane signaling by phase separation. *Annu Rev Biophys*. 2019;48:465–494.

64. Mulyasasmita W, Lee JS, Heilshorn SC. Molecular-level engineering of protein physical hydrogels for predictive sol-gel phase behavior. *Biomacromolecules*. 2011;12(10): 3406–3411.

65. McSwiggen DT, Mir M, Darzacq X, Tjian R. Evaluating phase separation in live cells: Diagnosis, caveats, and functional consequences. *Genes Dev*. 2019;33(23–24): 1619–1634.

66. Iakoucheva LM, Radivojac P, Brown CJ, et al. The importance of intrinsic disorder for protein phosphorylation. *Nucleic Acids Res*. 2004;32(3):1037–1049.

67. Gu J, Isaji T, Xu Q, et al. Potential roles of N-glycosylation in cell adhesion. *Glycoconj J*. 2012;29(8–9):599–607.

68. Roskoski Jr R. The ErbB/HER family of protein-tyrosine kinases and cancer. *Pharmacol Res*. 2014;79:34–74.

69. Narasumani M, Harrison PM. Discerning evolutionary trends in post-translational modification and the effect of intrinsic disorder: Analysis of methylation, acetylation and ubiquitination sites in human proteins. *PLoS Comput Biol*. 2018;14(8), e1006349.

70. Appadurai R, Uversky VN, Srivastava A. The structural and functional diversity of intrinsically disordered regions in transmembrane proteins. *J Membr Biol*. 2019;252 (4–5):273–292.

71. Ullrich A, Schlessinger J. Signal transduction by receptors with tyrosine kinase activity. *Cell*. 1990;61(2):203–212.

72. Phillips DR, Prasad KS, Manganello J, Bao M, Nannizzi-Alaimo L. Integrin tyrosine phosphorylation in platelet signaling. *Curr Opin Cell Biol*. 2001;13(5):546–554.

73. Cowan KJ, Law DA, Phillips DR. Identification of shc as the primary protein binding to the tyrosine-phosphorylated beta 3 subunit of alpha IIbeta 3 during outside-in integrin platelet signaling. *J Biol Chem*. 2000;275(46):36423–36429.

74. Wong EV, Schaefer AW, Landreth G, Lemmon V. Involvement of p90rsk in neurite outgrowth mediated by the cell adhesion molecule L1. *J Biol Chem*. 1996;271 (30):18217–18223.

75. Dickson TC, Mintz CD, Benson DL, Salton SR. Functional binding interaction identified between the axonal CAM L1 and members of the ERM family. *J Cell Biol*. 2002;157(7):1105–1112.

76. Kitajima Y, Aoyama Y, Seishima M. Transmembrane signaling for adhesive regulation of desmosomes and hemidesmosomes, and for cell-cell datachment induced by pemphigus IgG in cultured keratinocytes: Involvement of protein kinase C. *J Investig Dermatol Symp Proc*. 1999;4(2):137–144.

77. Dellambra E, Prislei S, Salvati AL, et al. Gene correction of integrin beta4-dependent pyloric atresia-junctional epidermolysis bullosa keratinocytes establishes a role for beta4 tyrosines 1422 and 1440 in hemidesmosome assembly. *J Biol Chem*. 2001;276 (44):41336–41342.

78. Matsuoka Y, Yamada T, Seishima M, Hirako Y, Owaribe K, Kitajima Y. Transient translocation of hemidesmosomal bullous pemphigoid antigen 1 from cytosol to membrane fractions by 12-O-tetradecanoylphorbol-13-acetate treatment and Ca2 + −switch in a human carcinoma cell line. *J Dermatol Sci*. 2001;27(3):206–214.

79. Chen CW, Huang LY, Liao CF, Chang KP, Chu YW. GasPhos: Protein phosphorylation site prediction using a new feature selection approach with a GA-aided ant Colony system. *Int J Mol Sci*. 2020;21(21):7891.

80. Linnemann D, Raz A, Bock E. Differential expression of cell adhesion molecules in variants of K1735 melanoma cells differing in metastatic capacity. *Int J Cancer.* 1989;43(4):709–712.

81. Miura M, Kobayashi M, Asou H, Uyemura K. Molecular cloning of cDNA encoding the rat neural cell adhesion molecule L1. Two L1 isoforms in the cytoplasmic region are produced by differential splicing. *FEBS Lett.* 1991;289(1):91–95.

82. Kamiguchi H, Long KE, Pendergast M, et al. The neural cell adhesion molecule L1 interacts with the AP-2 adaptor and is endocytosed via the clathrin-mediated pathway. *J Neurosci.* 1998;18(14):5311–5321.

83. Ohno H, Fournier MC, Poy G, Bonifacino JS. Structural determinants of interaction of tyrosine-based sorting signals with the adaptor medium chains. *J Biol Chem.* 1996;271 (46):29009–29015.

84. Long KE, Asou H, Snider MD, Lemmon V. The role of endocytosis in regulating L1-mediated adhesion. *J Biol Chem.* 2001;276(2):1285–1290.

85. Schaefer AW, Kamei Y, Kamiguchi H, et al. L1 endocytosis is controlled by a phosphorylation-dephosphorylation cycle stimulated by outside-in signaling by L1. *J Cell Biol.* 2002;157(7):1223–1232.

86. Hegyi H, Tompa P. Intrinsically disordered proteins display no preference for chaperone binding in vivo. *PLoS Comput Biol.* 2008;4(3): e1000017.

87. French DL, Coller BS, Usher S, et al. Prenatal diagnosis of Glanzmann thrombasthenia using the polymorphic markers BRCA1 and THRA1 on chromosome 17. *Br J Haematol.* 1998;102(2):582–587.

88. French DL, Seligsohn U. Platelet glycoprotein IIb/IIIa receptors and Glanzmann's thrombasthenia. *Arterioscler Thromb Vasc Biol.* 2000;20(3):607–610.

89. Liu J, Das M, Yang J, et al. Structural mechanism of integrin inactivation by filamin. *Nat Struct Mol Biol.* 2015;22(5):383–389.

90. Favier M, Bordet JC, Favier R, et al. Mutations of the integrin alphaIIb/beta3 intracytoplasmic salt bridge cause macrothrombocytopenia and enlarged platelet alpha-granules. *Am J Hematol.* 2018;93(2):195–204.

91. Fransen E, Schrander-Stumpel C, Vits L, Coucke P, Van Camp G, Willems PJ. X-linked hydrocephalus and MASA syndrome present in one family are due to a single missense mutation in exon 28 of the L1CAM gene. *Hum Mol Genet.* 1994;3(12): 2255–2256.

92. Vits L, Van Camp G, Coucke P, et al. MASA syndrome is due to mutations in the neural cell adhesion gene L1CAM. *Nat Genet.* 1994;7(3):408–413.

93. Fransen E, Van Camp G, Vits L, Willems PJ. L1-associated diseases: Clinical geneticists divide, molecular geneticists unite. *Hum Mol Genet.* 1997;6(10):1625–1632.

94. Weller S, Gartner J. Genetic and clinical aspects of X-linked hydrocephalus (L1 disease): Mutations in the L1CAM gene. *Hum Mutat.* 2001;18(1):1–12.

95. Yamasaki M, Thompson P, Lemmon V. CRASH syndrome: Mutations in L1CAM correlate with severity of the disease. *Neuropediatrics.* 1997;28(3):175–178.

96. Lermyte F. Roles, characteristics, and analysis of intrinsically disordered proteins: A Minireview. *Life (Basel).* 2020;10(12):320. https://doi.org/10.3390/life10120320.

97. Kiritsi D, Kern JS, Schumann H, Kohlhase J, Has C, Bruckner-Tuderman L. Molecular mechanisms of phenotypic variability in junctional epidermolysis bullosa. *J Med Genet.* 2011;48(7):450–457.

98. Vaisanen L, Has C, Franzke C, et al. Molecular mechanisms of junctional epidermolysis bullosa: Col 15 domain mutations decrease the thermal stability of collagen XVII. *J Invest Dermatol.* 2005;125(6):1112–1118.

99. Has C, South A, Uitto J. Molecular therapeutics in development for epidermolysis bullosa: Update 2020. *Mol Diagn Ther.* 2020;24(3):299–309.

100. Vacic V, Markwick PR, Oldfield CJ, et al. Disease-associated mutations disrupt functionally important regions of intrinsic protein disorder. *PLoS Comput Biol.* 2012;8(10): e1002709.
101. Sutto L, Gervasio FL. Effects of oncogenic mutations on the conformational free-energy landscape of EGFR kinase. *Proc Natl Acad Sci U S A.* 2013;110(26): 10616–10621.
102. Mitsudomi T, Yatabe Y. Epidermal growth factor receptor in relation to tumor development: EGFR gene and cancer. *FEBS J.* 2010;277(2):301–308.
103. Shan Y, Eastwood MP, Zhang X, et al. Oncogenic mutations counteract intrinsic disorder in the EGFR kinase and promote receptor dimerization. *Cell.* 2012;149 (4):860–870.
104. Forbes SA, Bindal N, Bamford S, et al. COSMIC: Mining complete cancer genomes in the catalogue of somatic mutations in cancer. *Nucleic Acids Res.* 2011;39(Database issue): D945–D950.
105. Lynch TJ, Bell DW, Sordella R, et al. Activating mutations in the epidermal growth factor receptor underlying responsiveness of non-small-cell lung cancer to gefitinib. *N Engl J Med.* 2004;350(21):2129–2139.
106. Keppel TR, Sarpong K, Murray EM, Monsey J, Zhu J, Bose R. Biophysical evidence for intrinsic disorder in the C-terminal tails of the epidermal growth factor receptor (EGFR) and HER3 receptor tyrosine kinases. *J Biol Chem.* 2017;292(2):597–610.
107. Singh I, Singh S, Verma V, Uversky VN, Chandra R. In silico evaluation of the resistance of the T790M variant of epidermal growth factor receptor kinase to cancer drug Erlotinib. *J Biomol Struct Dyn.* 2018;36(16):4209–4219.
108. Suda K, Onozato R, Yatabe Y, Mitsudomi T. EGFR T790M mutation: A double role in lung cancer cell survival? *J Thorac Oncol.* 2009;4(1):1–4.
109. Okamoto K, Sako Y. Single-molecule Forster resonance energy transfer measurement reveals the dynamic partially ordered structure of the epidermal growth factor receptor C-tail domain. *J Phys Chem B.* 2019;123(3):571–581.
110. Moeller A, Lee SC, Tao H, et al. Distinct conformational spectrum of homologous multidrug ABC transporters. *Structure.* 2015;23(3):450–460.
111. Bernado P, Svergun DI. Structural analysis of intrinsically disordered proteins by small-angle X-ray scattering. *Mol Biosyst.* 2012;8(1):151–167.
112. Jensen MR, Ruigrok RW, Blackledge M. Describing intrinsically disordered proteins at atomic resolution by NMR. *Curr Opin Struct Biol.* 2013;23(3):426–435.
113. Pryor Jr EE, Wiener MC. A critical evaluation of in silico methods for detection of membrane protein intrinsic disorder. *Biophys J.* 2014;106(8):1638–1649.
114. Meng F, Uversky VN, Kurgan L. Comprehensive review of methods for prediction of intrinsic disorder and its molecular functions. *Cell Mol Life Sci.* 2017;74(17):3069–3090.
115. Galea CA, High AA, Obenauer JC, et al. Large-scale analysis of thermostable, mammalian proteins provides insights into the intrinsically disordered proteome. *J Proteome Res.* 2009;8(1):211–226.
116. Galea CA, Pagala VR, Obenauer JC, Park CG, Slaughter CA, Kriwacki RW. Proteomic studies of the intrinsically unstructured mammalian proteome. *J Proteome Res.* 2006;5(10):2839–2848.
117. Qin J, Vinogradova O, Gronenborn AM. Protein-protein interactions probed by nuclear magnetic resonance spectroscopy. *Methods Enzymol.* 2001;339:377–389.
118. Clore GM, Gronenborn AM. The two-dimensional transferred nuclear Overhauser effect. *J Magn Reson.* 1982;48:402–417.
119. Marsh JA, Forman-Kay JD. Structure and disorder in an unfolded state under non-denaturing conditions from ensemble models consistent with a large number of experimental restraints. *J Mol Biol.* 2009;391(2):359–374.

120. Mittag T, Marsh J, Grishaev A, et al. Structure/function implications in a dynamic complex of the intrinsically disordered Sic1 with the Cdc4 subunit of an SCF ubiquitin ligase. *Structure*. 2010;18(4):494–506.

121. Murthy AC, Fawzi NL. The (un)structural biology of biomolecular liquid-liquid phase separation using NMR spectroscopy. *J Biol Chem*. 2020;295(8):2375–2384.

122. Freedberg DI, Selenko P. Live cell NMR. *Annu Rev Biophys*. 2014;43:171–192.

123. Rambo RP, Tainer JA. Characterizing flexible and intrinsically unstructured biological macromolecules by SAS using the Porod-Debye law. *Biopolymers*. 2011;95(8):559–571.

124. Brosey CA, Tainer JA. Evolving SAXS versatility: Solution X-ray scattering for macromolecular architecture, functional landscapes, and integrative structural biology. *Curr Opin Struct Biol*. 2019;58:197–213.

125. Svergun DI. Restoring low resolution structure of biological macromolecules from solution scattering using simulated annealing. *Biophys J*. 1999;76(6):2879–2886.

126. Svergun DI, Petoukhov MV, Koch MH. Determination of domain structure of proteins from X-ray solution scattering. *Biophys J*. 2001;80(6):2946–2953.

127. Grant TD. Ab initio electron density determination directly from solution scattering data. *Nat Methods*. 2018;15(3):191–193.

128. West GM, Chien EY, Katritch V, et al. Ligand-dependent perturbation of the conformational ensemble for the GPCR beta2 adrenergic receptor revealed by HDX. *Structure*. 2011;19(10):1424–1432.

129. Majumdar A, Mukhopadhyay S. Fluorescence depolarization kinetics to study the conformational preference, structural plasticity, binding, and assembly of intrinsically disordered proteins. *Methods Enzymol*. 2018;611:347–381.

130. Mukhopadhyay S. The dynamism of intrinsically disordered proteins: Binding-induced folding, amyloid formation, and phase separation. *J Phys Chem B*. 2020;124(51): 11541–11560.

131. Clister T, Mehta S, Zhang J. Single-cell analysis of G-protein signal transduction. *J Biol Chem*. 2015;290(11):6681–6688.

132. LeBlanc SJ, Kulkarni P, Weninger KR. Single molecule FRET: A powerful tool to study intrinsically disordered proteins. *Biomolecules*. 2018;8(4):140. https://doi.org/10.3390/biom8040140.

133. Gomes G-NW, Krzeminski M, Namini A, et al. Integrating multiple experimental data to determine conformational ensembles of an intrinsically disordered protein. *bioRxiv*. 2020. https://doi.org/10.1101/2020.02.05.935890.

CHAPTER FIVE

Molecular simulations of IDPs: From ensemble generation to IDP interactions leading to disorder-to-order transitions

Hebah Fatafta[a], Suman Samantray[a,b], Abdallah Sayyed-Ahmad[c], Orkid Coskuner-Weber[d], and Birgit Strodel[a,e,*]

[a]Institute of Biological Information Processing (IBI-7: Structural Biochemistry), Forschungszentrum Jülich, Jülich, Germany
[b]AICES Graduate School, RWTH Aachen University, Aachen, Germany
[c]Department of Physics, Birzeit University, Birzeit, Palestine
[d]Molecular Biotechnology, Turkish–German University, Sahinkaya Caddesi, Istanbul, Turkey
[e]Institute of Theoretical and Computational Chemistry, Heinrich Heine University Düsseldorf, Düsseldorf, Germany
*Corresponding author: e-mail address: b.strodel@fz-juelich.de

Contents

Abstract

Intrinsically disordered proteins (IDPs) lack a well-defined three-dimensional structure but do exhibit some dynamical and structural ordering. The structural plasticity of IDPs indicates that entropy-driven motions are crucial for their function. Many IDPs undergo function-related disorder-to-order transitions upon by their interaction with specific binding partners. Approaches that are based on both experimental and theoretical tools enable the biophysical characterization of IDPs. Molecular simulations provide insights into IDP structural ensembles and disorder-to-order transition mechanisms. However, such studies depend strongly on the chosen force field parameters and simulation techniques. In this chapter, we provide an overview of IDP characteristics, review all-atom force fields recently developed for IDPs, and present molecular dynamics-based simulation methods that allow IDP ensemble generation as well as the characterization of disorder-to-order transitions. In particular, we introduce metadynamics, replica exchange molecular dynamics simulations, and also kinetic models resulting from Markov State modeling, and provide various examples for the successful application of these simulation methods to IDPs.

1. Introduction to IDPs

1.1 Definition of IDPs

Proteins are central key components in catalyzing the chemical reactions in living systems. Before the turn of the century, some protein characteristics have led us to think that proteins function only when they are folded into their right structures.[1] The central dogma of molecular biology states that the DNA encodes the genetic information which is transcribed into messenger RNA and then translated into an amino acid sequence that folds into a protein. The biophysical mechanisms which govern how an amino acid sequence folds into the correct three-dimensional protein structure are still not fully understood.[2] The widely accepted protein–structure paradigm dominated scientific minds for more than 100 years. However, despite this paradigm, intrinsically dynamic and flexible and biologically active proteins were detected during the modern history of protein science because the genomic era that began at the end of the 20th century gave scientists access to complete genome sequences.[3] Scientists noted that some of the amino acid sequences were not expected to fold into globular protein structures.[4] In parallel, experimental investigations began to uncover samples of crucial proteins and domains that were incompletely structured or completely disordered in solution, yet remained biologically functional.[3] As a consequence, the concept "intrinsically disordered proteins" was introduced at the turn of the century.[4] The crucial hypothesis was formulated and it stated that intrinsically disordered proteins (IDPs), instead of being rare exceptions, represent a unique and a very broad class of proteins.[4] The currently accepted meaning of disorder embraces protein regions or proteins that are biologically active, but are dynamically flexible conformational ensembles either at the secondary and/or tertiary structure level.[5] IDPs and intrinsically disordered protein regions (IDPRs) exist as dynamical ensembles resembling "protein clouds" whereby the atom positions and backbone Ramachandran angles vary significantly with time, without obtaining specific equilibrium values.[6] A hallmark of IDPs is a marked bias in their amino acid composition, which includes a relatively small proportion of hydrophobic and aromatic residues, but a relatively large proportion of polar and charged amino acid residues.[7] The numbers of identified IDPs and IDPRs are ever increasing. All IDPs and IDPRs identified to date are collected in the DisProt database (https://disprot.org/), which to date contains ≈1600 nonambiguous IDPs and ≈3500 IDPRs.[8]

1.2 Functions of IDPs

The abundance and functional significance of IDPs in eukaryotes is known.[9] In the mid-1990s, experimental measurements of regulatory proteins and bioinformatics studies of the genome sequences that were just emerging revealed that disordered regions are very common in eukaryotic proteins.[10] The occurrence of IDPRs of significant size that contain more than 50 amino acid residues is common in functional proteins.[11] The existence of functional IDPs, such as polypeptide hormones, has been noted for many years and IDPs were detected in intact cells in nuclear magnetic resonance (NMR) experiments.[12] Functions of IDPs include transcription and translation regulation, cellular signal transduction, the storage of small molecules, protein phosphorylation, and self-assembly regulation, such as ribosome and bacterial flagellum.[13] IDPRs can function as chaperones for RNA molecules and for other proteins, indicating that IDPRs exert their function by binding to (mis)folded proteins or RNA molecules, thereby providing a function related to recognition elements and/or unfolding and loosening of kinetically trapped folding intermediates.[14] In Fig. 1 the different functions of IDPs and IDPRs are summarized.

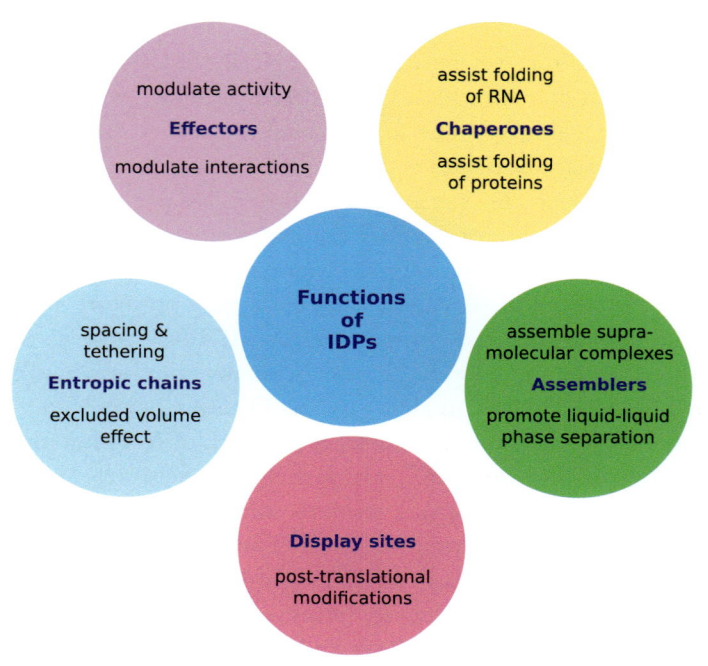

Fig. 1 Overview of the different functions of IDPs.

1.2.1 Folding upon binding

One of the main functions of IDPs/IDPRs is to facilitate the binding with other partners, such as other proteins, DNA, RNA, or small molecules.[6] In fact, IDPs/IDPRs are promiscuous binders which interact invariably with various partners through varying binding scenarios. They are capable of forming static, semistatic, fuzzy, and dynamic complexes.[15] IDPs/IDPRs are capable of undergoing transitions to more ordered states upon binding with their targets, that is, they undergo a coupled binding-and-folding processes. The degree of folding and resulting structures can vary depending on the binding partner.[6] In fact, the binding modes that IDPs can acquire are highly diverse, even for an individual IDP. Such binding plasticity can also include dynamic complexes in which IDPs rapidly switch between different binding modes that are mostly devoid of a folded structure.[16] In these binding scenarios, protein functionality is proposed to commonly originate from disorder-to-order transitions.[17]

Based on bioinformatics analysis, Uversky and coworkers introduced the idea that certain protein structural elements mediate the binding events of IDPRs. They called these elements "molecular recognition elements" (MoREs) or "molecular recognition features" (MoRFs), which are usually short regions that undergo coupled binding and folding within a longer region of disorder.[18–20] Thus, MoRFs mediate both molecular recognition and binding with simultaneous folding. Uversky et al. established a MoRFs database that distinguishes three basic types of MoRFs: α-MoRFs, β-MoRFs, and ι-MoRFs, which form α-helices, β-strands, and irregular secondary structure upon binding, respectively.[19] They further showed that MoRFs and their binding partners feature a significantly different residue composition and geometric and physicochemical properties as compared to the binding interfaces found in homodimers, heterodimers, and antigen–antibody complexes.[21] Moreover, MoRFs are found across the three domains of life, i.e., Eukaryota, Bacteria, and Archaea, with a similar abundance and amino acid composition, yet with an elevated level of disorder in Eukaryota.[22]

An example for a disorder-to-order transition is provided by the cAMP-regulated transcription factor CREB, which contains a kinase-inducible transcriptional-activation domain (KID) that is intrinsically disordered, but folds into a pair of orthogonal helices upon binding to the target domain of its interaction partner, the CREB-binding protein (CBP)[23] (Fig. 2). Coupled binding and folding may involve few amino acid residues like in the mentioned domain of CREB, or they can involve entire protein domains. For instance, the N-terminal region of the DNA-fragmentation factor, a 45 kDa subunit with 116 amino acid residues, is disordered in solution; however, upon forming a heterodimeric complex with a 40 kDa subunit of the DNA-fragmentation factor, it folds into an ordered globular structure.[24] Another examples involves the transcriptional coactivators PARALOGUE p300 and CBP, which modify both chromatin and transcription factor via acetyltransferase activity and function as scaffolds for the assembly and recruitment of transcriptional processes.[25] About 2442 amino acid residues of the CBP (more than half of the sequence) are intrinsically disordered. Furthermore, the binding of IDPs with their targets is usually regulated by covalent modifications that leads to biological switches, and this example can be detected in the CBP/p300 system.[25] Such a process requires binding to more than just one target, such as binding to a modifying enzyme and to a receptor. Due to differences in conformational requirements for binding to different targets, this binding may be facilitated by the presence of disordered structures in the protein.

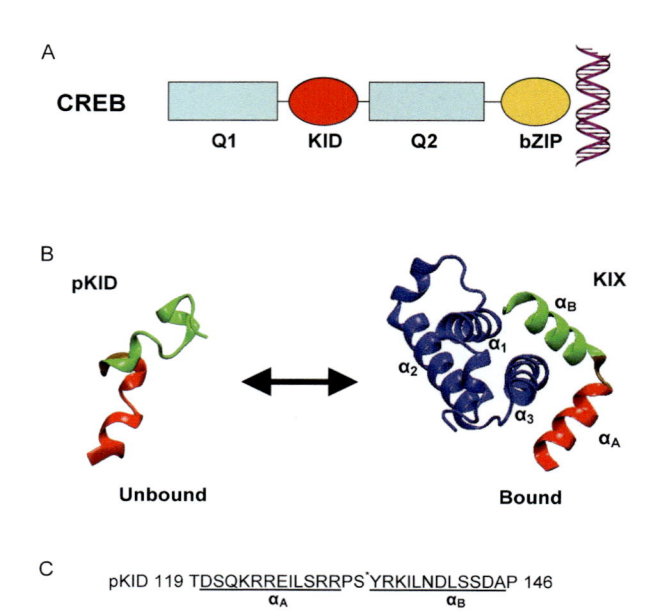

Fig. 2 Disorder-to-order transition of CREB. (A) Domain structure of CREB, consisting of two glutamine-rich domains Q1 and Q2, which are separated by the intrinsically disordered KID. On the C-terminal side the DNA-binding domain basic region/leucine zipper (bZIP) is located. (B) Phosphorylated KID (pKID) folds into a pair of orthogonal helices (*red*: helix αA; *green*: helix αB) upon binding to CBP (*blue*), while pKID is disordered in its unbound state. (C) The sequence of the structured part of pKID is shown with residues in helices αA and αB underlined. *Reproduced with permission Turjanski A, Gutkind JS, Best R, Hummer G. Binding-induced folding of a natively unstructured transcription factor. PLoS Comput. Biol. 2008; 4:e1000060. Copyright 2021, PLOS.*

1.2.2 Signaling and regulation processes

IDPs possess physical characteristics that give rise to an exquisite level of cellular signaling processes control.[10] These characteristics are their flexibility that allows IDPs to interact with different targets under varying physiological conditions, their accessible sites for posttranslational modifications, their capability of conserved sequence motif utilization for mediating binding interactions, the presence of small recognition elements that fold upon binding to a partner, and their ability to bind to partners with high specificity but moderate affinity, which leads to spontaneous and fast dissociation and thus termination of a signal.[10] They also possess kinetic advantages in signaling because of their fast association rates, which enable signals to be rapidly turned on.[26] During signaling functions, IDPs bind transiently to various partners within the dynamic regulatory networks, which respond quantitatively and precisely to cellular signals and provide complex information processing.[10] Such molecular interactions are dynamic and transient, meaning IDPs are capable of exchanging binding partners, and they compete for binding to central hub proteins that are present even in small amounts.[10] An important aspect of central hub IDPs is that they can associate with many partners, often involving different IDPRs to do so.[27–29] The interactions can be further modified by posttranslational modifications, which enables IDPs to function as switches and rheostats.[10]

Not all IDPs undergo folding transitions for performing their biological functions. Some IDPs and IDPRs appear to function as highly dynamic and unstructured linkers between disordered or globular interaction domains, while other IDPs remain disordered after binding with their targets and form "fuzzy" complexes.[6] Such flexible binding interactions can have biological functions, such as enhancing target binding affinities, modulating allosteric interactions, and mediating pathway crosstalk through ternary complex formations with other binding partners.[6] It was shown that IDPS and IDPRs are accessible for posttranslational modifications, estimating that there may be one million instances of peptide interaction motifs within IDPRs of the human proteome when we take posttranslational modifications into account.[30] This remarkable number highlights the key role of IDPs in cellular signaling and regulation, and helps to gain insights into their functional diversity. Modifications of IDPs by varying methylases, acetylases, kinases, or other enzymes can yield different signaling outputs, which increases the complexity of signaling pathways.[30]

1.2.3 IDPs in evolution

In contrast to structured proteins, IDPs and IDPRs present an overall increased rate of evolution. The sequence conservation modes of disordered regions provide the possibility that analysis of the similarity and difference determinations along their lines of function will provide a better knowledge of the common features of IDPs and IDPRs in biological processes.[31] Indeed, disregarding evolutionary relationships would separate molecular function from its developmental constraints. While there are numerous IDPs characterized in the animal kingdom and human diseases (see below), the studies of IDPs in plants or still in its infancy. Nonetheless, some studies identified IDPs in plants as well. For instance, late embryogenesis abundant (LEA) proteins were first identified in cotton seeds and represent one of the most famous examples of IDP-mediated stress responses in plants.[32] These proteins were shown to play key roles in plant responses to salinity, freezing, heat, drought, and desiccation.[32] Experimental and computational studies suggested that a key characteristic of LEA family proteins is their partial or total lack of stable structure, which equips them with high flexibility.[33] Some bacterial or plant LEA proteins also have folded domains which are involved in the desiccation response and may originate from *Archaea*.[34] LEA proteins have diverse functionalities, ranging from being chaperons for proteins and membranes to being sequesters of radicals and metal ions. Besides plant organisms, LEA proteins were also detected in other desiccation-tolerant organisms including bacteria, fungi, and invertebrates, which suggests a common mechanism across different life forms.[35] Even though LEA proteins do not exist in mammalian genomes, AfrLEA2 transfected into human HepG2 cells enhanced desiccation tolerance in the vicinity of intracellular trehalose, and, after rehydration, gave rise to increased membrane integrity.[36]

1.2.4 Phase separation of IDPs

Proteins that possess low-complexity and/or prion-like sequences are capable of promoting phase separation for forming membrane-less organelles within the nucleoplasm or cytoplasm, proposing that they are capable of contributing in a regulated manner to their compartmentalization.[37] Almost all membrane-less compartments contain a large amount of IDPs and their characteristics were proposed to be essential for liquid–liquid phase separation, and thus membrane-less compartment formation.[38] The high flexibility provided by IDPs may provide the needs for the dynamic behavior of membrane-less compartments. Furthermore, IDP-related protein-protein interactions may permit the reversible and spontaneous formation of sufficient, local protein concentrations to initiate liquid–liquid phase separation, which enables membrane-less compartments to rapidly form on demand, fuse, shear, exchange their content and disassemble, and therefore to concentrate proteins and biochemical reactions at specific location when required.[38] These protein-protein interactions are proposed to be mediated by repetitive sequence elements, resulting in overall IDP sequence simplicity. A known example for a membrane-less compartment-forming IDP is the essential pyrenoid component 1 (EPYC1), which is an indispensable part of the pyrenoid, a membrane-less compartment in the chloroplast of many algae which concentrates carbon fixation machinery components to increase its efficiency.[39]

1.3 IDPs in diseases

When misexpressed, mismodified, misprocessed, and/or dysregulated, IDPs and IDPRs are prone to engage in promiscuous and unwanted interactions and thus are associated with the development of various diseases.[40] Conformational or protein-folding diseases are divided into two classes. The first class includes errors in the genetic blueprint which leads to misfolded proteins that affect the function as well.[41] Examples of this first class include p53 and specific alterations in diseases such as cystic fibrosis and sickle cell anemia. The second class is through the formation of multi-molecular structures or plaques with the property of modifying normal cell function.[41] Such alterations, known as amyloidosis, are detected in severe diseases, such as Alzheimer's disease, Parkinson's disease, Creutzfeld-Jakob disease, and type II diabetes.

1.3.1 IDPs in cancer

Examples of IDPs in cancer are A-fetoprotein (AFP), p53, and BRCA-1. AFP is an oncofetal glycoprotein which has 590 residues and is produced by fetal liver and yolk sac cells.[42] Usually, AFP levels correlate with fetal maturity and these are very high during embryo development, while decreasing right after birth.[43] An increase in AFP level in adult sera is an evidence of various pathologies. Therefore, AFP is a biomarker that indicates cancer or fetal abnormality development. Cancer researchers studied the interaction between p53 and Mdm2 extensively due to the central role of p53 in tumor suppression through apoptosis.[44] Tumor protein p53, which is also called "the Guardian of the Genome," is a transcription factor which targets genes involved in the regulation of cell cycle and apoptosis.[45] Therefore, a loss of p53 function is a significant factor in the development of cancer. The differential response of breast cancer type 1 susceptibility protein (BRCA-1) to various types of DNA damage is an example of an IDP that is capable of signal-flow modifications.[46] This IDP is involved in various signaling processes, such as DNA damage response, cell-cycle checkpoint control, transcription, oncogenesis, apoptosis, tumor suppression, and stress response. Concomitant with its central role in DNA damage response, BRCA-1 is at the center of a variety of cancers.[47] Furthermore, prostate-associated gene 4 (PAGE4) is an IDP implicated in prostate cancer.[48] Specifically, PAGE 4 is a remarkably prostate-specific cancer/testis antigen which is highly upregulated in the human fetal prostate and its disease states, but not in the adult normal gland, and it functions as a stress–response protein to suppress reactive oxygen species as well as prevent DNA damage.[48] According to a disorder propensity analysis by bioinformatics tools, PAGE4, although an IDP, possesses several regions with an increased tendency to order (most significantly residues 13–19 and 86–92, and to a lesser degree 49–61). NMR measurements also showed that PAGE4 has metastable secondary structure elements.[12]

1.3.2 IDPs in amyloid diseases

Amyloid-β (Aβ) and α-synuclein (αS) are examples of two IDPs that are at the center of neurodegenerative diseases, such as Alzheimer's and Parkinson's diseases.[49,50] Aβ represents peptides of 36 to 43 amino acid residues, and it is the main component of the amyloid plaques that are detected in the brains of Alzheimer's disease patients. These IDPs derive from the amyloid precursor protein (APP) via cleavage by both β- and γ-secretases. Aβ peptides can oligomerize to form flexible soluble oligomers which may exist in several forms.[50] It has been proposed that specifically misfolded Aβ oligomers can induce other Aβ peptides to also take the misfolded form in oligomerization, which leads to a chain reaction as in a prion infection.[49] These oligomers are toxic to nerve cells and misfolded Aβ can induce misfolding in tau protein in Alzheimer's disease.[51] The neuronal protein αS is an IDP that is encoded by the *SNCA* gene in humans and it has 140 amino acid residues.[49] Although the function of αS is not fully understood, studies proposed that it plays a role in restricting the mobility of synaptic vesicles, consequently attenuating synaptic vesicle recycling and neurotransmitter release.[49] However, αS is prone to aggregation, forming insoluble fibrils in pathological conditions, such as Parkinson's disease, dementia with Lewy bodies, or multiple system atrophy. The pathological deposition of the misfolded prion protein (PrP) into its aggregated form causes prion diseases, which are called transmissible spongiform encephalopathies (TSEs).[52] TSEs range from chronic wasting disease of mule deer and elk to Creutzfeld–Jakob disease in humans.[52] About 100 amino acid residues in the N-terminal region of PrP are unstructured, while the C-terminal region is folded into an α-helical structure and is stabilized by a single disulfide bond.[53]

Hirudin and thrombin are example proteins with IDPRs that are at the center of cardiovascular diseases. Native hirudin contains an N-terminal region that is stabilized by three disulfide bonds and a C-terminal region, which is highly disordered (residues 50 to 65).[54] Hirudin is a thrombin-specific inhibitor with 65 amino acid residues, an anticoagulant protein which occurs in the salivary glands of the medial leech *Hirudo medicinalis*.[55] Thrombin is a blood protein that is involved in clotting and coagulation.[56] As a serine protease, thrombin cleaves bonds after Arg and Lys and converts fibrinogen into fibrin.[57] In a complex with thrombomodulin, thrombin activates protein C that inhibits coagulation.[55] Thrombin has 295 residues and is produced from prothrombin that has 579 residues. Thrombin has a light chain that comprises 36 amino acid residues and a heavy chain with 259 amino acid residues. The light chain of thrombin is disordered, while the heavy chain is an ordered protein region.[58] As a final example, amylin, that is an IDP of 37 amino acid residues and central to type II diabetes, shall be mentioned.[59] In addition to insulin, pancreatic β-cells produce the peptide amylin, which is also called islet amyloid polypeptide (IAPP). Amylin possesses several functions linked to the regulation of energy metabolism. One hallmark of type II diabetes is the deposition of amyloid fibrils in the islets of Langerhans. Amylin is the major component of these deposits. Amylin is an IDP of 37 amino acid residues.[59]

1.4 Biophysical characterization of IDPs

To address the challenge of understanding the structure and dynamics of IDPs/IDPRs, several biophysical characterization techniques have been utilized. These include solid and solution NMR spectroscopy, X-ray fiber diffraction and X-ray crystallography, Fourier transform infrared (FTIR) spectroscopy, X-ray absorption spectroscopy (XAS), circular dichroism (CD), thioflavin-S and thioflavin-T fluorescence, Congo red binding and bire-fringence, single-molecule fluorescence resonance energy transfer (smFRET) spectroscopy, scanning tunneling microscopy (STM), transmission electron microscopy (TEM), scanning transmission microscopy (STEM), electron spin resonance microscopy (ESR), and atomic force microscopy (AFM). In the following, we will survey the usage of some of these techniques in characterizing IDPs, with a focus on Aβ and αS as representative examples.

1.4.1 Solution NMR spectroscopy

Aβ and αS structures are modulated by solvent and crowding effects.[60] They undergo fast dynamic conformational changes in monomeric and oligomeric states and are prone to rapid aggregation processes. Insights into these structures have been provided by ^{1}H, ^{13}C, and ^{15}N NMR experiments.[61] In particular, changes in chemical shift values, cross peaks, line widths, and the nuclear Overhauser effect (NOE) are used in characterizing the monomer-to-oligomer transition. However, due to the heterogeneous nature of $A\beta$ and αS and the lack of sufficiently high concentrations of these IDPs, it is extremely difficult to fully study the structures of monomeric and oligomeric $A\beta$ and αS by these experiments. Nonetheless, monomeric and oligomeric $A\beta$ and αS structures have been presented in the literature using solution NMR experiments.[62,63]

Importantly, NMR techniques are constantly evolving to improve the characterization of IDPs. For instance, Blackledge and coworkers characterized the long-range order and local features in IDPs by a combination of paramagnetic relaxation enhancements and residual dipolar couplings.[64] In addition, Kaderavek and coworkers introduced a 3D-HNCO-based two-field NMR tool for protein backbone amide longitudinal relaxation rate measurements. Using this new tool, they overcame the limitations of high-resolution relaxometry and two-field NMR measurement in getting high-resolution structures of IDPs. They studied the C-terminal domain of RNA polymerase from *Bacillus subtilis*.[65] Interestingly, computer simulations using various force field parameters have taken chemical shift values from solution NMR studies as reference data sets in evaluating the accuracy of available force field parameters for IDPs.[66]

1.4.2 Solid-state NMR spectroscoply

Solution NMR is a key tool in the studies of IDPs/IDPRs; however, IDPs/IDPRs are not limited to solution and are also found in nonsoluble systems, such as fibrils and membrane proteins. Solid-state NMR measurements are conducted for investigating nonsoluble proteins.[67] The absence of isotropic tumbling in the solid state implies that dipole-dipole couplings and chemical shift anisotropies are not averaged, resulting in broadening of the line width, and hence lower resolutions. However, cross polarization (CP), high power proton decoupling, and magic angle spinning (MAS) help to obtain high-resolution structures.[68] Solid-state NMR investigations have been conducted on truncated and full-length IDPs. For example, Baldus and coworkers studied the interactions of disordered regions of microtubule-associated proteins with the dynamic microtubule surface by solid-state NMR measurements.[69] Zweckstetter and coworkers studied the tau structures in Alzheimer's disease by combining solid-state NMR measurements with cryogenic electron microscopy (cryo-EM).[70] They showed that the sequence KVAVVRT, which is the most hydrophobic patch within the proline-rich region, loses its flexibility upon formation of amyloid fibrils.

Heise and coworkers developed a novel approach to study the conformational ensemble of IDPs by applying dynamic nuclear polarization-enhanced solid-state NMR spectroscopy of sparsely isotope-labeled αS in frozen solution to obtain snapshots of the structural ensemble by exploiting the inhomogeneously broadened line shapes.[71]

1.4.3 ESR spectroscopy

ESR experiments measure the energy levels of unpaired electrons in an externally applied magnetic field. ESR detects the spin signals of electrons while NMR measures the spin signals of atomic nuclei.[68] Starker sensitive measurements can be expected from ESR rather than NMR spectroscopy because the energy splitting between the electron spins is larger than that of nuclear spins in a given magnetic field. This makes the background noise very low. However, unpaired electrons usually do not occur in biological samples. Therefore, extrinsic probes called spin labels have to be introduced into the protein. Such spin labels are usually nitroxide derivatives with a stable unpaired electron and a functional group allowing attachment to the protein. This is often achieved via covalent linking to a cysteine residue, which, if necessary, can be introduced into the protein by mutation. ESR measurements have been conducted for studying IDPs. Irie and coworkers reported the mechanism of Met35 oxidation by a phenoxy radical at Tyr10 position and measured the distance between Tyr10 and Met35 in Aβ by site-directed spin labeling ESR spectroscopy and proposed that Aβ42 is more toxic than Aβ40.[72] Moreover, Eliezer and coworkers demonstrated the usefulness of pulse ESR spectroscopy in the measurements of intramolecular distances in αS bound to detergent and lysophospholipid micelles.[73] Additionally, Marsh and coworkers characterized the association of αS with lipid membranes by ESR measurements.[74]

1.4.4 X-ray crystallography and absorption spectroscopy

A useful tool for the atomic-level characterization of structures is the usage of diffraction patterns of X-ray radiation. Molecules most suitable for such analysis usually exist in a highly ordered state as present in crystals. Unfortunately, these measurements are not applicable to IDPs as they cannot be crystallized due to their intrinsic flexibility.[68] X-ray fiber diffraction is alternatively used on pulverized and lower quality crystalline species to extract more information about the structure of IDPs, such as Aβ or αS.[68] Additionally, X-ray absorption spectra provide knowledge about the local environment and electronic state around heavy atoms. A specific quantum of energy is absorbed and electrons are excited and ejected from their orbitals when a high-energy X-ray beam hits the species. In XAS spectra, the absorption coefficients of the heavy atoms are plotted against the incident X-ray energy. These spectra are usually classified into two regions: X-ray absorption near edge structures (XANES) and extended X-ray absorption fine structure (EXAFS).[68] Such measurements were conducted in the studies of oxidative stress and transition metal ions with $A\beta$ or αS.[75,76]

1.4.5 CD spectroscopy

CD measurements have been extensively applied to IDPs such as $A\alpha$ or αS since they allow to determine the secondary structure properties as well as conformational transitions under a variety of conditions.[68] For instance, CD measurements were useful in identifying PAGE4 as an IDP since the CD spectra showed that the PAGE4 polypeptide chain contains no significant α-helical or β-strand secondary structural elements over a temperature range from 278 to 298 K, as evidenced by low ellipticity values in the 215–230 nm region.[77]

1.4.6 Light spectroscopic techniques

Identification of aggregated IDPs (such as $A\beta$ or αS) in tissue samples is usually conducted using Congo red staining. Congo red binding with amyloid results in a unique blue-green birefringence under cross-polarized light and enables therewith the visual characterization of amyloid formation.[68] Furthermore, FRET experiments are used to measure the energy transfer between a donor and an acceptor when they are in resonance state. However, its efficiency depends strongly on the distance between the fluorophores. Such measurements have been used in the studies for determining inter- and intramolecular distances, IDP misfolding, and disordered protein-membrane interactions.[68] For instance, FRET was used along with molecular dynamics (MD) simulations in the studies of monomeric $A\beta$ structures.[78]

1.4.7 Microscopy techniques

Different microscopy techniques such as TEM, STEM, and AFM are used to gain structural information on IDPs from images produced.[68] In TEM, negative staining, rotary shadowing, and cryo–electron microscopy are used to exchange the image and contrast. Conversely, STEM utilizes a field emission gun that delivers a sub-nanometer beam of 100 kV electrons on species.

The image of the species is produced as the focused beam moves step by step over the specimen as have been shown in characterizing homogeneity and structures of Aβ or αS.[68] Finally, AFM measurements were used in IDP assembly studies to investigate aggregate heterogeneity and biophysical properties.[68,79]

1.4.8 Small-angle X-ray scattering

Small-angle X-ray scattering (SAXS) is a small-angle scattering technique which is sensitive to nanoscale differences in the electron density of a sample. When X-rays travel through the material, their elastic scattering at small angles is recorded. This allows to determine the size, shape, and oligomeric state of biological macromolecules on a size scale of one to several hundred nanometers. The method can be applied to the samples in aqueous solution without any label or crystallization being required, which is an important advantage over other methods. For proteins, SAXS can resolve the structure including transition state structures and conformational changes, and resolve protein complex or aggregate formation, as well as ligand binding. It allows to identify flexible parts within a protein, which makes this method attractive for IDPs. For instance, SAXS has been applied in the studies of Aβ structural changes induced by transition metals, such as divalent copper ion.[80] Moreover, SAXS imaging can have the potential to image Aβ plaques in vivo in the brain without tracers for assessment of Alzheimer's disease.[81]

2. Force fields for IDPs

Biomolecular force fields (FFs) are represented by an energy function that provide models for the potential energy surface. Proteins and other biomolecules are predominantly simulated using two ways: using either an atomistic or a coarse-grained FF. Many FFs have been tailored for simulating the structural dynamics of folded proteins, and for a long time it was assumed that they are equally applicable to IDPs. However, as several FF benchmarks revealed, this is not the case.[82–89] It turned out to be difficult to accurately simulate the structural ensembles of IDPs or IDPRs, as they do not fold into a well-defined three-dimensional structure under physiological conditions and instead populate a dynamic conformational ensemble of rapidly interconverting structures. It was found that a good balance between protein-protein and protein-water interaction parameters is key for a good description of IDP ensembles. Small changes to this subtle balance will either produce overly compact IDP structures where protein-protein interactions are overestimated, excessively aggregation-prone proteins due to under-estimated protein-water interactions, or, if the latter are overestimated, extremely soluble IDP states that avoid protein-protein contacts. As accurate FFs are needed for the reliable generation of IDP ensembles, numerous FFs have been developed in recent times. They are usually based on existing FFs and different strategies were applied during their reparameterization, which will be explained in the following.

2.1 Definition of Force Fields

A model for the potential energy of a molecular system is composed of mathematical functions and associated constant parameters. The interactions between the particles of the system can be classified into two categories: bonded and nonbonded interactions. As an example, we provide the potential energy function which is used by the various CHARMM FFs[90]:

$$
\begin{aligned}
U(\mathbf{q}) = & \sum_{bonds} k_b (b - b_0)^2 + \sum_{angles} K_\theta (\theta - \theta_0)^2 \\
& + \sum_{dihedrals} V_\varphi (1 + \cos(n\varphi - \delta)) + \sum_{impropers} k_\omega (\omega - \omega_0)^2 \\
& + \sum_{i,j\ (LJ)} \epsilon_{ij}^{min} \left[\left(\frac{R_{min,ij}}{r_{ij}} \right)^{12} - 2 \left(\frac{R_{min,ij}}{r_{ij}} \right)^6 \right] + \sum_{i,j\ (Coul.)} \frac{q_i q_j}{4\pi\epsilon_0 r_{ij}} \\
& + \sum_{Urey-Bradley} k_{UB} (s - s_0)^2 + \sum_{CMAP} U_{CMAP}(\phi, \psi)
\end{aligned}
\tag{1}
$$

Here, $\mathbf{q}$ denotes the conformation of the system consisting of N atoms with coordinates $\mathbf{q} = (q_{1,x}, q_{1,y}, q_{1,z}, q_{2,x}, q_{2,y}, q_{2,z}, \ldots, q_{N,x}, q_{N,y}, q_{N,z})$. The bonded energy terms describe bond stretching around the equilibrium values b_0 with force constants k_b, angle bending around equilibrium angles θ_0 and with force constants k_θ, torsions around bonds as characterized by the dihedral angles φ, periodicity n, shift δ, and energy barrier V_φ, and out-of-plane bending, also called improper torsion, with the minimum at ω_0 and force constant k_ω. The nonbonded interactions contain Lennard-Jones and Coulomb potentials for interacting particles i and j. The Lennard-Jones (LJ) potential is a 12–6 potential, where the repulsive $1/r^{12}$ term describes the Pauli repulsion at short distances of the interacting particles due to overlapping electron orbitals, and the attractive $1/r^6$ term describes attractions arising from dispersion forces, which are also called van der Waals (vdW) interactions. The distance between the two interacting particles is given by r_{ij}, $\epsilon_{ij}^{\min}$ is the depth of the potential well, $R_{\min,ij}$ is the distance at which the particle-particle LJ potential energy is minimal and can be calculated from the van der Waals radii of the particles i and j. The Coulomb potential models the electrostatic interactions between the partial charges q_i and q_j of atoms i and j with distance r_{ij} between them, where ϵ_0 is the vacuum permittivity. The terms described thus far, which are summarized in Fig. 3, are common to all all-atom biomolecular force fields as found in the AMBER,[91] CHARMM,[90] OPLS-AA,[92] and GROMOS[93] packages. Though, depending on the force field, small differences to Eq. (1) can occur, such as that $\cos(\theta)$ is used for defining the harmonic potential describing angle bending.

In the CHARMM FFs, two correction terms are added to the potential energy. The Urey-Bradley (UB) term is used to improve the description of

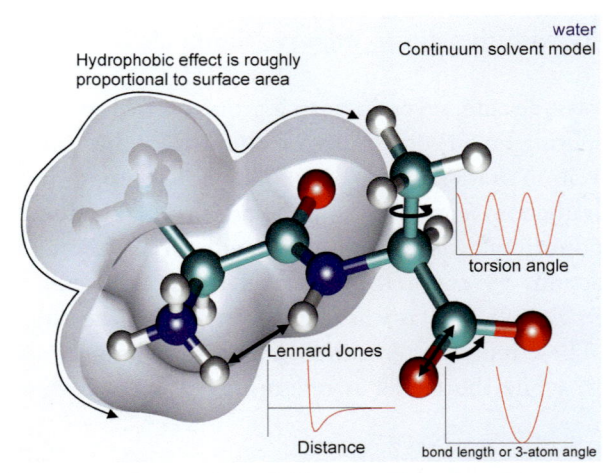

Fig. 3 Contributions in all-atom force fields. The interactions between the atoms are divided into bonded and nonbonded interactions. Harmonic potentials are used to describe the vibrations of bonds and bond-angle bending, while periodic functions are needed for modeling the torsion around bonds. The nonbonded interactions are between atoms that are separated by at least three bonds or between atoms of different molecules. They arise from charge-charge interactions as described by the Coulomb potential, and from hydrophobic interactions as well as repulsive interactions if two atoms get too close to each other, which are collectively modeled by the Lennard-Jones potential. The water around a protein or other biomolecules can be modeled explicitly (not shown) using typical water models, such as TIP3P or TIP4P, or using an implicit solvent model. Explicit modeling of the water molecules gives usually better results, especially in the case of IDPs. *Reproduced with permission (https://en.wikipedia. org/wiki/Force_field_(chemistry)).*

angle bending, where s is the distance between the first and third atoms that define a bond angle. However, most force fields do not include Urey-Bradley terms, and also in CHARMM FFs, no new UB terms were added in the past since the only advantage of these terms is the better reproduction of subtleties in vibrational spectra. However, the goal of classical MD simulations seldomly is the calculation of infrared spectra. In fact, in most of the MD simulations of proteins the bond lengths are anyhow restrained to their equilibrium values in order to allow an increase in the time step used for the integration of the equations of motions. Moreover, many of the vibrations, especially those involving hydrogen bonds, would require a quantum-mechanical description for proper modeling as classical simulations reach their limit of validity here. The second correction term is called CMAP, which is a grid-based correction and accounts for the correlation between the backbone dihedral angles ϕ and ψ. Unlike the UB term, the CMAP correction has gained in popularity and was included in other FFs too, especially with the aim to improve the modeled ensembles of IDPs (see below).

Over the last decade, several research groups aimed at developing better FFs for IDPs, using Eq. (1) or a similar equation as starting point. Two major reparameterization strategies have been followed: first, optimization of dihedral angle parameters to provide better descriptions of the tendency of IDPs to adopt random coil conformations; second, strengthening the protein-water interactions to counteract the preference of proteins to collapse into a molten globule state as was seen with many of the common protein force fields. These reparameterization strategies are briefly explained in the next sections, while the reader is referred to a recent review[89] for more details.

2.2 Optimization of dihedral parameters

In many of the recently developed FFs for IDPs, adjustments of dihedral-angle parameters, especially those of the backbone dihedrals ϕ and ψ, were made.[84,94,95] The reason behind this strategy is that many of the common FFs overestimate propensities for α-helix and β-sheet formation in IDPs, as was recently reviewed by Mu et al.[89] One remedy to resolve this secondary structure bias is to use dihedral-angle data of coil-like fragments in the training sets of FFs.[96] This approach was employed for AMBER03* and AMBER99SB*[96] which are based on AMBER03[97] and AMBER99SB,[98] respectively. However, AMBER03* overestimates and AMBER99SB* underestimates the helical content with respect to their respective pre-decessing FF. For OPLS-AA/M[99] and OPLS3[100] a similar refitting approach was followed, including the reparameterization of the side-chain dihedrals, and as training set data from *ab initio* torsional energy scanning of blocked dipeptides was used. However, their ability to produce good IDP ensembles beyond proline dipeptides and glycine tripeptides remains to be shown. In the CHARMM family of FFs, CHARMM22*[101] also a resulted from refitting dihedral angles based on CHARMM22[102]. In our analyses of various FFs with regard to their ability to produce Aβ ensembles in agreement with the data obtained from NMR and fluorescence spectroscopy, CHARMM22* yielded acceptable results[87,103]; in our earlier study of these two[103] it was even the best-performing FF. In the latter study, AMBER99SB-UCB turned out to be by far best suited for modeling Aβ.[87] This FF is based on AMBER99SB/TIP4P-Ew and includes modified backbone torsion parameters[104] and optimized protein–solvent Lennard-Jones parameters.[105] A special reparamterization approach for the dihedral angles was adopted in AMBER99SB-UCB, since it was limited to ϕ', which defines the torsion about the C-N-Cα-Cβ atoms, in order to only shift the equilibrium between the β and PPII states, but leave the α-helical state unaffected.[104] The mentioned FFs employed a universal refitting strategy for dihedral parameters. An alternative approach is to optimize these parameters in a residue-specific manner. This approach was adopted for RSFF1,[106] which is based on OPLS/AA, and RSFF2,[107] which derives from AMBER ff99SB. In both cases the dihedral distributions from a protein coil library were used as the training set.

2.3 Adding CMAP corrections

The majority of the CHARMM protein FFs that are based on CHARMM22* include CMAP corrections.[108,109] The CMAP residue-specific correction for backbone dihedral parameters is a grid-based energy correction map dependent on the (ϕ, ψ) distribution of the backbone dihedrals of the protein residues. The two-dimensional (ϕ, ψ) angle distribution per residue is evenly divided into 24 × 24 bins with a 15° step size between neighboring bins. The dihedral free energy for bin i is given by

$$\Delta G_i^{\text{sim}} = RT \ln \left(\frac{N_i}{N_{\text{max}}} \right) \tag{2}$$

where N_i is the number a dihedral angles falling into the bin in question and N_{max} refers to the total number of dihedral data in the sampling. The CMAP correction for each bin is provided as difference of the experimental database value (ΔG_i^{exp}) and the current simulation value from Eq. (2):

$$U_i^{CMAP} = \Delta G_i^{exp} - \Delta G_i^{sim} \tag{3}$$

Thus, the energy added is the larger, the larger the deviation from the database value is. Since a step size of 15° yields only 576 bins, nearest-neighbor[110] or cubic interpolation[111] is applied to generate a continuous energy-correction surface so that U_{CMAP} can be determined for any conformation and allow the computation of forces.

The CMAP method was first applied in CHARMM22/CMAP[109], which is also known as CHARMM27 and is based on CHARMM22.[102] However, CHARMM27 did not produce convincing results for IDPs as demonstrated for α-synuclein, where the helical conformation was overestimated, and generally fails to generate a stable hairpin structure. Therefore, the CMAP approach was revised for CHARMM36.[112] This newer FF performs generally better for IDPs; however, left-handed helices tend to be overpopulated.[82] To overcome this and other shortcomings, CHARMM36m was developed and claimed to be particularly suited for IDPs.[94] However, the FF benchmark by Robustelli et al. showed that, while CHARMM36m performs well for folded proteins, for many of the IDPs, it does not produce convincing results.[84] This excludes Aβ, for which CHARMM36m produced acceptable results,[87] which are better than those obtained with AMBER99SB-disp developed by Robustelli et al.[84] Chen and coworkers picked up the CMAP idea and implemented it into various FFs with the aim to improve the modeling of IDPs. They augmented this approach by deriving CMAP potentials for all 20 standard amino acids, instead of applying a universal CMAP correction in Eq. (3). They implemented their CMAP corrections into various FFs, yielding AMBER14IDPS,[113] CHARMM36IDPS,[114,115] and OPLSIDPSFF.[116] Another extension of the CMAP idea is the CMAP energy correction map based on a three-dimensional distribution of dihedrals that includes side-chain dihedrals, which was implemented into RSFF2[107] and yielded RSFF2C.[117] This FF was demonstrated to provide good models for IDPs, IDPRs, and also folded proteins.[117]

2.4 Refining protein–water interactions

Between protein and water, electrostatic and vdW interactions can occur. The short-range vdW interactions in particular influence the compactness of the simulated IDP conformations, which can be assessed by the radius of gyration (R_g) or the end-to-end distance and be compared to the corresponding values obtained from SAXS or FRET experiments.[118,119] Since the observation was that the common FFs in general produce too compact IDP conformations, one strategy for the development of IDP-appropriate FFs is to increase the vdW interactions between protein and water. The first approach along this line was realized by Best et al. who uniformly scaled the LJ interactions between protein and water by a factor of 1.1 in AMBER03, leading to AMBER03WS, which recovered the correct dimensions of IDPs or unfolded proteins in their simulations.[120] Shaw and coworkers approached the problem by increasing the ϵ_O value of the oxygen atom of water in the TIP4P water model, yielding TIP4P-D where "D" stands for dispersion as the authors aimed at providing a correct description of the water dispersion interactions. However, while this water model indeed improved the R_g values of some simulated IDPs, it also caused some α-helices to unfold and overestimated the R_g of several longer IDPs.[121] The recently developed AMBER99SB-disp force field is based on the TIP4P-D water model and increased the dispersion interactions of water even further, in addition to refining the backbone dihedral potentials and LJ interactions between backbone carbonyl oxygen atoms and backbone amide hydrogen atoms.[84] Along with CHARMM36m an alternative TIP3P water model was published.[94] Opposite to TIP4P-D where ϵ_O of water was modified, in CHARMM36m MacKerell and coworkers changed the LJ well depth parameter ϵ_H of the water hydrogen atoms while the oxygen LJ parameters and the water–water interactions were maintained. The rationale behind altering the ϵ_H and not the ϵ_O value is that by changing the water oxygen atom LJ parameters one would affect its effective size based on the repulsive r^{-12} term. Since the water hydrogen atoms have a very small vdW radius, their repulsive LJ term is basically unaffected when modifying their ϵ_H parameter. For the IDPs CHARMM36mW—the name that our group decided to give to CHARMM36m with the modified water model[122]—performed better than CHARMM36m in the study by MacKerell and coworkers. Head-Gordon and coworkers followed a different approach and adjusted the LJ parameters of the amino acids on atom type basis with the aim to reproduce experimental solvation free energies of a diverse set of 34 molecules, instead of uniformly scaling the vdW interactions between protein and water.[105] This modification along with the one affecting the backbone torsion parameters[104] (see above) resulted in AMBER99SB-UCB. In our recent study, this FF turned out to be the most suitable one for Aβ[87] while for peptide aggregation, we identified CHARMM36m(W) as the best FF.[122] All force fields that modified the LJ parameters of the water oxygen atom as well as AMBER99SB-UCB failed to simulate the aggregation of A$\beta(16 - 22)$, which is in clear disagreement with experiment.[88,122,123]

2.5 Polarizable force fields

A limitation of the FFs discussed thus far, and which can most likely not be fully compensated by refitting of dihedral parameters and tuning of LJ protein-water interactions, is that they are fixed-charged models. The accuracy of the FFs for IDPs can be improved by refining of electrostatic and hydrogen-bonding interactions.[124–126] Instead of using fixed atomic charges on amino-acid residues, which were derived for model compounds and are assumed to work under all circumstances, the charge distribution should be determined on the fly as it depends on the neighboring residues and degree of solvent interactions. This requirement can be met by polarizable FFs which explicitly incorporate electronic polarizability.[126,127] This is commonly achieved via fluctuating charge methods (CHARMM-FQ model), Drude oscillators (CHARMM-Drude model),[128] and the multipole expansions method (AMOEBA force field). For the CHARMM Drude FF it was demonstrated that it can simulate the experimentally relevant folding kinetics of small IDP-typical sequences.[129] A disadvantage of polarizable FFs is their higher computational costs compared to nonpolarizable FFs, largely limiting their application in the field of molecular simulations of IDPs, which usually require long simulation times. To overcome this problem, new strategies are developed such as usage of Gaussian models to treat the electrostatic interactions in AMBER FFs.[130] The future will show whether polarizable FFs will be applied more often to IDPs and IDPRs, especially if computationally more efficient polarizable FFs that make use of hardware developments should become available. For a comprehensive review of polarizable FFs the reader is referred to Ref. 131.

2.6 Self-learning algorithms

Machine learning/artificial intelligence (ML/AI) approaches can be implemented to optimize FF parameters against experimental observables like small-angle X-ray and neutron scattering (SAXS and SANS, respectively) intensities in the ForceBalance algorithm developed by Wang et al.[132] The optimization machinery in ForceBalance-SAS (small-angle scattering) method[133] is based on prediction and fitting with ensemble-averaged properties obtained from SANS and SAXS experiments with derivable gradients and Hessians with respect to user-input FF parameters of choice. Though one could systematically improve on the initial set of FF parameters to a large extent, the approach requires longer simulations to optimize the finer scales of atom-pair interaction to develop transferable optimized FF parameters.[134] The other limitation of ML approaches-based FF design is that it is heavily dependent on the data used during training. For a review of ML techniques for the sampling of IDPs, the reader is referred to Ref. 135.

2.7 Coarse-grained models

An alternative to computationally demanding all-atom FFs are coarse-grained FFs. Some of these force fields were suggested to be particularly suited for IDPs. One of them is AWSEM-IDP, which is based on AWSEM and was developed by P. G. Wolynes and G. A. Papoian, and has been successfully applied to IDP aggregation problems.[136,137] Another coarse-grained FF suited for IDP simulations is the MOFF potential model,[138] which is based on the maximum entropy algorithm to reproduce ensemble-averaged properties of experimental observables. The OPEP coarse-grained protein model developed by Derreumaux and coworkers has been applied to a wide range of applications, including $A\beta$ simulations and amyloid fibril formation.[139] Thirumalai and coworkers used a CG model for intrinsically disordered proteins (termed as self-organized polymer-intrinsically disordered protein [SOP-IDP] model)[140] to characterize the conformational ensembles of $A\beta40$ and $A\beta42$.[141] Shea and coworkers developed a CG approach using a field-theoretic simulation approach that allowed them to compute the complete phase diagram for liquid–liquid phase separation of IDPs.[142]

3. Simulation-based IDP ensemble generation and characterization

Generating 3D structures of the highly dynamic IDPs is of great importance for better understanding of their function. Therefore, structural information of IDPs at the atomic level is highly demanded.[143] In this regard, as mentioned earlier different experimental techniques have been utilized to characterize IDP structures, including NMR spectroscopy,[144] SAXS and SANS,[145,146] and smFRET.[147,148] Due to their dynamic nature, IDPs exist as ensemble of extended or partially folded states with a rapid exchange between these states. Most of the mentioned experimental methods lack the spatial and temporal resolutions to provide detailed local structural information of the protein, such as local residue contacts, side-chain orientations, the life time of contacts, or the secondary structure content. The readouts obtained from these methods are commonly static or an average over structures. Molecular dynamics (MD) simulations are able to provide the missing information and complement experiments of IDPs.[149–153] The simulation of the molecular motion is realized by the numerical solution of the classical Newtonian dynamic equations:

$$F_{i,a}(\mathbf{q}) = -\frac{\partial U(\mathbf{q})}{\partial r_{i,a}} = m_i \frac{dv_{i,a}}{dt} \tag{4}$$

where $a = x$, y or z, $F_{i,a}$ is the force acting on particle i in direction a, $U(\mathbf{q})$ is the potential energy defined in Eq. (1), m_i is the particle's mass, and $v_{i,a}$ its velocity in direction a. MD simulations enable exploring the conformational space of IDPs and provide an atomic description of their structure and dynamics (if an all-atom FF is used for $U(\mathbf{q})$). They also give information about the exchange rate between different conformations and the interactions between the solvent and ions.

However, significant challenges limit the application of standard MD simulations to study IDP structure and dynamics. These challenges include: (i) At the time scales that are accessible by conventional MD simulations, a peptide or protein often remains trapped in an energy basin and seldomly overcomes relevant energy barriers. This restricts the study of slow dynamic processes that occurs on the (sub-)millisecond or longer time scale. Since typical simulation time steps in all-atom MD simulations are on the order of femtoseconds, $\gtrsim 10^{12}$ time steps are needed to observe such slow dynamics. (ii) Even if a longer simulation can be managed, the analysis of simulation data is not trivial. The simulation trajectory tracks the Cartesian coordinate of the atoms present in the system and thus might contain millions of data points in tens of thousands of dimensions.

In this section, we will introduce different approaches-based on MD simulations that were developed to enhance the conformational-space sampling of IDPs and their dynamics[154,155] (Section 3.1). We will also survey some of the novel analysis methods that have been developed to reduce the complexity of MD data[156] (Section 3.2).

3.1 Enhanced sampling

To reduce the computational requirements of conventional MD simulations and improve the sampling efficiency, several enhanced sampling MD algorithms have been developed.[157,158] In particular, methods that are used to enhance the conformational sampling of IDPs can be separated into two classes: (i) collective-variable (CV)-based methods[159] such as metadynamics (MetaD) (Fig. 4A),[160] and (ii) non-CV-based methods such as replica exchange molecular dynamics (REMD).[161–164] Both classes achieve efficient sampling of IDPs and yield conformational ensembles and transitions in quantitative agreement with experiments.

3.1.1 Collective variables and free energy

Collective variables are widely used in enhanced sampling schemes[158,160] to describe the slow dynamics in many processes of interest. This is achieved by using a mapping from the $3N$ dimensional atomic coordinates $\mathbf{q}$ of the simulated system to N_{CV} collective variables $\mathbf{s}(\mathbf{q})$. The equilibrium distribution of the CV that gives the probability of observing the system at a specific CV point $\mathbf{s}$ is given by:

$$p_0(\mathbf{s}) = \langle \delta[\mathbf{s} - \mathbf{s}(\mathbf{q})] \rangle \tag{5}$$

where δ is the Dirac delta function and $\langle \cdot \rangle$ denotes averaging over the equilibrium probability distribution of $\mathbf{q}$. $p_0(\mathbf{s})$ can be utilized to compute the free energy of the system using

$$F(\mathbf{s}) = -k_B T \log[p_0(\mathbf{s})] \tag{6}$$

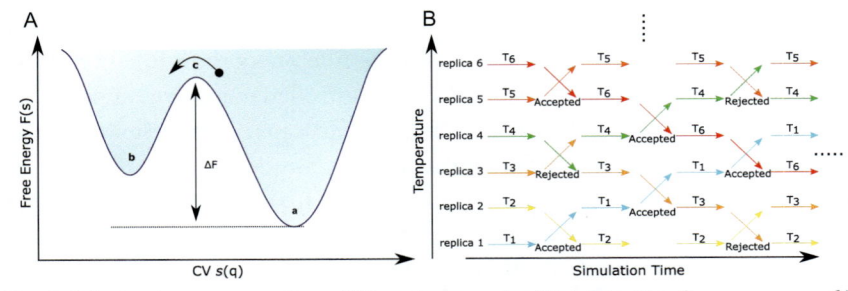

Fig. 4 Schematic representation of (A) metadynamics (MetaD) with a free energy profile of two energy basins (**a** and **b**) separated by an energy barrier ΔF at **c**. Filling the wells with a bias potential (*light blue*) allows the system to transition between a and b by crossing c. (B) Presentation of the REMD method for a system with six replicas simulated at different temperatures from T_1 (lowest T) to T_6 (highest T). Exchange attempts between neighbored replicas are marked by arrows.

where k_B is the Boltzmann's constant and T is the temperature.

In CV-based methods such as MetaD, a bias potential $V(s)$ is added to allow the sampling of the CV space and help to overcome the energy barriers that separate two or more states in the configuration space. The free energy of the system can then be calculated as

$$F(s) = -k_B T \log [p(s)] - V(s) \tag{7}$$

where $p(s)$ is the sampled distribution of $s(q)$.

In general, it is difficult to determine the proper set of collective variables that provide the best sampling efficiency and capture the states of the simulated system. Most CVs are usually chosen based on the nature of the investigated problem. For example, in studying binding/unbinding events the center of mass distance between the molecules can be used, while in studying order/disorder transitions, the number of contacts among residues is more suitable. Better choices of CV need experience and validation. Some considerations should be taken into account when picking the CV set: (i) The number of CVs should be limited to reduce the computational effort and to avoid ending up with a high-dimensional CV space to explore which is not trivial. (ii) CVs should be dynamically meaningful; capable to distinguish between conformations (initial, intermediate, and final conformations). (iii) CVs should include all the slow modes of the system to ensure a reasonable convergence of the bias potential to $F(s)$.

3.1.2 Metadynamics

Metadynamics is an enhanced sampling method originally developed by Parrinello and coworkers.[165] It is widely applied for the calculation of free energies and accelerating rare-event sampling in complex biomolecular systems. The idea of MetaD[166] is to fill the free energy minima of a metastable state with bias potentials (Eq. 7) in a controlled manner to enhance the exploration of other states in the energy landscape. This scheme can be achieved by running the MD simulation with a modified Hamiltonian for which a history-dependent bias potential $V(s(q), t)$ is added. This bias potential is a function of $s(q)$, and can be built as a sum of Gaussian potentials deposited within the CV space to push the system toward sampling unexplored configurations:

$$V(s,t) = \sum_{k\tau < t} W(k\tau) \exp \left(-\sum_{i=1}^{N_{CV}} \frac{(s_i - s_i(q(k\tau)))^2}{2\sigma_i^2} \right) \tag{8}$$

where σ_i is the width of the Gaussian function for the ith collective variable, $W(k\tau)$ is the height of the Gaussian at the simulation time $t = k\tau$, which is constant in the case of standard metadynamics, and τ is the deposition rate of the Gaussian functions. Ultimately, the bias potential converges in the long-time limit to the negative of the free energy as a function of the collective variables,

$$V(\mathbf{s}, t \to \infty) = -F(\mathbf{s}) + C \qquad (9)$$

where C is a constant. The latter equation implies that the deposited bias potential is optimal to enable transition events as it flattens the biased energy landscape. In standard MetaD, the height of the added Gaussian potentials is constant during the simulation. Therefore the estimated free energy landscape oscillates when converging toward the real free energy profile. This limitation can be overcome by utilizing well-tempered MetaD[167,168] that rescales the height of the added Gaussian potentials such that it is decreasing with time. In addition to well-tempered MetaD, different variants of standard MetaD have been implemented to enhance its sampling efficiency, such as parallel-tempering MetaD,[169,170] multiple-walkers MetaD,[171] and bias-exchange MetaD.[172] However, an outstanding drawback of MetaD is the possibility of driving the system into a physically irrelevant region of the space. Its accuracy is strongly dependent on the proper choice of CVs and the proper selection of the Gaussian parameters $W(k\tau)$ and σ.[166,173]

MetaD has been applied to study the binding of IDPs[174] and also the free energy landscape of IDPs.[175]

3.1.3 Replica exchange molecular dynamics

REMD is one of the non CV-based methods that alleviates the problem of prior knowledge of the system under study as required by CV-based methods. The idea of REMD is to run several replicas of the same system, yet at different temperatures. Exchanges between the temperatures or the configurations of neighboring replicas are attempted every few time steps (Fig. 4B).[176] This way, configurations that are accessible at high temperatures are exchanged with those sampled at a lower temperature. This process enhances the conformational sampling of the system and allows to accurately compute its thermodynamic properties. REMD has been widely applied in the study of conformational ensemble of IDPs. Its first application was by Sgourakis et al. to explore the differences in the conformation accessible to the hydrated Aβ monomers (Aβ40 and Aβ42).[177] Similarly, it was then employed to understand the conformational preferences of the histone tails (highly flexible N- or C-terminal) and its effect on the binding affinity to linker DNA.[178] It has been further utilized to understand the interplay of residual structure and conformational fluctuations in coupled binding and folding of IDPs applied to transcription coactivator CREB.[179,180] Miller et al. studied the conformational ensemble of several IAPP variants.[181] REMD simulations have been also used to investigate how posttranslational modifications affect the conformational ensemble of different IDPs such as tau, hIAPP,[182], and KIDS.[183]

However, this approach is computationally demanding as the number of replicas grows with the square root of system size. A more efficient version of REMD, called Hamiltonian replica exchange MD (HREMD), was developed by Bussi et al.[184] In this approach, the different replicas of the system evolve according to different Hamiltonians. It is rendered to be more efficient than REMD due to the lower number of replicas needed. Different types of HREMD approaches have been developed, which mainly differ by the modifications made to the Hamiltonian in the different replicas. In the approach of Bussi et al.[184] the system is divided into hot (H) and cold (C) regions, and Hamiltonian of the H-region is modified in order to accelerate the sampling. For a system of N_{rep} replicas with $\mathbf{q}_i$ being the coordinates of the ith replica, the ensemble probability is defined as

$$p(\mathbf{q}_1) \times \ldots \times p(\mathbf{q}_{N_{rep}}) \propto \exp\left(\frac{-U_1(\mathbf{q}_1)}{k_B} \cdot \ldots \cdot \frac{-U_{Nrep}(\mathbf{q}_{Nrep})}{k_B T}\right) \quad (10)$$

Thus, in HREMD the energy and not the temperature is modified; instead all replicas are usually simulated at the same temperature. The Hamiltonian of the system in the H-region will be changed depending on a scale factor λ such that only the force field terms contributing to noteworthy energy barriers are scaled, whereas the Hamiltonian of the C-region is kept unperturbed. In particular, the charges of the atoms in the H-region are scaled by a factor of $\sqrt{\lambda}$, while the Lennard-Jones parameters ϵ are scaled by a factor of λ. The proper dihedral potential is scaled by a factor of λ or $\sqrt{\lambda}$ depending on if both the first and fourth atoms or only one of them is in the H-region.

Considering that the temperature and energy are related as shown in Eq. (10) (for example, scaling the energy to its half is equivalent to doubling the temperature), it is plausible to think of the HREMD scheme as simulating each region at an "effective temperature" of $\frac{T}{\lambda}, \frac{T}{\sqrt{\lambda}}$ and T for interactions inside the H-region, between the H- and the C- regions, and inside the

C-region, respectively. The scaling factor λ ranges from 1 (for the unmodified system) to minimal 0 (infinite temperature or zero interaction in the hot region). In praxis, the lowest λ value is usually chosen as 0.6 or 0.5. The probability for exchanges between replicas should satisfy the Metropolis criterion to ensure detailed balance as a condition to obtain the correct ensemble for the Hamiltonian of interest:

$$p(\mathbf{q}_i \leftrightarrow \mathbf{q}_j) = \min\left[1, \exp\left(\frac{U_i(\mathbf{q}_i) - U_i(\mathbf{q}_j)}{k_B T} + \frac{U_j(\mathbf{q}_j) - U_j(\mathbf{q}_i)}{k_B T}\right)\right] \quad (11)$$

Though enhanced sampling techniques yield the relevant thermodynamic state of the studied system and allow the calculation of the corresponding thermodynamic properties accurately, estimating the transition rate among different states in the configuration space remains a limiting step.

HREMD has been successfully applied in the study of different IDPs, including Aβ,[164,185] Aβ fragments,[186] the disordered N-terminal of c-Src kinase,[154] histatin 5 (24 residues), and Sic 1 (92 residues). Furthermore, HREMD has also been coupled with a CG force field to further enhance the sampling efficiency.[187]

Enhanced sampling methods are implemented in different MD programs, such as the Colvars module in NAMD[188] that enables MetaD simulations and is flexible in defining CV using Tcl scripts. Similarly, the Plumed plugin[189] together with GROMACS[190] enables running HREMD and MetaD simulation.

3.2 Long-time methods for MD simulations

A major advance in the realm of alternative methods to analyze MD data has been made by the proposal of transition matrices and discretization of the phase space. This approach improves the MD sampling efficiency by widening its focus to include, in addition to identifying the energy minima, the simulation and definition of informative transition pathways. The transition process between substates is often described by a memoryless master equation,

$$\frac{d\mathbf{p}(t)}{dt} = \mathbf{K}\mathbf{p}(t) \quad (12)$$

where $\mathbf{p}(t)$ is a column vector contains the probability of finding the system in each of its m states at time t, $\mathbf{K}$ is the rate matrix with its elements, and K_{ij}

is the rate constants of transitions from state i to state j. In a Marokovian process, the system dynamics can be described by a discrete time transition matrix $\mathbf{T}(\tau)$ as explained below.

3.2.1 Markov state models

Markov state models (MSMs) derived from MD data are kinetic models used to define the kinetically relevant states in the configuration space and the transitions (memoryless jumps) between them.[191–193] In other words, MSMs transform the MD trajectory from structures over time to a human readable network of macrostates that enables extracting the hidden kinetics in the high-dimensional MD simulation data. The output of MSMs is Markovian, i.e., the transition to the next state is only affected by the current state and not the past history; thus it is memoryless.

In a Markovian process, the system dynamics can be described by a discrete time transition matrix $\mathbf{T}(\tau)$ with its entries T_{ij} representing the probability of finding the system in state j at time $t + \tau$ knowing that it was in state i at time t. In accordance with Eq. (12) this can be expressed as

$$\mathbf{p}((k+1)\tau) = \mathbf{T}(\tau)\mathbf{p}(k\tau) \tag{13}$$

where τ is the lag time selected by the user to determine $\mathbf{T}(\tau)$. Both Eqs. (12) and (13) give equivalent result at $t = k\tau$, and are related to each other by $\mathbf{T}(\tau) = \exp(\tau\mathbf{K})$. The focus of MSMs is on the transition matrix $\mathbf{T}(\tau)$, which describes the transitions between substates, with its eigenvectors $\mathbf{u}_i$ and corresponding eigenvalues λ_i. They are a measure for the relaxation time of a process described by the eigenvector in question, i.e., $\lambda_i \rightarrow 0$ is the fast mode corresponding to fast thermal fluctuations while $\lambda_i = 1$ represents the stationary distribution. The implied timescale t_i^* of a transition mode i is given by

$$t_i^* = -\frac{\tau}{\ln \lambda_i} \tag{14}$$

For equilibrium molecular dynamics, all λ_i are positive, and there exists a unique equilibrium (or stationary) distribution π that fulfills detailed balance:

$$\pi_i T_{ij} = \pi_j T_{ji} \tag{15}$$

The first step in constructing MSMs is to discretize the configuration space based on a suitable distance metric selected according to the scientific question of interest. For example, in studies interested in conformational dynamics or protein folding, the backbone atom coordinates can be used to define the conformations. This is followed by clustering, such as k-means clustering to define the microstates. Even though the idea is to identify the kinetically stable states, initially geometrical clustering is used to define microstates based on which the subsequent kinetic partitioning of the configuration space is performed. The transition matrix described above defines the kinetic relationship between these microstates. The quality of the resulting MSM depends on the chosen lag time. It should be long enough so that the system is memoryless (Markovian) while it should be short enough to resolve the different dynamic processes. Whether the models satisfy Markovianity can be assessed by the Chapman-Kolmogorov test[194]:

$$\mathbf{T}(k\tau) = \mathbf{T}^k(\tau) \tag{16}$$

Finally, the MSM can be further optimized by coarse-graining it, for which macrostates are defined and the microstates assigned to them. This yields so-called Hidden Markov Models.

A number of software packages are available to construct MSMs from MD trajectories of biomolecular systems and validate the resulting models. The most popular ones are PyEMMA (http://emma-project.org)[195] and MSMBuilder (http://msmbuilder.org/).[196] Recent applications of MSM-based methods in the realm of IDPs have shed light on the mechanisms of fibril formation for amyloid-β (Aβ), human islet amyloid polypeptide (hIAPP), and other intrinsically disordered peptides.[197,198] It further gives insight into the thermodynamics and kinetics of IDPs, such as the amyloid-β peptide,[87,199] and are useful in studying the effect of posttranslational modification (e.g., phosphorylated kinase[200]) on the conformational kinetics and structural ensembles of IDPs.

3.2.2 Transition networks

In contrast to MSMs, TNs rely only on geometry to define the transition pathways.[201] In this approach, the configuration space is discretized based on clustering using certain descriptors (f_i) that map the geometrical conformations $\mathbf{q}_t$ to a state $S(t)$ which is a composition of different f_i:

$$S(t) = [f_1(\mathbf{q}_t), f_2(\mathbf{q}_t), ..., f_n(\mathbf{q}_n)] \tag{17}$$

The conformations of an MD trajectory are mapped to the time sequence of states $S(t)$, which are then used for building the TNs. The key aim of f_i is to encode the process under investigation the best while avoiding network complexity. These descriptors can be measures that give information about the protein secondary structure, the number of electrostatics or hydrophobic contacts, or molecular shape. Importantly, the selected descriptors should be limited to a combination of f_i with minimum correlation among them and maximum sensitivity to the data to avoid increasing the complexity of TNs. For example, if two descriptors are strongly affecting each other (i.e., highly correlated), using both of them will increase the complexity of the generated TN without a major information gain. Thus, using only one of them simplifies the resulting TN and, at the same time, not too much information will be lost. The descriptor sensitivity can be inferred from whether the theoretically possible values of f_i are indeed sampled (which is good) or only limited to a small range (which would be insufficient).

After discretizing the configuration according to Eq. (17), the transition matrix can be built by counting the number of transitions between different sates $S(t)$ using certain time lag (spacing) between MD frames. The spacing between successive MD frames should be chosen such that it is not too large in order not to lose important transitions, but also not too small as larger conformational transitions require a certain amount of time. However, unlike to MSM building, there is no quantitative measure such as the convergence of the implied time scales in MSMs that allows to assess whether the selected lag time for TN building is a good choice.

The Python notebook ATRANET (Automated TRAnsition NETwork) (https://github.com/strodel-group/ATRANET) can be used to generate TNs.[201] It was recently employed for characterizing the amyloid-aggregation pathways of $A\beta(16-22)$[201] and full-length $A\beta42$.[202]

4. Modeling the disorder-to-order transition of IDPs

An important characteristic of IDPs is their ability to modulate interactions with different binding partners. This is achieved by limited structural ordering and significant collective motions elucidated in specific function-related disorder-to-order transitions. In this section we will review a few studies where such disorder-to-order transitions were studied by means of molecular simulations.

4.1 Conformational switching of PAGE4

A typical example of such phenomena is observed in the well-studied prostate-associated gene 4 (PAGE4). This IDP is a stress-response protein that interacts with multiple protein partners, including c-Jun that binds to

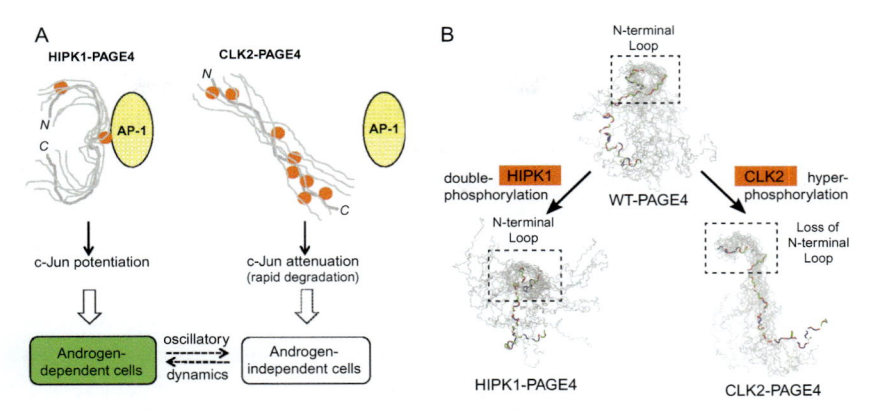

Fig. 5 PAGE4 (dys)functions and conformational dynamics. (A) The stress–response kinase HIPK1 phosphorylates PAGE4 at Ser9 and Thr51 (*orange circles*), resulting in a relatively compact PAGE4 ensemble that can potentiate c-Jun and in turn leads to androgen-receptor transactivation. In contrast, the kinase CLK2 hyperphosphorylates PAGE4 at eight different Ser/Thr residues (*orange circles*), including Ser9 and Thr51, leading to a more random-like PAGE4 ensemble that attenuates c-Jun transactivation and is likely to be degraded rapidly. Differential phosphorylation of PAGE4 by HIPK1 and CLK2 results in oscillations of the levels of HIPK1-PAGE4, CLK2-PAGE4, and CLK2. (B) The influence of the differential phosphorylation on the structural ensembles of the resulting PAGE4 variants was characterized by simulations. N-terminal loops are present in WT- and HIPK1-PAGE4, while no loops are formed in CLK2-PAGE4, giving rise to more extended and random coil CLK2-PAGE4 structures. *Panel (A): Reproduced with permission Kulkarni P, Jolly MK, Jia D, et al. Phosphorylation-induced conformational dynamics in an intrinsically disordered protein and potential role in phenotypic heterogeneity. Proc. Natl. Acad. Sci. U.S.A. 2017; 114(13):E2644–E2653. Copyright 2021, United States National Academy of Sciences. Panel (B): Figures showing the structural ensembles are reproduced with permission Lin X, Kulkarni P, Bocci F, et al. Structural and dynamical order of a disordered protein: molecular insights into conformational switching of PAGE4 at the systems level. Biomolecules. 2019; 9(2):77. Copyright 2021, Elsevier.*

c-Fos to form the activator protein-1 complex (AP-1), homeodomain-interacting protein kinase 1 (HIPK1) that phosphorylates PAGE4 at two different phosphorylation sites to enhance c-Jun activity, and CDC-like kinase 2 (CLK2) that hyper-phosphorylates PAGE4 and thereby inhibits c-Jun activity[203] (Fig. 5A).

Lin et al. performed molecular simulations using the coarse-grained model AWSEM to study the conformational and dynamical transitions observed in the various physiologically relevant phosphorylated forms of PAGE4.[205,206] The free energy profiles constructed from their simulations show a shift in the degree of collapse of PAGE4 upon phosphorylation, which is also confirmed by SAXS and smFRET experiments.[204]

In particular, the calculated average radius of gyration of HIPK1-PAGE4 is 32.1 Å, while that of CLK2-PAGE4 is 41.8 Å. The expansion of the latter is manifested in changing populations with different secondary structures and collective motions. Namely, CLK2-PAGE4 exhibits a reduced propensity for forming turns that might be associated with the loss of binding affinity of PAGE4 with the AP-1 complex, while both wild-type (WT-) PAGE4 and HIPK1-PAGE4 are observed to form a stable N-terminal loop in HIPK1-PAGE4 contacting the central acidic region (residues 43–62) of the protein (Fig. 5B). This loop formation imposes a structural order over the underlying disordered dynamics of PAGE4. The ordering was also reflected in the regularity of collective motions. Such ordered characteristics are directly associated with the functional interactions of WT-PAGE4, HIPK1-PAGE4 and CLK2-PAGE4 with the AP-1 signaling axis. The authors argue that the modulation of the structural ordering is crucial for sculpting its interactions. Even for a protein such as PAGE4 that has been identified as a near-complete random coil by bioinformatics studies, Lin et al. detected underlying structural features that confer on it the ability to respond differently to various levels of posttranslational phosphorylation. On a local scale, a cracking motion features an order–disorder–order transition that is utilized by some proteins for lowering the barrier of conformational transition, or to facilitate allosteric transitions. On a global scale, a complete structural rearrangement of the system may be required to deliver the functional parts of a protein into its target location. The plasticity of protein structures facilitates such transitions, but it is the ordering of the structure that enables its functions. In this study, Lin et al concluded that PAGE4 and its interacting partners emphasize the role that conformational dynamics of proteins can play in rewiring regulators networks which influences phenotypic plasticity. Such plasticity may be related to translational noise but their analysis suggested that conformational noise may also modulate cell fate. In summary, their simulations[205,206] provided detailed insights into structural and dynamic encoding in the sequence of PAGE4 and how these are modulated by different extents of phosphorylation via kinases HIPK1 and CLK2.

4.2 Order-disorder transition in CREB

Structural transitions and binding interaction patterns associated with CREB are largely influenced by the phosphorylated serine residue of the kinase inducible domain (KID)[207,208] (Fig. 2). The main function of phosphorylated KID is to increase the binding affinity between CREB and its coactivator CREB binding protein (CBP) via binding with the kinase-inducible domain interacting (KIX) domain of CBP.[209] The KID is characterized as an IDP, and its conformational dynamics has been extensively studied using both NMR experiments and MD simulations. The first simulation of the folding-upon-binding of the KID to the KIX domain of the CBP was achieved by Turjanski et al.[210] They used a coarse-grained Gō-type model and performed Langevin dynamics simulations, which were scrutinized using transition-path and Φ-value analyses. They found that the binding transition state resembles the unstructured state in solution, implying that CREB becomes structured only after binding to the CBP. Therefore, it was concluded that this particular order-disorder transition agrees to an induced-fit mechanism and does not follow a mechanism in which prestructured conformations that exist in the unbound state in solution bind to its interaction partner.

About ten years later, Liu et al. performed unbiased MD simulations using the latest generation all-atom FF for IDPs, AMBER14IDPS,[113] to improve the sampling efficiency of disordered regions of KID.[211] They concluded that in comparison to other IDP-specific FFs, AMBER14IDPS generated more diverse disordered conformers. Coupled with that, the secondary chemical shifts produced from the AMBER14IDPS simulations agree best with experimental results with a low RMSD of 0.76 ppm. From their computational studies, they could segregate the residence times of conformers sampled during the folding transitions of KID. They reported that the order-disorder-order transition of the native α-helical configuration of KID follows a two-state process starting with the tertiary transition followed by the helical transition. The residue-interaction networks reveal transition stages where one can observe the decline of order-disorder transitions a.k.a. "folding" and the restart of disorder-order transitions a.k.a. "refolding." Enhanced formation of expanded structures was observed during the last stage of simulation noted by the higher values of the radius of gyration, which can be linked to the refolding transition state. Also, the conformers of KID during the refolding state can be classified as a pre-molten globule state, a subset of the random coil state which features partly disordered and helical or β-sheet conformers and is very well in agreement with experimental observations. The residue networks also helped identify the important residues responsible for the binding interactions of KID with KIX. The residues of KID, namely Ile127, Lys128, and Lys141, are considered the hub nodes in the dynamical networks, and from clinical studies they are confirmed to play a major role in the binding kinetics with KIX due to the posttranslational phosphorylation. The conformational dynamics and order–disorder–order kinetics predicted from the MD simulations of KID with ff14IDPSFF confirmed that this FF is accurate enough to resolve the structure–function paradigm of IDPs.

4.3 Disorder-to-order transition in the H1 linker histone protein

Another example for the disorder-to-order transitions of IDPs is found in the H1 linker histone, which was studied using enhanced sampling techniques by Sridhar et al.[212] H1 is a partially disordered chromatin-binding protein that binds to nucleosomes near the entry/exit site of the linker DNA, interacts with ≈20 bp of the DNA linkers, and critically influences chromatin organization. In mammals, eleven variants of H1 have been identified (H1.0–H1.10), which are of different characteristics, such as chromatin compacting capabilities or expression timing. All of them are composed of the highly conserved domain, the globular domain (a helix of 75 residues), the unstructured N-terminal domain (NTD) (≈45 residues), and an almost unstructured C-terminal domain (CTD) (≈100 residues). Both the NTD and the CTD are strongly varied among the H1 variants.

Sridhar et al. employed both parallel-tempering metadynamics (PT-MetaD) and temperature-REMD (TREMD) at the atomistic scale using CHARMM36m and AMBER99SB-ILDN as force fields to study how the conformational landscape of the NTD of H1 subtypes (H1.0, H1.1, H1.2) changes upon charge neutralization or binding to DNA. They found that the NTD of all H1 subtypes is unstructured, but adopting compact conformations in solution. However, the NTD was found to undergo a disorder-to-order transition yielding a helical conformation upon charge neutralization or binding to DNA, with a stronger effect being seen in the latter case. Furthermore, the degree of helicity is subtype-dependent and correlates with the experimentally observed DNA-binding affinity of H1 subtypes. In particular, H1.0 and H1.2 emerged as optimal for helical folding due to a higher charge density per unit length and the low number of the helix-breaking residues Pro/Gly in their NTD. The disorder-to-order transition in H1.0 and H1.2 leads to an amphipathic helical conformation that orients the basic charged residues of NTD on one side of the helix, fostering the interactions with the DNA grooves. This explains the higher nucleosome binding affinity of the H1.0 subtype as its NTD forms a particularly stable amphipathic helix that increases the stability of the DNA-NTD complex, helping to glue H1 to the nucleosome.

4.4 Disorder-to-order transition in the IDPR of sortase

Sortase is a transpeptitase found in Gram–positive bacteria which involves a disordered loop that undergoes a disorder–to–order transition upon binding of both a peptide and a calcium ion (Fig. 6A). This transition is accompanied by the motion of another loop, which is therefore called dynamic loop, while the two β-sheets are relatively rigid. Moritsugu et al. performed enhanced sampling simulation at the atomistic scale and in explicit solvent for conformational sampling of the disordered loop and dynamic loop in the ligand-free enzyme as well as in three bound forms with i) a peptide, ii) a calcium ion, iii) both peptide and Ca^{2+}.[213] By studying these four systems, their aim was to elucidate the influence of the binding of the peptide and/or a calcium ion to the free-energy landscape of sortase. As simulation method they used multiscale essential sampling (MSES).[213] This sampling method combines an all–atom description of the protein and its surrounding solvents (MM) and coarse-grained models for selected states of the protein (CG), leading to following Hamiltonian:

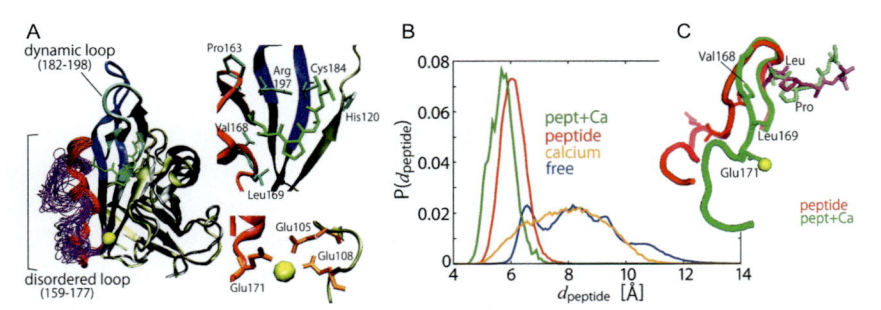

Fig. 6 Disorder-order transition in sortase. (A) (*left*) NMR solution structures of sortase in its ligand-free (PDB code 1IJA; *violet, blue,* and *gray*) and ligand-bound state (PDB code 2KID; *red, cyan,* and *olive*). The disordered and dynamic loops are indicated. The bound peptide is shown with *green sticks* and the calcium ion as *yellow sphere*. (*right*) Close-up views of the peptide binding pocket and the Ca^{2+} binding region. (B) Binding of the peptide causes the binding site at sortase to become more compact, as monitored by the distance between Pro163 of the disordered loop and Arg197 of the dynamic loop (denoted as $d_{peptide}$). Binding of both peptide and Ca^{2+} causes the most compact formation (*green*), while the ligand-free state adopts more open states (*blue*). Binding of Ca^{2+} only resembles the ligand-free state (*yellow*), while binding of the peptide alone causes quite compact conformations (*red*). (C) Binding of Ca^{2+} in addition to the peptide completes the disorder-to-order transition in the disordered loop by inducing a 3_{10}-helix formation in that loop, which can be seen by comparing the conformation shown in *green* vs the one in *red*. *Figures shown in the panels are reproduced with permission Moritsugu K, Terada T, Kidera A. Disorder-to-order transition of an intrinsically disordered region of sortase revealed by multiscale enhanced sampling. J. Am. Chem. Soc. 2012; 134(16):7094–7101. Copyright 2021, American Chemical Society.*

$$H = H_{MM} + H_{CG} + k_{MMCG}(\theta(\mathbf{q}_{MM}) - \mathbf{q}_{CG})^2 \qquad (18)$$

The last term in this equation defines the coupling between the MM and CG models with coordinates $\mathbf{q}_{MM}$ and $\mathbf{q}_{CG}$, respectively, with $\theta(\mathbf{q}_{MM})$ as a projection of $\mathbf{q}_{MM}$ onto the CG space and k_{MMCG} as coupling constant. For H_{MM}, Amber ff03 and TIP3P were used, while H_{CG} was defined as a mixture of two elastic networks, one for the bound structure as determined by NMR spectroscopy,[214] and the other one for the unbound state, for which a structure was generated for a prior high-temperature MD simulation. In order to determine the FES for the unbiased systems without coupling to the CG model, HREMD simulations were performed where the replicas differed in the k_{MMCG} values ranging from zero (for the unbiased system) and large values for enforcing switches between the bound and unbound states. For the ligand-free state, 20×100 ns replicas were run, while it was 12×50 ns for each of the three bound states.

For the ligand-free sortase, large flexibility in both the disordered and the dynamic loop was observed. However, a certain correlation between their motions was identified when they are close to each other, adopting a conformation similar to the peptide-bound one. This correlation originates from an interaction between Pro163 in the N-terminal part of the disordered loop and Arg197 in the dynamic loop. The motion of the C-terminal part of the disordered loop, on the other hand, is anticorrelated to the motion of the dynamic loop. Already in the ligand-free state, sortase has a significant propensity to adopt a conformation close to the one of the bound form, "as if it is ready to bind ligand molecules" as concluded by Moritsugu et al.[213] The binding of the peptide and/or a calcium ion drastically reduced the conformational freedom of both loops, where binding of the peptide has a larger ordering effect than Ca^{2+} has. However, only when both peptide and Ca^{2+} are bound, the disorder-to-order transition is completed (Fig. 6B). This suggests the presence of an allosteric effect resulting from Ca^{2+} binding, which agrees with the experimental data that revealed that the presence of Ca^{2+} increases the peptide binding affinity by eight times.[215] The MSES simulations elucidated that Glu171 binds to the Ca^{2+} ion, which, when the peptide is also bound, causes a partial folding of a 3_{10}–helix in the disordered loop, and, in turn, gives rise to a tighter hydrophobic binding between the peptide and the disordered loop (Fig. 6C). It was therefore concluded that the N–terminal part of the flexible loop moves together with the dynamic loop to form the peptide binding site, which occurs even in the ligand-free state with a population shift upon peptide binding while the C-terminal part shows an induced-fit behavior following the binding of a calcium ion.

5. Summary and outlook

After providing a short introduction into the peculiarities of IDPs at the start of this chapter, we concentrated on molecular simulations of IDPs in the remainder. Here, our focus was on MD simulations in conjunction with all-atom force fields. The critical evaluation of such force fields with regard to their ability to correctly model the structural ensemble and dynamics of IDPs revealed that the standard force fields originally developed for folded proteins are usually not applicable to IDPs. In particular, these force fields tend to produce overly compact and often too folded IDP conformations. To overcome this limitation, several research groups developed force fields aimed at solving these problems by either optimizing backbone torsion parameters, adding CMAP corrections and/or strengthening the van der Waals interactions between protein and water. Some of the resulting force fields are specifically designed for IDPs, whereas others, like AMBER99SB-disp and CHARMM36m, were demonstrated to be suitable for both folded proteins and IDPs. The other important ingredient to simulating IDPs is the sampling method. Within this context, we concentrated on MD simulation techniques and presented metadynamics and replica exchange molecular dynamics (REMD) as approaches that are often applied to enhance the sampling of IDPs. This is demonstrated in the last section of this chapter where we discussed several IDPs that undergo disorder-to-order transitions upon binding to a specific interaction partner. These transitions were simulated at the atomistic level using a metadynamics, REMD, or a closely related approach. In addition to these techniques we also presented two ways of network model building, which are very useful in analyzing the MD date of IDPs in order to elucidate relevant transition pathways. One of them are the increasingly popular MSMs, which are kinetic models of the process under study that, due to their coarse-grained nature, are, readily humanly understandable. Transition networks, on the other hand, are not only based on conformational clustering only but also provide mechanistic insight into transition processes.

Future developments will most likely see more applications of machine learning techniques in the realm of IDP simulations. The integration of information obtained from various experimental techniques commonly applied to IDPs into the simulations will be another important aspect for gaining further access to atomistic details of IDPs. Many force field improvements for better models of IDPs have already been determined. Further advances in this regard will most likely be made if polarizable force fields become more common in the simulations of IDPs, which, however, will only happen if the computational cost of these force fields compared to nonpolarizable force fields gets further reduced. And as with more or less all simulations of proteins, longer simulations and on larger scales, so that the biological environment of IDPs or their interaction partners can be considered in the simulation setup, will also be of tremendous importance to access slower processes and in more realistic environments.

Acknowledgments

H.F., A.S.-A., and B.S. acknowledge funding for this project from the Palestinian-German Science Bridge financed by the German Federal Ministry of Education and Research (BMBF).

References

1. Orengo CA, Todd AE, Thornton JM. From protein structure to function. *Curr Opin Struct*. 1999;9(3):374–382.
2. Dill KA, Ozkan SB, Shell MS, Weikl TR. The protein folding problem. *Annu Rev Biophys*. 2008;37:289–316.
3. Oldfield CJ, Dunker AK. Intrinsically disordered proteins and intrinsically disordered protein regions. *Annu Rev Biochem*. 2014;83:553–584.
4. Dunker AK, Lawson JD, Brown CJ, et al. Intrinsically disordered protein. *J Mol Graph*. 2001;19(1):26–59.
5. Uversky VN. Intrinsically disordered proteins from A to Z. *Int J Biochem Cell Biol*. 2011;43(8):1090–1103.
6. Uversky VN. A decade and a half of protein intrinsic disorder: biology still waits for physics. *Protein Sci*. 2013;22(6):693–724.
7. Habchi J, Tompa P, Longhi S, Uversky VN. Introducing protein intrinsic disorder. *Chem Rev*. 2014;114(13):6561–6588.
8. Hatos A, Hajdu-Soltész B, Monzon AM, et al. Disprot: intrinsic protein disorder annotation in 2020. *Nucleic Acids Res*. 2019;48(D1):D269–D276.
9. Basile W, Salvatore M, Bassot C, Elofsson A. Why do eukaryotic proteins contain more intrinsically disordered regions? *PLoS Comput Biol*. 2019;15(7):e1007186.
10. Wright PE, Dyson HJ. Intrinsically disordered proteins in cellular signalling and regulation. *Nat Rev Mol Cell Biol*. 2015;16(1):18–29.
11. Uversky VN. The alphabet of intrinsic disorder: II. Various roles of glutamic acid in ordered and intrinsically disordered proteins. *Intrinsically Disord Proteins*. 2013;1(1): e24684.
12. Kosol S, Contreras-Martos S, Cede no C, Tompa P. Structural characterization of intrinsically disordered proteins by NMR spectroscopy. *Molecules*. 2013;18(9): 10802–10828.

13. Jakob U, Kriwacki R, Uversky VN. Conditionally and transiently disordered proteins: awakening cryptic disorder to regulate protein function. *Chem Rev.* 2014;114(13): 6779–6805.

14. Hegyi H, Tompa P. Intrinsically disordered proteins display no preference for chaperone binding in vivo. *PLoS Comput Biol.* 2008;4(3):e1000017.

15. Tompa P, Fuxreiter M. Fuzzy complexes: polymorphism and structural disorder in protein-protein interactions. *Trends Biochem Sci.* 2008;33(1):2–8.

16. Uversky VN. Multitude of binding modes attainable by intrinsically disordered proteins: a portrait gallery of disorder-based complexes. *Chem Soc Rev.* 2011;40(3): 1623–1634.

17. Makepeace KAT, Brodie NI, Popov KI, et al. Ligand-induced disorder-to-order transitions characterized by structural proteomics and molecular dynamics simulations. *J Proteomics.* 2020;211:103544.

18. Oldfield CJ, Cheng Y, Cortese MS, Romero P, Uversky VN, Dunker AK. Coupled folding and binding with α-helix-forming molecular recognition elements. *Biochemistry.* 2005;44(37):12454–12470.

19. Mohan A, Oldfield CJ, Radivojac P, et al. Analysis of molecular recognition features (MoRFs). *J Mol Biol.* 2006;362(5):1043–1059.

20. Cheng Y, Oldfield CJ, Meng J, Romero P, Uversky VN, Dunker AK. Mining α-helix-forming molecular recognition features with cross species sequence alignments. *Biochemistry.* 2007;46(47):13468–13477.

21. Vacic V, Oldfield CJ, Mohan A, et al. Characterization of molecular recognition features, MoRFs, and their binding partners. *J Proteome Res.* 2007;6(6):2351–2366.

22. Yan J, Dunker AK, Uversky VN, Kurgan L. Molecular recognition features (MoRFs) in three domains of life. *Mol Biosyst.* 2016;12(3):697–710.

23. Dyson HJ, Wright PE. Role of intrinsic protein disorder in the function and interactions of the transcriptional coactivators CREB-binding protein (CBP) and p300. *J Biol Chem.* 2016;291(13):6714–6722.

24. Dyson HJ, Wright PE. Intrinsically unstructured proteins and their functions. *J Biol Chem.* 2005;6(3):197–208.

25. Chen J, Li Q. Life and death of transcriptional co-activator p300. *Epigenetics.* 2011; 6(8):957–961.

26. Liu Z, Huang Y. Advantages of proteins being disordered. *Protein Sci.* 2014;23 (5):539–550.

27. Oldfield CJ, Meng J, Yang JY, Yang MQ, Uversky VN, Dunker AK. Flexible nets: disorder and induced fit in the associations of p53 and 14-3-3 with their partners. *BMC Genom.* 2008;9(1):1–20.

28. Hsu WL, Oldfield C, Meng J, et al. Intrinsic protein disorder and protein-protein interactions. In: Altman RB, Dunker AK, Hunter L, Murray TA, Klein TE, eds. Biocomputing 2012. World Scientific; 2012:116–127.

29. Hsu WL, Oldfield CJ, Xue B, et al. Exploring the binding diversity of intrinsically disordered proteins involved in one-to-many binding. *Protein SCI.* 2013;22(3):258–273.

30. Darling AL, Uversky VN. Intrinsic disorder and posttranslational modifications: the darker side of the biological dark matter. *Front Genet.* 2018;9:158.

31. Lieutaud P, Ferron F, Uversky AV, Kurgan L, Uversky VN, Longhi S. How disordered is my protein and what is its disorder for? A guide through the "dark side" of the protein universe. *Intrinsically Disord Proteins.* 2016;4(1):e1259708.

32. Wallmann A, Kesten C. Common functions of disordered proteins across evolutionary distant organisms. *Int J Mol Sci.* 2020;21(6):2105.

33. Artur MAS, Rienstra J, Dennis TJ, Farrant JM, Ligterink W, Hilhorst H. Structural plasticity of intrinsically disordered LEA proteins from Xerophyta schlechteri provides protection in vitro and in vivo. *Front Plant Sci.* 2019;10:1272.

34. Mertens J, Aliyu H, Cowan DA. LEA proteins and the evolution of the WHy domain. *Appl Environ Microbiol.* 2018;84(15).
35. Tunnacliffe A, Hincha DK, Leprince O, Macherel D. Lea proteins: versatility of form and function. In: Lubzens E, Cerda J, Clark M, eds. Dormancy and Resistance in Harsh Environments. Springer; 2010:91–108.
36. Li S, Chakraborty N, Borcar A, Menze MA, Toner M, Hand SC. Late embryogenesis abundant proteins protect human hepatoma cells during acute desiccation. *Proc Natl Acad Sci USA.* 2012;109(51):20859–20864.
37. Franzmann TM, Alberti S. Prion-like low-complexity sequences: key regulators of protein solubility and phase behavior. *J Biol Chem.* 2019;294(18):7128–7136.
38. Uversky VN. Intrinsically disordered proteins in overcrowded milieu: membrane-less organelles, phase separation, and intrinsic disorder. *Curr Opin Struct Biol.* 2017; 44:18–30.
39. Cuevas-Velazquez CL, Dinneny JR. Organization out of disorder: liquid-liquid phase separation in plants. *Curr Opin Plant Biol.* 2018;45:68–74.
40. Kulkarni P, Uversky VN. Intrinsically disordered proteins in chronic diseases. *Biomolecules.* 2019;9:147.
41. Mendoza-Espinosa P, García-González V, Moreno A, Castillo R, Mas-Oliva J. Disorder-to-order conformational transitions in protein structure and its relationship to disease. *Mol Cell Biochem.* 2009;330(1):105–120.
42. Anzai H, Kazama S, Kiyomatsu T, et al. Alpha-fetoprotein-producing early rectal carcinoma: a rare case report and review. *World J Surg Oncol.* 2015;13(1):1–5.
43. Sell S. Alpha-fetoprotein, stem cells and cancer: how study of the production of alpha-fetoprotein during chemical hepatocarcinogenesis led to reaffirmation of the stem cell theory of cancer. *Tumor Biol.* 2008;29(3):161–180.
44. Shi D, Gu W. Dual roles of MDM2 in the regulation of p53: ubiquitination dependent and ubiquitination independent mechanisms of MDM2 repression of p53 activity. *Genes Cancer.* 2012;3(3–4):240–248.
45. Beckerman R, Prives C. Transcriptional regulation by p53. *Cold Spring Harb Perspect Biol.* 2010;2(8):a000935.
46. Wu J, Lu LY, Yu X. The role of BRCA1 in DNA damage response. *Protein Cell.* 2010;1(2):117–123.
47. Roy R, Chun J, Powell SN. BRCA1 and BRCA2: different roles in a common pathway of genome protection. *Nat Rev Cancer.* 2012;12(1):68–78.
48. Zeng Y, He Y, Yang F, et al. The cancer/testis antigen prostate-associated gene 4 (PAGE4) is a highly intrinsically disordered protein. *J Biol Chem.* 2011;286(16): 13985–13994.
49. Coskuner O, Uversky VN. Intrinsically disordered proteins in various hypotheses on the pathogenesis of Alzheimer's and Parkinson's diseases. *Prog Mol Biol Transl Sci.* 2019;166:145–223.
50. Nguyen PH, Ramamoorthy A, Sahoo BR, et al. Amyloid oligomers: a joint experimental/computational perspective on Alzheimer's Disease, Parkinson's disease, type II diabetes, and amyotrophic lateral sclerosis. *Chem Rev.* 2021;121:2545–2647.
51. Kayed R, Lasagna-Reeves CA. Molecular mechanisms of amyloid oligomers toxicity. *J Alzheimers Dis.* 2013;33:S67–S78.
52. Kovacs GG, Budka H. Prion diseases: from protein to cell pathology. *Am J Pathol.* 2008;172(3):555–565.
53. Ostapchenko VG, Makarava N, Savtchenko R, Baskakov IV. The polybasic N-terminal region of the prion protein controls the physical properties of both the cellular and fibrillar forms of PrP. *J Mol Biol.* 2008;383(5):1210–1224.

54. Huang Y, Zhang Y, Zhao B, et al. Structural basis of RGD-hirudin binding to thrombin: Tyr 3 and five C-terminal residues are crucial for inhibiting thrombin activity. *BMC Struct Biol*. 2014;14(1):1–9.

55. Cheng B, Liu F, Guo Q, et al. Identification and characterization of hirudin-HN, a new thrombin inhibitor, from the salivary glands of Hirudo nipponia. *PeerJ*. 2019;7:e7716.

56. Palta S, Saroa R, Palta A. Overview of the coagulation system. *Indian J Anaesth*. 2014; 58(5):515.

57. Weisel JW, Litvinov RI. Fibrin formation, structure and properties. *Fibrous Proteins Struct Mechanisms*. 2017;82:405–456.

58. Carter ISR, Vanden Hoek AL, Pryzdial ELG, MacGillivray RTA. Thrombin a-chain: activation remnant or allosteric effector? *Thrombosis*. 2010;2010(18):416167.

59. Asthana S, Mallick B, Alexandrescu AT, Jha S. IAPP in type II diabetes: basic research on structure, molecular interactions, and disease mechanisms suggests potential intervention strategies. *Biochim Biophys Acta Mol Basis Dis*. 2018;1860(9):1765–1782.

60. Schreck SJ, Bridstrup J, Yuan JM. Investigating the effects of molecular crowding on the kinetics of protein aggregation. *J Phys Chem B*. 2020;124(44):9829–9839.

61. Phelan MM, Caamano-Gutierrez E, Gant MS, Grosman RX, Madine J. Using an NMR metabolomics approach to investigate the pathogenicity of amyloid-beta and alpha-synuclein. *Metabolics*. 2017;13(12):151–159.

62. Karamanos KT, Kalverda PA, Thompson SG, Redford ES. Mechanisms of amyloid formation revealed by solution NMR. *Prog Nucl Magn Reson Spectros*. 2015;88(89): 86–104.

63. Liang B, Tamm LK. Solution NMR of SNAREs, complexin and α-synuclein in association with membrane-mimetics. *Prog Nucl Magn Reson Spectrosc*. 2018;105:41–53.

64. Salmon L, Nodet G, Ozenne V, et al. Nmr characterization of long-range order in intrinsically disordered proteins. *J Am Chem Soc*. 2010;132(24):8407–8418.

65. Jaseňáková Z, Zapleta V, Padrta P, Zachrdla M, et al. Boosting the resolution of low-field 15N relaxation experiments on intrinsically disordered proteins with triple-resonance NMR. *J Biomol NMR*. 2020;74(2–3):139–145.

66. Caliskan M, Mandaci YS, Uversky NV, Coskuner-Weber O. Secondary structure dependence of amyloid-β(1–40) on simulation techniques and force field parameters. *Chem Biol Drug Des*. 2021;97(5):1100–1108.

67. Loquet A, El Mammeri N, Stanek J, et al. 3D structure determination of amyloid fibrils using solid-state NMR spectroscopy. *Methods*. 2018;138:26–38.

68. Akbayrak YI, Caglayan IS, Uversky NV, Coskuner-Weber O. Current challenges and limitations in the studies of intrinsically disordered proteins in neurodegenerative diseases by computer simulations. *Curr Alzheimer Res*. 2020;17(9):805–818.

69. Luo Y, Xiang S, Hooikaas JP, et al. Direct observation of dynamic protein interactions involving human microtubules using solid-state NMR spectroscopy. *Nat Commun*. 2020;11(1):18–19.

70. Savastano A, Jaipuria G, Andreas L, Mandelkow E, Markus Zweckstetter M. Solid-state NMR investigation of the involvement of the P2 region in tau amyloid fibrils. *Sci Rep*. 2020;10(1):21210–21218.

71. Uluca B, Viennet T, Petrović D, et al. DNP-Enhanced MAS NMR: a tool to snapshot conformational ensembles of α-synuclein in different states. *Biophys J*. 2018; 114(7):1614–1623.

72. Kazuma M, HIdeyuki H, Masuda Y, Ohigashi H, Irie K. Distance measurement between Tyr10 and Met35 in amyloid beta by site-directed spin-labeling ESR spectroscopy: implications for the stronger neurotoxicity of Abeta42 than Abeta40. *ChemBioChem*. 2007;8(18):2308–2314.

73. Borbat P, Ramlall FT, Freed HJ, Eliezer D. Inter-helix distances in lysophospholipid micelle-bound alpha-synuclein from pulsed ESR measurements. *J Am Chem Soc.* 2006;128(31):10004–10005.

74. Ramakrishnan M, Jensen PH, Marsh D. Association of α-synuclein and mutants with lipid membranes: spin-label ESR and polarized IR. *Biochemistry.* 2006;45(10): 3386–3395.

75. Alies B, Sasaki I, Proux O, et al. Zn impacts Cu coordination to amyloid-β, the Alzheimer's peptide, but not the ROS production and the associated cell toxicity. *Chem Commun.* 2013;49(12):1214–1216.

76. Shearer J, Szalai V. The amyloid-beta peptide of Alzheimer's disease binds Cu(I) in a linear bis-his coordination environment: insight into a possible neuroprotective mechanism for the amyloid-beta peptide. *J Am Chem Soc.* 2008;31(130):17826–17835.

77. Zeng Y, He Y, et al. The cancer/testis antigen prostate-associated gene 4 (PAGE4) is a highly intrinsically disordered protein. *J Biol Chem.* 2011;286(16):13985–13994.

78. Meng F, Bellaiche M, Kim JY, Zerze HG, Best BR, Sung Chung H. Highly disordered amyloid-β monomer probed by single-molecule fret and md simulation. *Biophys J.* 2018;114(4):870–884.

79. Kodera N, Noshiro D, et al. Structural and dynamics analysis of intrinsically disordered proteins by high-speed atomic force microscopy. *Nat Nanotechnol.* 2021;16(2): 181–189.

80. Ryan T, Kirby N, Pham C, et al. Small angle X-ray scattering analysis of Cu(2+)-induced oligomers of the Alzheimer's amyloid β peptide. *Metallomics.* 2015;7(3): 536–543.

81. Choi M, Dahal E, Badano A. Feasibility of imaging amyloid in the brain using small-angle x-ray scattering. *Biomed Phys and Eng Exp.* 2020;7:015008.

82. Rauscher S, Gapsys V, Gajda MJ, Zweckstetter M, de Groot BL, Grubmüller H. Structural ensembles of intrinsically disordered proteins depend strongly on force field: a comparison to experiment. *J Chem Theory Comput.* 2015;11:5513–5524.

83. Carballo-Pacheco M, Strodel B. Comparison of force fields for Alzheimer's A β42: a case study for intrinsically disordered proteins. *Protein Sci.* 2017;26(2):174–185.

84. Robustelli P, Piana S, Shaw DE. Developing a molecular dynamics force field for both folded and disordered protein states. *Proc Natl Acad Sci USA.* 2018;115(21): E4758–E4766.

85. Man VH, He X, Derreumaux P, et al. Effects of all-atom molecular mechanics force fields on amyloid peptide assembly: the case of Aβ 16–22 dimer. *J Chem Theory Comput.* 2019;15(2):1440–1452.

86. Rahman MU, Rehman AU, Liu H, Chen HF. Comparison and evaluation of force fields for intrinsically disordered proteins. *J Chem Inf Model.* 2020;60(10):4912–4923.

87. Paul A, Samantray S, Anteghini M, Khaled M, Strodel B. Thermodynamics and kinetics of the amyloid-β peptide revealed by Markov state models based on MD data in agreement with experiment. *Chem Sci.* 2021;12(19):6652–6669.

88. Strodel B. Amyloid aggregation simulations: challenges, advances and perspectives. *Curr Opin Struct Biol.* 2021;67:145–152.

89. Mu J, Liu H, Zhang J, Luo R, Chen HF. Recent force field strategies for intrinsically disordered proteins. *J Chem Inf Model.* 2021;61(3):1037–1047.

90. Brooks BR, Bruccoleri RE, Olafson BD, States DJ, Swaminathan S, Karplus M. CHARMM: a program for macromolecular energy, minimization, and dynamics calculations. *J Comput Chem.* 1983;4(2):187–217.

91. Bayly CI, Merz KM, Ferguson DM, et al. A second generation force field for the simulation of proteins, nucleic acids, and organic molecules. *J Am Chem Soc.* 1995; 117(19):5179–5197.

92. Jorgensen WL, Tirado-Rives J. The OPLS potential functions for proteins. Energy minimizations for crystals of cyclic peptides and crambin. *J Am Chem Soc*. 1988; 110(6):1657–1666.

93. Scott WRP, Hünenberger PH, Tironi IG, et al. The GROMOS biomolecular simulation program package. *J Phys Chem A*. 1999;103(19):3596–3607.

94. Huang J, Rauscher S, Nawrocki G, et al. CHARMM36m: an improved force field for folded and intrinsically disordered proteins. *Nat Methods*. 2016;14(1):71–73.

95. Piana S, Robustelli P, Tan D, Chen S, Shaw DE. Development of a force field for the simulation of single-chain proteins and protein-protein complexes. *J Chem Theory Comput*. 2020;16(4):2494–2507.

96. Best RB, Hummer G. Optimized molecular dynamics force fields applied to the helix-coil transition of polypeptides. *J Phys Chem B*. 2009;113(26):9004–9015.

97. Duan Y, Wu C, Chowdhury S, et al. A point-charge force field for molecular mechanics simulations of proteins based on condensed-phase Quantum mechanical calculations. *J Comput Chem*. 2003;24(16):1999–2012.

98. Hornak V, Abel R, Okur A, Strockbine B, Roitberg A, Simmerling C. Comparison of multiple amber force fields and development of improved protein backbone parameters. *Proteins*. 2006;65(3):712–725.

99. Robertson MJ, Tirado-Rives J, Jorgensen WL. Improved peptide and protein torsional energetics with the OPLS-AA force field. *J Chem Theory Comput*. 2015;11(7): 3499–3509.

100. Harder E, Damm W, Maple J, et al. OPLS3: a force field providing broad coverage of drug-like small molecules and proteins. *J Chem Theory Comput*. 2016;12(1):281–296.

101. Piana S, Lindorff-Larsen K, Shaw DE. How robust are protein folding simulations with respect to force field parameterization? *Biophys J*. 2011;100(9):L47–L49.

102. MacKerell Jr AD, Bashford D, Bellott MLDR, et al. All-atom empirical potential for molecular modeling and dynamics studies of proteins. *J Phys Chem B*. 1998; 102(18):3586–3616.

103. Carballo-Pacheco M, Strodel B. Comparison of force fields for Alzheimer's A: a case study for intrinsically disordered proteins. *Protein Sci*. 2017;26(2):174–185.

104. Nerenberg PS, Head-Gordon T. Optimizing protein-solvent force fields to reproduce intrinsic conformational preferences of model peptides. *J Chem Theory Comput*. 2011;7 (4):1220–1230.

105. Nerenberg PS, Jo B, So C, Tripathy A, Head-Gordon TL. A new protein and water force field combination for reproducing solvation free energies. *J Phys Chem B*. 2012;116(15):4524–4534.

106. Jiang F, Zhou CY, Wu YD. Residue-specific force field based on the protein coil library. RSFF1: modification of OPLS-AA/L. *J Phys Chem B*. 2014;118(25): 6983–6998.

107. Zhou CY, Jiang F, Wu YD. Residue-specific force field based on protein coil library. RSFF2: modification of AMBER ff99SB. *J Phys Chem B*. 2015;119(3):1035–1047.

108. MacKerell AD, Feig M, Brooks CL. Improved treatment of the protein backbone in empirical force fields. *J Am Chem Soc*. 2004;126(3):698–699.

109. Mackerell AD, Feig M, Brooks CL. Extending the treatment of backbone energetics in protein force fields: limitations of gas-phase quantum mechanics in reproducing protein conformational distributions in molecular dynamics simulation. *J Comput Chem*. 2004;25(11):1400–1415.

110. Olivier R, Hanqiang C. Nearest neighbor value interpolation. *Int J Adv Comput Sci Appl*. 2012;3(4).

111. Taylor LR, Press WH, Flannery BP, Teukolsky SA, Vetterling WT. Numerical recipes: the art of scientific computing. *J Anim Ecol*. 1987;56(1):374.

112. Huang J, Mackerell AD. CHARMM36 all-atom additive protein force field: Validation based on comparison to NMR data. *J Comput Chem*. 2013;34(25): 2135–2145.

113. Song D, Luo R, Chen HF. The IDP-specific force field ff14IDPSFF improves the conformer sampling of intrinsically disordered proteins. *J Chem Inf Model*. 2017; 57(5):1166–1178.

114. Liu H, Song D, Lu H, Luo R, Chen HF. Intrinsically disordered protein-specific force field CHARMM36IDPSFF. *Chem Biol Drug Des*. 2018;92(4):1722–1735.

115. Liu H, Song D, Zhang Y, Yang S, Luo R, Chen HF. Extensive tests and evaluation of the CHARMM36IDPSFF force field for intrinsically disordered proteins and folded proteins. *Phys Chem Chem Phys*. 2019;21(39):21918–21931.

116. Yang S, Liu H, Zhang Y, Lu H, Chen H. Residue-specific force field improving the sample of intrinsically disordered proteins and folded proteins. *J Chem Inf Model*. 2019;59(11):4793–4805.

117. Kang W, Jiang F, Wu YD. Universal implementation of a residue-specific force field based on CMAP potentials and free energy decomposition. *J Chem Theory Comput*. 2018;14(8):4474–4486.

118. Huang J, MacKerell AD. Force field development and simulations of intrinsically disordered proteins. *Curr Opin Struct Biol*. 2018;48:40–48.

119. Chen J, Wu Y. Recent development of atomistic force fields and simulations of intrinsically disordered proteins. *Sci Sin Chim*. 2020;50(10):1320–1332.

120. Best RB, Zheng W, Mittal J. Balanced protein-water interactions improve properties of disordered proteins and non-specific protein association. *J Chem Theory Comput*. 2014;10(11):5113–5124.

121. Piana S, Donchev AG, Robustelli P, Shaw DE. Water dispersion interactions strongly influence simulated structural properties of disordered protein states. *J Phys Chem B*. 2015;119(16):5113–5123.

122. Samantray S, Yin F, Kav B, Strodel B. Different force fields give rise to different amyloid aggregation pathways in molecular dynamics simulations. *J Chem Info Model*. 2020;60(12):6462–6475.

123. Carballo-Pacheco M, Ismail AE, Strodel B. On the applicability of force fields to study the aggregation of amyloidogenic peptides using molecular dynamics simulations. *J Chem Theory Comput*. 2018;14(11):6063–6075.

124. Kaminski GA, Stern HA, Berne BJ, et al. Development of a polarizable force field for proteins via ab initio quantum chemistry: first generation model and gas phase tests. *J Comput Chem*. 2002;23(16):1515–1531.

125. Baldwin R, Baker DJ. *Peptide Solvation and H-Bonds*. Elsevier; 2006.

126. Friesner RA. Modeling polarization in proteins and protein-ligand complexes: methods and preliminary results. In: Baldwin R, Baker DJ, eds. *Peptide Solvation and H-bonds*. vol. 72. Elsevier; 2005:79–104.

127. Lopes PEM, Huang J, Shim J, et al. Polarizable force field for peptides and proteins based on the classical drude oscillator. *J Chem Theory Comput*. 2013;9(12):5430–5449.

128. Huang J, Mackerell AD. Induction of peptide bond dipoles drives cooperative helix formation in the (AAQAA)3 peptide. *Biophys J*. 2014;107(4):991–997.

129. Lemkul JA, Huang J, MacKerell AD. Induced dipole-dipole interactions influence the unfolding pathways of wild-type and mutant amyloid β-peptides. *J Phys Chem B*. 2015;119(51):15574–15582.

130. Wei H, Qi R, Wang J, Cieplak P, Duan Y, Luo R. Efficient formulation of polarizable Gaussian multipole electrostatics for biomolecular simulations. *J Chem Phys*. 2020; 153(11):114116.

131. Lemkul JA. Chapter One–Pairwise-additive and polarizable atomistic force fields for molecular dynamics simulations of proteins. In: Strodel B, Barz B, eds. *Computational Approaches for Understanding Dynamical Systems: Protein Folding and Assembly*. Progress in Molecular Biology and Translational Science; vol. 170. Academic Press; 2020:1–71.

132. Wang LP, Martinez TJ, Pande VS. Building force fields: an automatic, systematic, and reproducible approach. *J Phys Chem Lett*. 2014;5(11):1885–1891.

133. Demerdash O, Shrestha UR, Petridis L, Smith JC, Mitchell JC, Ramanathan A. Using small-angle scattering data and parametric machine learning to optimize force field parameters for intrinsically disordered proteins. *Front Mol Biosci*. 2019;6:64.

134. Laury ML, Wang LP, Pande VS, Head-Gordon T, Ponder JW. Revised parameters for the AMOEBA polarizable atomic multipole water model. *J Phys Chem B*. 2015;119(29):9423–9437.

135. Ramanathan A, Ma H, Parvatikar A, Chennubhotla SC. Artificial intelligence techniques for integrative structural biology of intrinsically disordered proteins. *Curr Opin Struct Biol*. 2021;66:216–224.

136. Davtyan A, Schafer NP, Zheng W, Clementi C, Wolynes PG, Papoian GA. AWSEM-MD: Protein structure prediction using coarse-grained physical potentials and bioinformatically based local structure biasing. *J Phys Chem B*. 2012; 116(29):8494–8503.

137. Wu H, Wolynes PG, Papoian GA. AWSEM-IDP: a Coarse-Grained force field for intrinsically disordered proteins. *J Phys Chem B*. 2018;122(49):11115–11125.

138. Latham AP, Zhang B. Maximum entropy optimized force field for intrinsically disordered proteins. *J Chem Theory Comput*. 2020;16(1):773–781.

139. Sterpone F, Melchionna S, Tuffery P, et al. The OPEP protein model: from single molecules, amyloid formation, crowding and hydrodynamics to DNA/RNA systems. *Chem Soc Rev*. 2014;43(13):4871–4893.

140. Baul U, Chakraborty D, Mugnai ML, Straub JE, Thirumalai D. Sequence effects on size, shape, and structural heterogeneity in intrinsically disordered proteins. *J Phys Chem B*. 2019;123(16):3462–3474.

141. Chakraborty D, Straub J, Thirumalai D. Differences in the free energies between the excited states of Aβ40 and Aβ42 monomers encode their aggregation propensities. *Proc Natl Acad Sci USA*. 2020;117(33):19926–19937.

142. McCarty J, Delaney KT, Danielsen SPO, Fredrickson GH, Shea JE. Complete phase diagram for liquid-liquid phase separation of intrinsically disordered proteins. *J Phys Chem Lett*. 2019;10(8):1644–1652.

143. Van Der Lee R, Buljan M, Lang B, et al. Classification of intrinsically disordered regions and proteins. *Chem Rev*. 2014;114(13):6589–6631.

144. Dyson HJ, Wright PE. Perspective: the essential role of NMR in the discovery and characterization of intrinsically disordered proteins. *J Biomol NMR*. 2019;73(12):651–659.

145. Cordeiro TN, Herranz-Trillo F, Urbanek A, et al. Small-angle scattering studies of intrinsically disordered proteins and their complexes. *Curr Opin Struct Biol*. 2017; 42:15–23.

146. Mansouri AL, Grese LN, Rowe EL, et al. Folding propensity of intrinsically disordered proteins by osmotic stress. *Mol Biosyst*. 2016;12(12):3695–3701.

147. LeBlanc SJ, Kulkarni P, Weninger KR. Single molecule FRET: A powerful tool to study intrinsically disordered proteins. *Biomolecules*. 2018;8(4):140.

148. Schuler B, Soranno A, Hofmann H, Nettels D. Single-molecule FRET spectroscopy and the polymer physics of unfolded and intrinsically disordered proteins. *Annu Rev Biophys*. 2016;45:207–231.

149. Chong SH, Chatterjee P, Ham S. Computer simulations of intrinsically disordered proteins. *Ann Rev Phys Chem*. 2017;68:117–134.
150. Wang W. Recent advances in atomic molecular dynamics simulation of intrinsically disordered proteins. *Phys Chem Chem Phys*. 2021;23(2):777–784.
151. Battisti A, Tenenbaum A. Molecular dynamics simulation of intrinsically disordered proteins. *Mol Simul*. 2012;38(2):139–143.
152. Best RB. Computational and theoretical advances in studies of intrinsically disordered proteins. *Curr Opin Struct Biol*. 2017;42:147–154.
153. Kasahara K, Terazawa H, Takahashi T, Higo J. Studies on molecular dynamics of intrinsically disordered proteins and their fuzzy complexes: a mini-review. *Comput Struct Biotechnol J*. 2019;17:712–720.
154. Shrestha UR, Juneja P, Zhang Q, et al. Generation of the configurational ensemble of an intrinsically disordered protein from unbiased molecular dynamics simulation. *Proc Natl Acad Sci USA*. 2019;116(41):20446–20452.
155. Shrestha UR, Smith JC, Petridis L. Full structural ensembles of intrinsically disordered proteins from unbiased molecular dynamics simulations. *Commun Biol*. 2021;4(1):1–8.
156. Bhattacharya S, Lin X. Recent advances in computational protocols addressing intrinsically disordered proteins. *Biomolecules*. 2019;9(4):146.
157. Abrams C, Bussi G. Enhanced sampling in molecular dynamics using metadynamics, replica-exchange, and temperature-acceleration. *Entropy*. 2014;16(1):163–199.
158. Yang YI, Shao Q, Zhang J, Yang L, Gao YQ. Enhanced sampling in molecular dynamics. *J Chem Phys*. 2019;151(7):070902.
159. Paul S, Nair NN, Vashisth H. Phase space and collective variable based simulation methods for studies of rare events. *Mol Simul*. 2019;45(14–15):1273–1284.
160. Bussi G, Laio A. Using metadynamics to explore complex free-energy landscapes. *Nat Rev Phys*. 2020;2(4):200–212.
161. Affentranger R, Tavernelli I, Di Iorio EE. A novel hamiltonian replica exchange md protocol to enhance protein conformational space sampling. *J Chem Theory Comput*. 2006;2(2):217–228.
162. Roe DR, Bergonzo C, Cheatham IIITE. Evaluation of enhanced sampling provided by accelerated molecular dynamics with Hamiltonian replica exchange methods. *J Phys Chem B*. 2014;118(13):3543–3552.
163. Samantray S, Cheung DL. Effect of the air-water interface on the conformation of amyloid beta. *Biointerphases*. 2020;15(6):061011.
164. Fatafta H, Poojari C, Sayyed-Ahmad A, Strodel B, Owen MC. Role of oxidized gly25, gly29, and gly33 residues on the interactions of Aβ1–42 with lipid membranes. *ACS Chem Neurosci*. 2020;11(4):535–548.
165. Laio A, Parrinello M. Escaping free-energy minima. *Proc Natl Acad Sci USA*. 2002;99 (20):12562–12566.
166. Barducci A, Bonomi M, Parrinello M. Metadynamics. *Wiley Interdiscip Rev Comput Mol Sci*. 2011;1(5):826–843.
167. Barducci A, Bussi G, Parrinello M. Well-tempered metadynamics: a smoothly converging and tunable free-energy method. *Phys Rev Lett*. 2008;100(2):020603.
168. Branduardi D, Bussi G, Parrinello M. Metadynamics with adaptive gaussians. *J Chem Theory Comput*. 2012;8(7):2247–2254.
169. Bussi G, Gervasio FL, Laio A, Parrinello M. Free-energy landscape for β hairpin folding from combined parallel tempering and metadynamics. *J Am Chem Soc*. 2006;128 (41):13435–13441.
170. Deighan M, Bonomi M, Pfaendtner J. Efficient simulation of explicitly solvated proteins in the well-tempered ensemble. *J Chem Theory Comput*. 2012;8(7): 2189–2192.

171. Raiteri P, Laio A, Gervasio FL, Micheletti C, Parrinello M. Efficient reconstruction of complex free energy landscapes by multiple walkers metadynamics. *J Phys Chem B*. 2006;110(8):3533–3539.

172. Piana S, Laio A. A bias-exchange approach to protein folding. *J Phys Chem B*. 2007;111 (17):4553–4559.

173. Miao Y, McCammon JA. Unconstrained enhanced sampling for free energy calculations of biomolecules: a review. *Mol Simul*. 2016;42(13):1046–1055.

174. Han M, Xu J, Ren Y, Li J. Simulation of coupled folding and binding of an intrinsically disordered protein in explicit solvent with metadynamics. *J Mol Graph Model*. 2016;68:114–127.

175. Granata D, Baftizadeh F, Habchi J, et al. The inverted free energy landscape of an intrinsically disordered peptide by simulations and experiments. *Sci Rep*. 2015; 5(1):1–15.

176. Sugita Y, Okamoto Y. Replica-exchange molecular dynamics method for protein folding. *Chem Phys Lett*. 1999;314(1–2):141–151.

177. Sgourakis NG, Yan Y, McCallum SA, Wang C, Garcia AE. The Alzheimer's peptides $A\beta 40$ and 42 adopt distinct conformations in water: a combined MD/NMR study. *J Mol Biol*. 2007;368(5):1448–1457.

178. Potoyan DA, Papoian GA. Energy landscape analyses of disordered histone tails reveal special organization of their conformational dynamics. *J Am Chem Soc*. 2011;133 (19):7405–7415.

179. Zhang W, Ganguly D, Chen J. Residual structures, conformational fluctuations, and electrostatic interactions in the synergistic folding of two intrinsically disordered proteins. *PLoS Comput Biol*. 2012;8(1):e1002353.

180. Knott M, Best RB. A preformed binding interface in the unbound ensemble of an intrinsically disordered protein: evidence from molecular simulations. *PLoS Comput Biol*. 2012;8(7):e1002605.

181. Miller C, Zerze GH, Mittal J. Molecular simulations indicate marked differences in the structure of amylin mutants, correlated with known aggregation propensity. *J Phys Chem B*. 2013;117(50):16066–16075.

182. Zerze GH, Mittal J. Effect of O-linked glycosylation on the equilibrium structural ensemble of intrinsically disordered polypeptides. *J Phys Chem B*. 2015;119(51):15583–15592.

183. Ganguly D, Chen J. Atomistic details of the disordered states of KID and pKID. Implications in coupled binding and folding. *J Am Chem Soc*. 2009;131(14):5214–5223.

184. Bussi G. Hamiltonian replica exchange in GROMACS: a flexible implementation. *Mol Phys*. 2014;112(3–4):379–384.

185. Liao Q, Owen MC, Bali S, Barz B, Strodel B. $A\beta$ under stress: the effects of acidosis, Cu^{2+}-binding, and oxidation on amyloid β-peptide dimers. *Chem Commun*. 2018;54 (56):7766–7769.

186. Itoh SG, Okumura H. Oligomer formation of amyloid-β (29–42) from its monomers using the hamiltonian replica-permutation molecular dynamics simulation. *J Phys Chem B*. 2016;120(27):6555–6561.

187. Liu X, Gong X, Chen J. Accelerating atomistic simulations of proteins using multiscale enhanced sampling with independent tempering. *J Comput Chem*. 2021;42(5):358–364.

188. Phillips JC, Braun R, Wang W, et al. Scalable molecular dynamics with NAMD. *J Comput Chem*. 2005;26(16):1781–1802.

189. Bonomi M, Branduardi D, Bussi G, et al. PLUMED: a portable plugin for free-energy calculations with molecular dynamics. *Comput Phys Commun*. 2009;180(10): 1961–1972.

190. Abraham MJ, Murtola T, Schulz R, et al. Gromacs: high performance molecular simulations through multi-level parallelism from laptops to supercomputers. *SoftwareX*. 2015;1:19–25.

191. Chodera JD, Noé F. Markov state models of biomolecular conformational dynamics. *Curr Opin Struct Biol*. 2014;25:135–144.

192. Schwantes CR, McGibbon RT, Pande VS. Perspective: Markov models for long-timescale biomolecular dynamics. *J Chem Phys*. 2014;141(9):09B201_1.

193. Pande VS, Beauchamp K, Bowman GR. Everything you wanted to know about markov state models but were afraid to ask. *Methods*. 2010;52(1):99–105.

194. Prinz JH, Wu H, Sarich M, et al. Markov models of molecular kinetics: generation and validation. *J Chem Phys*. 2011;134(17):174105.

195. Scherer MK, Trendelkamp-Schroer B, Paul F, et al. PyEMMA 2: A software package for estimation, validation, and analysis of Markov models. *J Chem Theory Comput*. 2015;11(11):5525–5542.

196. Harrigan MP, Sultan MM, Hernández CX, et al. Msmbuilder: statistical models for biomolecular dynamics. *Biophys J*. 2017;112(1):10–15.

197. Lin YS, Bowman GR, Beauchamp KA, Pande VS. Investigating how peptide length and a pathogenic mutation modify the structural ensemble of amyloid beta monomer. *Biophys J*. 2012;102(2):315–324.

198. Qiao Q, Bowman GR, Huang X. Dynamics of an intrinsically disordered protein reveal metastable conformations that potentially seed aggregation. *J Am Chem Soc*. 2013;135(43):16092–16101.

199. Löhr T, Kohlhoff K, Heller G, Camilloni C, Vendruscolo M. A kinetic ensemble of the alzheimer's Aβ peptide. *Nature Comput Sci*. 2021;1:71–78.

200. Stanley N, Esteban-Martín S, De Fabritiis G. Kinetic modulation of a disordered protein domain by phosphorylation. *Nat Commun*. 2014;5(1):1–8.

201. Illig AM, Strodel B. Performance of markov state models and transition networks on characterizing amyloid aggregation pathways from md data. *J Chem Theory Comput*. 2020;16(12):7825–7839.

202. Fatafta H, Khaled M, Sayyed-Ahmad A, Strodel B. Amyloid-β peptide dimers undergo a random coil to β-sheet transition in the aqueous phase but not at the neuronal membrane. *bioRxiv*. 2021: 2020-12. https://doi.org/10.1101/2020.12.31.424964.

203. Rajagopalan K, Qiu R, Mooney SM, et al. The stress-response protein prostate-associated gene 4, interacts with c-jun and potentiates its transactivation. *Biochim Biophys Acta Mol Basis Dis*. 2014;1842(2):154–163.

204. Kulkarni P, Jolly MK, Jia D, et al. Phosphorylation-induced conformational dynamics in an intrinsically disordered protein and potential role in phenotypic heterogeneity. *Proc Natl Acad Sci USA*. 2017;114(13):E2644–E2653.

205. Lin X, Kulkarni P, Bocci F, et al. Structural and dynamical order of a disordered protein: molecular insights into conformational switching of PAGE4 at the systems level. *Biomolecules*. 2019;9(2):77.

206. Lin X, Kulkarni P, Bocci F, et al. Structural and dynamical order of a disordered protein: molecular insights into conformational switching of page4 at the systems level. *Biomolecules*. 2019;9(2):77.

207. Parker D, Ferreri K, Nakajima T, et al. Phosphorylation of CREB at Ser-133 induces complex formation with CREB-binding protein via a direct mechanism. *Mol Cell Biol*. 1996;16(2):694–703.

208. Parker D, Rivera M, Zor T, et al. Role of secondary structure in discrimination between constitutive and inducible activators. *Mol Cell Biol*. 1999;19(8):5601–5607.

209. Zor T, Mayr BM, Dyson HJ, Montminy MR, Wright PE. Roles of phosphorylation and helix propensity in the binding of the KIX domain of CREB-binding protein by constitutive (c-Myb) and inducible (CREB) activators. *J Biol Chem*. 2002; 277(44):42241–42248.

210. Turjanski A, Gutkind JS, Best R, Hummer G. Binding-induced folding of a natively unstructured transcription factor. *PLoS Comput Biol*. 2008;4:e1000060.

211. Liu H, Guo X, Han J, Luo R, Chen HF. Order-disorder transition of intrinsically disordered kinase inducible transactivation domain of CREB. *J Chem Phys*. 2018; 148(22):225101.

212. Sridhar A, Orozco M, Collepardo-Guevara R. Protein disorder-to-order transition enhances the nucleosome-binding affinity of H1. *Nucleic Acids Res*. 2020;48(10): 5318–5331.

213. Moritsugu K, Terada T, Kidera A. Disorder-to-order transition of an intrinsically disordered region of sortase revealed by multiscale enhanced sampling. *J Am Chem Soc*. 2012;134(16):7094–7101.

214. Suree N, Liew CK, Villareal VA, et al. The structure of the staphylococcus aureus sortase-substrate complex reveals how the universally conserved lpxtg sorting signal is recognized. *J Biol Chem*. 2009;284(36):24465–24477.

215. Naik MT, Suree N, Ilangovan U, et al. Staphylococcus aureus Sortase A Transpeptidase: calcium promotes sorting signal binding by altering the mobility and structure of an active site loop. *J Biol Chem*. 2006;281(3):1817–1826.

CHAPTER SIX

Target-binding behavior of IDPs via pre-structured motifs

Do-Hyoung Kim and Kyou-Hoon Han*

Korea Research Institute of Bioscience and Biotechnology, Daejeon, South Korea
*Corresponding author: e-mail address: khhan600@kribb.re.kr

Contents

Abstract

Pre-Structured Motifs (PreSMos) are transient secondary structures observed in many intrinsically disordered proteins (IDPs) and serve as protein target-binding hot spots. The prefix "pre" highlights that PreSMos exist a priori in the target-unbound state of IDPs as the active pockets of globular proteins pre-exist before target binding. Therefore, a PreSMo is an "active site" of an IDP; it is not a spatial pocket, but rather a secondary structural motif. The classical and perhaps the most effective approach to understand the function of a protein has been to determine and investigate its structure. Ironically or by definition IDPs do not possess *structure* (here *structure* refers to tertiary structure only). *Are IDPs then entirely structureless?* The PreSMos provide us with an atomic-resolution answer to this question. For target binding, IDPs do not rely on the spatial pockets afforded by tertiary or higher structures. Instead, they utilize the PreSMos possessing particular conformations that highly presage the target-bound conformations. PreSMos are recognized or captured by targets via conformational selection (CS) before their conformations eventually become stabilized via structural induction into more ordered bound structures. Using PreSMos, a number of, if not all, IDPs can bind targets following a sequential pathway of CS followed by an induced fit (IF). This chapter presents several important PreSMos implicated in cancers, neurodegenerative diseases, and other diseases along with discussions on their conformational details that mediate target binding, a structural rationale for unstructured proteins.

Abbreviations

4EBP	eIF4E binding protein
AD	Alzheimer's disease
AF-1	activation function–1 domain
BMRB	biological magnetic resonance bank
CBD	C-terminal basic domain
CD	circular dichroism
CFTR	cystic fibrosis transmembrane conductance regulator
CR	conserved region
CREB	cAMP-response element binding protein
CS	conformational selection
CU	completely unstructured
DBD	DNA-binding domain
eIF4G	eukaryotic translation initiation factor 4G
Flg M	flagella M
HBV	hepatitis B virus
hGR	human glucocorticoid receptor
HPVs	human papillomaviruses
ICMRBS	International Conference on Magnetic Resonance in Biological Systems
IDR	intrinsically disordered region
IF	induced fit
KID	kinase inducible domain
KIX	kinase-inducible domain interacting
LBD	ligand binding domain
MIA	maternal immune activation
MoRF	molecular recognition feature
MT	microtubule
MU	mostly unstructured
NCBD	nuclear coactivator binding domain
NMR	nuclear magnetic resonance
NOE	nuclear Overhauser effect

OD	oligomerization domain
PD	Parkinson's disease
PHF	paired helical filament
PPII	polyproline II
PRE	paramagnetic relaxation enhancement
PreSMo	pre-structured motif
PSE	pre-formed structural element
RAN	repeat-associated non–AUG
RDC	residual dipolar coupling constant
REMD	replica exchange molecular dynamics
RPA	replication protein A
SLiM	short linear motif
SSP	secondary structure propensity
SUMO	small ubiquitin-like modifier
SUSP4	SUMO specific protease 4
TAD	transactivation domain
TAF	TATA box binding protein associated factor
TAT	transactivator of transcription
Tau1c	core domain of tau1
TAZ	transcriptional adaptor zinc-binding domain
TF	transcription factor
TFE	trifluoroethanol
TMAO	trimethylamine N-oxide
TrNOE	transferred NOE
VP16	virion protein 16
αS	alpha-synuclein

1. Introduction

Intrinsically disordered proteins (IDPs) are novel proteins that do not, under non-denaturing conditions, form uniquely-defined three-dimensional (3-D) structures.[1] Interests in these peculiar proteins arise because they are still capable of carrying out diverse biological functions despite lacking 3-D structures.[2,3] Reports on these unorthodox proteins began appearing in the 1990s with various names such as unstructured proteins, largely unstructured proteins, intrinsically unstructured proteins, natively unfolded proteins, inherently unstructured proteins, intrinsically unfolded proteins, natively unstructured proteins, natively disordered proteins, inherently disordered proteins, etc.[2–11] After 2013 a consensus was reached on the name as intrinsically disordered proteins when a dedicated journal titled Intrinsically Disordered Proteins was launched.[1] Interests in IDPs have been very keen because their existence not only contradict the decades-old paradigm in protein science, 3-D structure = function, but also they are associated with fatal diseases such as cancers, prion diseases, Alzheimer's disease (AD), Parkinson's disease (PD) and so on,[2-4,12] strongly suggesting that meaningful effort is needed to understand what IDPs are and how they function. IDPs and intrinsically disordered regions (IDRs) in IDPs[13] are distinct from the phenomenon of protein disorder. The latter has been known for decades and usually refers to short segments (typically consisting of less than 20 residues) that link secondary structures in globular proteins[14] even though there are exceptions to this definition such as the disorder exhibited by heme pockets.[15] On the other hand, IDPs/IDRs contain long disordered regions (minimally 40 residues and often encompassing hundreds of amino acid residues), which is the essence that makes IDPs novel.[3] Massive bioinformatics predictions have shown that IDPs/IDRs occupy as much as one half of entire human proteome,[16] strengthening the assertion that without clear understanding of IDPs an accurate and comprehensive research paradigm in protein science would not be possible, not to mention effectively managing the associated diseases.

2. Target-binding of IDPs

Target-binding of proteins is a multi-faceted process that depends on several experimental variables such as ionic strength, pH, temperatures, etc.[17] Two mutually exclusive mechanisms, induced fit (IF) and conformational selection (CS), were proposed for target binding of globular proteins half a century ago.[18] In Fig. 1, we describe how these two models might be applied for IDPs. While this picture is simplistic since the actual binding landscape is much more complex[24] it suffices to deliver an overall concept. An IDP exists as an ensemble of conformers where partially structured conformers and fully unstructured ones containing no secondary structures coexist. In principle a large number of different conformers separated by low energy barriers may exist, but it is probably more realistic to view that the conformational ensemble consists of only a limited number of conformers as shown experimentally.[19,20] Different encounter complexes form depending on which conformer is engaged in initial interaction. To become a final bound structure partially structured conformers need a partial structural tightening whereas the fully unstructured ones would have to undergo a full (100%) structural induction. A more elegant picture can be found in fig. 1 of Wang et al.[25] In the case of globular proteins, it appears that CS is favored over IF.[18,26] For IDPs, however, there are not yet enough experimental data to conclude which mechanism describes target binding better. It needs to be noted that even for globular proteins the mechanism varies from one protein

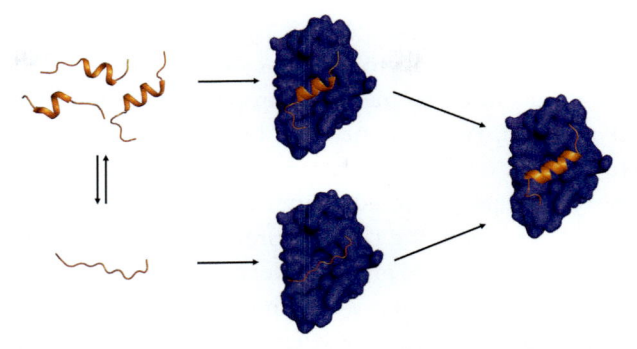

Fig. 1 A schematic diagram showing two possible mechanisms for IDP-target binding. (*Left column*) An unbound IDP or a sufficiently long IDR (>40 residues) exists in a conformational ensemble where a completely unstructured state (bottom) is in equilibrium with more than one pre-structured conformers having transient helices of slightly different helical lengths (top). The fraction of the pre-structured state has been shown to vary from one IDP to another (10%–70%).[3] The three pre-structured helices with slightly different helical lengths were experimentally observed by Flexible-Meccano calculations using residual dipolar couplings in the case of the NTAIL of Sendai virus nucleoprotein[19] and 4EBP1.[20] (*Right column*) A transient pre-structured helix seen in the target-unbound state becomes a stable helix upon target binding, which was observed in the case of p53TAD/mdm2,[2,21] KID/KIX,[22] and 4EBP1/4EBP2 with eIF4E.[20,23] A peptide with a sequence of the pre-structured helix in p53TAD or in 4EBP1 is used as a basis for anti-cancer peptide therapeutics design. (*Middle column*) Two potential encounter complexes (EC). The top EC would be formed by a conformational selection (CS) of a PreSMo by a target while the bottom EC would be produced in an induced fit (IF) mechanism. The top CS path is likely to be more favorable in terms of entropy than the IF path since the bottom EC would have to face a larger entropic penalty to form a stable helix during or upon target binding. A more elegant figure can be found in Ref.[25] *Reproduced with permission from Kim DH, Han KH. Transient secondary structures as general target-binding motifs in intrinsically disordered proteins. Int J Mol Sci. 2018;19(11):3614. https://doi.org/10.3390/ijms19113614.*

to another. The same would be true for IDPs. An essential point that needs to be clarified when considering the target binding behavior of IDPs is to check first if IDPs are fully unstructured or not. If an IDP is fully unstructured containing no conformers resembling the bound structure, a full IF is an inevitable mechanism that has been popular in the IDP field for a good period. But if not, e.g., because transient structures resembling the bound structure exist in free IDPs, other possibilities open up. Regardless of which mechanism we are dealing with it is important to remember that it is the structure (or absence of structure) of a protein which plays a pivotal role in target binding. In globular proteins, tertiary and quaternary structures maintain the overall topology, which in turn render the target–binding regions or pockets to be properly presented for the target. As this has been the textbook dogma for decades the new view that a protein does not necessarily need to form a 3-D structure to function was met with strong skepticism when the notion of IDPs first appeared. Here one needs to be reminded that no matter how unconventional IDPs may be they are fundamentally just proteins after all; they are heteropolymers composed of amino acids just like globular proteins (for the moment we ignore the fact that proteins can be post-translationally modified by various functional groups). Since proteins including IDPs manifest their function by interacting with other biomolecules such as proteins, peptides, nucleic acids, lipids, carbohydrates, metals, lipids, and small ligands including drug molecules,[27–29] an immediate query regarding IDPs arises as to how a long stretch of amino acid residues with no specific structural content could interact with targets or be associated with diseases. More specifically, one must answer whether these "unstructured" proteins or long IDRs would still abide by the golden rule of structural biology, i.e., structure governs function, even though their binding to targets may not rely on 3-D structures. Alternatively, one could hypothesize that IDPs follow truly exquisite rules on their own since they are so novel.

Two main lines of thoughts have been tackling the above issues during the last two decades. The first is called the Completely Unstructured (CU) view, meaning that no structural elements whatsoever are present within free IDPs down to the level of secondary structures.[30–32] According to this line of thought IDPs, being so unorthodox, would use a target–binding mechanism where disorder itself dictates binding with no structural factors involved. This school has relied on negative information, i.e., the lack of 3-D structures, to explain the functions of IDPs rather than trying to ask if there may be some elements present within free IDPs that contribute to functions. In this school, IDPs are viewed like water that would shape itself as the shape of a container it is contained; it is argued that IDPs become structured or gain structure *only* upon target binding.[33,34] Three key reports that appeared just before the turn of the century, the early days of IDP research when not much was known about IDPs, strongly supported this CU view. The IDPs/IDRs studied in these investigations were described to be completely unstructured in the unbound state, i.e., random coils. The CU view naturally advocated an IF mechanism for IDP–target binding that involved *full* (100%) structural induction, e.g., from a random coil to a helix.[30–32] A limitation of this view has been that it does not provide a basis for the specificity of the initial recognition between an IDP and a target protein; specificity is assumed to reside within the inherent physicochemical properties of individual amino acids.

Very surprisingly, all the IDPs/IDRs examined in these key reports were found in later studies to contain transient secondary structural elements,[3] i.e., pre-structured motifs, seriously questioning the argument for the CU view. For example, in the case of the transactivation domain (TAD) of VP16 a short hTAF$_{II}$31-binding peptide was reported to be fully unstructured in the free state and then to be induced into a helix upon target binding.[31] This segment, however, was shown to form a $\sim$20% pre-structured helix in the free state when placed in the full-length VP16 TAD[35,36] (see Section 3.3 for more details). If a target binding segment in a free IDP is already pre-structured, e.g., in a helical conformation and the bound structure is a helix, one cannot expect a full structural induction involving a random coil-to-helix transition.[37] Intuitively, it would be thermodynamically unfavorable for the pre-structured segment to unwind before target binding and then to refold upon target binding. Above all, it is debatable whether this CU view could be rationalized noting that the structural state of even fully denatured proteins cannot be described by the so-called random coils where not only tertiary but also secondary structures are absent.[38–40] Nevertheless, this CU view has dominated the IDP field for a long time since few influential NMR structural biology groups[32,34] and a leading IDP bioinformatics group supported this view,[16,41] culminating in a special editor's article on IDPs in Nature[42] as if the CU view were the only applicable concept to IDPs, failing to acknowledge the existence and functional significance of PreSMos. For nearly two decades, numerous articles on IDPs used the terms such as disorder $\rightarrow$ order transition,[43] coupled folding and binding[34,44] and an induced fit (IF) involving a coil $\rightarrow$ helix folding transition, all underlining that IDPs become structured *only* upon binding. It is true that a certain degree of structural induction for the target-binding segments of IDPs/IDRs must occur since proteins are malleable by nature, but is it a coil-to-helix transition type of folding? After two decades of research on IDPs, it would be pertinent to ask if the early fascination with the novel concept of "unstructuredness" have blinded many in the field from considering other factors than the absence of tertiary or higher structures that might be contributing to the functions of IDPs.

The second school is the Mostly Unstructured (MU) view which acknowledges the presence of pre-structured motifs (PreSMos), the minimally-looking-yet-significant structures observed in free IDPs.[2,3,20,45,46] Experimental evidence has accumulated during the last two decades, demonstrating that dozens of free IDPs are pre-populated with one or more PreSMos and that these motifs having particular conformations are the specificity determinants for IDP-target binding.[3,45,46] This school therefore argues that *structure does matter* for IDPs' function as in globular proteins except that "structure" here refs only to secondary structures. Utilizing PreSMos, IDPs obey the structural complementarity rule extremely well (Fig. 2), the dogma established so solidly in structural biology for decades. It should be mentioned that early articles reporting the MU view received strong rejection from the IDP field which was dominated by the CU view at the time. The first and comprehensive article on the MU view made a timely appearance in 2012 and contributed significantly to the balanced progress of the IDP field by challenging the acceptance of the CU view which was formulated based on an inappropriate or premature interpretation of NMR spectral data.[3] The statement that IDP-target binding involves a coil → helix folding transition was fully invalidated by the discovery of PreSMos no matter which mechanism the binding follows. The same nullification goes true for the expression that IDPs/IDRs become structured only upon binding because the *PreSMos are structures*, activated secondary structures no matter how insignificant they may appear. Above all, they are clearly detected experimentally. The PreSMos are the "prevailing conformations" forming hydrophobic cores that were proposed by Tsai et al.[8] During the past two decades more than 120 PreSMos have been reported.[3,45,46]

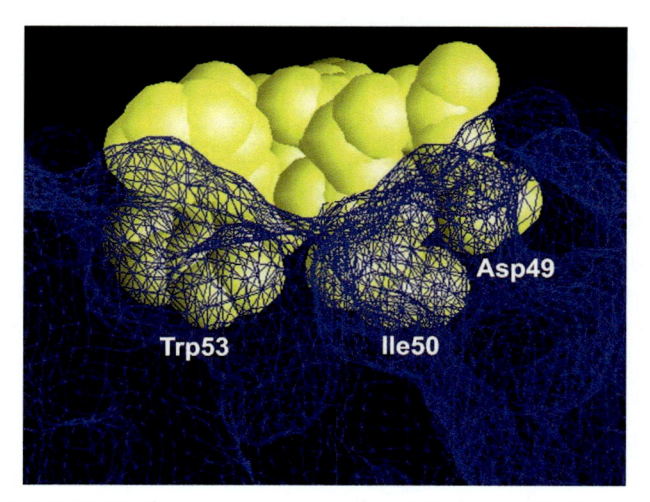

Fig. 2 The turn II PreSMo having a sequence of DIEQW observed in the transactivation domain of unbound p53 (yellow)[47] forms a complex with mdm2 (blue mesh), demonstratings that binding of a PreSMo to a target protein abides by the shape complementarity rule. The hydrophobic groove of mdm2 optimally accommodates the hydrophobic face formed by the Ile-Trp hydrophobic pair. A more elegant figure demonstrating the shape complementarity between a PreSMo and a target can be found in fig. 3 of Ref. 3

The MU view, supporting the idea that the pre-populated transient secondary structures in free IDPs formed by target-binding segments enable binding of IDPs with specific targets, has contributed to a balanced progress of the IDP field by proposing that the full induced fit is not the only option for the mechanism of IDP-target binding. What can be envisioned with the MU view is that formation of a transition state or an encounter complex which would be produced at the very early stage of IDP-target interaction by long-range electrostatic attraction or by random Brownian collisions ends with a specific recognition of a particular conformation of a PreSMo by a target. In this manner, the binding specificity originates from "structure," i.e., the pre-structured conformation. The IDP-target binding process will wrap up with a partial structural induction or tightening of a PreSMo's conformation into a more ordered or stable bound structure. The extent up to which a structural induction occurs will depend on the degree of pre-population a particular PreSMo has. For many PreSMos whose pre-population is $\sim$30% a maximum of $\sim$70% induction should be possible in principle. It is not clear why most PreSMos are populated only at such a level. For some reason too high or too low pre-population does not help IDPs to perform their function. In the original proposals of IF and CS with globular proteins the two pathways were proposed as the limiting models. In the case of IDPs the circumstance becomes flexible and the sequential involvement of the two in a concerted manner composed of CS followed by IF (CS-IF in short) is possible. This sort of CS-IF mechanism has been already perceived early on.[18] In short, to the question of which comes first, binding or folding,[17] the MU view argues that folding comes first; formation of a PreSMo is a partial folding.

One subtle feature in the description of IDP-target binding process which does not seem to be fully appreciated is that the binding involves structural changes at two different levels, global and local. The common expressions such as a disorder $\rightarrow$ order transition and binding coupled folding[34,44] for the most part are referring to the global-level large displacement of all the atoms in IDPs/IDRs, a type of structural change unlikely to be observed for globular proteins since most atoms in globular proteins will experience only minor displacements upon target binding although there are exceptions. The latter type, on the other hand, pertains to what happens locally to the atoms belonging to the PreSMo-forming residues that make direct contacts with targets via, for example, hydrophobic interactions. An expression like fuzzy complexes[48] can thus be valid only when one is referring to the global topological aspect but is not applicable for the PreSMo-forming segment which, once bound, forms a stable secondary structure in nearly all cases. If a long IDP/IDR that contains flanking segments at the two ends of a PreSMo is used the flanking segments, i.e., the non-PreSMo residues, could have a significant motional freedom even after target binding as reflected in several NMR parameters.[49]

It is important to remember that no matter how strongly the PreSMos' conformations presage their target-bound conformations the presence of a PreSMo per se is not a sufficient condition for CS.[50,51] Determining the exact role of a PreSMo in the target binding of an IDP requires careful measurement of both thermodynamic and kinetic parameters associated with binding for a sufficient number of MU type IDPs/IDRs under a variety of experimental conditions that take into to account of not only the aforementioned factors, but also length of IDRs, relative concentrations of IDP vs ligands and so on. Experimental determination of target binding mechanisms is challenging. One of the techniques that can be used to study the binding mechanism is the NMR relaxation dispersion experiment.[52,53] Finding an optimal experimental condition and fulfilling the hardware requirements for this experiment are both non-trivial. An in-silico approach using molecular dynamics simulation may become a reasonable alternative.[54,55]

3. The pre-structured motif (PreSMo)

The detailed definition of PreSMos,[3] the different ways of how investigators reported them[45] along with the significance of PreSMos as active target-binding sites in IDPs have already been discussed. Following will describe other aspects of PreSMos such as the historical circumstances surrounding the formulation of the PreSMo hypothesis and a few more subtle facets about PreSMos.

3.1 NMR and PreSMos

Multi-dimensional high-resolution NMR is a robust technique that allows one to measure several independent parameters which can be taken as evidence for the presence of secondary structures in proteins.[56] When used in combination with molecular dynamics simulation, NMR can produce ensemble conformations of IDPs at an atomic resolution based upon residual dipolar coupling constants (RDCs)[19,40,57,58] and long-range distance restraints derived from paramagnetic relaxation enhancement (PRE) experiments.[9,59–62] The unique information one can get from NMR measurements is the fraction of transient secondary structures in proteins.[3] For example, secondary structure propensity (SSP) scores that are derived from chemical shifts provide accurate fraction of a partially populated secondary structure in IDPs.[63] Using the SSP scores one can quickly tell that the PreSMos in IDPs are ~30% pre-populated on the average.[3] Several other NMR parameters such as interproton NOEs, heteronuclear NOEs,[3]J coupling constants, relaxation times, temperature coefficients of backbone amide protons, and hydrogen exchange rates can be used along with the SSP scores to precisely delineate the location of a PreSMo on a per-residue basis.[2,3,45] Although not common, the temperature coefficients of the carbonyl chemical shifts can also be used to detect PreSMos.[64] While using the NMR chemical shifts is an efficient and reliable way for detecting PreSMos and for secondary structures of proteins in general one needs to remember that use of different reference chemical shifts may lead to different conclusions. This may be particularly true for PreSMos since they are transient. Fig. 3 shows the SSP scores calculated for four MU type IDPs using three different chemical shift references. One can find that the conclusions based upon different references agree to a good extent. Although the degree of pre-populations seems to vary to some extent the conclusion on the presence and the locations of PreSMos does not change.

Practically all (>99%) PreSMos are detected by the high-resolution NMR technique. In lieu of PreSMos a generic term of transient (secondary) structures has been commonly used in the NMR literature to indicate the pre-structuring phenomenon,[3,19,45,68-70] which is fully reasonable since the PreSMos are transient secondary structural elements after all. However, the specific term of the pre-structured motif was coined since it explicitly depicts the transient pre-structuring in IDPs whereas the generic term of transient secondary structure could be used for any proteins and their segments or for non-proteins such as nucleic acids. In an effort to accentuate the ability of certain segments of IDPs to form transient pre-structured motifs the acronym PreSMo was first introduced during a lecture at the 2008 International Conference on Magnetic Resonance in Biological Systems (ICMRBS), one of the largest bio-magnetic resonance conferences dealing with NMR protein structural biology. The new terminology of PreSMo was documented in 2012 only when a statistically significant number of PreSMos were compiled and the functional significance of such pre-structuring became evident. A recommendable exercise would be to introduce a new jargon only when doing so provides a clear benefit of clarifying a novel concept especially in a newly developing field. The first article reporting the presence of PreSMos just used a generic expression of "local structural elements" without introducing a new terminology since it was not clear if the pre-structuring in free IDPs would

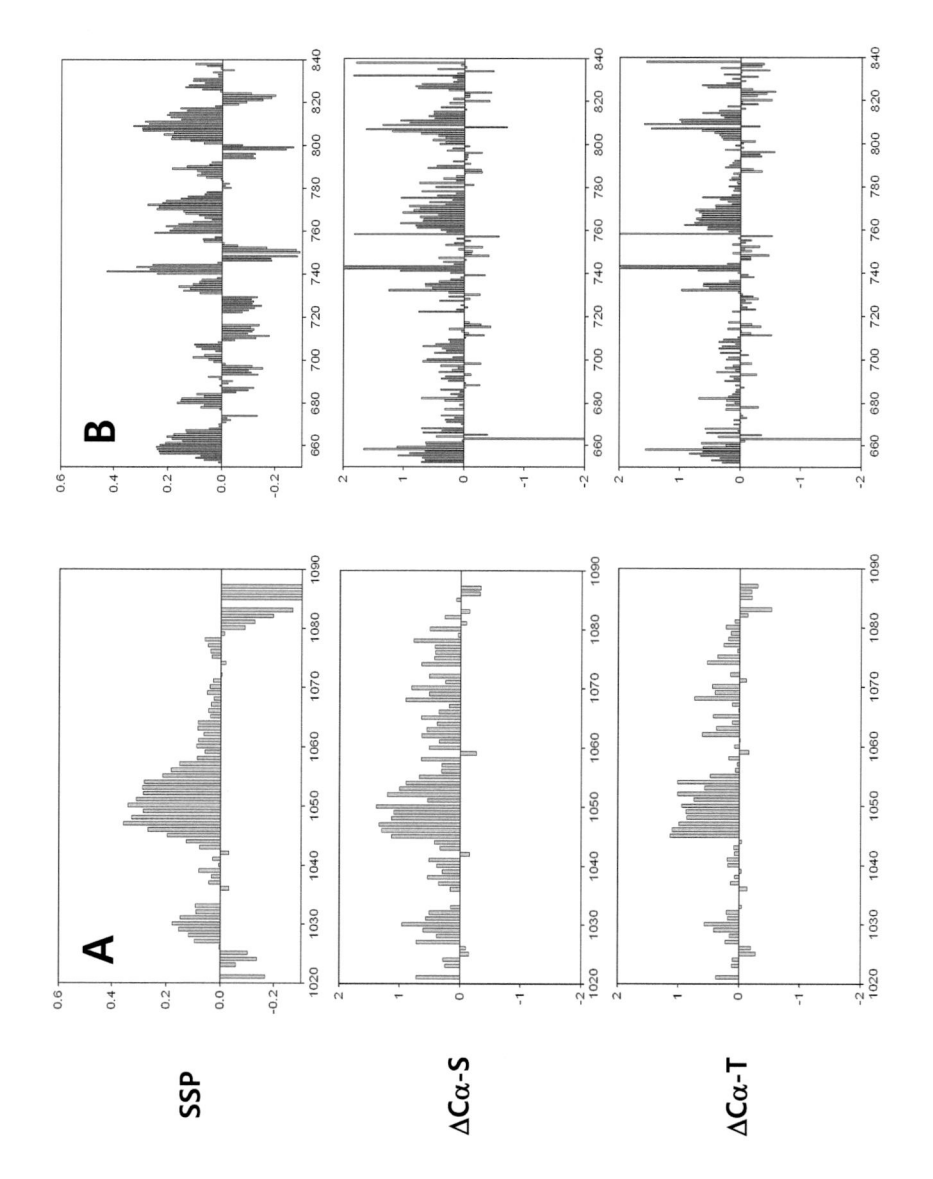

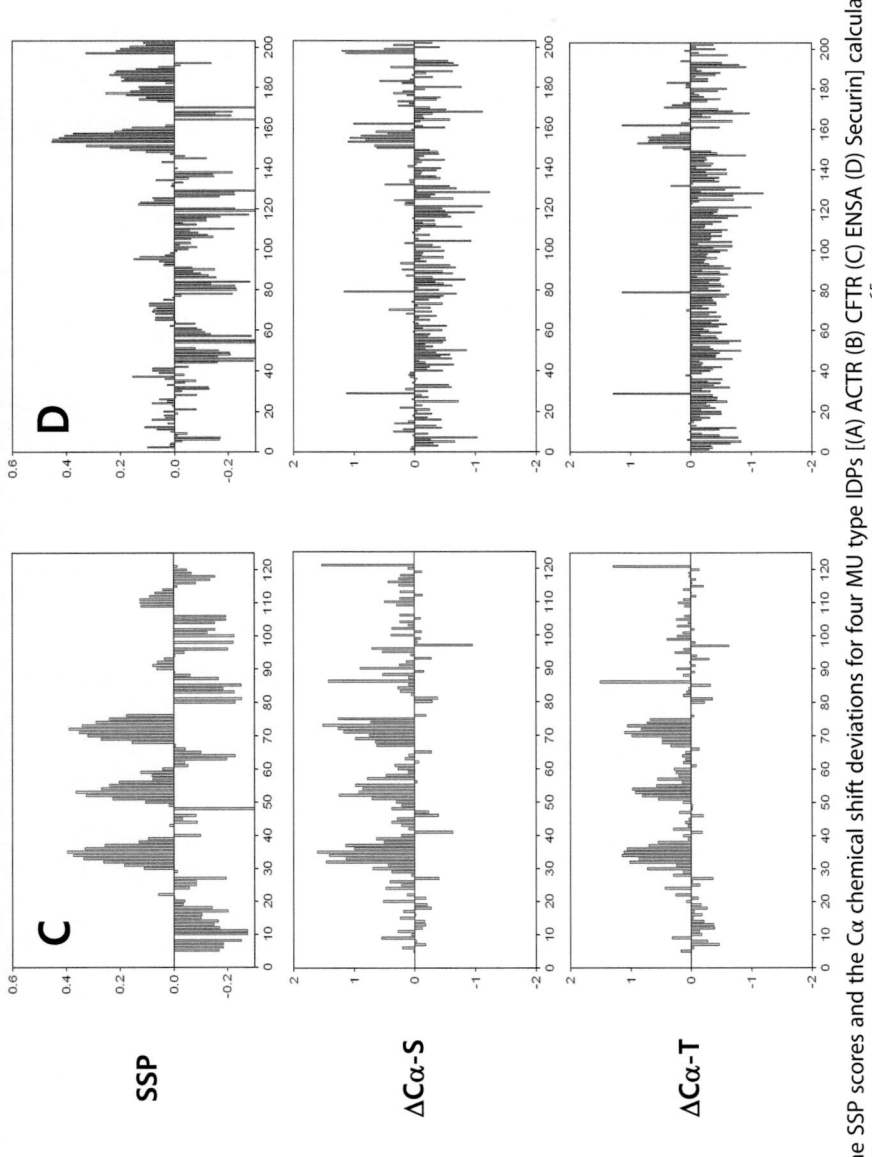

Fig. 3 The SSP scores and the Cα chemical shift deviations for four MU type IDPs [(A) ACTR (B) CFTR (C) ENSA (D) Securin] calculated using three different random-coil chemical shift references. (Top panel) SSP calculated using Neal et al.[65] (Middle panel) ΔCα-S calculated using Schwarzinger et al.[66] (Bottom panel) ΔCα-T calculated using Tamiola et al.[67] While the degree of pre-structuring somewhat varies depending upon which reference is used the locations of the PreSMos are same. *Reproduced with permission from Kim DH, Lee J, Mok KH, Lee JH, Han KH. Salient features of monomeric alpha-synuclein revealed by NMR spectroscopy. Biomolecules. 2020;10(3):428–442. https://doi.org/10.3390/biom10030428.*

be a general feature.[2] During the last two decades, numerous different yet similar terms were used in many NMR reports on IDPs to describe essentially the same phenomenon of pre-structuring in IDPs/IDRs (see Table 1). In retrospect, it may appear that too many terms have proliferated the IDP field for the same phenomenon. Nonetheless, this was inevitable in a way since the terms describing the pre-structuring phenomenon were not clearly acknowledged in the early days of IDP research.

Table 1 A list of various terms used to describe the pre-structuring phenomenon (transient structures) in IDPs.

Year	Terms used	References
1998	Partially collapsed, favoring helical conformations	71
1998	Nascent helical segments, intrinsic helical propensities	22
1998	Helix-forming segment	72
1999	Short two-turn-helix, another potential helix	73
1999	A preformed nascent helix	74
2000	Local structural elements, pre-existing minimal secondary structures, local structural order	2
2000	Local structural elements, long alpha-helix	75
2000	Preformed transient helices	76
2000	Transient, preordering	77
2000	Transiently ordered states, structural preordering	78
2001	A region with a preference for helical conformation, a nascent or transient helix.	79
2002	Pre-organized helical structure, marginally stable helical structure	80
2004	A short stretch which is structured transiently	81
2004	Nascent helix, regions with helical propensity	26
2006	Small islands of secondary structures	82
2007	Pre-structured motifs	83
2008	Partially folded, preferentially populates helical conformers, nascent helix formation	19
2008	Minimal ordering of short linear motifs	49

Table 1 A list of various terms used to describe the pre-structuring phenomenon (transient structures) in IDPs.—cont'd

Year	Terms used	References
2009	Residual secondary structural elements	36
2011	Preferred a-helical organization, local structural preformation	84
2012	Transiently ordered regions	85
2012	Dynamic local structure, distinct elements of secondary structure, local helical structural element	86
2012	Two helices that are transiently structured	70
2012	Regions that transiently adopt a-helical structures	87
2012	Weakly populated helical conformations	88
2013	Transient secondary structure	89
2015	Pre-structured helix	20
2016	Transient helical pre-structured motifs	90
2017	Pre-structured motif, transient helices	91
2017	Helix pre-structured motifs	92

3.2 The helix controversy

The discovery of the very first helix in an unbound transcription factor, i.e., the helix PreSMo detected in the p53 transactivation domain, had a huge impact to the field of transcription as illustrated below. For several years since the mid-1980s, several reports, both speculative as well as experimental, on TFs/TADs persistently suggested that TFs/TADs, albeit "unstructured," must have some specificity determinants that are responsible for transcriptional activity.[93] In one study employing a wild type GAL4 activation domain and its scrambled mutant with no transcriptional activity Giniger and Ptashne concluded that the mutant was inactive because its helix-forming propensity was compromised.[94] In another study Leuther et al., argued using CD data that it should be a β-sheet that is important for activity.[95] A keen debate went on whether target-free acidic TADs should possess an amphipathic helix as the specificity determinant for activity, a debate named "a helix controversy"[3] (see Table 2). Yet, no direct experimental evidence based on atomic-resolution structures was available. And it is the NMR investigation on the of p53 TAD which resolved this controversy by presenting a helix PreSMo as a specificity determinant for target (mdm2) binding.[2]

Table 2 Helix controversy and a list of excerpts from the articles that addressed the issue.

Year	Title/argument	References
1987	*A new class of yeast transcriptional activators* "… adjacent to these turns, α-helices composed of hydrophobic residues in many cases. Most of our activating sequences also have short segments that could form amphiphilic helices, that is, *α-helices* with charged residues on one side and hydrophobic residues on the other. Whether any of these features is important remains to be tested."	96
1987	*Transcription in yeast activated by a putative amphipathic α helix linked to a DNA binding unit* "The synthetic peptide is acidic and should it form *an α-helix*, that helix would be amphipathic, having one hydrophilic face bearing the acidic residues, and one hydrophobic face. When expressed in yeast, the artificial protein bearing this peptide efficiently activates the GAL1 gene which is ordinarily activated by GAL4. An otherwise identical protein with the novel 15 amino acids in a scrambled order, and which is thus unable to form an amphipathic structure, does not activate GAL1 transcription."	94
1988	*How eukaryotic transcriptional activators work* "One peptide (AH) was designed so that, should it form *an α-helix*, that structure would be amphipathic, bearing negatively charged residues along one surface and hydrophobic residues along another. The second peptide (SH) comprised identical amino acids, but in scrambled order. The construct bearing the first peptide activated transcription in yeast, whereas that bearing the second peptide did not. Deletion analysis of GCN4 also suggests that an α-helix could be an important component of an activating region"	97
1988	*GAL4-VP16 is an unusually potent transcription activator* "… probably at least in part is α-helix."	98
1991	*Critical structural elements of the VP16 transcriptional activation domain* "A putative amphipathic alpha helix did not appear to be an important structural component of the activation domain."	99
1993	*Pattern of aromatic and hydrophobic amino acids critical for one of two subdomains of the VP16 transcriptional activator* "… although neither is likely to be an amphipathic helix" & "Giniger and Ptashne proposed that acidic activation domains form amphipathic α-helices" & "our results are also inconsistent with a model of VP16 as an amphipathic α-helix"	100

Table 2 Helix controversy and a list of excerpts from the articles that addressed the issue.—cont'd

Year	Title/argument	References
1993	*The acidic activation domains of the GCN4 and GAL4 proteins are not α helical but form β sheets* "AH does have α-helical potential" & "However, there is no direct evidence that AH is indeed helical"	101
1993	*Genetic evidence that an activation domain of GAL4 does not require acidity and may form a β sheet* "… this region is not unstructured or α helical, but its function may require a β sheet."	95
1993	*Structure(?) and function of acidic transcription activators* "From work on other systems, it is not clear whether the β structure or any other defined structure is important for activation function."	102

Interesting to note is that this seminal manuscript had been rejected for years before it finally was able to see the light. Around the time this report was published, few other pioneering NMR reports on non-transcription factor type IDPs appeared, demonstrating that certain segments in these types of IDPs tended to be pre-structured.[71,77,78]

3.3 Helix vs random coil

From a technical point, detecting PreSMos or transient structures in proteins by NMR is not formidably challenging. Given this, it calls for a question as to why there were only few articles reporting the PreSMos between the mid-1990s and the mid-2000s, when PreSMo reports began to more or less proliferate.

The first reason seems to be related to the preoccupation many NMR investigators had on the narrow chemical shift dispersion which seemed to have played a considerable role in a negative sense.[25,30,31] Narrow chemical shift dispersion in NMR is taken as a default evidence for "no structure" or a random-coil state. Consequently, those NMR investigators who encountered narrow chemical shift dispersion in their spectra made statements that the IDP/IDR under study was "unstructured" or random-coil like without considering a possibility that the "unstructured" proteins might have some sort of structural elements other than 3-D structures such as transient secondary structures. A subtle point, *a random coil exhibits narrow chemical shift dispersion, and so do helices*, was missed. Most (>80%) PreSMos are helices.[3,45,46] Under the narrow chemical shift dispersion transient helices were hiding. Devil is in the details.

3.4 Confidence level

The second reason for the paucity of the reports on the PreSMos in the early days of IDP research seems to be related to the mindset or the level of confidence as illustrated in Fig. 4. Given the transient nature of PreSMos an investigator would find a PreSMo with rather good confidence if pre-structuring tendency is clearly visible, i.e., if the degree of pre-population of a PreSMo is $>\sim 10\%$. However, if the degree of pre-population of a PreSMo is lower it would become difficult for her/him to detect a PreSMo unless she/he has a belief that a PreSMo is present or is determined to find it. It is interesting that a neophyte NMR student may have a better chance of finding a PreSMo in this type of situation since a neophyte tends to have more meticulous attitude toward data analysis (leaving no NMR peaks unchecked no matter how small they are) than an experienced investigator preoccupied with educated guesses.

The third reason is also related to the confidence level but in a slightly different sense. It is challenging for researchers to publish results that contradict the dominant flow of a particular field. This reality seemed to have affected a good number of early IDP investigators, making them present only weak argument for the existence of PreSMos. When the uncomfortable idea of IDPs itself was facing strong skepticism in the late 1990s, the even more queer result that an IDP/IDR in its free state was not fully unstructured, but contained pre-existing motifs in contrast to the predominant belief, was another novel concept difficult for many to accept. The MU view was an

Fig. 4 Believe it or not. A white cat is sitting on snow. Finding a cat in this picture is like finding a PreSMo when the degree of pre-population of a PreSMo is low.

additional subtle eccentricity that even many IDP researchers found it difficult to accept let alone typical protein scientists. There seemed to be two reasons which made the publication of the heralding NMR work on the p53 TAD difficult. One was associated with an extremely meticulous peer review[103] due to the enormous impact the report would create if published and the other was related to the biological importance per se of the p53 protein itself. Around the mid-2000s bioinformatics analyses on IDPs suggested that there are many segments in IDPs which could potentially form what were described as MoRF[104] and PSE[105] or SLiMs.[106] Even though these terms were not based on experimental observation of the pre-structured motifs in target-free IDPs they made a decent contribution to letting many consider a possibility that perhaps IDPs are not CU, judging from the fact that several PreSMo reports began to appear since the mid-2000s. The term MoRF has been used interchangeably with PreSMos.[19,70,107] In the recent years, steady and strong literature is being published supporting the MU view over the CU view.

3.5 The formulation of the PreSMo hypothesis

An important, and perhaps the most significant, factor that enabled the birth of the PreSMo concept with a great confidence was the availability of the X-ray structure of the mdm2-bound p53 TAD peptide composed of the residues 15–29.[21] These residues overlapped precisely with those that form a helix PreSMo in free p53 TAD. A hypothesis was born that the transient helix present in the unbound p53 TAD should be recognized by mdm2 to become a stable helix shown in the X-ray structure. This is an excellent example demonstrating how the X-ray crystallography contributed to the IDP field, i.e., by providing the target-bound structure of a PreSMo, even though the technique cannot provide direct structural information on IDPs themselves. The insignificant-looking transient helix and the mdm2-bound helix were a lead to a goldmine of PreSMos (>120 PreSMos). The X-ray crystallography also contributed to the proposals of MoRF and PSE since these terms were proposed based on the X-ray structures of the complexes between globular targets and the supposedly fully disordered fragments of IDPs. No target-bound structures were available as in the case of p53 TAD for other IDPs/IDRs studied in the early days such as tau1c in human glucocorticoid receptor (hGR),[5] FlgM,[71] amyloid precursor protein,[77] and ribosomal protein S4.[78] An early NMR work on a transactivation domain KID found the presence of two transient helices in free KID[22] and its target bound structure was known.[30] But this work did not lead to the proposl of a hypothesis similar to the PreSMo hypothesis. It is educational that two other NMR studies on p53 TAD overlooked the PreSMos.[108,109] This illustrates how, especially in the early days of IDP research, difficult it was to look into the significance of the transient features in IDPs. A similar preoccupation seemed to have played a role in the early NMR study on the VP16 TAD where the protein was described to be CU despite that the NMR data indicated presence of transient structures.[110]

4. PreSMos in transcriptional and translational factors

IDPs take up ~50% of the entire human proteome when a criterion of 40 residues for the IDR length is used[16] and ~60% of the human transcription/translation factors are IDPs,[111] which explains why cancers represent the diseases IDPs are most frequently associated with. In this section we will describe four transcription factors, p53 transactivation domain, minimal core activation domain of tau1 in human glucocorticoid receptor, the transactivation domain of VP16 from herpes simplex virus, c-Myc and a translation factor 4EBP1/2. It will be evident at the end of this section that all these IDPs/IDRs use the hydrophobic face of their amphipathic helix PreSMos to bind to the hydrophobic grooves in target proteins.

4.1 p53

p53 is a well-known transcriptional factor/activator associated with ~50% of human cancers.[112] It is an inherent tumor suppressor as its cognomen "guardian of the genome" implies.[113] This protein is active as a tetramer whose monomer is composed of 393-residues. Structurally p53 is a modular protein like many other transcriptional factors consisting of four independent functional domains/regions, a disordered transcriptional activation domain (TAD) at its N-terminus, two globularly well-structured domains, a DNA-binding domain (DBD) and an oligomerization domain (OD) that enables the formation of a tetramer, and finally a short disordered C-terminal basic domain (CBD) enriched with lysines.[114] Therefore, p53 is a hybrid protein where two structured domains (DBD and OD) and two intrinsically disordered regions (TAD and CBD) coexist.

For its importance in cancer biology, a great deal of efforts went into determining the high-resolution structure of a full p53 in the 1990s, but all ended up in vain since no single crystals suitable for X-ray crystallographic investigations could be obtained since p53 is an IDP containing disordered regions. Unfortunately, the disordered nature of p53 was not fully recognized at the time, probably letting many X-ray structural biologists experience deep frustration. Determination of the structure of a full p53 by NMR was not possible either since a full p53 monomer with 393 residues was too large for a complete resonance assignment and structure determination at the time even with $^{13}C/^{15}N$ double isotope labeling. An NMR spectrum of a full p53 monomer would be a mixture of resonances from the ordered domains and the disordered regions having different signal intensities and linewidths, making resonance assignment impossible. In the mid–1990s a divide and conquer strategy was successfully applied to produce the high–resolution structures of the globular domains of p53, i.e., DBD and OD, both having high helix content.[115–117] On the other hand, structural knowledge for the remaining disordered parts was not available until the turn of a century; the disordered nature of the N–terminal TAD and the C–terminal CBD was revealed by NMR[2] and by predictions, respectively. The NMR work on the disordered TAD of p53 was an important milestone both in the IDP field and in the transcription field, rendering p53 one of the most thoroughly studied IDPs in later years.

The TAD of p53 contains three PreSMos, a helix and two turns (turn I and turn II) (Fig. 5). It is common that IDPs/IDRs have multiple PreSMos.[2,3,45,46] In addition, none of these PreSMos are stable or 100% populated secondary

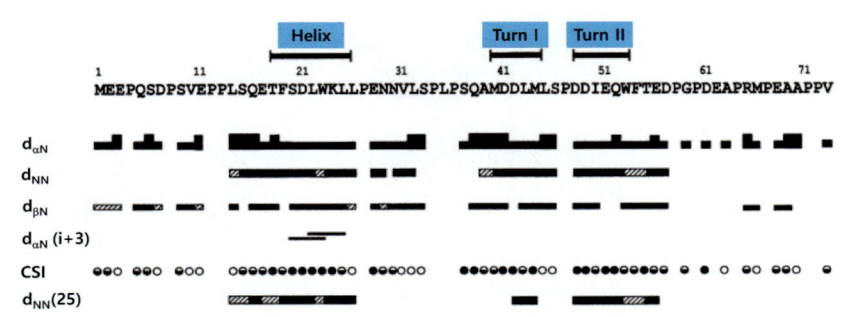

Fig. 5 Three PreSMos (Helix, Turn I and Turn II) are detected in the transactivation domain of the 73-residue full-length p53 based upon the observed the interproton NOEs and chemical shift index (CSI). *Thickness of bars* is proportional to NOE intensities (strong, medium, and weak) that are measured from an NOESY-HSQC spectrum obtained at 5 °C. The hatched portion in the sequential d_{NN} NOEs indicates an ambiguity regarding the presence of the particular NOE due to resonance overlap. The CSI values (1, 0, and −1) are represented by *open, half-filled*, and *filled circles*, respectively. A similar pattern of the sequential d_{NN} NOEs is obtained at 25 °C (shown at the *bottom*). Locations of an amphipathic helix (helix I) and two turns (turns I and II) are marked above the amino acid sequence. *Reproduced with permission from Lee H, Mok KH, Muhandiram R, et al. Local structural elements in the mostly unstructured transcriptional activation domain of human p53. J Biol Chem. 2000;275(38):29426–29432. https://doi.org/10.1074/jbc.M003107200.*

structures; PreSMos are transient.[3] The helix PreSMo that is ~30% pre-populated is an amphipathic helix formed by the residues 18–26 whose three bulky hydrophobic sidechains of F19, 23W and 26L are facing one side optimally primed for binding to the hydrophobic pockets of target proteins.[21,47,118] Two turns are 5%–15% pre-populated and formed by the residues 40–44 (turn I, MDDLM) and 48–53 (turn II, DIEQW), respectively. Turn II contains three hydrophobic residues. All three PreSMos of p53 TAD interact with several targets such as mouse double minute 2 (mdm2), p62/Tfb1 subunit of TFIIH, the single-stranded DNA-binding protein (replication protein A, RPA), TAZ1, KIX, TAZ2, and NCBD.[21,118–123] For example, the helix and the turn II PreSMos bind with mouse double minute 2 (mdm2),[47] the turn II with p62/Tfb1 subunit of TFIIH,[120] both turn I and turn II with RPA[119] and so on, demonstrating that the promiscuity of p53 TAD has a structural basis (Fig. 6).

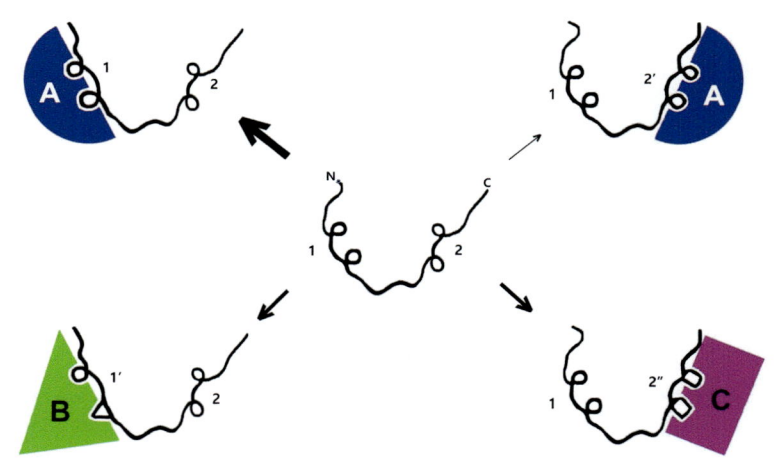

Fig. 6 A cartoon explaining the promiscuity of an IDP possessing two PreSMos (1 and 2) that interact with three target proteins. The PreSMo 1 binds target A in a conformation (e.g., a helix) that is similar to what is has (a transient helix) in the free state. The PreSMo 2 also binds to the binding pocket in target A, but by slightly adjusting its conformation (structural induction) to a helix in order to bind to the same binding pocket in target A where PreSMo 1 binds. The binding of PreSMo 2 is weaker than that of PreSMo 1 as indicated by a small arrow. This type of binding is observed for interaction of p53 TAD and mdm2[2,21] where mdm2 is target A and the PreSMos 1 and 2 are the helix PreSMo and the turn II PreSMo in p53 TAD, respectively. For binding with another target protein B the PreSMo 1 slightly adjusts its conformation to become 1′. TAZ2 may qualify as target B since it binds to the helix PreSMo of p53 TAD.[118] Binding of target C is achieved via the PreSMo 2 which adjusts its conformation to 2″. Replication protein A (RPA) qualifies as target C since two turns (turn I and turn II) were shown to become a helix upon binding to RPA.[119] Note that none of the interactions starts from a completely unstructured state (random coil).

The promiscuity of IDPs is an important facet which has not been thoroughly explained yet. The CBD of p53 was used as a model to explain promiscuity in a recent (7th ed.) version of the Lehninger's biochemistry textbook.[124] The CBD of p53, consisting of 33 residues, barely qualifies for an IDR. For some reason, perhaps due to the disorder prediction result, many seem to have been led to a notion that this IDR is CU. The argument used in the Lehninger's illustration relies on the structures of a few peptides derived from this IDR with lengths of 11–22 residues. However, the structural states of these peptides in their target-free state were not investigated to warrant such a CU nature. Careful examination of the structural features of these peptides must be carried out. Since they are short peptides, they are likely to exhibit no significant structures in the free state despite that the full–length p53 CBD is not fully unstructured (see Section 3.3 for further discussion). Besides, the studied peptides are not composed of the wildtype amino acids, containing modified residues such as acetylated Lys 382. A small modification of an amino acid in a short peptide can influence the structure to a significant degree. The validity of discussion on promiscuity based on these peptides must be checked since the result of a preliminary NMR experiment on this IDR indicates that p53 CBD appears to be MU. It has several d_{NN} type NOE crosspeaks typical of the NOESY spectra of the MU type IDPs (Fig. 7). No d_{NN} type NOE crosspeaks are observed in a NOESY spectrum of a CU type.

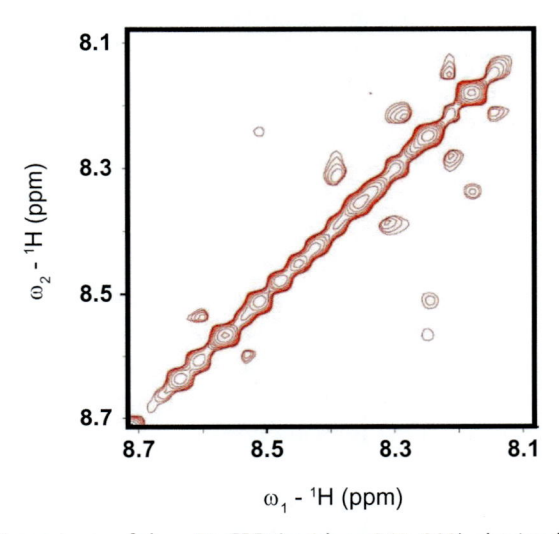

Fig. 7 A NOESY spectrum of the p53 CBD (residues 361–393) obtained with a mixing time of 150 ms using 1.6 mM protein, in 20 mM NaOAc-d_3, 50 mM NaCl, 5 mM DTT, 90% H_2O, 10% D_2O at pH 5.3. To avoid introducing any artifacts no symmetrization of the spectrum is done. Several d_{NN} type NOEs are clearly visible indicating the existence of transient structures.

The N-terminal domain of mdm2 forms a globular structure that provides a hydrophobic groove which binds to the helix and turn II PreSMos of p53 TAD.[47] This groove in fact can accommodate many hydrophobic compounds. Blockage of this groove produces anti-cancer effects by restoring the tumor suppressive activity of p53. This principle was utilized by Hoffmann-La Roche for developing the Nutlin family anti-cancer compounds among which RG7112 is under clinical trials.[125] Other potential anti-cancer drugs such as benzodiazepines,[126] indole-2-carboxylic acid derivatives[127] and pyrrolidines are also being developed based upon this feature. Non-small-molecule compounds such as a peptidomimetic derivative of the amphipathic helix of p53 TAD called a stapled helix was also designed.[128] For example, ALRN_6924 designed by Aileron Therapeutics is currently under clinical trials as a novel chemoprotective medicine to treat and protect healthy cells in patients with cancer that harbors p53 mutations to reduce or eliminate chemotherapy-induced side effects.[25]

4.2 Tau1c

Human glucocorticoid receptor is a steroid hormone-activated transcriptional factor that belongs to the nuclear receptor superfamily. By translocating itself from cytoplasm to the nucleus upon ligand binding hGR regulates various target genes involved in cell differentiation, inflammation, glucose homeostasis, etc.[129] This receptor is an important therapeutic target since mutations in hGR are associated with several diseases including autoimmune diseases, cancers and Cushing's syndrome. As a member of the nuclear receptor superfamily hGR has an N-terminal activation function-1 domain (AF-1), a central DNA binding domain, and a C-terminal ligand binding domain (LBD).[130] LBD in fact is a ligand-dependent activation function that forms the second activation function-2 domain (AF-2) of hGR. Of the two activation domains in hGR the N-terminal domain is a long domain consisting of residues 77–262 (186-residues) and is named tau1. The C-terminal activation domain is called tau2.[131] It was found that the minimal core domain of tau1 (named tau1c) composed of the residues 187–244 is responsible for ~70% of the transcriptional activity of the full tau1 domain.[132]

Tau1c provides us with an excellent model system which can be used to systematically explore the relationship between function (transcriptional activity) and the structural features embedded in the PreSMos since the activity profiles for a large number of change-of-function mutants are available.[133] Structurally, tau1c is an IDR and unstructured in an aqueous solution,[5] which is fully consistent with the well-known fact that transactivation domains are intrinsically unstructured.[3] To understand the structural features in tau1c that are important for transcriptional activity an NMR study was carried out at the dawn of IDP research, which revealed that three regions in tau1c had tendency to form helices.[5] Unfortunately, the experiments were carried out under a helix-inducing solvent, trifluoroethanol (TFE). Interpreting the biological relevance of any secondary structures observed under secondary-structure-inducing alcoholic solvents including TFE is not straightforward since observed structures may be an induced phenomenon rather than reflecting the true in-vivo structures. Using other types of solvents, such as trimethylamine N-oxide (TMAO), may help to reveal the structural features of tau1c that might be relevant to transcriptional activity as TMAO was reported to reveal true structures slightly better than alcoholic solvents.[134] In order to eliminate the ambiguity in data interpretation arising from the use of non-aqueous solvents and noting the definition of a PreSMo, i.e., transient secondary structures obtained in aqueous buffers are the biologically meaningful pre-structured motifs, Kim et al. carried out a whole new set of heteronuclear multi-dimensional NMR experiments on tau1c in an aqueous buffer solution at neutral pH.[92] Results showed that tau1c formed three helix PreSMos that are ~25% pre-populated, the locations of which overlapped to a good extent with those obtained in TFE (Fig. 8). PreSMos were observed in several target-unbound TADs as the key elements for transcriptional activity,[2,3,46,86] suggesting that the three helix PreSMos found in tau1c might be the critical determinants for the transcriptional activity.

The first helix PreSMo (Helix 1) in tau1c detected in aqueous solution is formed by the residues 185–202, the second (Helix 2) by residues 206–225, and the third (Helix 3) by residues 232–244. The helices observed under TFE were Helix I (189–201), Helix II (215–226) and Helix III (234–239), respectively. While the three helix PreSMos detected in the new study overlap with the helices observed under TFE, differences exist in all three helices; the first residue in Helix I was 189S, but that in Helix 1 is 185T. This difference is related to the fact that the starting residue of hGR tau1c in the current study is 181V while that in the previous work was 187D. Another notable difference is seen at the N-terminus of Helix 2. Helix 2 is 9-residues longer than Helix II at its N-terminus; it starts at residue 206K whereas Helix II starts at 215S. The 213W in tau1c is known to be important for activity.[133] In the new data, 213W belongs to Helix II, i.e., it is a part of a helix PreSMo, which explains why it is important for activity (see below). In the previous study, it was described to be a part of a loop region. Another difference is found in Helix 3 whose C-terminus is longer by five residues than Helix III.

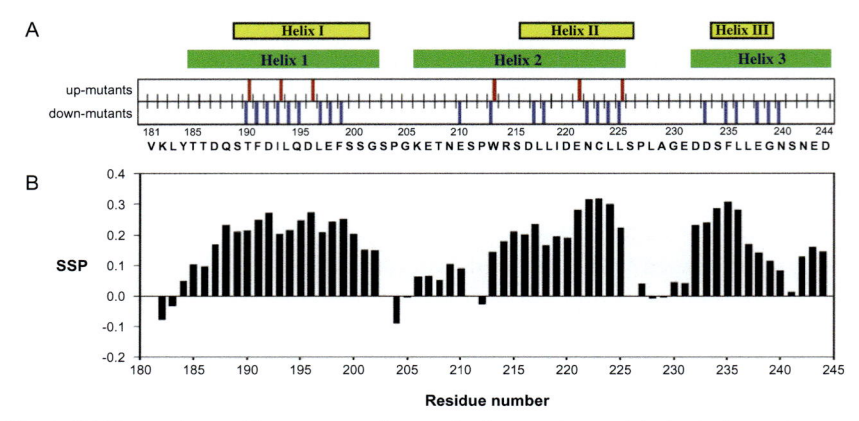

Fig. 8 (A) The amino acid sequence of tau1c is shown along with the helices observed under TFE (yellow box) and in aqueous solution (green box). Red and blue vertical bars indicate up mutants (gain-of-function) and down mutants (loss-of-function), respectively. Note that all the residues causing either up or down activities belong to the helix PreSMos. In the new report by Kim et al.[92] the functionally important 213W is a part of Helix 2. (B) The SSP scores of tau1c. Three helix PreSMos have the SSP scores of ~25%. *Reproduced with permission from Salamanova E, Costeira-Paulo J, Han KH, et al. A subset of functional adaptation mutations alter propensity for α-helical conformation in the intrinsically disordered glucocorticoid receptor tau1core activation domain. Biochim Biophys Acta Gen Subj 2018;1862(6):1452–1461. https://doi.org/10.1016/j.bbagen.2018.03.015.*

An important point revealed by the new investigation was that nearly all the change-of-function mutations (either gain-of-function or loss–of-function) belong to the helix PreSMos, establishing a solid correlation between the activity and the PreSMos in a helical conformation as shown in Fig. 8. Notably, the residue at 212 is a proline in Helix 2. Since a proline is a well-known helix-breaker 212P could introduce a kink just prior to 213W, which would make Helix 2 a helix-kink-helix motif. In tau1c two prolines, 204P and 226P, are flanking the C-termini of the two helix PreSMos, Helix 1, and Helix 2. Prolines frequently flank the helix PreSMos in IDPs and act as an activity switch by controlling the length of helix PreSMos that are important for activity[135] (Fig. 9). Borcheds et al. applied the proline mutation strategy to the helix PreSMo of p53 TAD and found that the increased helicity by mutation of 27P to Ala resulted in stronger binding to mdm2 and increased activity of p53,[136] supporting the conformational basis of activity provided by a helix PreSMo.

TAZ2 in CBP/p300 is a target of tau1c.[133] The binding site for tau1c in TAZ2 is hydrophobic like the hydrophobic groove in mdm2 that binds to the hydrophobic face of the amphipathic helix PreSMo in p53 TAD

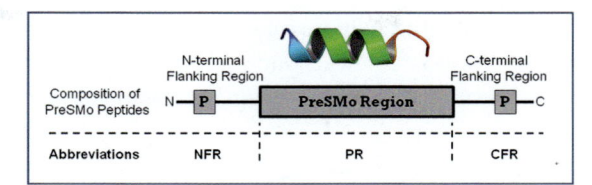

Fig. 9 A schematic diagram showing that PreSMos are flanked by prolines. *Reproduced with permission from Lee C, Kalmar L, Xue B, et al. Contribution of proline to the pre-structuring tendency of transient helical secondary structure elements in intrinsically disordered proteins. Biochim Biophys Acta 2014;1840(3):993–1003. https://doi.org/ 10.1016/j.bbagen.2013.10.042.*

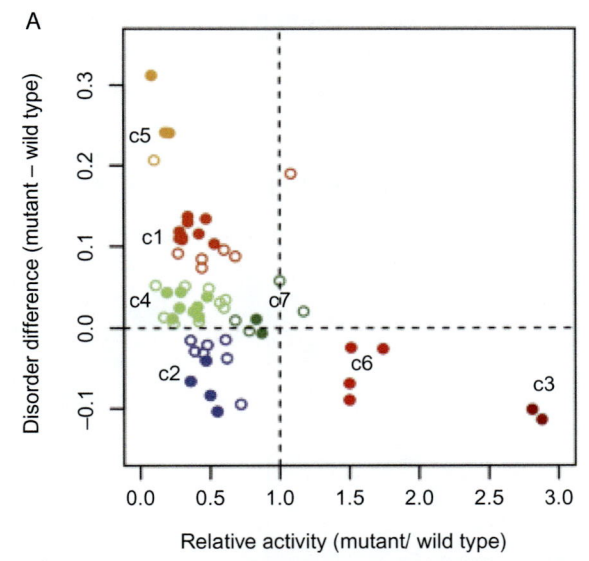

Fig. 10 The seven clusters of change-of-function mutants of tau1c. Among the seven, two c3 and c6, show gain of function. *Reproduced with permission from Salamanova E, Costeira-Paulo J, Han KH, et al. A subset of functional adaptation mutations alter propensity for α-helical conformation in the intrinsically disordered glucocorticoid receptor tau1core activation domain. Biochim Biophys Acta Gen Subj. 2018;1862(6):1452–1461. https://doi.org/10.1016/j.bbagen.2018.03.015.*

composed of 19F, 23W and 26L.[21,47] In a recent study Salamanova et al. partitioned the 60 change-of-function mutants of tau1c into seven clusters based upon their disorder prediction scores and the gene activation activities[137] (Fig. 10). The mutants showed either increased or decreased gene activation activity. Interestingly, gain-of-function mutations affecting

Helix 1 were associated with the increased stability of the helix. In particular, the mutants in the clusters 3 and 6 such as D196Y, E221F, T190F and T190Y exhibited a clear correlation between increased activity and decreased disorder or increased helicity.

An elaborate account for the activity changes in the mutant tau1c can be made based on the helicity of PreSMos. Fig. 11 shows a helical wheel diagram (A) and the cylindrical surface models of Helix 1 (B-D). In Fig. 11B one finds a triad of three bulky hydrophobic residues (191F, 194L and 197L shown in a cyan box) that are important for the activity of tau1c. Losing one of these hydrophobic residues significantly decreases the activity by 70% as can be seen in the L197E_30 mutant (according to the designation used in Salamanova et al.[137] the last number in this notation is the percentile activity of a mutant relative to the wild type). Fig. 11C shows what happens if the residue 196D is replaced by Tyr. It is fascinating that another triad of hydrophobic residues

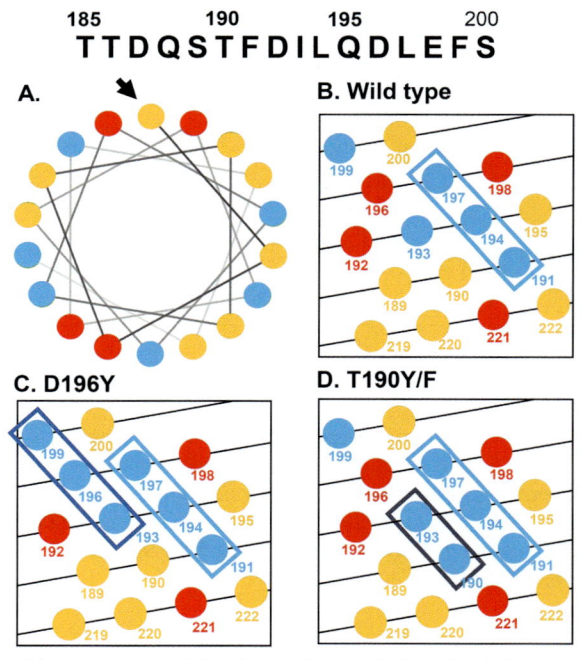

Fig. 11 The sky-blue drops in a bloody pool. (A) a helix wheel diagram for Helix 1 of tau1c. The arrow indicates the first residue. (B) Three hydrophobic residues are aligned in a box in cyan like a cluster of oil drops in a pool of hydrophilic residues (red for charged residues and orange for polar residues). (C) The D196Y mutation introduces another triad of hydrophobic residues (blue box). (D) The T190Y/F mutation makes a pair of hydrophobic residues (dark blue box).

(shown in a blue box) is formed by 193I, 196Y and 199F with an introduction of Tyr (an aromatic sidechain) at the position 196. Therefore, the mutation is expected to provide tau1c with a higher, if not doubled, probability for TAZ2 binding. The result is a dramatic gain of activity by 181% (D196Y_281). Fig. 11D shows what happens in the T190Y or T190F mutant. Even though one does not get an additional triad these mutants have a pair of two consecutive hydrophobic residues (a dark–blue box), 190Y/193I or 190F/193I. Due to this hydrophobic pair the TAZ2 binding capacity is likely to increase to a certain extent. As expected, activity increased by 50% (T190Y_150 and T190F_150). A short segment containing two hydrophobic residues that are three residues apart cannot form an amphipathic helix. But such a pair can form an amphipathic turn with a hydrophobic face on one side which can bind to the hydrophobic groove of TAZ2, but with a weaker affinity than a triad. This type of weakened binding for a turn was observed in the case of the turn II PreSMo of p53 TAD with a sequence of DIEQW. The Ile-Trp pair in this turn PreSMo formed a hydrophobic face that was capable of binding to the hydrophobic groove of mdmd2 where the amphipathic helix PreSMo of p53 TAD binds.[21,47] The binding strength of the turn II PreSMo, however, was weaker by an order of magnitude than that of the helix PreSMo. The mutants T190Y and T190F displayed ~50% higher activities than the wild type with a help of a pair of hydrophobic residues. This pair must have increased binding affinity of Helix 1 to TAZ2.

A similar explanation based on hydrophobicity is applicable to Helix 2 as shown in Fig. 12. Fig. 12A shows that Helix 2 in the wild type does not have a hydrophobic triad like Helix 1. Nevertheless, it has an interesting pseudo–quartet of hydrophobic residues formed by two pairs of hydrophobic residues, 212P-213W and 217L-218L (cyan box). A tryptophan has a large hydrophobic sidechain and is often critically involved in protein recognition. A well-known example is the 23W of the p53 helix PreSMo,[2,21] removal of which abolishes the binding of mdm2 with p53. The 213W in tau1c is critical for activity. Removal of 213W in Helix 2 eliminates the pseudo–quartet as shown in Fig. 12B, which is expected to lower the activity. Two such examples are the W213G_42 or W213R_30 mutants; activity went down by more than 50%. The position 221 in Helix 2 appears to play an extremely crucial role. This position is Glu in the wild type. Replacement of 221E by a hydrophobic residue Phe is dramatically increases the activity by 188% (E221F_288). Fig. 12C reveals a remarkable fact that the E221F mutant can have three triads of hydrophobic residues (purple boxes): the first by 218L, 221F and 224L, the second by 213W, 217L and 221F and

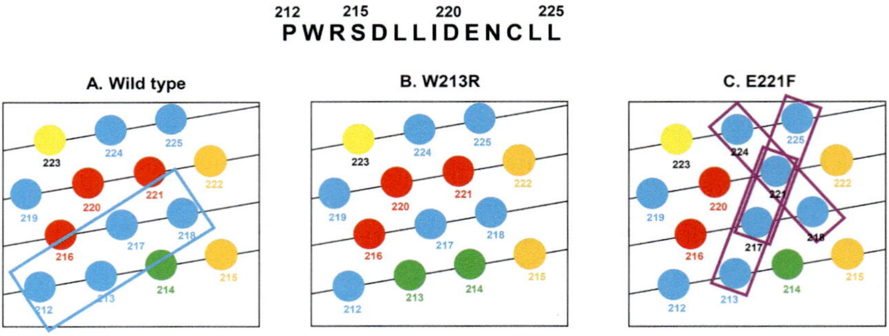

Fig. 12 (A) In the wildtype no triad of hydrophobic residues is found. Instead, it has a pseudo-quartet of hydrophobic residues outlined by a cyan box. A yellow circle is used for 223C. (B) Three triads of hydrophobic residues (purple boxes) found in the gain-of-function mutant E221F_288. Activity goes up by 188% relative to the wildtype. (C) Replacing 213W in the wildtype by non-hydrophobic residues (G or R) as in two mutants W213G_42 and W213R_30 would disrupt the pseudo-quartet of hydrophobic residues, lowering the activity. A green circle is used for 213R.

the third by 217L, 221F and 225L. The TAZ2 binding affinity of Helix 2 with 221F is expected to be much higher than any helix equipped with only one triad of hydrophobic residues. Note that E221F_288 has an even higher activity than D196Y_281 although the difference might be within the limit of measurement errors. Above results are neatly compatible with a reasoning that tau1c equipped with two or three triads of hydrophobic residues may slide around the hydrophobic groove of TAZ2 to produce a higher binding affinity for TAZ2.

The above logic has explained why four gain-of-functions mutants (two in the cluster c3 and two in the cluster c6) have higher activities than the wildtype. There are two gain-of-function mutants in the cluster c6 whose activities are not accounted for yet. The I193F_151 mutant has a larger aromatic sidechain Phe than the wildtype Ile at the position 193. This position in the wildtype forms a hydrophobic pair 193I-194L (Fig. 11B). In the mutant the pair turns into a somewhat stronger hydrophobic pair 193F-194L producing a somewhat higher activity. Similarly, the L225F_174 mutant has a slightly stronger hydrophobic pair 224L-225F than the wildtype 224L–225L (Fig. 12) that increased the activity by 74%. The hydrophobic interactions also explain loss-of-function mutants as well. For example, any replacement of 191F into a charged residue will reduce the activity as is demonstrated in the case of F191D_28 and F191E_29. Requirement for a large hydrophobic residue at the position 191 seems to be very stringent. Even a small reduction in the hydrophobicity at this position, for example, replacing Phe by Ile or Val decreases the activity significantly as shown in F191I_38 and F191V_48.

In summary, the new investigation by Salamanova et al. showed that it was the "helical conformation" of the helix PreSMos which was critical for activity, not the amino acid types. Only a helical conformation allows the formation of a triad, the pseudo-quartet and a pair of hydrophobic residues. This study also pointed to that the PreSMo stabilization (i.e., helicity increase) appears to be the main mechanism by which adaptive mutations can increase the activity of tau1c. From an evolutionary point of view, PreSMo stabilization or destabilization could be the strategies for functional adaptation. Tau1c of hGR is one of the IDRs that was studied very early in IDP research[5] when the keen helix controversy was still going on regarding the question of whether transactivation domains should contain some sort of specificity determinants that mediate target binding. However, it was difficult to obtain a clear explanation for the activity changes in tau1c mutants for a long time due to absence of an adequate structural rationale. The study by Salamanova et al. provided a realistic answer in a quite systematic manner using the PreSMo concept. In fact, tau1c is the best example so far that demonstrated the PreSMo-function relationship. Obtaining a more elaborate and truly quantitative PreSMo-function relationship for tau1c should be possible if the activity profiles for a few more selected mutants, e.g., a double mutant E221F/W213G, are examined along with their target binding mode with TAZ2.

4.3 VP16 TAD

VP16 TAD is the transactivation domain of the virion protein 16 (VP16) which is one of proteins that herpes simplex virus encodes. Even though it is of viral origin we describe it here since it is a well-known transactivation domain that was used as a model system in the helix controversy from the early days of IDP research.[98–100] The TAD in VP16 is located at its C-terminus (residues 412–490) and was classified as an acidic activation domain containing 17 Asp and 5 Glu.[138] It exhibits a potent transcriptional activity and is responsible for activation of the viral immediate early genes; it strongly transactivates relevant genes when fused to the DNA-binding domains of other transactivators.[98] Like many other activation domains (E2F1, TREB5, c-Jun, ESX, HSF-1, MyoD, NF kB p65, NF-IL6,

NHAT1, TEF-1, ALL1, HFV Bel-1, HIF-1α and p53 TAD) VP16 TAD also contains the so-called positionally-conserved hydrophobic residues such as Leu439, Phe442, Leu444, Phe479 and Leu483 as shown in Fig. 13. Mutation of any of these hydrophobic residues critically influences the transcriptional activity.[139] Many helix PreSMos in the activation domains are amphipathic with their hydrophobic faces that are involved in binding to the hydrophobic grooves in target proteins. The positionally-conserved hydrophobic residues are positioned optimally to form amphipathic helices.[47] It is ironical that the hydrophobic residues which are known to be depleted in IDPs are the critical factors for activity.

In 1997, the year a critical article on the random-coil nature of kinase inducible domain (KID) was published, another report titled "Induced Alpha-helix in the VP16 Activation Domain upon Binding to a Human TAF" was published.[31] This report described the structural nature of the hTAF$_{II}$31-binding region in VP16 TAD and claimed that the fragment consisting of the residues 472–483 acquired a helical conformation upon binding to its target hTAF$_{II}$31 based on the results of a transferred NOE (TrNOE) experiment. The TrNOE technique is designed to provide the structural information of a small ligand that has a bound memory which can be detected in its NOESY spectrum if its free and target-bound structure is in fast exchange.[140] The 17-residue peptide (residues 469–485) used in this experiment was reported to be completely unstructured in the free state, which was used as a basis for proposing a random coil to α-helix induced fit transition this peptide was supposed to undergo upon binding.

Fig. 13 The positionally-conserved hydrophobic residues in transactivation domains that form amphipathic helices. *Reproduced with permission from Chi SW, Lee SH, Kim DH, et al. Structural details on mdm2-p53 interaction. J Biol Chem. 2005;280 (46):38795–38802. https://doi.org/10.1074/jbc.M508578200.*

This conclusion is in line with that made for KID of CREB which was described to undergo a coil → helix folding transition.[30] This VP16 TAD article became one of the key reports supporting the CU view.

The PreSMo hypothesis states that the target-binding segments form PreSMos. Hence, suspicion arose if the fragment of VP16 TAD reported to be CU in the above article might form a PreSMo since it was the hTAF$_{II}$31-binding segment. Two later studies demonstrated that this putatively CU segment was pre-structured as a helix ($\sim$25% pre-populated)[35,36] when it was examined as an intact segment in a full 79-residue VP16 TAD. Jonker et al. showed that two regions (residues 431–449 and 467–487) formed helix PreSMos.[35] The region encompassing the residues 430–450 interacts with the human transcriptional coactivator positive cofactor 4 (PC4) while the helix PreSMo formed by the residues 460–490 bind to the general transcription factor TFIIB. Kim et al. showed that the residues 424–433, 442–446, 465–467 and 472–479 formed transient helices.[36] The last of the four composed of the residues 472–479 is the binding motif to hTAF$_{II}$31. These studies are another clear demonstration that the basis of transcriptional activity lies in the pre-structuring ability of the target-binding residues not in the physicochemical properties per se of amino acids.

Short peptides (<20 residues) free in aqueous solution are not able to form any discernible/detectable secondary structures unless they have an extremely strong inherent propensity to form a secondary structure.[141] This does not mean that the same residues constituting a short peptide would be in a completely unstructured conformation when they are present as a segment in a mother protein or in a full IDR like VP16 TAD. Within a mother protein or in a full domain the corresponding residues are supposed to experience what is known as the "tertiary effect" by flanking regions and may form a stable or at least transient secondary structure such as a transient helix. In the past, those who did not observe "significant" secondary structures in free peptides made an assumptive extrapolation to conclude that the segment would be CU even if it were placed within a mother protein or a full domain without actually carrying out an experiment with a mother protein or a full domain. The report on the VP16 TAD peptide is one example; the CU nature of the segment in a full IDP/IDR was claimed using the data obtained only with a short peptide. While it might as well be that the short VP16 peptide used was fully unstructured in the free state and indeed needed to undergo a coil → helix transition. But the same argument does not hold for binding between the full VP16 TAD and hTAF$_{II}$31 which is likely to require only a structural tightening of the target binding segment from a transient helix into a stable helix. Many unconfirmed speculations of this kind were common in the early-day IDP investigations.

4.4 c-Myc

The c-Myc oncoprotein (herein referred to as c-Myc) is a transcription factor that regulates many genes involved in apoptosis, differentiation, and proliferation.[142] The C-terminus of c-Myc heterodimerizes with the Max protein and forms a DNA-binding basic helix–loop–helix leucine-zipper motif. The transactivation domain in c-Myc is located at its N-terminus (residues 1–143) and is responsible for Myc-mediated transformation, differentiation, and apoptosis.[143] The c-Myc TAD is an important therapeutic target for inhibition of the heterodimerization. c-Myc is composed of two regions, Myc-1–88 (MBI) and Myc-92–167 (MBII).[144] Both regions are known to be intrinsically disordered, but are partially folded.[47,144,145] Andresen et al. discovered three transient structures within MBI using near-complete NMR resonance assignments.[70] Two helix PreSMos are formed by the residues 26–34 and 47–5, respectively and one β-turn by 20–23. Interesting to note is that these PreSMos are formed by the residues that are known to be conserved in Myc proteins. This is in line with the well-known fact that multiple PreSMos are formed by the positionally-conserved residues in many TADs as described above (see Fig. 13).

4.5 An eIF4E binding protein 1 (4EBP1)

Translation of eukaryotic mRNAs to proteins is a complex multi-step process that occurs through the formation of the eukaryotic translation initiation complex 4F (eIF4F) which is formed by a set of cap-binding protein, eukaryotic translation initiation factor 4E (eIF4E), a DEAD-Box RNA helicase, eukaryotic translation initiation complex 4A (eIF4A), and a eukaryotic translation initiation factor 4G (eIF4G).[146] eIF4G is a scaffold protein constituting the backbone of the eIF4F complex and interacts with eIF4E using a short consensus motif $YXXXXL\Phi$ (X is any amino acid and Φ is a hydrophobic residue). Since eIF4E binding proteins, inherent tumor suppressors in humans, also have this consensus motif 4EBPs and eIF4G compete each other for binding to eIF4E. Oncogenic transformation is associated with overexpression of eIF4E in tumor cells,[147,148] which can be down regulated by binding of 4EBPs. Binding of eIF4E with 4E-BP is phosphorylation-dependent, i.e., hyper-phosphorylation of 4EBPs disable their binding to eIF4E and leads to cancers.[149] In addition to cancers, eIF4E plays important in neurodevelopmental and neuropsychiatric disorders including depression, maternal immune activation (MIA), and repeat-associated non-AUG (RAN) translation since the function of eIF4E is controlled by ERK and mTOR signaling pathways.

The helix PreSMo detected in 4EBP1 is formed by the above consensus motif. This PreSMo carries a unique historical value in the development of the IDP research[20] because the original NMR report on 4EBP1 published at the dawn of the IDP research is one of the three reports that advocated the CU view, which ended up with making many take 4EBP1 as a paradigmatic CU type IDP.[16,32] Since 4EBP1 is composed of only 118 residues it may appear that a full NMR resonance assignment can be accomplished relatively easily with multidimensional high-resolution NMR methods. But obtaining a full NMR resonance assignment on a full-length 4EBP1 is very difficult due to heavy resonance overlap.[20] As a result, the first NMR report on this IDP was published without a full resonance assignment, but only with ~82% resonance assignment (the residues 4–10, 43–46 and 56–65 not assigned).[32] While such a level of resonance assignment for most proteins is usually sufficient to reveal the structural features important for function the situation on 4EBP1 was an exception since the residues for which the resonance assignment was missing happened to include the residues 56–65 that form the consensus motif critical for eIF4E binding. It took nearly 20 years before the realization that a full resonance assignment including the residues 56–65 needs to be obtained since these residues are target binding and are likely to form an important PreSMo. A new investigation using a ~70-residue 4EBP1 construct, named BP49, containing the consensus motif was carried out and a complete NMR resonance assignment was obtained including the PreSMo region.[20] A recent NMR study on 4EBP2, a homolog protein of 4EBP1, also discovered five regions with helical propensities, one of which encompassing the residues T50-N64 corresponds to the helix PreSMo formed by the consensus motif.[23] Had the PreSMo concept been available before the first NMR report on 4EBP1 with the message that NMR data on IDPs should be analyzed much more thoroughly than those on globular proteins an honest mistake of drawing a misleading conclusion from an incomplete resonance assignment data could have been avoided. The binding between eIF4E and the helix PreSMo of 4EBP1 is mainly mediated by hydrophobic interactions involving V69, L131 and I138 of eIF4E. Moerke et al.[150] discovered an inhibitor named 4EGI-1 that disrupts interaction of eIF4G-derived peptides with eIF4E and stabilizes the eIF4E/4E–BP1 interaction at the same time.

5. PreSMos in neurodegenerative IDPs

5.1 Alpha-synuclein

Alpha-synuclein (αS) a well-known IDP that is located at the nuclear envelope and at the presynaptic nerve terminals[151,152] and is responsible for the normal function and maintenance of synapses.[153] Clinically, it is strongly correlated with the pathogenesis of Parkinson's disease.[154] Neuronal fibrillar αS deposits known as Lewy bodies[155,156] is the diagnostic hallmark of PD being spherical protein inclusions found in the cytoplasm of nigral neurons in the brains of PD patients. For its clinical importance, the structural features of αS have been examined extensively. αS in fact is one of the IDPs whose structural features have been most extensively investigated. It has been revealed that the fibrillary aggregates of αS have a characteristic cross-β structure consisting of β-sheets, where the individual β-strands are perpendicular to the axis of the fibril.[157–159] These fibrillary aggregates are morphologically similar to the amyloid fibrils found in ADse neurotic plaques and in deposits associated with other amyloidogenic diseases.[160,161] The partially folded oligomeric intermediates that are formed along the αS fibril formation pathway are known to be cytotoxic.

αS is composed of 140 residues and its N-terminal region consisting of the residues 1–60 has a series of 11-amino acid repeats with a conserved KTKEGV motif that, upon binding to synthetic lipid vesicles or detergent micelles in vitro, adopts a highly helical conformation.[79,162,163] The mid-domain of αS is composed of the residues 61–95 and contains two additional KTKEGV repeats and the hydrophobic amyloidogenic NAC (non-amyloid-β component) region that is known to be involved in the formation of amyloid fibrils both in vitro and in vivo.[163] The highly acidic C-terminal region (residues 96–140) is responsible for the overall net negative charge. The critical early step in the fibrillation of αS is believed to involve a conformational transition from the native monomeric structure of αS into an aggregation-prone partially folded intermediate.[164–166] The truncation of the acidic C-terminus of αS accelerates fibril formation in vitro,[167] which suggested that the intermolecular electrostatic repulsions among the negatively charged C-terminal regions could be an important factor for slowing down αS aggregation. It also leads to clear phenotypes of the disease in mice.[168] Therefore, determining the structure of αS monomer became especially important and was carried out by several NMR laboratories under a variety of experimental conditions (see Table 3) with an aim to find out if αS monomer has any structural features that are associated with formation of oligomeric intermediates.

Table 3 A list of NMR studies on αS.

Year	Sample condition	BMRB	References
2001	~100 μM αS 100 mM NaCl, 10 mM Na$_2$HPO$_4$, pH 7.4, 283 K		27
2003	0.3 mM αS 20 mM sodium phosphate, 50 mM SDS, pH 7.4, 298 K	5744[a]	71
2006	1 mM human αS 20 mM phosphate, 0.5 mM EDTA, 200 mM NaCl, 10% D$_2$O, pH 6.5, 285.5 K	6968	72
2008	0.2 mM αS PBS buffer (8 mM Na$_2$HPO$_4$, 2 mM KH$_2$PO$_4$, 2.7 mM KCl, 137 mM NaCl), pH 7.4, 263 K		34
2009	0.65 mM αS 10 mM phosphate, 140 mM NaCl, pH 2.5, 10% D$_2$O, 288 K		73
2009	0.3 mM αS 20 mM NaOAc, 100 mM NaCl, 10% D$_2$O, pH 3.0 and pH 7.4, 288 K	16,342	74
2009	0.6 mM human αS, 20 mM Na$_2$HPO$_4$ (pH 6.0), 6% D$_2$O, 0.02% NaN$_3$, in phospholipids, 293 K		26
2010	0.6 mM human wildtype αS, mutants (A30P, E46K, A53T), 20 mM Na$_2$HPO$_4$ (pH 6.0), 6% D$_2$O, 0.02% NaN$_3$, in phospholipids, 293 K		75
2012	0.1 mM αS, 5 mM dioxane, 20 mM sodium phosphate buffer (pH 6), in phospholipids, 288 K		76
2013[b]	1.7 mM αS 10% D$_2$O, 90% H$_2$O, pH 6.2, 277 K	19,257	77
2013[c]	20 mM Tris-HCl (pH 7), 100 mM NaCl, 10% D$_2$O, 288 K		78
2014	0.35 mM αS, 20 mM Sodium phosphate (pH 6), 288 K		79
2014	50 μM αS, 5 × NaCl/Pi, 5% glycerol, 288 K		80
2014	0.3 mM αS 20 mM NaOAc, 100 mM NaCl, 10% D$_2$O, pH 3.0 and pH 7, 288 K		81
2015	0.5 mM, 0.7 mM, αS, H50Q, 10 mM sodium phosphate (pH 7.5), 100 mM NaCl, 5% D$_2$O, 0.01% NaN$_3$, 0.001% DSS, 283 K	25,227	82
2015	~0.43 mM (6.6 mg/mL) αS, 20 mM HEPES, 10% D$_2$O, pH 7.0, 277 K	26,557	83
2016[b]	0.4 mM human aS, 20 mM sodium phosphate, 150 mM NaCl, pH 7.0, 283 K		58

Continued

Table 3 A list of NMR studies on αS.—cont'd

Year	Sample condition	BMRB	References
2017[d]	75 μM αS PBS (1.76 mM KH_2PO_4), 10 mM Na_2HPO_4, 2.7 mM KCl, 137 mM NaCl, 0.02% NaN_3 pH 7.4, 310 K		84
2018	~1 mM αS, pH 5.0, 10% D_2O, 298 K		85
2018	1 mM human αS 200 mM NaCl, 0.5 mM EDTA, 20 mM phosphate pH 6.5	27,348	86

[a]BMRB 5744 was done for a folded αS, and hence is not used for calculation of SSP values.
[b]In cell NMR.
[c]No information available on the concentration of the protein.
[d]Aggregation inhibition experiment.
Only seven studies deposited the chemical shifts in the BMRB database as shown in the third column.

In general, NMR studies on globular proteins carried out under different sample conditions (protein concentration, buffer, temperature, pH, etc.) produce the structures with the same overall globular structures whose atoms have nearly the same spatial coordinates except for some minor differences at the sub-Angstrom level that originate from the rotational freedom of sidechains. The situation is similar in the NMR studies of IDPs; slightly different sample conditions do not significantly influence the results in terms of the presence and/or locations of PreSMos. For αS the situation has been somewhat different such that the results of different NMR studies do not fully agree with one another. αS is the only IDP studied so far that displays this kind of inconsistency. Several factors could contribute to this discrepancy as reviewed in a recent article.[169] Briefly, ensuring the monomeric state of a protein during NMR data acquisition becomes crucial when we deal with proteins that are prone to aggregation like αS. The pH of a sample is also an important factor since the charged states of the hydrophilic residues in IDPs are likely to influence the overall topology as well as the local conformations. For an IDP with an oligomerization tendency temperature can also have a significant effect on its conformation. Due to their structural homeostasis, the changes in the solvent buffer conditions do affect 3D structures of globular proteins. This may not be true for αS, as the counterion shielding of polar residues on the surface may significantly influence the conformations of IDPs. Finally, removal of residual detergents and lipids during the purification of αS is extremely important since both the normal function and the pathogenic role of αS are related to its interaction with lipid membranes. Incomplete removal can cause line broadening in NMR resonances.

In Fig. 14 the SSP scores and the chemical shift deviation of $C\alpha$ carbons calculated using three different random coil chemical shift references are shown for four BMRB deposited NMR results on αS. Even though the degree of pre-structuring and the locations of transient structures in αS vary to a certain extent depending upon the experimental conditions all the results point to that a helix PreSMo exists at the N-terminal region (encompassing the residues 10–30) of αS even though the degree of pre-population for this PreSMo varies slightly. This agrees with the early result reported by Eliezer et al.[79] describing that the $C\alpha$ chemical shifts for residues 6–37 in αS deviated more than $\sim$0.3 ppm from random chemical shifts, which indicates the existence of a helix PreSMo.

5.2 Tau

Tau protein is an IDP that plays a central role in the progression of Alzheimer's disease. Expressed in nerve axons the normal function of tau is to bind and stabilize microtubules (MTs).[170] Binding of tau to MTs modulates the transport of vesicles and organelles along MTs and supports the outgrowth of axons.[171,172] When tau becomes excessively phosphorylated in AD it can no longer bind MTs and partially folds or aggregates into an ordered b-structure, which produces intracellular abnormal protein deposits composed of paired helical filaments (PHFs).[173]

The human central nervous system has six isoforms of tau, the longest consisting of 441 amino acids. Being an IDP the structure of tau cannot be determined by x-ray crystallography and its structural features can be investigated only by NMR. However, because of its large size and of the fact that tau contains repeat sequences the overlap of resonances in the NMR spectrum of a full-length tau is severe. The degree of overlapping is $\sim$3 times worse than what one sees in the NMR spectrum of a globular protein composed of 731-residues (malate synthase G) which represents one of the largest proteins for which a complete resonance has been achieved. Therefore, all the previous structural studies on tau have used fragments of tau.[174–182]

An excellent NMR study was reported by Mukrasch et al., where they used a divide-and-conquer strategy to accomplish an almost complete (more than 98%) resonance assignment of a full-length tau.[183] What was taken advantage of was a feature unique to IDPs, a sum of the NMR spectra of segments of a long IDP is the same as that of a full-length IDP. They used three segments with overlapping sequences containing 185, 198 and 168 residues to ensure the correct assignment. Shown in Fig. 15 are

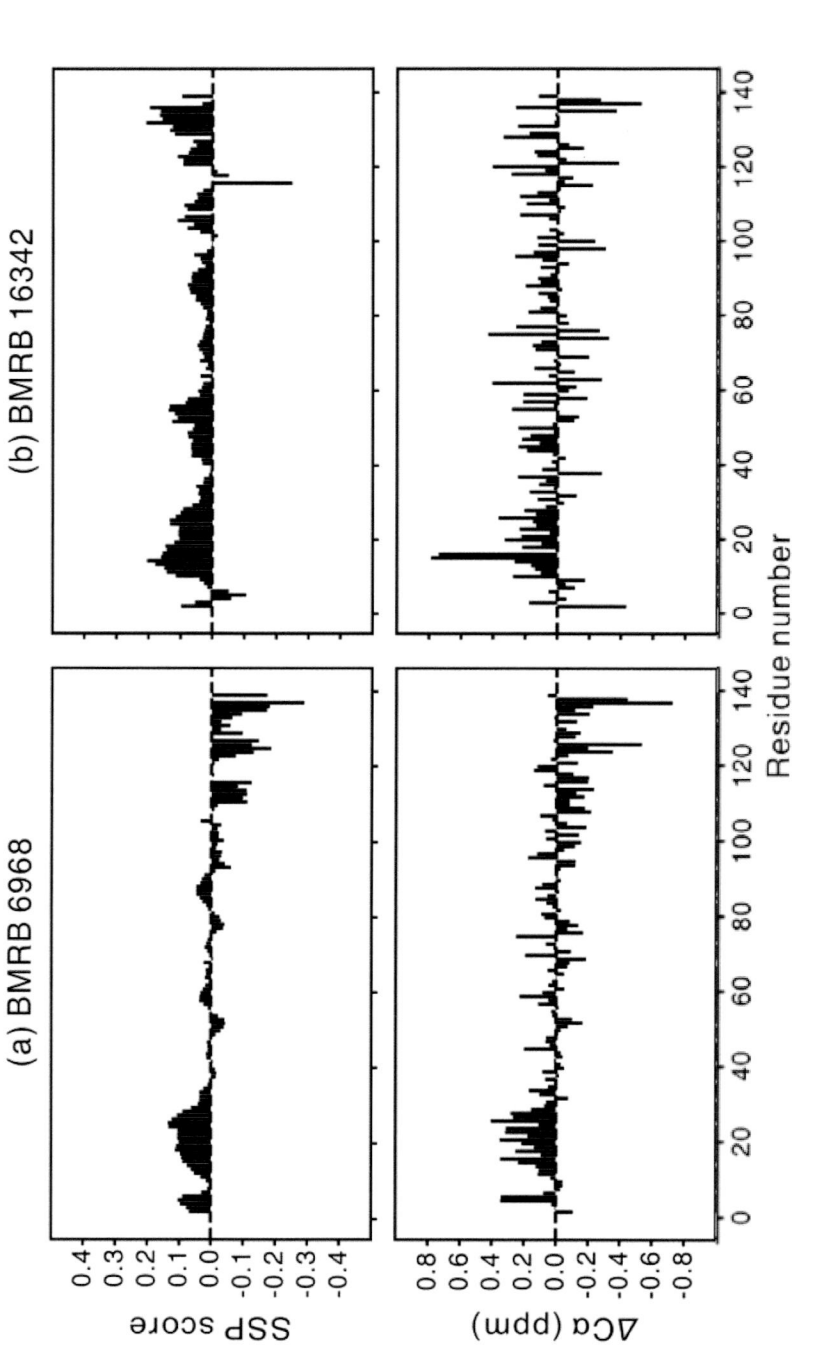

Fig. 14 Four SSP scores for a monomeric α-synuclein obtained in different experimental conditions. Although the scores vary all four agree on the presence of a helix PreSMo at the N-terminus. These results illustrate that the SSP scores which include several chemical shifts into calculation ($^{1}H\alpha$, $^{13}C\alpha$, $^{13}C\beta$ ^{13}CO, ^{15}N etc.) are a more reliable measure than the $\Delta C\alpha$ values alone. *Reproduced with permission from Csizmok V, Felli IC, Tompa P, et al. Structural and dynamic characterization of intrinsically disordered human securin by NMR spectroscopy. J Am Chem Soc. 2008; 130(50):16873–16879. https://doi.org/10.1021/ja805510b.*

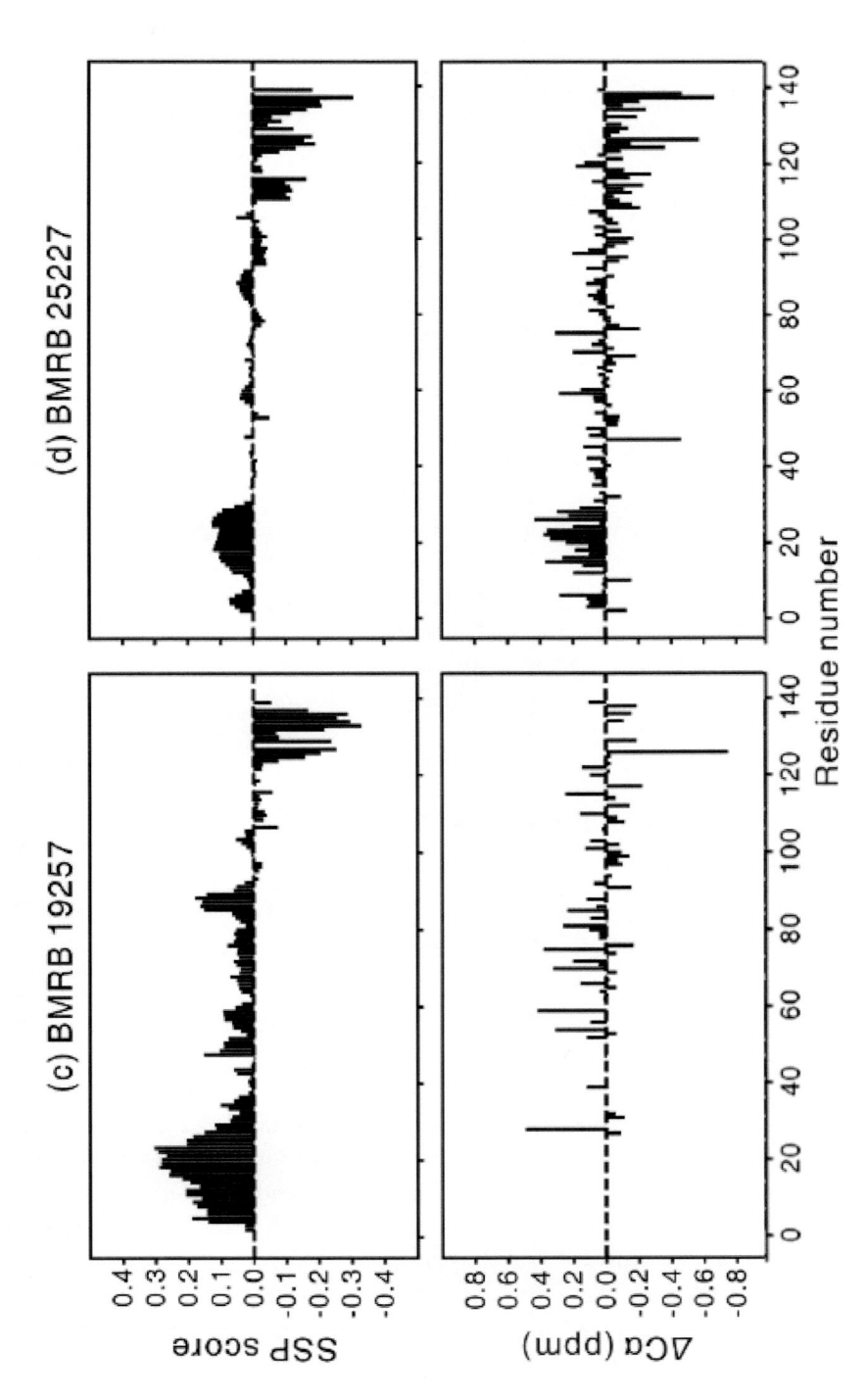

Fig. 14—Cont'd

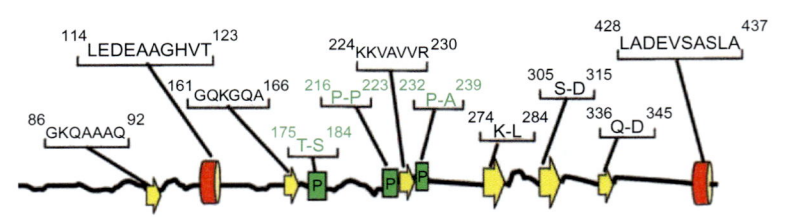

Fig. 15 The 11 PreSMos observed in Tau; two α-helices, three PPII helices, six β-strands. *Reproduced with permission from Mukrasch MD, Bibow S, Korukottu J, et al. Structural polymorphism of 441-residue tau at single residue resolution. PLoS Biol. 2009;7(2):e34. https:// doi.org/10.1371/journal.pbio.1000034.*

the 11 pre-structured motifs found in tau. Through this work tau has become not only the largest IDP whose NMR resonances are fully assigned, but also an IDP containing the largest number of PreSMos. This work has shown that the tau-MT binding relies on hydrophobic interactions that involve transiently populate β-structure and polyproline II helix.[184] It is also demonstrated that much of the N- and C-terminal domains of tau are highly flexible even when the repeat domain is attached to MTs.[170] This observation is related to the statement made above regarding the fuzzy complexes; certain portions composed of no-PreSMo-forming residues can be disordered even after target binding.

6. PreSMos in other IDPs

6.1 Viral IDPs

Microorganisms such as archaea, and bacteria contain ∼15% of IDPs. Viruses encode ∼20% IDPs as well. PreSMos have been observed in several viral IDPs.[16]

6.1.1 preS1

Hepatitis B virus (HBV) poses a significant health threat to millions of people worldwide, particularly in Africa, Asia, and certain parts of Europe.[185,186] chronic hepatitis B infection often develops into more serious symptoms such as cirrhosis, liver failure, and hepatocellular carcinoma.[187] Even though HBV vaccines were developed,[188–191] non-responders to these vaccines exist and escape mutants are found, requiring an effort to develop vaccines with better efficacy. Specific anti-HBV therapeutics are needed to completely eradicate HBV infection, since currently available drugs for HBV-related symptoms are not HBV specific, often causing side effects or inducing drug resistance. Anti–HBV agents based upon natural products or siRNAs are only partially successful.[192,193] The surface antigen of HBV consists of three envelope glycoproteins called the large, middle, and small proteins, named preS1, preS2, and S regions, respectively.[194] The large protein preS1 is the most abundant form found in the surface of infectious viral particles[194,195] and is believed to play the most critical role in the binding of HBV to hepatocytes as the outermost part of HBV particles.[196–199] PreS1 from the adr subtype HBV is an IDP composed of 119 residues.[194]

An NMR study on preS1 revealed that it contains several noncontiguous PreSMos.[83] Even though preS1 has 118 residues no transient pre-structuring was observed for the residues beyond 50D. The most prominent PreSMo encompasses residues 27–45 and contains two helical turn motifs composed of residues 32P-36A and 41P-45F, both of which are located within the putative HBD (residues 21–47) known to be important for hepatocyte binding. The PreSMo formed by residues 32P-36A partially overlaps with the putative hepatocyte binding fragment (Q29LDPAF34)[200] which is an immunodominant epitope recognized by several neutralizing monoclonal antibodies.[201] Additional pre-structuring tendency was observed for the residues 11–18, 22–25, 37–40, and 46–50. Kim et al.[202] reported that a 21-residue peptide (named 7524 BVS7) corresponding to the most prominent PreSMo potently inhibited HBV infection with an IC_{50} of ≈ 20 nM (Fig. 16).

6.1.2 C-terminal domain of Sendai virus nucleoprotein

Jensen et al. have shown that the partially folded C-terminal domain of Sendai virus nucleoprotein contains three helix PreSMos formed by the residues 476–488, 479–484, and 478–492 with the pre-population of 28%, 36%, and 11%, respectively.[19] In a more recent study, it was shown that the helix PreSMo (H2) formed by the residues 476–488 is the one that binds to target PX and stabilized in the bound state.[53]

6.1.3 E7

Cervical cancer, one of the leading causes of female death in developing countries. Human papillomaviruses (HPVs) are the primary agents causing cervical cancer. Despite the availability of anti-HPV vaccines,[203] effective pharmaceuticals are still needed for those who have already developed an infection with HPV. HPVs are classified into high- or low- risk types.[204] HPV16, classified as a high-risk alpha HPV is one of the most prevalent HPVs found in cervix carcinomas. Among eight proteins HPV16 encodes,

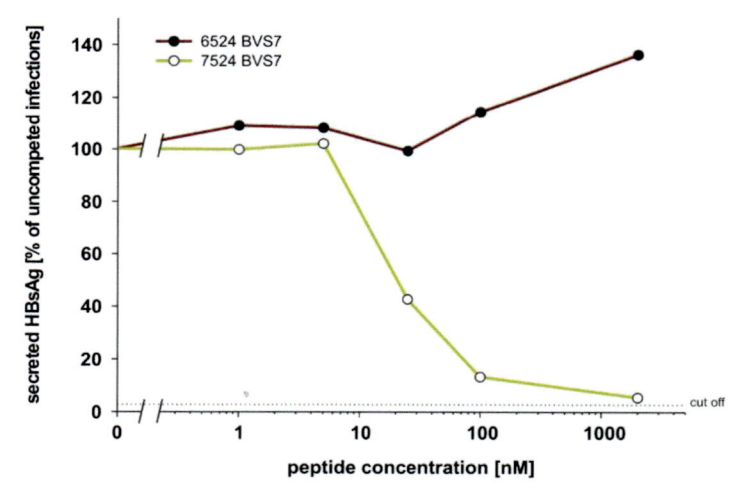

Fig. 16 The anti-viral activity of a 21-residue peptide (named 7524 BVS7) derived from the most prominent PreSMo in HBV preS1 shows an IC$_{50}$ of ≈ 20 nM. *Reproduced with permission from Kim DH, Yi N., Lee SH, et al. An anti-viral peptide derived from the preS1 surface protein of hepatitis B virus. BMB Reports. 2008;41(9): 640–644. http://doi.org/10.5483/BMBRep.2008.41.9.640.*

two oncoproteins, E6 and E7 are essential for the development of cancer. E7 is a small protein consisting only of 98 amino acid residues. Its primary structure can be divided into three conserved regions (CRs). Its N-terminus is an IDR with two CRs, CR1 (residues 2–15) and CR2 (residues 16–41). E7 is responsible for the transforming activity and promiscuously interacts with diverse cellular targets.[205,206] CR interacts with p600, Skp2, p300, and IRF-1, and C7 with HPV E2, Rb1, FHL2, TBP, CKII, p300, HIF-1a, and p21.

The structural characteristics of the N-terminal IDR (residues 1–46) of E7 (N-E7) were examined by high resolution NMR spectroscopy combined with replica exchange molecular dynamics (REMD) simulation.[90] Results based on the chemical shifts and the SSP scores show two helical PreSMos (residues 7–14 and 20–26) and a non-helical transient structure as shown in Fig. 17. The first helix is ~10% pre-populated and is an E2F mimic. The second is ~20% pre-populated and corresponds to the Rb-binding LXCXE segment. The C-terminal residues 30–46 has a potential to form a non-helical (β-type) PreSMo. A separate REMD simulations were carried out on N-E7 and revealed two helical PreSMos; the two calculated helical PreSMos obtained by REMD are slightly longer than the

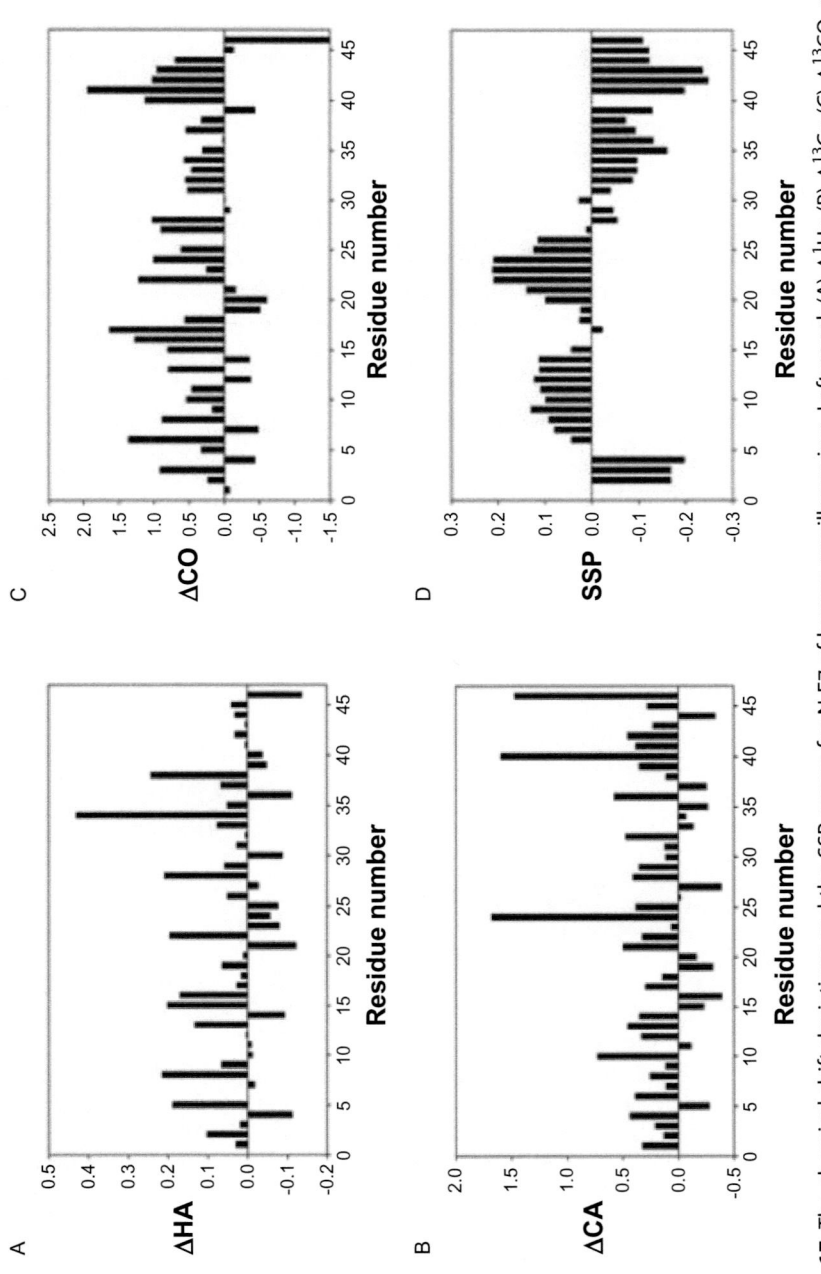

Fig. 17 The chemical shift deviations and the SSP scores for N-E7 of human papillomavirus. Left panel: (A) Δ^1Hα (B) Δ^{13}Cα (C) Δ^{13}CO, and (D) the SSP scores. *Reproduced with permission from Lee C, Kim DH, Lee SH, et al. Structural investigation on the intrinsically disordered N-terminal region of HPV16 E7 protein. BMB Rep. 2016;49(8):431–436. https://doi.org/10.5483/bmbrep.2016.49.8.021.*

ones detected by NMR as shown as ensemble structures in Fig. 18. The left is an ensemble superimposed along the first helix PreSMo (residues 7–14), and the right is an ensemble superimposed along the second helix PreSMo (residues 20–26).

6.1.4 HIV-1 Tat

To et al.[28] reported that the 121-residue HIV-1 transactivator of transcription (TAT) has five PreSMos formed by the residues 27–32 (helix 0), 41–59 (helix), 70–81 (β sheet), 93–99 (β sheet), 105–112 (β sheet), each pre-populated by ~20%, ~30%, ~25%, ~25%, ~10%, respectively.

6.2 Cystic fibrosis transmembrane conductance regulator (CFTR)

Baker et al. have carried out an NMR investigation on the intrinsically disordered regulatory region of a CFTR domain (residues 654–838) and found four α-helix PreSMos (the residues 654–668, 759–764, 766–776, and 801–817), and a β-strand (744–753). These PreSMos play an important role for interaction with the nucleotide-binding domain 1 of CFTR.[37]

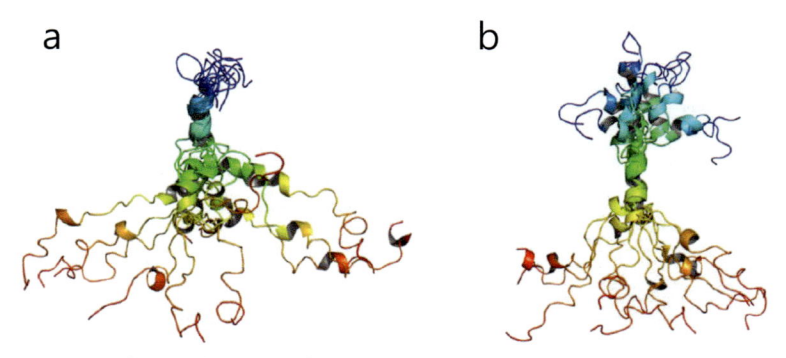

Fig. 18 Two replica exchange molecular dynamics (REMD) ensembles of N-E7. (A) An ensemble generated by aligning the residues 7–14 that form the first helical PreSMo. (B) An ensemble generated by superimposing the second helical PreSMo. All ensemble structures showed a high correlation with the SSP scores from NMR experiments (PCC > 0.65). The structures are color-coded from the N- to the C-terminus (blue → green → yellow → brown). *Reproduced with permission from Lee C, Kim DH, Lee SH, et al. Structural investigation on the intrinsically disordered N-terminal region of HPV16 E7 protein. BMB Rep. 2016;49(8):431–436. https://doi.org/10.5483/bmbrep.2016. 49.8.021.*

6.3 A p53 rescue motif in SUMO specific protease 4 (SUSP4)

SUMO is a small ubiquitin-like modifier. SUMO-specific protease SUSP4 is an inherent tumor suppressor that positively regulates p53 by promoting mdm2 self-ubiquitination.

Lee et al. discovered that SUSP 4 can restore the tumor suppressive activity of p53 by interfering with the binding of mdm2 to p53 TAD.[207] A subsequent NMR investigation on SUSP4 shows that it is the mid-domain (residues 201–300) of SUSP4 named P4-201 which binds to the hydrophobic pocket in mdm2 where the helix PreSMo of p53 TAD binds, resulting in release of p53 from the mdm2 attack. Even though P4-201 is 100-residues long it is an IDR. Detailed analysis of structural characteristics of P4-201 obtained by NMR and molecular dynamics simulation revealed that SUSP4 contains a PreSMo composed of 29 residues which consists of two transient helices connected by a hydrophobic linker.[91] Since this motif shows anti-cancer effects in various cancer cells it was named the p53 rescue motif. This study demonstrated that PreSMos may provide a new structure-assisted technology platform for drug design.

7. Summary and perspective

In this chapter we have returned to the basics and rekindled the fundamental question of why the unorthodox IDPs are functional at all. To answer this question we have delved deeply into the theme of how the unfolded IDPs can interact with specific targets and have reasoned out a *structural* rationale for *unstructured* proteins, the PreSMo-function relationship between the conformational characteristics of the PreSMos and the diverse IDPs' activities. This relationship is most clear for transcription factors and TADs. The activity critically depends on the helical conformation of the helix PreSMos. The connection between the detected transient structural features and the functions of the IDPs involved in neurodegenerative diseases, αS and tau, is not as clear. Yet, it is evident that the structural characteristics (α-helicity, PPII helix and β-structure) of the target binding segments in these IDPs are also important for target interactions. For αS and tau, unlike for the transcription activation domains, the atomic-resolution structural details of the PreSMo-binding interfaces are not fully known yet. In viral IDPs the knowledge gained so far is limited other than the fact that the target-binding segments of these IDPs form PreSMos that appear important for binding to relevant targets such as a hepatocyte receptor, PX, the nucleotide-binding domain 1 of CFTR, and mdm2 (p53 rescue motif). One exception is the C-terminal domain of Sendai virus nucleoprotein for which a thorough NMR relaxation dispersion experiment elegantly showed that it followed the CS-IF mechanism by forming an encounter complex between one of the three helix PreSMos detected and PX. Further investigations emulating this work would be valuable.

So, what is the verdict on the mechanism of IDP-target binding? The answer is up in the clouds as yet. The trend for the last two decades has been all CU and 100% induced fit, then consideration of the conformational selection pathway with the discovery of PreSMos, and finally leaning toward the enticing CS-IF sequential mechanism. As has been mentioned it took half a century to learn that CS seems to be favored slightly over IF for globular proteins that have been around us for a century. And, the mechanism varies from one protein to another. It perhaps is too ambitious to want an answer for IDPs that began to be noticed only ~20 years ago. However, despite the fact that the PreSMo-function relationship established so far is semi-quantitative it is encouraging that the concept can be utilized in the discovery of potential anti-viral and anti-cancer molecules. Further systematic studies taking into account of energetic details by measuring the kinetic and thermodynamic parameters associated with binding should be carried out for a sufficient number of IDPs and for their selective mutants suggested by the PreSMo concept. Availability of target-bound structures is essential to comprehend the atomic resolution picture of binding. Developing NMR techniques that allow detection of a low-level PreSMo should also be useful in this regard. While the NMR technique has made an unmatched contribution to the detection of PreSMos and to providing a reasonable PreSMo-activity relationship the technique may not be the most efficient tool for studying the target binding process of IDPs. Molecular dynamics simulation that can follow a microsecond level trajectory seems to be a promising alternative. Being able to detect a PreSMo just by disorder prediction algorithms would be nice.

It is not uncommon in the history of science that a new finding or concept has to go through a period of denial, rejection and challenges. An example is prion. It took decades before the idea that a non-bacterial, non-viral but protein-only can cause a disease. The PreSMo hypothesis is such an example. PreSMos initially appeared so insignificant that no duly attention was paid to them for years. The PreSMo hypothesis was even viewed as an uncomfortable outlier, but now is on the verge of becoming a putative theory providing the most coherent structural rationale for IDPs' diverse functions.

The fraction of IDPs containing PreSMos is $\sim$80% of all IDPs/IDRs structurally characterized so far and is expected to increase with time. Exploring the structural details of IDPs should be continued to accumulate the quantitative knowledge between the pre-structuring phenomenon in IDPs and function. This will help answer the lingering questions on how disordered proteins can function without globular structures and will provide the arsenals necessary for eradicating the IDP associated fatal diseases.

Acknowledgments

We wish to thank Christella Gordon, James A. Ferretti, Anthony P.H. Wright, Christian Griesinger, and Joan J. Han for carefully reading the manuscript and providing valuable suggestions.

References

1. Dunker AK, Babu MM, Barbar E, et al. What's in a name? Why these proteins are intrinsically disordered: Why these proteins are intrinsically disordered. *Intrinsically Disord Proteins*. 2013;1(1), e24157. https://doi.org/10.4161/idp.24157.
2. Lee H, Mok KH, Muhandiram R, et al. Local structural elements in the mostly unstructured transcriptional activation domain of human p53. *J Biol Chem*. 2000;275 (38):29426–29432. https://doi.org/10.1074/jbc.M003107200.
3. Lee SH, Kim DH, Han JJ, et al. Understanding pre-structured motifs (PreSMos) in intrinsically unfolded proteins. *Curr Protein Pept Sci*. 2012;13(1):34–54. https://doi.org/10.2174/138920312799277974.
4. James TL, Liu H, Ulyanov NB, et al. Solution structure of a 142-residue recombinant prion protein corresponding to the infectious fragment of the scrapie isoform. *Proc Natl Acad Sci USA*. 1997;94(19):10086–10091. https://doi.org/10.1073/pnas.94.19.10086.
5. Dahlman-Wright K, Baumann H, McEwan IJ, et al. Structural characterization of a minimal functional transactivation domain from the human glucocorticoid receptor. *Proc Natl Acad Sci USA*. 1995;92(5):1699–1703. https://doi.org/10.1073/pnas.92.5.1699.
6. Wright PE, Dyson HJ. Intrinsically unstructured proteins: re-assessing the protein structure-function paradigm. *J Mol Biol*. 1999;293(2):321–331. https://doi.org/10.1006/jmbi.1999.3110.
7. Uversky VN, Gillespie JR, Fink AL. Why are "natively unfolded" proteins unstructured under physiologic conditions? *Proteins*. 2000;41(3):415–427. https://doi.org/10.1002/1097-0134(20001115)41:3 < 415::aid-prot130 > 3.0.co;2-7.
8. Tsai CJ, Ma B, Sham YY, Kumar S, Nussinov R. Structured disorder and conformational selection. *Proteins*. 2001;44(4):418–427. https://doi.org/10.1002/prot.1107.
9. Bertoncini CW, Jung YS, Fernandez CO, et al. Release of long-range tertiary interactions potentiates aggregation of natively unstructured alpha-synuclein. *Proc Natl Acad Sci USA*. 2005;102(5):1430–1435. https://doi.org/10.1073/pnas.0407146102.
10. Fink AL. Natively unfolded proteins. *Curr Opin Struct Biol*. 2005;15(1):35–41. https://doi.org/10.1016/j.sbi.2005.01.002.
11. Dunker AK, Silman I, Uversky VN, Sussman JL. Function and structure of inherently disordered proteins. *Curr Opin Struct Biol*. 2008;18(6):756–764. https://doi.org/10.1016/j.sbi.2008.10.002.

12. Uversky VN, Davé V, Iakoucheva LM, et al. Pathological unfoldomics of uncontrolled chaos: intrinsically disordered proteins and human diseases. *Chem Rev.* 2014;114 (13):6844–6847. https://doi.org/10.1021/cr400713r.

13. We prefer the usage of a word "region" instead of domain since the latter tends to be implicitly tied with some form of structural features.

14. Many examples of protein disorder can be found as absence of x-ray diffraction density in the RCSB Protein Data Bank (https://www.rcsb.org).

15. Han KH, La Mar GN. Nuclear magnetic resonance study of the isotope exchange of the proximal histidyl ring labile protons in hemoglobin A. The exchange rates and mechanisms of individual subunits in deoxy and oxy-hemoglobin. *J Mol Biol.* 1986;189 (3):541–552. https://doi.org/10.1016/0022-2836(86)90323-2.

16. Dunker AK, Obradovic Z, Romero P, Garner EC, Brown CJ. Intrinsic protein disorder in complete genomes. *Genome Inform Ser Workshop Genome Inform.* 2000;11:161–171. https://doi.org/10.11234/gi1990.11.161.

17. Hammes GG, Chang YC, Oas TG. Conformational selection or induced fit: a flux description of reaction mechanism. *Proc Natl Acad Sci USA.* 2009;106(33): 13737–13741. https://doi.org/10.1073/pnas.0907195106.

18. Changeux JP, Edelstein S. Conformational selection or induced fit? 50 years of debate resolved. *F1000 Biol Rep.* 2011;3:19. https://doi.org/10.3410/B3-19.

19. Jensen MR, Houben K, Lescop E, Blanchard L, Ruigrok RW, Blackledge M. Quantitative conformational analysis of partially folded proteins from residual dipolar couplings: application to the molecular recognition element of Sendai virus nucleoprotein. *J Am Chem Soc.* 2008;130(25):8055–8061. https://doi.org/10.1021/ja801332d.

20. Kim DH, Lee C, Cho YJ, et al. A pre-structured helix in the intrinsically disordered 4EBP1. *Mol Biosyst.* 2015;11(2):366–369. https://doi.org/10.1039/c4mb00532e.

21. Kussie PH, Gorina S, Marechal V, et al. Structure of the MDM2 oncoprotein bound to the p53 tumor suppressor transactivation domain. *Science.* 1996;274(5289):948–953. https://doi.org/10.1126/science.274.5289.948.

22. Hua QX, Jia WH, Bullock BP, Habener JF, Weiss MA. Transcriptional activator-coactivator recognition: nascent folding of a kinase-inducible transactivation domain predicts its structure on coactivator binding. *Biochemistry.* 1998;37 (17):5858–5866. https://doi.org/10.1021/bi9800808.

23. Lukhele S, Bah A, Lin H, Sonenberg N, Forman-Kay JD. Interaction of the eukaryotic initiation factor 4E with 4E-BP2 at a dynamic bipartite interface. *Structure.* 2013;21 (12):2186–2196. https://doi.org/10.1016/j.str.2013.08.030.

24. Dogan J, Gianni S, Jemth P. The binding mechanisms of intrinsically disordered proteins. *Phys Chem Chem Phys.* 2014;16(14):6323–6331. https://doi.org/10.1039/c3cp54226b.

25. Wang H, Dawber RS, Zhang P, Walko M, Wilson AJ, Wang X. Peptide-based inhibitors of protein-protein interactions: biophysical, structural and cellular consequences of introducing a constraint. *Chem Sci.* 2021;12(17):5977–5993. https://doi.org/10.1039/d1sc00165e.

26. Lange OF, et al. Recognition dynamics up to microseconds revealed from an RDC-derived ubiquitin ensemble in solution. *Science.* 2008;269(41):25613–25620. https://doi.org/10.1016/S0021-9258(18)47294-8.

27. Thapar R, Mueller GA, Marzluff WF. The N-terminal domain of the Drosophila histone mRNA binding protein, SLBP, is intrinsically disordered with nascent helical structure. *Biochemistry.* 2004;43(29):9390–9400. https://doi.org/10.1021/bi036314r.

28. To V, Dzananovic E, McKenna SA, O'Neil J. The dynamic landscape of the full-length HIV-1 transactivator of transcription. *Biochemistry.* 2016;55(9):1314–1325. https://doi.org/10.1021/acs.biochem.5b01178.

29. Wells M, Tidow H, Rutherford TJ, et al. Structure of tumor suppressor p53 and its intrinsically disordered N-terminal transactivation domain. *Proc Natl Acad Sci USA.* 2008;105(15):5762–5767. https://doi.org/10.1073/pnas.0801353105.

30. Radhakrishnan I, Pérez-Alvarado GC, Parker D, Dyson HJ, Montminy MR, Wright PE. Solution structure of the KIX domain of CBP bound to the transactivation domain of CREB: a model for activator:coactivator interactions. *Cell.* 1997;91(6):741–752. https://doi.org/10.1016/s0092-8674(00)80463-8.

31. Uesugi M, Nyanguile O, Lu H, Levine AJ, Verdine GL. Induced alpha helix in the VP16 activation domain upon binding to a human TAF. *Science.* 1997;277(5330): 1310–1313. https://doi.org/10.1126/science.277.5330.1310.

32. Fletcher CM, Wagner G. The interaction of eIF4E with 4E-BP1 is an induced fit to a completely disordered protein. *Protein Sci.* 1998;7(7):1639–1642. https://doi.org/10.1002/pro.5560070720.

33. Oldfield CJ, Meng J, Yang JY, Yang MQ, Uversky VN, Dunker AK. Flexible nets: disorder and induced fit in the associations of p53 and 14-3-3 with their partners. *BMC Genomics.* 2008;9(suppl 1):S1. https://doi.org/10.1186/1471-2164-9-S1-S1.

34. Wright PE, Dyson HJ. Linking folding and binding. *Curr Opin Struct Biol.* 2009;19 (1):31–38. https://doi.org/10.1016/j.sbi.2008.12.003.

35. Jonker HR, Wechselberger RW, Boelens R, Folkers GE, Kaptein R. Structural properties of the promiscuous VP16 activation domain. *Biochemistry.* 2005;44(3):827–839. https://doi.org/10.1021/bi0482912.

36. Kim DH, Lee SH, Nam KH, Chi SW, Chang I, Han KH. Multiple hTAF(II)31-binding motifs in the intrinsically unfolded transcriptional activation domain of VP16. *BMB Rep.* 2009;42(7):411–417. https://doi.org/10.5483/bmbrep.2009.42.7.411.

37. Baker JM, Hudson RP, Kanelis V, et al. CFTR regulatory region interacts with NBD1 predominantly via multiple transient helices. *Nat Struct Mol Biol.* 2007;14(8):738–745. https://doi.org/10.1038/nsmb1278.

38. Neri D, Billeter M, Wider G, Wüthrich K. NMR determination of residual structure in a urea-denatured protein, the 434-repressor. *Science.* 1992;257(5076):1559–1563. https://doi.org/10.1126/science.1523410.

39. Baldwin RL, Zimm BH. Are denatured proteins ever random coils? *Proc Natl Acad Sci USA.* 2000;97(23):12391–12392. https://doi.org/10.1073/pnas.97.23.12391.

40. Bernadó P, Bertoncini CW, Griesinger C, Zweckstetter M, Blackledge M. Defining long-range order and local disorder in native alpha-synuclein using residual dipolar couplings. *J Am Chem Soc.* 2005;127(51):17968–17969. https://doi.org/10.1021/ja055538p.

41. Dunker AK, Lawson JD, Brown CJ, et al. Intrinsically disordered protein. *J Mol Graph Model.* 2001;19(1):26–59. https://doi.org/10.1016/s1093-3263(00)00138-8.

42. Chouard T. Structural biology: breaking the protein rules. *Nature.* 2011;471 (7337):151–153. https://doi.org/10.1038/471151a.

43. Uversky VN, Oldfield CJ, Dunker AK. Intrinsically disordered proteins in human diseases: introducing the D2 concept. *Annu Rev Biophys.* 2008;37:215–246. https://doi.org/10.1146/annurev.biophys.37.032807.125924.

44. Dyson HJ, Wright PE. Coupling of folding and binding for unstructured proteins. *Curr Opin Struct Biol.* 2002;12(1):54–60. https://doi.org/10.1016/s0959-440x(02)00289-0.

45. Kim DH, Han KH. Transient secondary structures as general target-binding motifs in intrinsically disordered proteins. *Int J Mol Sci.* 2018;19(11):3614. https://doi.org/10.3390/ijms19113614.

46. Kim DH, Han KH. PreSMo target-binding signatures in intrinsically disordered proteins. *Mol Cells.* 2018;41(10):889–899. https://doi.org/10.14348/molcells.2018.0192.

47. Chi SW, Lee SH, Kim DH, et al. Structural details on mdm2-p53 interaction. *J Biol Chem.* 2005;280(46):38795–38802. https://doi.org/10.1074/jbc.M508578200.

48. Fuxreiter M. Fuzziness in protein interactions-a historical perspective. *J Mol Biol.* 2018;430(16):2278–2287. https://doi.org/10.1016/j.jmb.2018.02.015.

49. Mittag T, Orlicky S, Choy WY, et al. Dynamic equilibrium engagement of a polyvalent ligand with a single-site receptor. *Proc Natl Acad Sci USA.* 2008;105(46):17772–17777. https://doi.org/10.1073/pnas.0809222105.

50. Rogers JM, Wong CT, Clarke J. Coupled folding and binding of the disordered protein PUMA does not require particular residual structure. *J Am Chem Soc.* 2014;136(14):5197–5200. https://doi.org/10.1021/ja4125065.

51. Crabtree MD, Borcherds W, Poosapati A, Shammas SL, Daughdrill GW, Clarke J. Conserved helix-flanking prolines modulate intrinsically disordered protein:target affinity by altering the lifetime of the bound complex. *Biochemistry.* 2017;56(18):2379–2384. https://doi.org/10.1021/acs.biochem.7b00179.

52. Sugase K, Dyson HJ, Wright PE. Mechanism of coupled folding and binding of an intrinsically disordered protein. *Nature.* 2007;447(7147):1021–1025. https://doi.org/10.1038/nature05858.

53. Schneider R, Maurin D, Communie G, et al. Visualizing the molecular recognition trajectory of an intrinsically disordered protein using multinuclear relaxation dispersion NMR. *J Am Chem Soc.* 2015;137(3):1220–1229. https://doi.org/10.1021/ja511066q.

54. Bucher D, Grant BJ, McCammon JA. Induced fit or conformational selection? The role of the semi-closed state in the maltose binding protein. *Biochemistry.* 2011;50(48):10530–10539. https://doi.org/10.1021/bi201481a.

55. Shaw DE, Maragakis P, Lindorff-Larsen K, et al. Atomic-level characterization of the structural dynamics of proteins. *Science.* 2010;330(6002):341–346. https://doi.org/10.1126/science.1187409.

56. Wuthrich K. *NMR of Proteins and Nucleic Acids.* New York: John Wiley and Sons; 1986.

57. Bernadó P, Blanchard L, Timmins P, Marion D, Ruigrok RW, Blackledge M. A structural model for unfolded proteins from residual dipolar couplings and small-angle x-ray scattering. *Proc Natl Acad Sci USA.* 2005;102(47):17002–17007. https://doi.org/10.1073/pnas.0506202102.

58. Marsh JA, Baker JM, Tollinger M, Forman-Kay JD. Calculation of residual dipolar couplings from disordered state ensembles using local alignment. *J Am Chem Soc.* 2008;130(25):7804–7805. https://doi.org/10.1021/ja802220c.

59. Lindorff-Larsen K, Kristjansdottir S, Teilum K, et al. Determination of an ensemble of structures representing the denatured state of the bovine acyl-coenzyme a binding protein. *J Am Chem Soc.* 2004;126(10):3291–3299. https://doi.org/10.1021/ja039250g.

60. Dedmon MM, Lindorff-Larsen K, Christodoulou J, Vendruscolo M, Dobson CM. Mapping long-range interactions in alpha-synuclein using spin-label NMR and ensemble molecular dynamics simulations. *J Am Chem Soc.* 2005;127(2):476–477. https://doi.org/10.1021/ja044834j.

61. Ganguly D, Chen J. Structural interpretation of paramagnetic relaxation enhancement-derived distances for disordered protein states. *J Mol Biol.* 2009;390(3):467–477. https://doi.org/10.1016/j.jmb.2009.05.019.

62. Mittag T, Forman-Kay JD. Atomic-level characterization of disordered protein ensembles. *Curr Opin Struct Biol.* 2007;17(1):3–14. https://doi.org/10.1016/j.sbi.2007.01.009.

63. Marsh JA, Singh VK, Jia Z, Forman-Kay JD. Sensitivity of secondary structure propensities to sequence differences between alpha- and gamma-synuclein: implications for fibrillation. *Protein Sci.* 2006;15(12):2795–2804. https://doi.org/10.1110/ps.062465306.

64. Bracken C, Carr PA, Cavanagh J, Palmer 3rd AG. Temperature dependence of intramolecular dynamics of the basic leucine zipper of GCN4: implications for the entropy of association with DNA. *J Mol Biol.* 1999;285(5):2133–2146. https://doi.org/10.1006/jmbi.1998.2429.

65. Neal S, Nip AM, Zhang H, Wishart DS. Rapid and accurate calculation of protein 1H, 13C and 15N chemical shifts. *J Biomol NMR*. 2003;26(3):215–240. https://doi.org/10.1023/a:1023812930288.

66. Schwarzinger S, Kroon GJ, Foss TR, Chung J, Wright PE, Dyson HJ. Sequence-dependent correction of random coil NMR chemical shifts. *J Am Chem Soc*. 2001;123(13):2970–2978. https://doi.org/10.1021/ja003760i.

67. Tamiola K, Acar B, Mulder FA. Sequence-specific random coil chemical shifts of intrinsically disordered proteins. *J Am Chem Soc*. 2010;132(51):18000–18003. https://doi.org/10.1021/ja105656t.

68. Csizmok V, Felli IC, Tompa P, Banci L, Bertini I. Structural and dynamic characterization of intrinsically disordered human securin by NMR spectroscopy. *J Am Chem Soc*. 2008;130(50):16873–16879. https://doi.org/10.1021/ja805510b.

69. Sung YH, Eliezer D. Residual structure, backbone dynamics, and interactions within the synuclein family. *J Mol Biol*. 2007;372(3):689–707. https://doi.org/10.1016/j.jmb.2007.07.008.

70. Andresen C, Helander S, Lemak A, et al. Transient structure and dynamics in the disordered c-Myc transactivation domain affect Bin1 binding. *Nucleic Acids Res*. 2012;40(13):6353–6366. https://doi.org/10.1093/nar/gks263.

71. Daughdrill GW, Hanely LJ, Dahlquist FW. The C-terminal half of the anti-sigma factor FlgM contains a dynamic equilibrium solution structure favoring helical conformations. *Biochemistry*. 1998;37(4):1076–1082. https://doi.org/10.1021/bi971952t.

72. Rudolph MG, Bayer P, Abo A, Kuhlmann J, Vetter IR, Wittinghofer A. The Cdc42/Rac interactive binding region motif of the Wiskott Aldrich syndrome protein (WASP) is necessary but not sufficient for tight binding to Cdc42 and structure formation. *J Biol Chem*. 1998;273(29):18067–18076. https://doi.org/10.1074/jbc.273.29.18067.

73. Geyer M, Munte CE, Schorr J, Kellner R, Kalbitzer HR. Structure of the anchor-domain of myristoylated and non-myristoylated HIV-1 Nef protein. *J Mol Biol*. 1999;289(1):123–138. https://doi.org/10.1006/jmbi.1999.2740.

74. Hazzard J, Südhof TC, Rizo J. NMR analysis of the structure of synaptobrevin and of its interaction with syntaxin. *J Biomol NMR*. 1999;14(3):203–207. https://doi.org/10.1023/a:1008382027065.

75. Zhao X, Georgieva B, Chabes A, et al. Mutational and structural analyses of the ribonucleotide reductase inhibitor Sml1 define its Rnr1 interaction domain whose inactivation allows suppression of mec1 and rad53 lethality. *Mol Cell Biol*. 2000;20(23):9076–9083. https://doi.org/10.1128/MCB.20.23.9076-9083.2000.

76. Zitzewitz JA, Ibarra-Molero B, Fishel DR, Terry KL, Matthews CR. Preformed secondary structure drives the association reaction of GCN4-p1, a model coiled-coil system. *J Mol Biol*. 2000;296(4):1105–1116. https://doi.org/10.1006/jmbi.2000.3507.

77. Ramelot TA, Gentile LN, Nicholson LK. Transient structure of the amyloid precursor protein cytoplasmic tail indicates preordering of structure for binding to cytosolic factors. *Biochemistry*. 2000;39(10):2714–2725. https://doi.org/10.1021/bi992580m.

78. Sayers EW, Gerstner RB, Draper DE, Torchia DA. Structural preordering in the N-terminal region of ribosomal protein S4 revealed by heteronuclear NMR spectroscopy. *Biochemistry*. 2000;39(44):13602–13613. https://doi.org/10.1021/bi0013391.

79. Eliezer D, Kutluay E, Bussell Jr R, Browne G. Conformational properties of alpha-synuclein in its free and lipid-associated states. *J Mol Biol*. 2001;307(4):1061–1073. https://doi.org/10.1006/jmbi.2001.4538.

80. Bienkiewicz EA, Adkins JN, Lumb KJ. Functional consequences of preorganized helical structure in the intrinsically disordered cell-cycle inhibitor p27(Kip1). *Biochemistry*. 2002;41(3):752–759. https://doi.org/10.1021/bi015763t.

81. Moncoq K, Broutin I, Craescu CT, Vachette P, Ducruix A, Durand D. SAXS study of the PIR domain from the Grb14 molecular adaptor: a natively unfolded protein with a transient structure primer? *Biophys J*. 2004;87(6):4056–4064. https://doi.org/10.1529/biophysj.104.048645.

82. Laptenko O, Prives C. Transcriptional regulation by p53: one protein, many possibilities. *Cell Death Differ*. 2006;13(6):951–961. https://doi.org/10.1038/sj.cdd.4401916.

83. Chi SW, Kim DH, Lee SH, Chang I, Han KH. Pre-structured motifs in the natively unstructured preS1 surface antigen of hepatitis B virus. *Protein Sci*. 2007;16 (10):2108–2117. https://doi.org/10.1110/ps.072983507.

84. Schedlbauer A, Gandini R, Kontaxis G, Paulmichl M, Furst J, Konrat R. The C-terminus of ICln is natively disordered but displays local structural preformation. *Cell Physiol Biochem*. 2011;28(6):1203–1210. https://doi.org/10.1159/000335852.

85. Zheng Z, Ma D, Yahr TL, Chen L. The transiently ordered regions in intrinsically disordered ExsE are correlated with structural elements involved in chaperone binding. *Biochem Biophys Res Commun*. 2012;417(1):129–134. https://doi.org/10.1016/j.bbrc.2011.11.070.

86. Lum JK, Neuweiler H, Fersht AR. Long-range modulation of chain motions within the intrinsically disordered transactivation domain of tumor suppressor p53. *J Am Chem Soc*. 2012;134(3):1617–1622. https://doi.org/10.1021/ja2078619.

87. Feuerstein S, Solyom Z, Aladag A, et al. Transient structure and SH3 interaction sites in an intrinsically disordered fragment of the hepatitis C virus protein NS5A. *J Mol Biol*. 2012;420(4–5):310–323. https://doi.org/10.1016/j.jmb.2012.04.023.

88. Gupta G, Qin H, Song J. Intrinsically unstructured domain 3 of hepatitis C virus NS5A forms a "fuzzy complex" with VAPB-MSP domain which carries ALS-causing mutations. *PLoS One*. 2012;7(6), e39261. https://doi.org/10.1371/journal.pone.0039261.

89. Nováček J, Janda L, Dopitová R, Žídek L, Sklenář V. Efficient protocol for backbone and side-chain assignments of large, intrinsically disordered proteins: transient secondary structure analysis of 49.2 kDa microtubule associated protein 2c. *J Biomol NMR*. 2013;56(4):291–301. https://doi.org/10.1007/s10858-013-9761-7.

90. Lee C, Kim DH, Lee SH, Su J, Han KH. Structural investigation on the intrinsically disordered N-terminal region of HPV16 E7 protein. *BMB Rep*. 2016;49(8):431–436. https://doi.org/10.5483/bmbrep.2016.49.8.021.

91. Kim DH, Lee C, Lee SH, et al. The mechanism of p53 rescue by SUSP4. *Angew Chem Int Ed Engl*. 2017;56(5):1278–1282. https://doi.org/10.1002/anie.201607819.

92. Kim DH, Wright A, Han KH. An NMR study on the intrinsically disordered core transactivation domain of human glucocorticoid receptor. *BMB Rep*. 2017;50 (10):522–527. https://doi.org/10.5483/bmbrep.2017.50.10.152.

93. Sigler PB. Transcriptional activation. Acid blobs and negative noodles. *Nature*. 1988;333(6170):210–212. https://doi.org/10.1038/333210a0.

94. Giniger E, Ptashne M. Transcription in yeast activated by a putative amphipathic alpha helix linked to a DNA binding unit. *Nature*. 1987;330(6149):670–672. https://doi.org/10.1038/330670a0.

95. Leuther KK, Salmeron JM, Johnston SA. Genetic evidence that an activation domain of GAL4 does not require acidity and may form a beta sheet. *Cell*. 1993;72(4):575–585. https://doi.org/10.1016/0092-8674(93)90076-3.

96. Ma J, Ptashne M. A new class of yeast transcriptional activators. *Cell*. 1987;51 (1):113–119. https://doi.org/10.1016/0092-8674(87)90015-8.

97. Ptashne M. How eukaryotic transcriptional activators work. *Nature*. 1988;335 (6192):683–689. https://doi.org/10.1038/335683a0.

98. Sadowski I, Ma J, Triezenberg S, Ptashne M. GAL4-VP16 is an unusually potent transcriptional activator. *Nature*. 1988;335(6190):563–564. https://doi.org/10.1038/335563a0.

99. Cress WD, Triezenberg SJ. Critical structural elements of the VP16 transcriptional activation domain. *Science.* 1991;251(4989):87–90. https://doi.org/10.1126/science. 1846049.

100. Regier JL, Shen F, Triezenberg SJ. Pattern of aromatic and hydrophobic amino acids critical for one of two subdomains of the VP16 transcriptional activator. *Proc Natl Acad Sci USA.* 1993;90(3):883–887. https://doi.org/10.1073/pnas.90.3.883.

101. Van Hoy M, Leuther KK, Kodadek T, Johnston SA. The acidic activation domains of the GCN4 and GAL4 proteins are not alpha helical but form beta sheets. *Cell.* 1993;72 (4):587–594. https://doi.org/10.1016/0092-8674(93)90077-4.

102. Hahn S. Structure(?) and function of acidic transcription activators. *Cell.* 1993;72 (4):481–483. https://doi.org/10.1016/0092-8674(93)90064-w.

103. The final rejection letter from an editor of the journal where the manuscript on the NMR study of p53 TAD [2] was submitted contained the following statement. "... that you may view these experiments as redundant to the results that you have already obtained, the reviewer was very clear in requesting a very rigorous demonstration of the putative helical structures in the unbound TAD. Indeed, since your work sharply contrasts with currently held beliefs, this is the admittedly very rigorous standard that we feel would be required for consideration of the paper for CELL. We certainly agree that the work is very interesting and I'm sure that you will have no problem in finding another journal in which to...".

104. Mohan A, Oldfield CJ, Radivojac P, et al. Analysis of molecular recognition features (MoRFs). *J Mol Biol.* 2006;362(5):1043–1059. https://doi.org/10.1016/j.jmb.2006.07.087.

105. Fuxreiter M, Simon I, Friedrich P, Tompa P. Preformed structural elements feature in partner recognition by intrinsically unstructured proteins. *J Mol Biol.* 2004;338 (5):1015–1026. https://doi.org/10.1016/j.jmb.2004.03.017.

106. Van Roey K, Uyar B, Weatheritt RJ, et al. Short linear motifs: ubiquitous and functionally diverse protein interaction modules directing cell regulation. *Chem Rev.* 2014;114(13):6733–6778. https://doi.org/10.1021/cr400585q.

107. Bourhis JM, Johansson K, Receveur-Bréchot V, et al. The C-terminal domain of measles virus nucleoprotein belongs to the class of intrinsically disordered proteins that fold upon binding to their physiological partner. *Virus Res.* 2004;99(2):157–167. https:// doi.org/10.1016/j.virusres.2003.11.007.

108. Bell S, Klein C, Müller L, Hansen S, Buchner J. p53 contains large unstructured regions in its native state. *J Mol Biol.* 2002;322(5):917–927. https://doi.org/10.1016/s0022-2836(02)00848-3.

109. Dawson R, Müller L, Dehner A, Klein C, Kessler H, Buchner J. The N-terminal domain of p53 is natively unfolded. *J Mol Biol.* 2003;332(5):1131–1141. https://doi.org/10.1016/j.jmb.2003.08.008.

110. O'Hare P, Williams G. Structural studies of the acidic transactivation domain of the Vmw65 protein of herpes simplex virus using 1H NMR. *Biochemistry.* 1992;31 (16):4150–4156. https://doi.org/10.1021/bi00131a035.

111. Minezaki Y, Homma K, Kinjo AR, Nishikawa K. Human transcription factors contain a high fraction of intrinsically disordered regions essential for transcriptional regulation. *J Mol Biol.* 2006;359(4):1137–1149. https://doi.org/10.1016/j.jmb.2006.04.016.

112. Sherr CJ. Principles of tumor suppression. *Cell.* 2004;116(2):235–246. https://doi.org/10.1016/s0092-8674(03)01075-4.

113. Lane DP. Cancer. p53, guardian of the genome. *Nature.* 1992;358(6381):15–16. https://doi.org/10.1038/358015a0.

114. Joerger AC, Fersht AR. Structural biology of the tumor suppressor p53. *Annu Rev Biochem.* 2008;77:557–582. https://doi.org/10.1146/annurev.biochem.77. 060806.091238.

115. Clore GM, Omichinski JG, Sakaguchi K, et al. High-resolution structure of the olig-omerization domain of p53 by multidimensional NMR. *Science*. 1994;265 (5170):386–391. https://doi.org/10.1126/science.8023159.

116. Jeffrey PD, Gorina S, Pavletich NP. Crystal structure of the tetramerization domain of the p53 tumor suppressor at 1.7 angstroms. *Science*. 1995;267(5203):1498–1502. https://doi.org/10.1126/science.7878469.

117. Cho Y, Gorina S, Jeffrey PD, Pavletich NP. Crystal structure of a p53 tumor suppressor-DNA complex: understanding tumorigenic mutations. *Science*. 1994;265 (5170):346–355. https://doi.org/10.1126/science.8023157.

118. Feng H, Jenkins LM, Durell SR, et al. Structural basis for p300 Taz2-p53 TAD1 binding and modulation by phosphorylation. *Structure*. 2009;17(2):202–210. https://doi.org/10.1016/j.str.2008.12.009.

119. Bochkareva E, Kaustov L, Ayed A, et al. Single-stranded DNA mimicry in the p53 transactivation domain interaction with replication protein A. *Proc Natl Acad Sci USA*. 2005;102(43):15412–15417. https://doi.org/10.1073/pnas.0504614102.

120. Di Lello P, Jenkins LMM, Jones TN, et al. Structure of the Tfb1/p53 complex: Insights into the interaction between the p62/Tfb1 subunit of TFIIH and the activation domain of p53. *Mol Cell*. 2006;22(6):731–740. https://doi.org/10.1016/j.molcel.2006.05.007.

121. Miller Jenkins LM, Feng H, Durell SR, et al. Characterization of the p300 Taz2-p53 TAD2 complex and comparison with the p300 Taz2-p53 TAD1 complex. *Biochemistry*. 2015;54(11):2001–2010. https://doi.org/10.1021/acs.biochem.5b00044.

122. Lee CW, Arai M, Martinez-Yamout MA, Dyson HJ, Wright PE. Mapping the interactions of the p53 transactivation domain with the KIX domain of CBP. *Biochemistry*. 2009;48(10):2115–2124. https://doi.org/10.1021/bi802055v.

123. Lee CW, Martinez-Yamout MA, Dyson HJ, Wright PE. Structure of the p53 transactivation domain in complex with the nuclear receptor coactivator binding domain of CREB binding protein. *Biochemistry*. 2010;49(46):9964–9971. https://doi.org/10.1021/bi1012996.

124. Nelson DL, Cox NM. *Lehninger Principles of Biochemistry*. 7th ed. New York: W.H. Freeman and Company; 2017.

125. Tovar C, Graves B, Packman K, et al. MDM2 small-molecule antagonist RG7112 activates p53 signaling and regresses human tumors in preclinical cancer models. *Cancer Res*. 2013;73(8):2587–2597. https://doi.org/10.1158/0008-5472.CAN-12-2807.

126. Parks DJ, Lafrance LV, Calvo RR, et al. 1,4-Benzodiazepine-2,5-diones as small molecule antagonists of the HDM2-p53 interaction: discovery and SAR. *Bioorg Med Chem Lett*. 2005;15(3):765–770. https://doi.org/10.1016/j.bmcl.2004.11.009.

127. Neochoritis CG, Wang K, Estrada-Ortiz N, et al. 2,30-Bis(10H-indole) heterocycles: new p53/MDM2/MDMX antagonists. *Bioorg Med Chem Lett*. 2015;25(24): 5661–5666. https://doi.org/10.1016/j.bmcl.2015.11.019.

128. Silva JL, Lima CGS, Rangel LP, et al. Recent synthetic approaches towards small molecule reactivators of p53. *Biomolecules*. 2020;10(4):635. https://doi.org/10.3390/biom10040635.

129. Rhen T, Cidlowski JA. Antiinflammatory action of glucocorticoids- -new mechanisms for old drugs. *N Engl J Med*. 2005;353(16):1711–1723. https://doi.org/10.1056/NEJMra050541.

130. Evans RM. The steroid and thyroid hormone receptor superfamily. *Science*. 1988;240 (4854):889–895. https://doi.org/10.1126/science.3283939.

131. Hollenberg SM, Evans RM. Multiple and cooperative trans–activation domains of the human glucocorticoid receptor. *Cell*. 1988;55(5):899–906. https://doi.org/10.1016/0092-8674(88)90145-6.

132. Dahlman-Wright K, Almlöf T, McEwan IJ, Gustafsson JA, Wright AP. Delineation of a small region within the major transactivation domain of the human glucocorticoid receptor that mediates transactivation of gene expression. *Proc Natl Acad Sci USA.* 1994;91(5):1619–1623. https://doi.org/10.1073/pnas.91.5.1619.

133. Almlöf T, Gustafsson JA, Wright AP. Role of hydrophobic amino acid clusters in the transactivation activity of the human glucocorticoid receptor. *Mol Cell Biol.* 1997;17 (2):934–945. https://doi.org/10.1128/MCB.17.2.934.

134. Baskakov IV, Kumar R, Srinivasan G, Ji YS, Bolen DW, Thompson EB. Trimethylamine N-oxide-induced cooperative folding of an intrinsically unfolded transcription-activating fragment of human glucocorticoid receptor. *J Biol Chem.* 1999;274(16):10693–10696. https://doi.org/10.1074/jbc.274.16.10693.

135. Lee C, Kalmar L, Xue B, et al. Contribution of proline to the pre-structuring tendency of transient helical secondary structure elements in intrinsically disordered proteins. *Biochim Biophys Acta.* 2014;1840(3):993–1003. https://doi.org/10.1016/j.bbagen.2013.10.042.

136. Borcherds W, Theillet FX, Katzer A, et al. Disorder and residual helicity alter p53-Mdm2 binding affinity and signaling in cells. *Nat Chem Biol.* 2014;10 (12):1000–1002. https://doi.org/10.1038/nchembio.1668.

137. Salamanova E, Costeira-Paulo J, Han KH, Kim DH, Nilsson L, Wright APH. A subset of functional adaptation mutations alter propensity for α-helical conformation in the intrinsically disordered glucocorticoid receptor tau1core activation domain. *Biochim Biophys Acta Gen Subj.* 2018;1862(6):1452–1461. https://doi.org/10.1016/j.bbagen.2018.03.015.

138. Blau J, Xiao H, McCracken S, O'Hare P, Greenblatt J, Bentley D. Three functional classes of transcriptional activation domain. *Mol Cell Biol.* 1996;16(5):2044–2055. https://doi.org/10.1128/MCB.16.5.2044.

139. Sullivan SM, Horn PJ, Olson VA, et al. Mutational analysis of a transcriptional activation region of the VP16 protein of herpes simplex virus. *Nucleic Acids Res.* 1998;26 (19):4487–4496. https://doi.org/10.1093/nar/26.19.4487.

140. Campbell AP, Sykes BD. The two-dimensional transferred nuclear Overhauser effect: theory and practice. *Annu Rev Biophys Biomol Struct.* 1993;22:99–122. https://doi.org/10.1146/annurev.bb.22.060193.000531.

141. Shoemaker KR, Kim PS, York EJ, Stewart JM, Baldwin RL. Tests of the helix dipole model for stabilization of alpha-helices. *Nature.* 1987;326(6113):563–567. https://doi.org/10.1038/326563a0.

142. Meyer N, Penn LZ. Reflecting on 25 years with MYC. *Nat Rev Cancer.* 2008;8 (12):976–990. https://doi.org/10.1038/nrc2231.

143. Kato GJ, Barrett J, Villa-Garcia M, Dang CV. An amino-terminal c-myc domain required for neoplastic transformation activates transcription. *Mol Cell Biol.* 1990;10 (11):5914–5920. https://doi.org/10.1128/mcb.10.11.5914-5920.1990.

144. Fladvad M, Zhou K, Moshref A, Pursglove S, Säfsten P, Sunnerhagen M. N and C-terminal sub-regions in the c-Myc transactivation region and their joint role in creating versatility in folding and binding. *J Mol Biol.* 2005;346(1):175–189. https://doi.org/10.1016/j.jmb.2004.11.029.

145. Burton RA, Mattila S, Taparowsky EJ, Post CB. B-myc: N-terminal recognition of myc binding proteins. *Biochemistry.* 2006;45(32):9857–9865. https://doi.org/10.1021/bi060379n.

146. Gingras AC, Raught B, Sonenberg N. eIF4 initiation factors: effectors of mRNA recruitment to ribosomes and regulators of translation. *Annu Rev Biochem.* 1999;68:913–963. https://doi.org/10.1146/annurev.biochem.68.1.913.

147. Gingras AC, Raught B, Gygi SP, et al. Hierarchical phosphorylation of the translation inhibitor 4E-BP1. *Genes Dev.* 2001;15(21):2852–2864. https://doi.org/10.1101/gad.912401.

148. Ruggero D, Montanaro L, Ma L, et al. The translation factor eIF-4E promotes tumor formation and cooperates with c-Myc in lymphomagenesis. *Nat Med.* 2004;10(5):484–486. https://doi.org/10.1038/nm1042.

149. Constantinou C, Clemens MJ. Regulation of the phosphorylation and integrity of protein synthesis initiation factor eIF4GI and the translational repressor 4E-BP1 by p53. *Oncogene.* 2005;24(30):4839–4850. https://doi.org/10.1038/sj.onc.1208648.

150. Moerke NJ, Aktas H, Chen H, et al. Small-molecule inhibition of the interaction between the translation initiation factors eIF4E and eIF4G. *Cell.* 2007;128(2):257–267. https://doi.org/10.1016/j.cell.2006.11.046.

151. Maroteaux L, Campanelli JT, Scheller RH. Synuclein: a neuron-specific protein localized to the nucleus and presynaptic nerve terminal. *J Neurosci.* 1988;8(8):2804–2815. https://doi.org/10.1523/JNEUROSCI.08-08-02804.1988.

152. Iwai A, Masliah E, Yoshimoto M, et al. The precursor protein of non-a beta component of Alzheimer's disease amyloid is a presynaptic protein of the central nervous system. *Neuron.* 1995;14(2):467–475. https://doi.org/10.1016/0896-6273(95)90302-x.

153. Burré J. The synaptic function of α-synuclein. *J Parkinsons Dis.* 2015;5(4):699–713. https://doi.org/10.3233/JPD-150642.

154. Poewe W, Seppi K, Tanner CM, et al. Parkinson disease. *Nat Rev Dis Primers.* 2017;3:17013. https://doi.org/10.1038/nrdp.2017.13.

155. Spillantini MG, Schmidt ML, Lee VM, Trojanowski JQ, Jakes R, Goedert M. Alpha-synuclein in Lewy bodies. *Nature.* 1997;388(6645):839–840. https://doi.org/10.1038/42166.

156. Braak H, Del Tredici K, Rüb U, de Vos RA, Jansen Steur EN, Braak E. Staging of brain pathology related to sporadic Parkinson's disease. *Neurobiol Aging.* 2003;24(2):197–211. https://doi.org/10.1016/s0197-4580(02)00065-9.

157. Kirschner DA, Abraham C, Selkoe DJ. X-ray diffraction from intraneuronal paired helical filaments and extraneuronal amyloid fibers in Alzheimer disease indicates cross-beta conformation. *Proc Natl Acad Sci USA.* 1986;83(2):503–507. https://doi.org/10.1073/pnas.83.2.503.

158. Serpell LC, Berriman J, Jakes R, Goedert M, Crowther RA. Fiber diffraction of synthetic alpha-synuclein filaments shows amyloid-like cross-beta conformation. *Proc Natl Acad Sci USA.* 2000;97(9):4897–4902. https://doi.org/10.1073/pnas.97.9.4897.

159. Tuttle MD, Comellas G, Nieuwkoop AJ, et al. Solid-state NMR structure of a pathogenic fibril of full-length human α-synuclein. *Nat Struct Mol Biol.* 2016;23(5):409–415. https://doi.org/10.1038/nsmb.3194.

160. Conway KA, Harper JD, Lansbury Jr PT. Fibrils formed in vitro from alpha-synuclein and two mutant forms linked to Parkinson's disease are typical amyloid. *Biochemistry.* 2000;39(10):2552–2563. https://doi.org/10.1021/bi991447r.

161. El-Agnaf OM, Jakes R, Curran MD, Wallace A. Effects of the mutations Ala30 to Pro and Ala53 to Thr on the physical and morphological properties of alpha-synuclein protein implicated in Parkinson's disease. *FEBS Lett.* 1998;440(1–2):67–70. https://doi.org/10.1016/s0014-5793(98)01419-7.

162. Uéda K, Fukushima H, Masliah E, et al. Molecular cloning of cDNA encoding an unrecognized component of amyloid in Alzheimer disease. *Proc Natl Acad Sci USA.* 1993;90(23):11282–11286. https://doi.org/10.1073/pnas.90.23.11282.

163. El-Agnaf OM, Irvine GB. Aggregation and neurotoxicity of alpha-synuclein and related peptides. *Biochem Soc Trans.* 2002;30(4):559–565. https://doi.org/10.1042/bst0300559.

164. Uversky VN, Li J, Fink AL. Evidence for a partially folded intermediate in alpha-synuclein fibril formation. *J Biol Chem.* 2001;276(14):10737–10744. https://doi.org/10.1074/jbc.M010907200.

165. Uversky VN, Fink AL. Conformational constraints for amyloid fibrillation: the importance of being unfolded. *Biochim Biophys Acta.* 2004;1698(2):131–153. https://doi.org/10.1016/j.bbapap.2003.12.008.

166. Wetzel R. For protein misassembly, it's the "I" decade. *Cell.* 1996;86(5):699–702. https://doi.org/10.1016/s0092-8674(00)80143-9.

167. Li J, Uversky VN, Fink AL. Conformational behavior of human alpha-synuclein is modulated by familial Parkinson's disease point mutations A30P and A53T. *Neurotoxicology.* 2002;23(4–5):553–567. https://doi.org/10.1016/s0161-813x(02)00066-9.

168. Hernandez AM, Silbern I, Geffers I, et al. Low-expressing synucleinopathy mouse models based on oligomer-forming mutations and C-terminal truncation of α-synuclein. *Front Neurosci.* 2021;15:643391. https://doi.org/10.3389/fnins.2021.643391.

169. Kim DH, Lee J, Mok KH, Lee JH, Han KH. Salient features of monomeric alpha-synuclein revealed by NMR spectroscopy. *Biomolecules.* 2020;10(3):428–442. https://doi.org/10.3390/biom10030428.

170. Sillen A, Barbier P, Landrieu I, et al. NMR investigation of the interaction between the neuronal protein tau and the microtubules. *Biochemistry.* 2007;46(11):3055–3064. https://doi.org/10.1021/bi061920i.

171. Weingarten MD, Lockwood AH, Hwo SY, Kirschner MW. A protein factor essential for microtubule assembly. *Proc Natl Acad Sci USA.* 1975;72(5):1858–1862. https://doi.org/10.1073/pnas.72.5.1858.

172. Drubin DG, Kirschner MW. Tau protein function in living cells. *J Cell Biol.* 1986;103(6 pt 2):2739–2746. https://doi.org/10.1083/jcb.103.6.2739.

173. Wischik CM, Novak M, Edwards PC, Klug A, Tichelaar W, Crowther RA. Structural characterization of the core of the paired helical filament of Alzheimer-disease. *Proc Natl Acad Sci USA.* 1988;85(13):4884–4888. https://doi.org/10.1073/pnas.85.13.4884.

174. Mukrasch MD, von Bergen M, Biernat J, et al. The "jaws" of the tau-microtubule interaction. *J Biol Chem.* 2007;282(16):12230–12239. https://doi.org/10.1074/jbc.M607159200.

175. Mukrasch MD, Biernat J, von Bergen M, Griesinger C, Mandelkow E, Zweckstetter M. Sites of tau important for aggregation populate {beta}-structure and bind to microtubules and polyanions. *J Biol Chem.* 2005;280(26):24978–24986. https://doi.org/10.1074/jbc.M501565200.

176. Eliezer D, Barré P, Kobaslija M, Chan D, Li X, Heend L. Residual structure in the repeat domain of tau: echoes of microtubule binding and paired helical filament formation. *Biochemistry.* 2005;44(3):1026–1036. https://doi.org/10.1021/bi048953n.

177. Mukrasch MD, Markwick P, Biernat J, et al. Highly populated turn conformations in natively unfolded tau protein identified from residual dipolar couplings and molecular simulation. *J Am Chem Soc.* 2007;129(16):5235–5243. https://doi.org/10.1021/ja0690159.

178. Barré P, Eliezer D. Folding of the repeat domain of tau upon binding to lipid surfaces. *J Mol Biol.* 2006;362(2):312–326. https://doi.org/10.1016/j.jmb.2006.07.018.

179. Andronesi OC, von Bergen M, Biernat J, et al. Characterization of Alzheimer's-like paired helical filaments from the core domain of tau protein using solid-state NMR spectroscopy. *J Am Chem Soc.* 2008;130(18):5922–5928. https://doi.org/10.1021/ja7100517.

180. Sibille N, Sillen A, Leroy A, et al. Structural impact of heparin binding to full-length Tau as studied by NMR spectroscopy. *Biochemistry.* 2006;45(41):12560–12572. https://doi.org/10.1021/bi060964o.

181. Landrieu I, Lacosse L, Leroy A, et al. NMR analysis of a Tau phosphorylation pattern. *J Am Chem Soc.* 2006;128(11):3575–3583. https://doi.org/10.1021/ja054656.

182. von Bergen M, Friedhoff P, Biernat J, Heberle J, Mandelkow EM, Mandelkow E. Assembly of tau protein into Alzheimer paired helical filaments depends on a local sequence motif ((306)VQIVYK(311)) forming beta structure. *Proc Natl Acad Sci USA.* 2000;97(10):5129–5134. https://doi.org/10.1073/pnas.97.10.5129.

183. Mukrasch MD, Bibow S, Korukottu J, et al. Structural polymorphism of 441-residue tau at single residue resolution. *PLoS Biol.* 2009;7(2), e34. https://doi.org/10.1371/journal.pbio.1000034.

184. Kadavath H, Hofele RV, Biernat J, et al. Tau stabilizes microtubules by binding at the interface between tubulin heterodimers. *Proc Natl Acad Sci U S A.* 2015;112 (24):7501–7506. https://doi.org/10.1073/pnas.1504081112.

185. Fung SK, Lok AS. Viral hepatitis in 2003. *Curr Opin Gastroenterol.* 2004;20(3):241–247. https://doi.org/10.1097/00001574-200405000-00008.

186. Wright TL. Introduction to chronic hepatitis B infection. *Am J Gastroenterol.* 2006;101 (suppl 1):S1–S6. https://doi.org/10.1111/j.1572-0241.2006.00469.x.

187. Blumberg BS, Millman I, Venkateswaran PS, Thyagarajan SP. Hepatitis B virus and hepatocellular carcinoma--treatment of HBV carriers with *Phyllanthus amarus*. *Cancer Detect Prev.* 1989;14(2):195–201. 2559794.

188. Stephenne J. Development and production aspects of a recombinant yeast-derived hepatitis B vaccine. *Vaccine.* 1990;8(suppl):S69–S73. discussion S79-80 https://doi.org/10.1016/0264-410x(90)90221-7.

189. Mahoney FJ, Woodruff BA, Erben JJ, et al. Effect of a hepatitis B vaccination program on the prevalence of hepatitis B virus infection. *J Infect Dis.* 1993;167(1):203–207. https://doi.org/10.1093/infdis/167.1.203.

190. Jilg W. Novel hepatitis B vaccines. *Vaccine.* 1998;16(suppl):S65–S68. https://doi.org/10.1016/s0264-410x(98)00300-4.

191. Poland GA, Jacobson RM. Clinical practice: prevention of hepatitis B with the hepatitis B vaccine. *N Engl J Med.* 2004;351(27):2832–2838. https://doi.org/10.1056/NEJMcp041507.

192. Saag MS. Emtricitabine, a new antiretroviral agent with activity against HIV and hepatitis B virus. *Clin Infect Dis.* 2006;42(1):126–131. https://doi.org/10.1086/498348.

193. Shin D, Kim SI, Kim M, Park M. Efficient inhibition of hepatitis B virus replication by small interfering RNAs targeted to the viral X gene in mice. *Virus Res.* 2006;119 (2):146–153. https://doi.org/10.1016/j.virusres.2005.12.012.

194. Heermann KH, Goldmann U, Schwartz W, Seyffarth T, Baumgarten H, Gerlich WH. Large surface proteins of hepatitis B virus containing the pre-s sequence. *J Virol.* 1984;52(2):396–402. https://doi.org/10.1128/JVI.52.2.396-402.1984.

195. Stibbe W, Gerlich WH. Structural relationships between minor and major proteins of hepatitis B surface antigen. *J Virol.* 1983;46(2):626–628. https://doi.org/10.1128/JVI.46.2.626-628.1983.

196. Neurath AR, Kent SB, Strick N, Parker K. Identification and chemical synthesis of a host cell receptor binding site on hepatitis B virus. *Cell.* 1986;46(3):429–436. https://doi.org/10.1016/0092-8674(86)90663-x.

197. Pontisso P, Ruvoletto MG, Gerlich WH, Heermann KH, Bardini R, Alberti A. Identification of an attachment site for human liver plasma membranes on hepatitis B virus particles. *Virology.* 1989;173(2):522–530. https://doi.org/10.1016/0042-6822(89)90564-3.

198. Ryu CJ, Cho DY, Gripon P, Kim HS, Guguen-Guillouzo C, Hong HJ. An 80-kilodalton protein that binds to the pre-S1 domain of hepatitis B virus. *J Virol.* 2000;74(1):110–116. https://doi.org/10.1128/jvi.74.1.110-116.2000.

199. De Falco S, Ruvoletto MG, Verdoliva A, et al. Cloning and expression of a novel hepatitis B virus-binding protein from HepG2 cells. *J Biol Chem*. 2001;276 (39):36613–36623. https://doi.org/10.1074/jbc.M102377200.

200. Paran N, Geiger B, Shaul Y. HBV infection of cell culture: evidence for multivalent and cooperative attachment. *EMBO J*. 2001;20(16):4443–4453. https://doi.org/10.1093/emboj/20.16.4443.

201. Küttner G, Kramer A, Schmidtke G, et al. Characterization of neutralizing anti-pre-S1 and anti-pre-S2 (HBV) monoclonal antibodies and their fragments. *Mol Immunol*. 1999;36(10):669–683. https://doi.org/10.1016/s0161-5890(99)00074-7.

202. Kim DH, Ni Y, Lee SH, Urban S, Han KH. An anti-viral peptide derived from the preS1 surface protein of hepatitis B virus. *BMB Rep*. 2008;41(9):640–644. https://doi.org/10.5483/bmbrep.2008.41.9.640.

203. Frazer I. Vaccines for papillomavirus infection. *Virus Res*. 2002;89(2):271–274. https://doi.org/10.1016/s0168-1702(02)00195-8.

204. Roman A, Munger K. The papillomavirus E7 proteins. *Virology*. 2013;445 (1–2):138–168. https://doi.org/10.1016/j.virol.2013.04.013.

205. Chemes LB, Glavina J, Faivovich J, de Prat-Gay G, Sánchez IE. Evolution of linear motifs within the papillomavirus E7 oncoprotein. *J Mol Biol*. 2012;422(3):336–346. https://doi.org/10.1016/j.jmb.2012.05.036.

206. Münger K, Basile JR, Duensing S, et al. Biological activities and molecular targets of the human papillomavirus E7 oncoprotein. *Oncogene*. 2001;20(54):7888–7898. https://doi.org/10.1038/sj.onc.1204860.

207. Lee MH, Lee SW, Lee EJ, et al. SUMO-specific protease SUSP4 positively regulates p53 by promoting Mdm2 self-ubiquitination. *Nat Cell Biol*. 2006;8:1424–1431. https://doi.org/10.1038/ncb1512.

CHAPTER SEVEN

The role of dancing duplexes in biology and disease

Heather M. Forsythe and Elisar Barbar*

Department of Biochemistry and Biophysics, Oregon State University, Corvallis, OR, United States
*Corresponding author: e-mail address: barbare@oregonstate.edu

Contents

Abstract

Across species, a common protein assembly arises: proteins containing structured domains separated by long intrinsically disordered regions, and dimerized through a self-association domain or through strong protein interactions. These systems are termed "IDP duplexes." These flexible dimers have roles in diverse pathologies including development of cancer, viral infections, and neurodegenerative disease. Here we discuss the role of disorder in IDP duplexes with similar domain architectures that bind hub protein, LC8. LC8-binding IDP duplexes are categorized into three groups: IDP duplexes that contain a self-association domain that is extended by LC8 binding, IDP duplexes that have no self-association domain and are dimerized through binding several copies of LC8, and multivalent LC8-binders that also have a self-association domain. Additionally, we discuss non-LC8-binding IDP duplexes with similar domain organizations, including the Nucleocapsid protein of SARS-CoV-2. We propose that IDP duplexes have structural features that are essential in many biological processes and that improved understanding of their structure function relationship will provide new therapeutic opportunities.

Progress in Molecular Biology and Translational Science, Volume 183
ISSN 1877-1173
https://doi.org/10.1016/bs.pmbts.2021.06.004

1. Introduction

The relationship between protein structure and function is evident. Just as the shapes of tools in a toolbox determine the job they can complete, the exact shape of a protein leads to its unique binding interactions and associated role within a cell. The specific shape-function codependence can be disrupted at times with even the slightest alteration in amino acid sequence, resulting in failure of protein function, frequently followed by disease development. In this light, studies and discussions of protein function often center around folded structural domain conservation[1–4]; however, the shape of a protein is not the sole determinant of its function. Instead, it is the shape combined with the flexibility and motions that the protein can sample that sets the protein's abilities, just as a metal ring and rubber band have the same circular shape, but only the rubber band can stretch around and bind together a bouquet of flowers. Key properties of intrinsic disorder are the ability to wrap around, reach toward, bind, and connect proteins together. Therefore, it is imperative that protein comparisons take into account not only structural conservation, but also unstructured conservation.

Across biological systems, an emerging new class of protein assembly arises: protein dimers brought together by either a strong dimerization domain or by binding to a dimeric protein, but leaving large regions of the protein intrinsically disordered. These intrinsically disordered protein (IDP) duplexes are essential for proper cellular function and pathology. IDP duplexes are implicated in viral assembly, transcription and replication,[5,6] and in eukaryotic cellular processes such as co-transcriptional gene silencing,[7] transcription,[8–10] intracellular transport,[10,11] and mitosis.[12] Proteins involved in such essential cellular functions are by association, implicated in diseases that arise when these functions are disrupted such as cancer,[13] or in the case of viral IDP duplexes, disease arises and persists when these IDP duplexes are not disrupted.[14] Here we review IDP duplexes dimerized by binding to the hub protein LC8, and also those that have similar structural features but do not require LC8 for its binding. All retain significant regions of disorder that allow for functional protein regulation.

2. LC8-facilitated dimerization in IDP duplexes

Dynein light chain 8 (LC8) is a dynamic hub protein that is highly conserved and functionally diverse.[15–18] LC8's binding events are well characterized in multiple systems and are largely predictable based on sequence

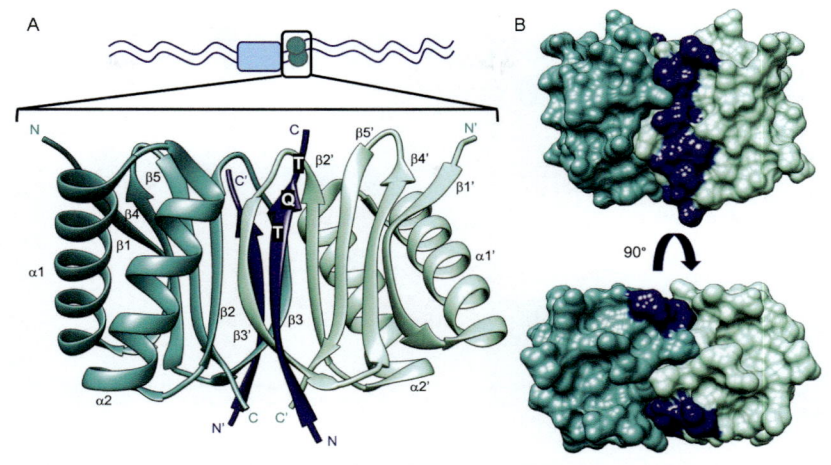

Fig. 1 An LC8 homodimer (teal) binds intrinsically disordered protein regions (dark blue) containing a "TQT" motif. (A) Two disordered regions of an IDP duplex interact with LC8, and LC8-binding induces formation of a beta strand. (B) The disordered binding partner fits within the LC8 homodimer groove. PDB: 3E2B.

and secondary structure.[15] A ~20 kDa homodimer in solution, LC8 binds intrinsically disordered regions (IDRs) of proteins containing a highly conserved "TQT" motif. IDRs pack in the LC8's highly conserved binding-groove at the dimer interface, with each chain making contacts with both protomers of LC8 (Fig. 1).[19] The binding interface of LC8's partners includes 8 IDR residues with a propensity to form a beta strand when packed against the beta sheet LC8 structure.[15,19] The LC8 binding site is recurrently preceded by a weakly predicted self-association domain as in Swallow that only forms a strong dimer upon LC8 binding demonstrating that LC8 binding is thermodynamically coupled to Swallow self-association (Fig. 1, Fig. 2).

Swallow (*Swa*) is a protein required for proper mRNA localization during drosophila oogenesis. Correct and specific mRNA localization is necessary so that translated proteins are asymmetrically distributed, creating cell polarity required for axial polarity, assembly of germ plasm, and neuronal development.[20] More specifically, Swa is known to have a role in restricting bicoid and huli-tai shao-adducin-like mRNA to the anterior pole of drosophila embryos, which functions to specify anterior cell fates during development.[21] The system of mRNA localization and subsequent cell differentiation during oogenesis is better understood within the model system of drosophila, but similar protein networks are present in mammalian systems as well.[22] Genetic mutations impacting RNA-binding proteins involved in mRNA localization are associated with diverse neurodevelopmental and

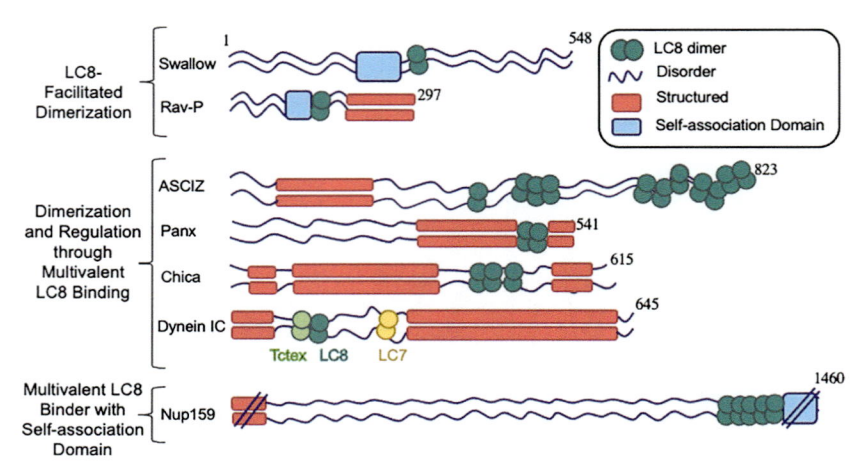

Fig. 2 LC8-binding IDP duplexes. LC8 (green circles) binding within IDP duplexes may play variable roles in different systems including, strengthening of an already existing self-association domain (top), dimerization and regulation through multivalent binding of LC8 (middle), or multivalent binding in IDP duplexes with self-association domains (bottom). LC8-binding IDP duplexes all share similar domain architecture, consisting of a combination of structured regions (rectangles) and disorder (dark blue line), dimerized through either a self-association domain, LC8 homodimer binding, or a combination of the two.

neuropsychiatric disorders, and neuromuscular and neurodegenerative diseases such as myotonic dystrophy, spinal muscular atrophy, and fragile X syndrome.[23]

Swa consists of a self-association domain located in the middle of long N- and C-terminal disordered tails (Fig. 2). Without LC8 and at physiological conditions, Swa's dimerization is relatively weak at 4 µM.[21,22,24] LC8-free Swa exists primarily as a flexible monomer, but has a region with a high propensity to form a coiled-coil, and is thought to weakly form a dimer through this coiled-coil, resulting in a monomer–dimer equilibrium (Fig. 3A). Swa also possesses an LC8-binding domain directly next to its self-association domain (Fig. 2).[19,21,24] Binding of the strong LC8 homodimer C-terminal of Swa's coiled-coil domain disrupts the monomer–dimer equilibrium, reinforcing the Swa dimer and promoting a stronger coiled-coil (Fig. 3A).[24] Mutations in Swa's coiled-coil domain results in bicoid mRNA uniformly dispersed throughout the oocyte cytoplasm. Other Swa mutants additionally result in improper localization of huli-tai shao-adducin-like mRNA, cytoskeleton abnormalities, and associated phenotypes in drosophila, suggesting that both the structured coiled-coil and the disorder must be conserved for proper Swa function.[22]

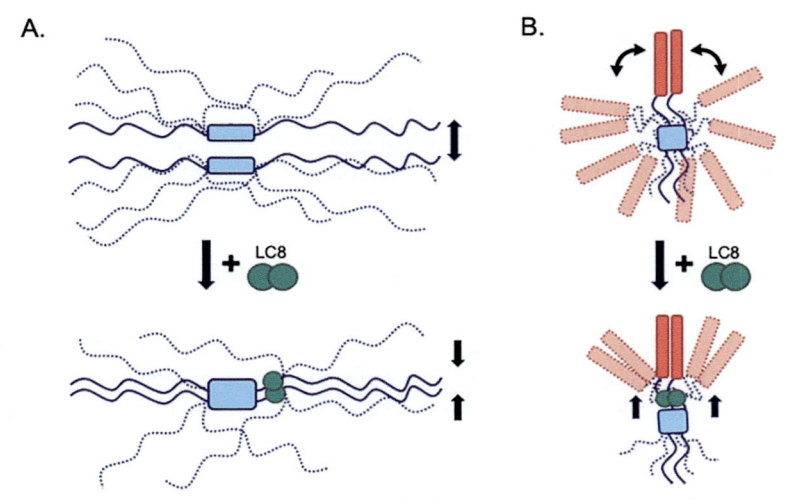

Fig. 3 LC8 binding is capable of facilitating (A) and extending (B) dimerization. (A) LC8 binding to Swallow stabilizes its dimer state. Swallow possesses a centrally located self-association region (light blue), through which, Swallow makes a weak dimer in solution (top). When LC8 is present, the strong homodimer holds two Swallow monomers together, stabilizing the IDP duplex through supporting the formation of the self-association domain's coiled-coil (bottom). (B) LC8 binding to Rav-P. In solution, the structured nucleotide binding domain of Rav-P (pink) is dynamic, able to sample many states surrounding the self-association domain (light blue) (top). Upon LC8 binding to the disordered domain separating the nucleotide and self-association domains, motion of the structured nucleotide binding domain becomes restricted, oriented away from the self-association domain and N-terminal disordered tail (bottom).

The coiled–coil domain of the Swa dimerization domain shows reduced packing at the coil center compared to the N- and C-terminal ends, creating a bulge that is thought to make dimerization less favorable than a more tightly packed coiled-coil, suggesting that this is the origin of Swa's monomer-dimer equilibrium present in LC8-free Swa.[22] Without LC8, Swallow also tends to aggregate at increased concentrations.[22] LC8 binding directly decreases the amount of flexibility in the region it binds to, but additionally, the organized dimerization irreversibly pushes coiled–coil formation such that the monomeric coiled–coil region is no longer available for non–specific interactions with other protein regions. The result is an IDP duplex that is most likely less prone to aggregation due to an overall decrease in entropy of Swa in solution.

Rabies disease, caused by Rabies lyssavirus (RAV), is preventable by vaccination but is otherwise lethal. One of the key proteins involved in assembly and replication of RAV is the rabies phosphoprotein (*Rav-P*). Rav-P exists as an IDP duplex (Fig. 2) and is a non-catalytic polymerase cofactor that functions within the rabies virus through its numerous nucleotide and protein connections; one of these connections being with LC8. LC8 interacts with Rav-P through its TQT motif and the interaction is critical for the success of the rabies virus.[6] The Rav-P-LC8 interaction, regulated by only a few residues, is significant enough to solely determine the lethality rate of the rabies virus. Mice exposed to rabies with the TQT of Rav-P mutated away, and the LC8-interaction consequently eliminated, showed a 100% survival rate, in contrast to those which received WT Rav-P and experienced 100% lethality.[14] This extreme phenotype observed is not attributed to a complete termination of the virus, but rather to an extremely attenuated viral infection, suggesting that Rav-P is still able to carry out its role, but does so far less efficiently.[6,14]

Rav-P's structured C-terminal domain's (CTD) connection with gRNA is mediated through its interaction with the rabies nucleocapsid protein, Rav-N, which encapsulates the gRNA, and Rav-P and stabilizes Rav-N's connection with RNA polymerase L. Rav-P likely organizes and prevents Rav-N from making undesired RNA interactions. Additionally, Rav-P also interacts with Complex I within the mitochondria, increasing reactive oxygen species (ROS) generation.[25] Localized to disordered residues 139–172 of Rav-P, this contact with Complex I is in direct competition with LC8, binding at residue 147. The disordered linkers of Rav-P also bind to ribosomal protein L9,[26] IRF-3,[27] STAT1,[28] the RNA protected by Rav-N,[29] PML,[30] and nuclear import and export factors.[31] Rav-P is also known to bind to BECN1,[32] HSP90, and Cdc37.[33] While this complex network of binding interactions remains to be fully understood, all are potentially influenced by changes in flexibility within the connecting linkers upon LC8 binding, or are directly competing with LC8 to occupy Rav-P.

Several IDP duplexes such as RavP already have strong self-association domains before binding LC8, suggesting that LC8's role in these systems is not only to dimerize the IDP, but to strengthen and/or lengthen the IDP's pre-existing structured region of dimerization. This additional rigidity decreases mobility and limits the number of conformational assemblies the duplex may sample. Increased restriction of IDP duplex movement likely makes orientation-dependent interactions with nucleotides or other proteins more favorable, as the interacting domain is more likely to be available for binding rather than making more distant transient or specific interactions. Because this fortified structure is controlled by LC8 binding and is not inherent to the IDP duplex's structure, the system may use LC8 as a regulator of function, enabling more or less flexibility depending on the availability of LC8 or other biomolecules.

LC8 is not required to dimerize Rav-P, but when LC8 interacts with Rav-P next to its strong dimerization domain, mobility of the RNA-binding CTD becomes more restricted with the additional structure, orienting the CTD away from the N-terminus (Fig. 3B).[6] LC8 binding to Rav-P places both CTDs of the IDP duplex in closer proximity to each other, increasing the likelihood that both can interact with Rav-N/RNA, and also decreasing the probability that the CTD will interfere with the N-terminal end's binding to Rav-L. The result is that LC8 binding of Rav-P increases polymerase activity and transcription rates through modulation of IDP duplex dynamics.[6]

3. Multivalent binding of LC8 along IDP duplexes

Several IDP duplexes contain multiple binding motifs for LC8 within their IDRs. LC8 multivalent binding in these IDP duplexes is responsible for functional dimerization of the protein, increased structure, and can result in an array of bound and unbound states. Multivalent LC8-binding introduces increased complexity to IDP duplex systems as the possibility of variable occupancy by LC8 arises. Additionally, multivalent binding combined with flexibility of LC8 binding partners introduces the potential for off-register binding, such that LC8 may bind to site A of a disordered partner on one side of the LC8 dimer, but attach to site B on the opposite side. Recent work shows that LC8 trends toward in-register binding and that this binding is regulated by binding site sequence along with the length of connecting linkers between sites.[34] Differential motif sequences, linker length, and number of LC8-binding sites may offer diverse mechanisms for regulation of multivalent LC8-binding IDP duplexes.[34]

ASCIZ (ATM substrate Chk2-interacting ZnF2+; ATMIN; ZNF822) is a protein consisting of a structured N-terminal zinc finger DNA-binding domain, and a C-terminal LC8-binding disordered tail (Fig. 2). ASCIZ is primarily found in the nucleus of a cell[8] and was first discovered as a DNA damage response protein, having a role in base excision repair.[35] The primary function of ASCIZ, however, appears to be related to its role as a transcription factor for hub protein, LC8.[9,36] Loss of functional ASCIZ results in severe developmental abnormalities, abnormally low levels of B cells, and death during embryogenesis in both mice and flies.[8] Since the interactome of LC8 includes processes such as cell division, viral transcription, and apoptosis, both LC8 and its transcription factor ASCIZ, are critical for cellular homeostasis and ultimately, organismal health and function.[15,16]

As ASCIZ contains no predicted self-association domain, dimerization of ASCIZ is driven entirely by multivalent LC8 binding. Human ASCIZ has 11 LC8 binding sites, but site occupancy is dynamic and variable, with two-to-four LC8 dimers being the predominant species when assessed by negative stain electron microscopy (Fig. 4).[8] ASCIZ remains in a dynamic equilibrium of variable LC8 occupancy, with LC8 occupancy acting as a sensor for LC8 concentration levels in the cell. LC8 is predicted to regulate its own transcription through its ASCIZ binding, dimerization, and increased structure, such that when more LC8 is present transcription is inhibited, and with less LC8, transcription increases (Fig. 4).[8] Based on this observation, increased flexibility is associated with increased transcription, suggesting that the long, disordered tail of ASCIZ is directly involved in transcription, likely through other binding interactions that no longer become accessible when bound to LC8. Another possibility is that decreased flexibility in an LC8-bound ASCIZ dimer brings two zinc-finger DNA-binding domains closer together, orienting them so that transcription is not possible, or is less likely to occur, as one physically inhibits the other's ability make contacts with DNA.

Panoramix (*panx*) is a protein involved in the piwi-interacting RNA pathway.[7] This pathway functions within animal gonads as an RNA-based immune system, controlling the expression of transposons by inducing

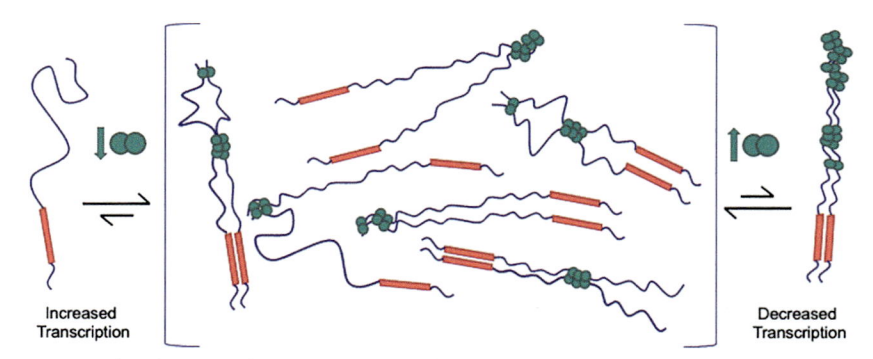

Fig. 4 Multivalent LC8 binding to ASCIZ. hsASCIZ contains 11 LC8-binding sites, but exists in equilibrium in variable bound states, with the majority of ASCIZ dimers bound to 2–4 LC8s. LC8 is thought to regulate its own transcription through ASCIZ interactions such that elevated levels of LC8 are associated with decreased levels of transcription, and decreased levels of LC8 with increased levels of transcription.

heterochromatin formation. Important for the transcriptional gene silencing of transposon insertions, this pathway works to maintain genomic integrity by Piwi's targeting of chromatin epigenetically marked for repression.[37] Defects within this pathway result in impaired transposon repression, loss of repressive histone modifications, elevated occupancy of RNA polymerase II at transposon insertions, and often sterility.[7] While the drosophila system is better understood, similar proteins and mechanisms are thought to function within mammalian gonads, protecting genomes from nascent RNA transcripts.

Panx is one of the proteins that assembles the panoramix-induced co-transcriptional silencing (PICTS) complex, functioning upstream of the machinery that modifies chromatin, but downstream of Piwi.[7] In somatic and germline cells of the ovary, Panx forms a complex with Nuclear Export Factor 2 (Nxf2) and its co-factor, NTF2-related export protein 1 (Nxt1) to achieve co-transcriptional silencing of transposons.[7]

In drosophila, the LC8 homolog (cut-up) is essential for PICTS's function of transcriptional gene silencing of transposon insertions. Panx consists of a long disordered N-terminus, followed by a structured domain consisting of the Nxf2/Nxt1 interaction region and the protein region responsible for regulating protein degradation, the degron. Following the structured degron is a shorter disordered region, and a structured C-terminus predicted to be alpha-helical (Figs. 2 and 5). Two LC8-binding sites are located in the disordered region C-terminal of the degron, facilitating dimerization of Panx, and by association, the PICTS complex.[7] This dimerization is key for the complex's transposon repression as when Panx is artificially dimerized, LC8 is no longer necessary for the PICTS complex to fulfill its function.[7]

Binding of LC8 to its two motifs restricts motion of the disordered linker and orients the predicted C-terminal helices away from the structured degron (Fig. 5). The largely disordered N-terminal region is the driving force for transcriptional silencing through mechanisms that remain unresolved, however, dimerization is required for transposon repression, suggesting that the mechanism is more efficient when the N-terminal disorder is in closer proximity to another N-terminal tail. As the disordered N-terminus is far less conserved than the more structured C-terminal half of Panx, this suggests that its role in gene silencing is not related to recruitment of specific binding partners, but rather it remains important that the region is flexible overall. In contrast, the structured C-terminus is the interaction site for all known cofactors.[7] Therefore, LC8-facilitated dimerization may facilitate the functionality of the PICTS complex by limiting motion, pushing the far C-terminus out of

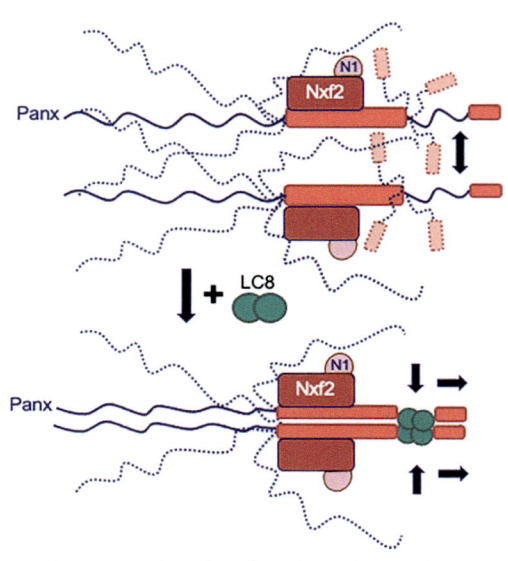

Fig. 5 LC8 binding to Panx complexed with Nxf2 and its co-factor, Nxt1. LC8-free Panx exists as a monomer in solution (top). When LC8 binds to Panx, this facilitates dimerization of not only Panx itself but of the larger PICTS complex, including Nxf2/Nxt1. Additionally, LC8 binding increase's structure and limits motion of the C-terminal end of Panx, making the region available for other necessary interactions.

the way, and making space for other functional binding interactions to be made along the structured C-terminal end of Panx.

The Chica spindle adaptor protein (*Chica*) is involved in asymmetric cortical localization of dynein. Chica is additionally involved in the orientation of spindle fibers, as cells that are depleted in Chica will not orient mitotic spindles correctly.[38] Through Chica's role in dynein localization and spindle organization, proper Chica function is related to somatic and germ cell replication, therefore, mutations that disrupt Chica's binding interactions or structure would likely result in chromosomal disorders or disease associated with improper cell replication.[38]

Chica has three helical regions with the rest of the protein consisting of a disordered coil. Within the disorder, Chica has four predicted TQT motifs but only three bind experimentally, with LC8:Chica binding stoichiometry shown to be 3:1 by isothermal titration calorimetry, and the least conserved motif shown by NMR to be the non-binder.[12] Disruption of the LC8-Chica interaction shows that this binding event is necessary for dynein to be localized to the cell cortex and to orient mitotic spindles correctly for mitosis.[38] LC8 binding induces dimerization of Chica, resulting in an IDP duplex, with the three site affinities varying from 0.4 μM to much weaker binding.[12]

Along with LC8, Chica is also known to interact with microtubules indirectly through its interaction with HMMR.[38] The Chica-LC8 complex and its functionality remains to be fully understood, however, it is possible that, as in the Panx-LC8 complex, LC8 functions to dimerize Chica and its associated HMMR and microtubules during mitosis. The stability of the LC8-bound Chica duplex is the same whether all three LC8 sites are available or only the first site is bound, suggesting that the role of the other two LC8-binding sites is not to promote dimerization but to fulfill other cellular functions of the Chica-LC8 complex.[12] Chica's disordered region is serine and threonine rich and Chica is known to be heavily phosphorylated in mitosis.[39] Likely sites of phosphorylation are located within and directly adjacent to LC8 binding sites, therefore, a potential function of additional LC8 binding is to physically compete with and inhibit kinases at variable stages of the cell cycle.

Cytoplasmic dynein is a motor protein responsible for most long-distance negative-end directed microtubule-based transport within human cells, consequently, proper dynein function is essential for diverse cellular functions. Dynein's many functions include mitotic spindle fiber assembly, axonal transport, and mutations within dynein motor components often result in neurodevelopmental and neurodegenerative disease.[11,40] Dynein has also been implicated in the transport, and therefore the infectivity and function of, viruses within a human cell.[41]

Dynein exists as a large, 1.6-MDa, IDP duplex made up of folded motor domains that function to attach to microtubules, and a more flexible cargo attachment domain. Essential to the cargo attachment domain of dynein is the intermediate chain (*IC*). Monomeric IC is made up of N-terminal alpha helices and a C-terminal beta-propeller, separated by a disordered linker.[42] In solved dynein structures, IC remains unresolved due to its flexible nature, and this disorder is responsible for dynein's cargo attachment, motor protein stability, and regulation of dynein activity broadly.[11,43–45] Dynein is dimerized through multivalent light chain binding interactions in the IC linker, including LC8, Tctex, and LC7 (Fig. 2).[42] LC8 binding to IC takes on the role of overcoming the entropic barrier, dimerizing two highly disordered monomers, while also increasing the favorability of LC7 and Tctex binding. This cooperative multivalent dimerization is termed, polybivalency.[46,47] Light chain binding and dimerization results in tight packing at protein interfaces but leaves the connecting linkers flexible.

4. Nup159: A multivalent LC8 binder with a self-association domain

Across species, the nuclear pore complex regulates transport of macromolecules between the cell cytoplasm and nucleus. Nuclear pore complexes are made up of the nucleoporin family of proteins, and numerous pathologies have links to these proteins. Overexpression of nucleoporin 88 is associated with tumor growth,[48] and other nucleoporins have been linked to cancer through their role in spindle assembly and aneuploidy arising through disrupted nucleoporin function.[49] Impaired nucleoporins are also involved in Amyotrophic Lateral Sclerosis (ALS) and frontotemporal dementia disease (FTD).[50,51]

Nup159 is a subunit at the cytoplasmic end of the yeast nuclear pore complex, which consists of ~30 proteins, many of which are largely intrinsically disordered.[52] Nup159 serves both as a structural component and as a binding site for transiently associated nuclear transport factors, and forms a nucleoporin subcomplex with Nup82 and Nsp1.[46] Nup159 contains an N-terminal beta-propeller, a C-terminal coiled-coil that functions as a self-association domain, and a disordered linker between the two.[46] This domain organization is similar to that of dynein IC, which has a C-terminal beta propeller region, an N-terminal alpha helix, and a connecting linker.

The long, disordered linker separating the two structured domains of Nup159 contains five LC8-binding sites directly N-terminal of the coiled-coil domain (Fig. 2). LC8 binding likely stabilizes favorable coiled-coil interactions, similar to that observed in Swa (Fig. 3A), and facilitates larger complex formation as observed in Panx (Fig. 5). Similar to what is observed in multivalent LC8-binding to ASCIZ, Nup159 is also observed to favor an intermediate number of three bound LC8 dimers rather than taking on full occupancy of five LC8s due to instability arising from unfavorable changes in entropy.[46]

Since optimal stability is achieved with only three tandem LC8 dimers, this suggests an additional role of LC8 is to provide increased rigidity with all five sites bound, and more flexibility with fewer LC8s.[46] Moreover, EM studies show that LC8-bound Nup159 forms a structured "bead-on-a-string" phenomenon, while a Nup159 construct engineered with fewer LC8 binding sites is too flexible to capture by EM, supporting this model.[46] Fine tuning of Nup159 flexibility vs rigidity may be critical in Nup159's binding interactions within the nuclear membrane and its nuclear pore complex role in facilitating macromolecule passage, adjusting flexibility as molecules move in and out of the nucleus. For example, increased concentrations of LC8 could result in a more rigid and closed structure, only allowing small, passive transport, while decreased LC8 concentrations could allow for the flexibility necessary to open up a nuclear pore complex to accommodate larger molecules that require mediated transport between the cytoplasm and nucleus.[52]

5. VSV-P: The role of disorder in a non-LC8 binding IDP duplex

Rav-P has a homologous protein that does not bind LC8, vesicular stomatitis virus phosphoprotein (*VSV-P*) (Fig. 6A).[6,53] Vesicular stomatitis livestock virus, like RAV, is a member of the Family Rhabdoviridae. In place of LC8 binding, VSV-P has an extended dimerization domain which structures the space that is otherwise bound by LC8 in Rav-P, resulting in a linker length between the nucleotide binding and self-association domains that is comparable to that of Rav-P (Fig. 6).[6,54] Due to its longer dimerization domain, VSV-P naturally restricts its nucleotide binding domain away

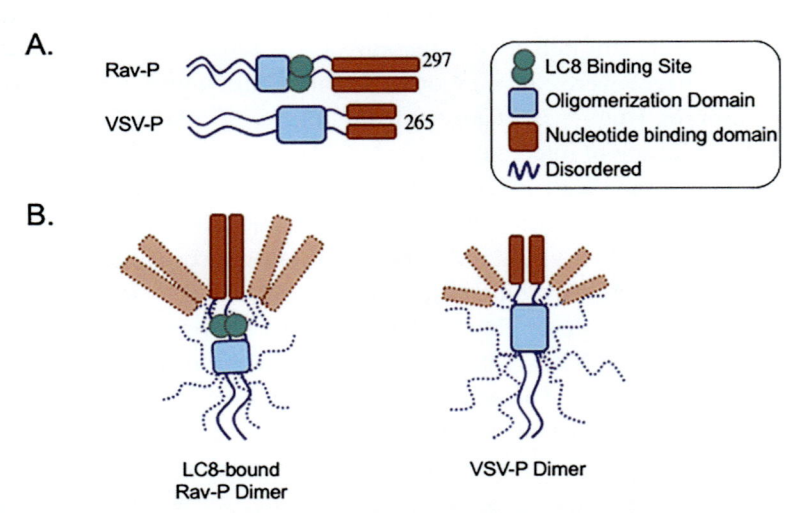

Fig. 6 VSV-P and Rav-P are homologous proteins of the Family Rhabdoviridae. (A) Rav-P and VSV-P share common domain organization, but only Rav-P binds LC8. (B) Rav-P contains an LC8-binding site within the disordered linker connecting the oligomerization and nucleotide binding domains that is necessary for restricting motion of the nucleotide binding domain and ultimately for viral transcription. Alternatively, the VSV-P linker is naturally restricted due to its extended oligomerization domain.

from the N-terminal region, the role that LC8-binding fulfills in Rav-P. Although VSV-P's linker is shorter than that of Rav-P, its disordered N-terminal tail is longer. Despite the length of the disordered N-terminal tail in VSV-P, the protein structure remains extended due to the high concentration of acidic residues that are charge repulsive.[53] This long N-terminal tail has a long reach, capable of extending more than five N protomers from the VSV-P binding site, and is therefore thought to capture the upstream polymerase, as well as deliver it downstream during replication.[53] RAV-P's ability to bind LC8, allowing for variable linker flexibility rather than a fixed orientation suggests that LC8 allows for multiple states of and, in its LC8-free form, a role beyond what non–LC8-binding protein VSV-P is capable of.[6]

6. The role of disorder and dimerization of the SARS-CoV-2 nucleocapsid phosphoprotein IDP duplex

COVID-19 disease, caused by the SARS-CoV-2 virus, has resulted in one of the most devastating public health crises of the century. A member of the human-infecting class of CoVs, β-coronavirus, SARS-CoV-2 causes viral pneumonia through its occupation of the lower respiratory system but can also have severe effects on other major organs, with some symptoms lasting months later.[55] A major concern regarding SARS-CoV-2 is its unusually high rate of infectivity, and as of March 2021, COVID19 disease has resulted in the death of over 2.6 million people worldwide, less than 2 years after the virus entered the human population.[56]

The SARS-CoV-2 virion is composed of four structural proteins, spike (S), membrane (M), envelope (E), and nucleocapsid (N). Protein S, M, and E combine to shape the viral membrane, surrounding the helical ribonucleocapsid (RNP). The nucleocapsid protein (*CoV-N*) is one of the most highly expressed proteins, with many copies binding and packaging the ~30 kb SARS-CoV-2 genomic RNA (gRNA), one of the largest genomes among RNA viruses.[57] Beyond virus assembly, packaging, and protecting the gRNA within the virion structure, the CoV-N also plays an important role in gRNA replication and transcription, as well as in disassembly of the virus during infection.[5,58] The role of CoV-N is therefore similar to that of Rav-P, but while Rabies contains two separate proteins fulfilling roles of negative-strand RNA protection, replication, and transcription, the rabies phosphoprotein (Rav-P) and rabies nucleocapsid

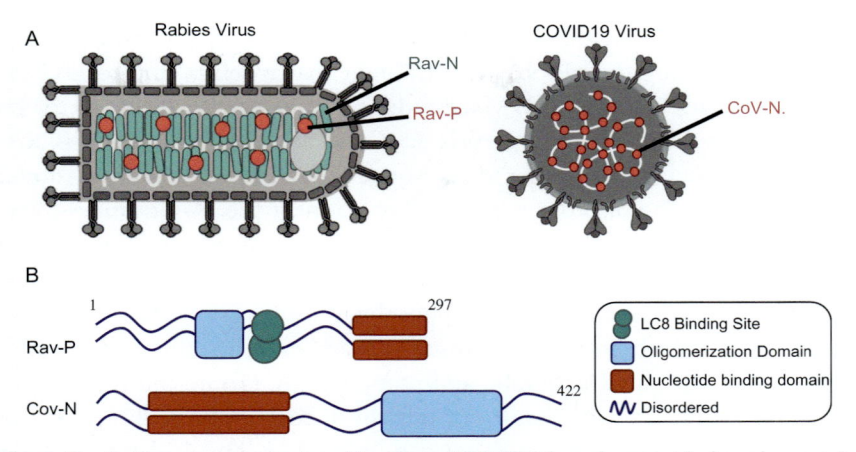

Fig. 7 The Rabies phosphoprotein (Rav-P) and COVID-19 nucleocapsid phosphoprotein (CoV-N) fulfill similar functions within a virus, as well as a similar domain architecture. (A) Both Rav-P and CoV-N play a role in packaging and protecting genomic RNA (gRNA), and consequently have roles in viral replication and transcription. In the rabies virus Rav-P protects gRNA with Rav-N, while in a COVID-19 virus, CoV-N binds gRNA alone. (B) Rav-P and CoV-N both exist as IDP duplexes, dimerized thought a self-association domain, with disordered linkers separating the origin of dimerization from the structured nucleotide binding domains.

(Rav-N),[6] SARS–CoV-2 has one protein, the nucleocapsid phosphoprotein (CoV–N) bound to its positive-strand RNA genome (Fig. 7A). Beyond the function, the domain architecture of Rav-P and CoV-N is also similar (Fig. 7B), as both proteins consist of two globular regions, an RNA-binding domain and a self-association domain. These two structured components are flanked by three regions of disorder thought to be involved in regulation of RNP complex formation along with numerous viral and host protein binding interactions.

In solution, CoV–N exists as an IDP duplex, dimerized through its strong self-association domain. In addition to its function in dimerization, this domain participates in RNA binding and is thought to participate in other protein binding events such as with CoV-M.[59,60] Additionally, the self-association domain is also the origin of higher order oligomer formation of CoV-N. CoV-N is known to form biomolecular condensates with viral RNA, and while the conditions and functionality of this phenomenon remain to be fully understood, there is increasing evidence that condensate formation and regulation of CoV-N's various binding interactions and functional resides within the protein's disorder.[61–63]

Along the SARS-CoV-2 genome, CoV-N binds multivalently, packaging and protecting the gRNA. It is thought that when bound to gRNA, CoV-N remains flexible as NMR binding studies with 1 kb of the 30 kb gRNA show that disorder persists in CoV-N when bound to RNA. Despite the fact that disorder does not appear to be involved in direct RNA binding, the disorder is required to connect the N-terminal RNA-binding domain to the dimeric C-terminal self-association domain for tighter binding affinity. One factor likely important for increasing gRNA binding affinity is the presence of phosphorylation sites within the disordered linker, which increase electrostatic interactions, making higher order oligomerization more favorable, and consequently promoting multivalent binding, as one CoV-N dimer recruits the next.[58] Within the packaged RNP flexibility likely remains, leaving disordered regions available for other protein binding interactions and post-translational modifications. Persisting flexibility within the RNP may be important for replication and transcription machinery to interact with gRNA without disrupting the entirety of RNP structure.

The middle-disordered linker of CoV-N contains a serine/arginine-rich region of phosphorylation (residues 184–196) that is thought to help regulate CoV-N's various functions and interactions, including transcription and translation of viral RNA and proteins, self-oligomerization, and its ability to bind nucleic acids and host and viral proteins.[64,65] CoV-N is phosphorylated at multiple serine residues during viral infection, and while intracellular CoV-N is phosphorylated by several kinases, the N-protein in extracellular virions is not phosphorylated, implicating host phosphatases in virion replication.[65] Furthermore, MD simulations have shown that even a single phosphorylation within the CoV-N linker can increase the number of inter- and intra-molecular protein contacts due to the formation of salt bridges between the phosphate groups and arginine side chains. This increase in salt bridges attenuates CoV-N RNA binding,[63] which is potentially the cause for the observation that phosphorylation of CoV-N in human cells alters CoV-N condensate formation.[63,66] Phosphorylated CoV-N, for example, binds to all seven human hub protein 14-3-3 isoforms, which are highly prevalent in human tissues where SARS-CoV-2 is abundant during infection.[65] This interaction is thought to be important for nucleocytoplasmic transport of CoV-N, and may also takeover human cellular pathways through sequestering 14-3-3.[65]

Liquid-liquid phase separation (LLPS) of CoV-N facilitates concentration of components necessary for replication, including RNA, nsp7, nsp8, and RNA-dependent RNA-polymerase RdRp,[63] and the intrinsically disordered N-terminal tail, as well as the linker are known to be important for LLPS.[62–64] In particular, the disordered N-terminal tail of CoV-N has been reported to be essential for CoV-N LLPS, as truncation of this IDR results in loss of CoV-N aggregation and droplet formation.[67] CoV-N forms different condensate types corresponding to its two primary functions: RNP formation and transcription.[64] Phosphorylation promotes the protein's transcriptional function likely by inhibiting stabilizing interactions necessary for organized RNP formation, causing CoV-N to form more dynamic, liquid-like condensates, as opposed to gel-like condensates containing discrete RNP particles as observed with unmodified CoV-N.[64] Therefore, disorder within the IDP duplex of CoV-N is important not only because its properties promote functional LLPS, but also because they contain the SR-rich region that allows phosphorylation events to switch CoV-N between its viral stages of virion assembly, replication, and transcription. Additionally, it is essential that CoV-N exists as a duplex rather than a monomer as dimerization improves RNA-binding, and is the seed for formation of higher order oligomers.

7. Future directions

The reversibly ordered multivalent LC8-IDP duplex systems, which play critical roles throughout the cell, are vastly under-studied compared to their prevalence and importance. Future studies require a host of innovative methods integrating multiscale computations with a range of experimental modalities to address key questions in the functional structural biology of these duplexes. This could include a combination of state-of-the-art NMR, novel negative stain electron microscopy, single-molecule characterization, biophysical and thermodynamic modeling, and fluorescence fluctuation analysis of live cells. As structural biology technologies and techniques continue to surpass previously held limitations, new and persisting questions may be answered to illuminate how LC8 ultimately affects IDP functional behavior.

Among many LC8-binding IDP duplexes, LC8 binding plays an essential role and disrupting this interaction would result in disease development or prevention. For example, loss of LC8 binding in ASCIZ does not produce viable life, and the loss of the LC8-Chica prevents dynein from localizing to the cell cortex and mitotic spindles are not oriented correctly for mitosis, but loss of LC8 binding to Rav-P would inhibit Rabies progression.

Furthermore, loss of LC8-binding in Panx would likely result in defects typical of piwi-interacting RNA pathway disruption such as impaired transposon repression, loss of repressive histone modifications, elevated occupancy of RNA polymerase II at transposon insertions, or sterility, as LC8 is needed to dimerize the PICTS complex. LC8 is also needed to stabilize dimerization of Swa, and without this, proper RNA localization would not take place, as mutations in Swa's coiled-coil domain result in delocalized RNA. With LC8's role in dimerization in mind, analysis of disease-associated mutations should take into account the LC8 binding motif as well as changes in disorder propensity surrounding and LC8 binding motif. It should further be considered that engineering of an LC8 binding site into a region of disorder has the potential to improve dimerization or to change protein functionality.

The role of LC8 in dimerization is increasingly clear, however, in multivalent IDP duplex systems, dynamic states appear to have diverse functions. As only one of Chica's three LC8-binding sites appears to be necessary for dimerization, the other two likely serve other functions not yet uncovered. Additionally, the long ASCIZ disordered tail binds up to 11 LC8s, but appears to exists in a complicated and diverse equilibrium of mostly two-to-four LC8 dimers bound. Improved understanding of the fine-tuning along multivalent LC8-binding disordered tails offers not only a better understanding of the role of disorder in IDP duplexes in disease, but may also offer opportunities for biotechnology and protein design advancements such that variable amounts of LC8 or disorder present could be used to mediate transcription of other protein functions attached to a long disordered tail.

Knowledge about the role of flexibility and dynamics in IDP duplex systems continues to be uncovered, offering new opportunities to target such systems. For example, it is the increased structure upon binding of LC8 to Rav-P that enables Rav-P to properly carry out its function. Therefore, drug therapies may be developed to bind Rav-P at the site of LC8 binding, thereby preventing life-threatening LC8 interactions. Knowing that Rav-P must be a dimer also presents an opportunity to disrupt rabies virus through disruption of Rav-P dimerization.

Reversible protein phosphorylation is a common method of regulating protein function and physiological effects within a cell. Polar residues serine, arginine, tyrosine, and threonine are common within intrinsically disordered protein regions and are often phosphorylated across the proteome, acting as a functional switch. IDP duplexes known to be regulated by phosphorylation have the potential to be further targeted through physical binding to and blocking of the phosphorylation site, or through prompting increased structure in the region of phosphorylation, making the residue unavailable for phosphorylation. For example, LC8 binding to Chica's serine-threonine rich region is thought to physically compete with and inhibit kinases at variable stages of the cell cycle.

Continued understanding of the disorder within IDP duplexes offers the potential for new strategies for targeting diseases as diverse as the prevalence of IDP duplexes in biology. For example, uncovering that dimerization is essential to the function of CoV-N and therefore for SARS-CoV-2 infectivity means that RNA sequences coding for the dimerization domain could be replaced with sequences coding for regions of disorder, or the dimerization domain could be removed completely. Similarly, the decreased disorder of VSV-P compared to Rav-P reveals that increasing linker length of VSV-P could likely disrupt function of this virus. This knowledge could be harnessed to insert a longer region of disorder into VSV-P, making the protein too flexible to function. Finally, knowledge of how phosphorylation regulates an IDP duplex could be used to mutate away key phosphorylated residues.

8. Conclusion

In summary, IDP duplexes are extremely prevalent in biology and essential in regulating diverse cellular processes such as cell division, cellular transport, transcription, viral replication, phase-separation, and gene silencing. Since IDP duplex architecture and functions are conserved across species, it is imperative to think of protein structural conservation in terms of not only conserved structural domains, but also in terms of unstructured components, length of intrinsically disordered regions, regions of post-translational modifications, dimerization state, and multivalent interactions.

Acknowledgments

We acknowledge support from the National Science Foundation (Awards 1617019 and 2034446).

References

1. Ingles-Prieto A, et al. Conservation of protein structure over four billion years. *Structure*. 2013;21(9):1690–1697.
2. Zhang S, et al. Conservation and variation of the hepatitis E virus ORF2 capsid protein. *Gene*. 2018;675:157–164.
3. Walls AC, et al. Structure, function, and antigenicity of the SARS-CoV-2 spike glycoprotein. *Cell*. 2020;**181**(2):281–292. e6.

4. Andreani J, Quignot C, Guerois R. Structural prediction of protein interactions and docking using conservation and coevolution. *Wiley Interdiscip Rev: Comput Mol Sci.* 2020;10(6): e1470.

5. Cong Y, et al. Nucleocapsid protein recruitment to replication-transcription complexes plays a crucial role in coronaviral life cycle. *J Virol.* 2020;94(4):e01925–19.

6. Jespersen NE, et al. The LC8-RavP ensemble structure evinces a role for LC8 in regulating lyssavirus polymerase functionality. *J Mol Biol.* 2019;431(24):4959–4977.

7. Eastwood EL, et al. Dimerisation of the PICTS complex via LC8/Cut-up drives co-transcriptional transposon silencing in Drosophila. *Elife.* 2021;10, e65557.

8. Clark S, et al. Multivalency regulates activity in an intrinsically disordered transcription factor. *Elife.* 2018;7:e36258.

9. Jurado S, et al. ATM substrate Chk2-interacting Zn2 + finger (ASCIZ) is a bi-functional transcriptional activator and feedback sensor in the regulation of dynein light chain (DYNLL1) expression. *J Biol Chem.* 2012;287(5):3156–3164.

10. Becker JR, et al. The ASCIZ-DYNLL1 axis promotes 53BP1-dependent non-homologous end joining and PARP inhibitor sensitivity. *Nat Commun.* 2018;9(1):1–12.

11. Reck-Peterson SL, et al. The cytoplasmic dynein transport machinery and its many cargoes. *Nat Rev Mol Cell Biol.* 2018;19(6):382–398.

12. Clark S, et al. The anchored flexibility model in LC8 motif recognition: insights from the Chica complex. *Biochemistry.* 2016;55(1):199–209.

13. Lou J, et al. Spatiotemporal dynamics of 53BP1 dimer recruitment to a DNA double strand break. *Nat Commun.* 2020;11(1):1–11.

14. Tan GS, et al. The dynein light chain 8 binding motif of rabies virus phosphoprotein promotes efficient viral transcription. *Proc Natl Acad Sci U S A.* 2007;104(17):7229–7234.

15. Jespersen N, et al. Systematic identification of recognition motifs for the hub protein LC8. *Life Sci Alliance.* 2019;2(4):e201900366.

16. Barbar E. Dynein light chain LC8 is a dimerization hub essential in diverse protein networks. *Biochemistry.* 2008;47(2):503–508.

17. Jespersen N, Barbar E. Emerging features of linear motif-binding Hub proteins. *Trends Biochem Sci.* 2020;45(5):375–384.

18. Rapali P, et al. DYNLL/LC8: A light chain subunit of the dynein motor complex and beyond. *FEBS J.* 2011;278(17):2980–2996.

19. Benison G, Karplus PA, Barbar E. Structure and dynamics of LC8 complexes with KXTQT-motif peptides: Swallow and dynein intermediate chain compete for a common site. *J Mol Biol.* 2007;371(2):457–468.

20. Weil TT, et al. Distinguishing direct from indirect roles for bicoid mRNA localization factors. *Development.* 2010;137(1):169–176.

21. Kidane AI, et al. Structural features of LC8-induced self-association of swallow. *Biochemistry.* 2013;52(35):6011–6020.

22. Loening NM, Barbar E. Structural characterization of the self-association domain of swallow. *Protein Sci.* 2021;30(5):1056–1063.

23. Wang ET, et al. Dysregulation of mRNA localization and translation in genetic disease. *J Neurosci.* 2016;36(45):11418–11426.

24. Wang L, et al. Dynein light chain LC8 promotes assembly of the coiled-coil domain of swallow protein. *Biochemistry.* 2004;43(15):4611–4620.

25. Kammouni W, et al. Rabies virus phosphoprotein interacts with mitochondrial complex I and induces mitochondrial dysfunction and oxidative stress. *J Neurovirol.* 2015;21(4):370–382.

26. Li Y, et al. Rabies virus phosphoprotein interacts with ribosomal protein L9 and affects rabies virus replication. *Virology.* 2016;488:216–224.

27. Chelbi-Alix MK, et al. Rabies viral mechanisms to escape the IFN system: The viral protein P interferes with IRF-3, Stat1, and PML nuclear bodies. *J Interferon Cytokine Res.* 2006;26(5):271–280.

28. Vidy A, Chelbi-Alix M, Blondel D. Rabies virus P protein interacts with STAT1 and inhibits interferon signal transduction pathways. *J Virol.* 2005;79(22):14411–14420.

29. Albertini AA, et al. Crystal structure of the rabies virus nucleoprotein-RNA complex. *Science.* 2006;313(5785):360–363.

30. Blondel D, et al. Rabies virus P and small P products interact directly with PML and reorganize PML nuclear bodies. *Oncogene.* 2002;21(52):7957–7970.

31. Oksayan S, et al. A novel nuclear trafficking module regulates the nucleocytoplasmic localization of the rabies virus interferon antagonist, *P protein. J Biol Chem.* 2012;287 (33):28112–28121.

32. Liu J, et al. BECN1-dependent CASP2 incomplete autophagy induction by binding to rabies virus phosphoprotein. *Autophagy.* 2017;13(4):739–753.

33. Xu Y, et al. The co-chaperone Cdc37 regulates the rabies virus phosphoprotein stability by targeting to Hsp90AA1 machinery. *Sci Rep.* 2016;6(1):1–15.

34. Reardon PN, et al. The dynein light chain 8 (LC8) binds predominantly "in-register" to a multivalent intrinsically disordered partner. *J Biol Chem.* 2020;295(15): 4912–4922.

35. McNees CJ, et al. ASCIZ regulates lesion-specific Rad51 focus formation and apoptosis after methylating DNA damage. *EMBO J.* 2005;24(13):2447–2457.

36. Anjos-Afonso F, et al. Perturbed hematopoiesis in mice lacking ATMIN. *Blood.* 2016;128(16):2017–2021.

37. Fabry MH, et al. piRNA-guided co-transcriptional silencing coopts nuclear export factors. *Elife.* 2019;8, e47999.

38. Dunsch AK, et al. Dynein light chain 1 and a spindle-associated adaptor promote dynein asymmetry and spindle orientation. *J Cell Biol.* 2012;198(6):1039–1054.

39. Santamaria A, et al. The spindle protein CHICA mediates localization of the chromokinesin Kid to the mitotic spindle. *Curr Biol.* 2008;18(10):723–729.

40. Htet ZM, et al. LIS1 promotes the formation of activated cytoplasmic dynein-1 complexes. *Nat Cell Biol.* 2020;22(5):518–525.

41. Scherer J, Yi J, Vallee RB. Role of cytoplasmic dynein and kinesins in adenovirus transport. *FEBS Lett.* 2020;594(12):1838–1847.

42. Nyarko A, Barbar E. Light chain-dependent self-association of dynein intermediate chain. *J Biol Chem.* 2011;286(2):1556–1566.

43. Morgan JL, Song Y, Barbar E. Structural dynamics and multiregion interactions in dynein-dynactin recognition. *J Biol Chem.* 2011;286(45):39349–39359.

44. Jie J, Löhr F, Barbar E. Interactions of yeast dynein with dynein light chain and dynactin: general implications for intrinsically disordered duplex scaffolds in multiprotein assemblies. *J Biol Chem.* 2015;290(39):23863–23874.

45. Gill SR, et al. Dynactin, a conserved, ubiquitously expressed component of an activator of vesicle motility mediated by cytoplasmic dynein. *J Cell Biol.* 1991;115 (6):1639–1650.

46. Nyarko A, et al. Multiple recognition motifs in nucleoporin Nup159 provide a stable and rigid Nup159-Dyn2 assembly. *J Biol Chem.* 2013;288(4):2614–2622.

47. Hall A, et al. Redox-dependent dynamics of a dual thioredoxin fold protein: evolution of specialized folds. *Biochemistry.* 2009;48(25):5984–5993.

48. Brustmann H, Hager M. Nucleoporin 88 expression in normal and neoplastic squamous epithelia of the uterine cervix. *Ann Diagn Pathol.* 2009;13(5):303–307.

49. Itoh G, et al. Nucleoporin Nup188 is required for chromosome alignment in mitosis. *Cancer Sci.* 2013;104(7):871–879.

50. Aizawa H, et al. Impaired nucleoporins are present in sporadic amyotrophic lateral sclerosis motor neurons that exhibit mislocalization of the 43-kDa TAR DNA-binding protein. *J Clin Neurol.* 2019;**15**(1):62.

51. Chou C-C, et al. TDP-43 pathology disrupts nuclear pore complexes and nucleocytoplasmic transport in ALS/FTD. *Nat Neurosci.* 2018;21(2):228–239.

52. Heisel KA, Krishnan VV. NMR based solvent exchange experiments to understand the conformational preference of intrinsically disordered proteins using FG-nucleoporin peptide as a model. *Pept Sci*. 2014;102(1):69–77.

53. Leyrat C, et al. Ensemble structure of the modular and flexible full-length vesicular stomatitis virus phosphoprotein. *J Mol Biol*. 2012;423(2):182–197.

54. Green TJ, et al. Structure of the vesicular stomatitis virus nucleoprotein-RNA complex. *Science*. 2006;313(5785):357–360.

55. Carfi A, Bernabei R, Landi F. Persistent symptoms in patients after acute COVID-19. *JAMA*. 2020;324(6):603–605.

56. World Health Organization. *WHO Coronavirus Disease (COVID-19) Dashboard*. Geneva: World Health Organization; 2020.

57. Bar-On YM, et al. Science forum: SARS-CoV-2 (COVID-19) by the numbers. *Elife*. 2020;9, e57309.

58. Chang CK, et al. The SARS coronavirus nucleocapsid protein—forms and functions. *Antiviral Res*. 2014;103:39–50.

59. Hurst KR, et al. A major determinant for membrane protein interaction localizes to the carboxy-terminal domain of the mouse coronavirus nucleocapsid protein. *J Virol*. 2005;79(21):13285–13297.

60. Lu S, et al. *The SARS-CoV-2 Nucleocapsid Phosphoprotein Forms Mutually Exclusive Condensates With RNA and the Membrane-Associated M Protein*. bioRxiv; 2020.

61. Cubuk J, et al. *The SARS-CoV-2 Nucleocapsid Protein Is Dynamic, Disordered, and Phase Separates with RNA*. BioRxiv; 2020.

62. Iserman C, et al. Genomic RNA elements drive phase separation of the SARS-CoV-2 nucleocapsid. *Mol Cell*. 2020;80(6):1078–1091. e6.

63. Savastano A, et al. Nucleocapsid protein of SARS-CoV-2 phase separates into RNA-rich polymerase-containing condensates. *Nat Commun*. 2020;11(1):1–10.

64. Carlson CR, et al. Phosphoregulation of phase separation by the SARS-CoV-2 N protein suggests a biophysical basis for its dual functions. *Mol Cell*. 2020;80(6): 1092–1103. e4.

65. Tugaeva KV, et al. The mechanism of SARS-CoV-2 nucleocapsid protein recognition by the human 14-3-3 proteins. *J Mol Biol*. 2021;433(8):166875.

66. Perdikari TM, et al. SARS-CoV-2 nucleocapsid protein phase-separates with RNA and with human hnRNPs. *EMBO J*. 2020;39(24), e106478.

67. Wang J, et al. SARS-CoV-2 nucleocapsid protein undergoes liquid–liquid phase separation into stress granules through its N-terminal intrinsically disordered region. *Cell Discov*. 2021;7(1):1–5.

CHAPTER EIGHT

Intrinsic disorder in protein kinase A anchoring proteins signaling complexes

Mateusz Dyla[a,b] and Magnus Kjaergaard[a,b,c],*

[a]Department of Molecular Biology and Genetics, Aarhus University, Aarhus, Denmark
[b]The Danish Research Institute for Translational Neuroscience (DANDRITE), Nordic EMBL Partnership for Molecular Medicine, Aarhus, Denmark
[c]The Danish National Research Foundation Center for Proteins in Memory (PROMEMO), Aarhus, Denmark
*Corresponding author: e-mail address: magnus@mbg.au.dk

Contents

Abstract

Protein kinase A (PKA) is regulated by a diverse class of anchoring proteins known as AKAPs that target PKA to subsets of its activators and substrates. Recently, it was reported that PKA can remain bound to its regulatory subunit after activation in contrast to classical model of activation-by-dissociation. This implies that PKA remains bound to the AKAPs and its substrates, and thus suggest many phosphorylation reactions occur while PKA is physically connected to its substrate. Intra-complex reactions are sensitive to the architecture of the signaling complex, but generally concentration independent. We show that most AKAPs have long intrinsically disordered regions, and suggest that they represent an adaptation for intra-complex phosphorylation. Based on polymer models of the disordered proteins, we predict that the effective concentrations of tethered substrates range from the low millimolar range to tens of micromolar. Based on recent models for intra-complex enzyme reactions, we suggest that the structure of the AKAP signaling complex is likely to be source of allosteric regulation of PKA signaling.

An astonishing number of signaling pathways converge on activation of protein kinase A (PKA). A wealth of G-protein coupled-receptors activate an adenylate cyclase to produce the second messenger cyclic AMP (cAMP). The main target of cAMP is PKA, which is activated to phosphorylate a range of protein substrates with remarkably little crosstalk between pathways.[1] Accordingly, PKA is involved in pathways controlling as diverse physiological functions as heart rate,[2] energy metabolism[1] and memory.[3] How does specificity emerge? The signaling fidelity of PKA cannot be explain by the substrate specificity of the catalytic domain. PKA phosphorylates a simple linear motif that, relative to the phosphorylated residue, consists of positive residues in position -3 and -2 and a hydrophobic residue in position $+1$.[4] The broad intrinsic substrate specificity means that most proteins have potential PKA phosphorylation sites. Nevertheless, PKA is able to achieve signaling fidelity, because its activity if tightly spatially regulated by a class of anchoring proteins known as AKAPs.[5]

AKAPs are a highly diverse family of proteins united by their ability to bind the dimerization-and-docking domain (D/D) of the regulatory subunits of PKA. AKAPs bind the regulatory subunit of PKA via a $\sim$20 residue amphipathic α-helix that docks in a groove of the D/D domain.[6,7] Beyond this helical motif, there is little sequence conservation, and AKAPs are thus defined by a functional trait rather than a shared evolutionary history. AKAPs thus have widely different molecular architectures and interact with widely different proteins, which allow them to compartmentalize PKA signaling by targeting the kinase to a certain cellular location. As cAMP signaling is tightly restricted on the nanoscale,[8,9] the AKAP thus targets PKA to activation by the subset of receptors found in that compartment. Furthermore, as the AKAP also recruits substrates of PKA substrates, it links activation by a certain receptor to phosphorylation of certain substrates, and thus creating signaling specificity from an otherwise promiscuous kinase.[10]

Cyclic AMP activates PKA by binding to the CNB domain in the regulatory subunit. The canonical model for PKA model suggests that cAMP binding leads to dissociation of the catalytic subunit from the regulatory subunit. As the regulatory subunit of PKA binds across the catalytic cleft of the kinase domain, subunit release also leads to disinhibition of the kinase. As the AKAP only binds to the D/D domain of the regulatory subunit of PKA, dissociation of the catalytic subunit also releases it from the subset of substrates that it is tethered to, and is thus free to phosphorylate a broader range of substrates in the local environment. Recently, however, an alternative model was proposed, where full dissociation only occurs at high cAMP levels.[11] At moderate cAMP levels, the holoenzyme adopts a conformation

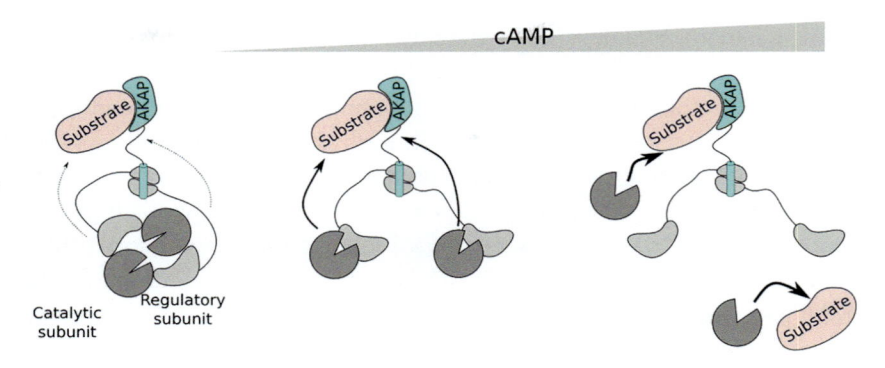

Fig. 1 Generalized model of the complex between a PKA holoenzyme and an AKAP as a function of cAMP stimulation. In the absence of cAMP, the catalytic and regulatory subunits form a compact complex with low enzymatic activity that is linked to a tethered substrate via the AKAP. At moderate cAMP concentration, the holoenzyme opens up to allow moderate catalytic activity, while the kinase remains tethered to the substrate. While the catalytic subunit is tethered to the regulatory subunit, it can only phosphorylate substrates within the reach dictated by the anchoring and regulatory proteins, and is biased toward intra-complex reactions by the high effective concentrations of the tethered substrates. At high cAMP levels, the catalytic subunit is released from the regulatory subunit and the anchoring protein. Under these conditions, the enzyme is free to diffuse and is thus no longer restricted to the immediate vicinity of the anchoring point, or to favor intra-complex reactions.

where the kinase is active while it remains bound to the regulatory subunit and thus also to the AKAP (Fig. 1). PKA also has a low basal phosphorylation activity in the absence of cAMP, which is sufficient to phosphorylate tethered substrates such as the R–subunit[12] and associated AKAPs.[13] In addition to the compact holo-enzyme structures solved by crystallography, electron microscopy has revealed much more open and dynamic conformation of the PKA holoenzyme, where the kinase searches its surroundings for substrates.[11,13,14] Modification of the linkers of the regulatory subunit has shown that the basal phosphorylation rate is highly sensitive to the dynamic ensemble of the holoenzyme.[13]

1. Tethered phosphorylation

PKA is enzymatically active without dissociating from the regulatory subunit at low cAMP stimulation. This has mechanistic consequences for how substrate specificity of PKA is regulated. When catalytic subunit dissociates, PKA phosphorylation occurs in the normal Michaelis–Menten regime, where the rate depends on the substrate concentration. In this regime, AKAP targets the enzyme to a sub-cellular environment and a subset of activating receptors. When the catalytic subunit is active while bound to the regulatory subunit, the enzymatic domain is physically connected to a subset of substrates including the regulatory subunit, the AKAP and any substrates bound to the AKAP. The physically associated substrates occur at a high local abundances, and the kinase is biased toward substrates in its complex. In addition to targeting the kinase to sub-cellular environment and receptors, the AKAP also targets kinase activity to specific substrates. Intra-molecular reactions are generally concentration independent, and are instead regulated by the connection between the reactants. Similarly, high-affinity tethering of PKA will approach the concentration independent regime, where the reaction rate is determined by the molecular architecture of the complex.[15] Low-affinity tethering will likely result in a reaction jointly controlled by the architecture of the complex and the substrate concentration.[16]

How does the architecture of signaling complex affect intra-complex phosphorylation reactions? We have had good kinetic descriptions of untethered enzymatic reactions for more than a century,[17] but no comparable equation has emerged for intra-complex enzymatic reactions until recently.[15] We believe this challenge stems from the challenge of describing the complex architecture quantitatively: It is straight-forward to change the concentration of a substrate in an untethered reaction, but how does one systematically change the architecture of a signaling complex? Similarly, substrate concentrations can readily be compared between different enzyme systems, but how can we quantitatively compare enzyme tethering architectures as that varies as widely as PKA-AKAP signaling complexes? Concentrations determine the encounter rate between enzyme and substrate in untethered systems. In tethered systems, the encounter rate is determined by the architecture of the complex and can be expressed as an effective concentration, C_{eff}. C_{eff} corresponds to the concentration of free substrate that would give the same encounter rate as the tethered substrate. To a first approximation, the connection between enzyme and substrate can thus be reduced to the effective concentration it enforces. Effective concentrations can be measured directly in competition experiments,[18,19] or estimated from physics based models[20,21] most recently through the web server "C_{eff} Calculator."[22] Effective concentrations thus provide a quantitative measure of how closely linked tethered substrates are in signaling complexes.

The kinetics of intra-complex reactions are complicated by the fact that the tethered substrate will be kinetically different from all other substrates.

For stable complexes, the reaction may effectively be single turnover and concentration-independent, whereas weaker complexes can dissociate and bind another substrate and thus undergo several rounds of transiently tethered phosphorylation. Upon a brief burst of cAMP production, PKA will thus be biased toward phosphorylating tethered substrates. In many pathways PKA acts as a switch that only needs to phosphorylate few substrates per kinase to exert a biological function, e.g., activation or inactivation of a receptor. This means that for tethered kinases it is often more relevant to consider the time needed for the first enzymatic turnover rather than "substrates per unit time" as for steady-state reaction. Thus tethered kinase reactions are often best understood in terms of single-turnover or pre-steady-state kinetics.

We recently investigated how the architecture of the anchoring protein regulates tethered PKA reactions in a model system.[15] The model system consisted of the catalytic domain of PKA tethered to a substrate via two disordered linkers of variable length joined by a coiled-coil. The length of the disordered linkers provided a way of continuously tuning the effective concentration of the tethered substrate. Tethering increased the single turnover phosphorylation rate by more than 100-fold compared to the untethered rate. The phosphorylation rate was concentration independent, but depended strongly on both the sequence motif and the length of the linker. Tethered reactions could be described by the following equation:

$$k_{tet} = \frac{k_{phos} C_{eff}}{K_d + C_{eff}} \tag{1}$$

where k_{tet} is the observed single-turnover rate constant, k_{phos} is the rate of phospho-transfer, C_{eff} is the effective concentration of the tethered substrate and K_d is the dissociation constant for substrate binding to the kinase. K_d and k_{phos} account for the differences between substrates, and are approximately equal to K_M and k_{cat} except for near consensus substrates. C_{eff} captures the architecture of the PKA-AKAP signaling complexes. Eq. (1) provides a framework for analyzing pre-steady state phosphorylation reactions in PKA-AKAP signaling complexes (Fig. 2).

2. Intrinsic disorder in PKA-AKAP signaling complexes

Intra-complex reactions require a high degree of flexibility of the kinase anchoring proteins. A rigid PKA-AKAP signaling complexes with fixed interdomain orientation would leave little room for the kinase to

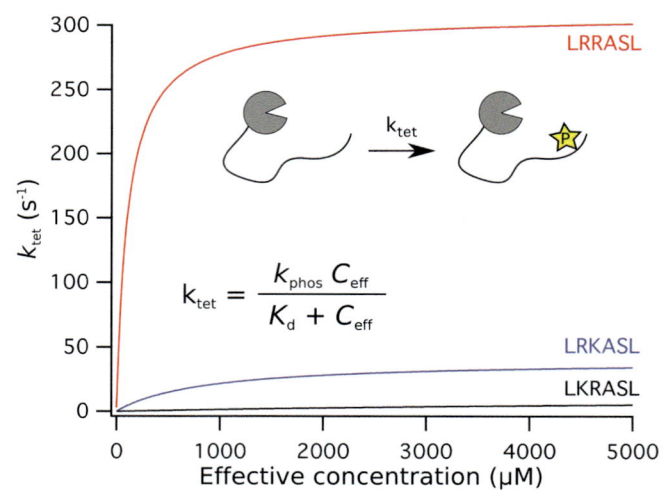

Fig. 2 Simulated single-turnover phosphorylation kinetics for three substrates. The effective concentration (C_{eff}) is defined by the connection between the catalytic domain and the substrate, and can be highly variable. Simulation parameters are based on experimental parameters[15]: LRRASL ($k_{phos} = 307\ s^{-1}$, $K_d = 142\ \mu M$), LRRASL ($k_{phos} = 40\ s^{-1}$, $K_d = 837\ \mu M$) and LRRASL ($k_{phos} = 14.1\ s^{-1}$, $K_d = 8400\ \mu M$).

search for substrates. A rigidly tethered kinase would thus rely on the diffusion of the substrate to initiate contact, and would be unable to process tethered substrates. Signaling events where PKA remains associated to the regulatory subunit pose increased demands on the flexibility of the PKA-AKAP signaling complexes. In protein complexes, flexibility relies on intrinsically disordered regions.

In intrinsically disordered proteins (IDPs) and regions (IDRs), the functional state lacks a fixed three-dimensional structure. As much as 20% of all eukaryotic proteins contain at least one IDR,[23] but IDPs/IDRs are especially enriched in proteins involving, e.g., signaling and regulation.[24] Functional annotation of IDRs in the DisProt database suggests that acting as a disordered linker is the most common function of IDRs.[25] Intrinsically disordered regions are highly accessible to enzymes, and are thus overrepresented in phosphorylation sites.[26] IDPs can be recognized by their amino acid composition[27] as hydrophilic and secondary structure–disrupting amino acids are more common, and hydrophobic amino acids are less common.[28] Due to this sequence bias, IDRs can be predicted from sequences alone with a steadily increasing accuracy.[29] In the following, we will focus on predictions made by the recently developed predictor ODiNpred.[30]

The PKA holoenzyme structure is inherently flexible, which can be attributed to flexible linkers between the dimerization and docking (D/D) domain and the CNB domains.[11,13,31,32] ODiNPred clearly identify these IDRs and confirm the experimental observation that the length of the linkers vary between subunits as RI subunits have ~35 disordered residues whereas RII have ~90 (Fig. 3). Note, that ODiNPRED is able to distinguish between the true flexible linker, and the linker proximal region that is not part of the conserved CNB domain but serves as an allosteric regulator.[33]

3. The prevalence of intrinsic disorder in AKAPs

In addition to the PKA holoenzyme, the AKAP also form a part of the relevant signaling complex. There are sporadic reports of intrinsic disorder in AKAPs,[13,14,34] but due to the low sequence conservation such conclusions cannot be extrapolated to AKAPs as a class. To investigate the prevalence of intrinsic disorder in AKAPs, we used ODiNpred[30] to predict the frequency and distribution of IDRs. As AKAPs are defined functionally rather than by sequence similarity, it is not straight forward to delimit which proteins should be included in such an analysis. For consistency, we restricted our analysis to the consensus variant of human proteins annotated as AKAPs in UNIPROT. This results in a list of 16 proteins ranging in length from ~100 to 4000 residues (Table 1).

To survey the role of prevalence of IDRs in AKAPs, we performed disorder predictions on the AKAPs identified above (Figs. 4–7). To map IDRs on the architecture of the complex, we furthermore identified the known or predicted binding helix for the D/D domain and identified conserved domains from PFAM.[35] All AKAPs tested contained predicted IDRs. The fraction of IDR residues varied from ~17% to 92% with a highly variable distribution through-out the sequence. This re-iterates the point, that unlike many other protein classes AKAPs do not have a unifying molecular architecture. The most disordered AKAPs such as AKAP5 (which is known as AKAP79 in humans or AKAP150 in mouse) and AKAP12 are predicted to be almost entirely disordered. Short regions (<30 residues) of predicted order embedded in IDRs are unlikely to represent stably folded domains, but usually indicate the presence of a short linear motif as these are enriched in hydrophobic residues compared to other IDRs.[36] Other AKAPs have an intermediate degree of predicted disorder such as AKAP1 and AKAP7γ. These proteins contains conserved, folded domains, and the

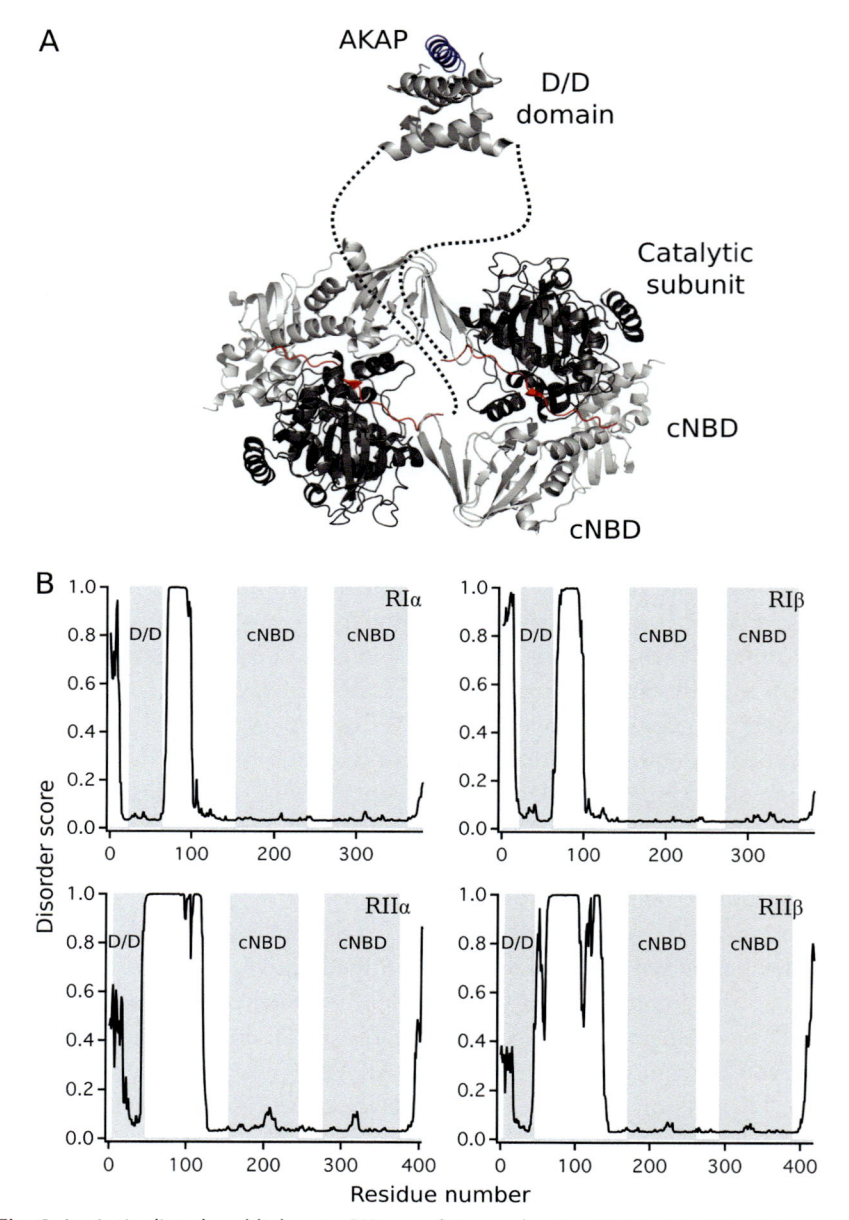

Fig. 3 Intrinsic disordered linkers in PKA regulatory subunits. (A) Model of the tetrameric holoenzyme in an inhibited state. The catalytic subunit (C) in dark gray and the regulatory subunit (R) is shown in light gray with the inhibitory sequence is highlighted in red. The position of the disordered linkers are displayed as dashed lines. The R2:C2 structure is based on (PDB: 3TNP),[56] and D/D:AKAP structure based on (PDB: 2HWN).[7] D/D = dimerization and docking, cNBD = cyclic nucleotide binding domain. (B) Disorder predictions for each of the four human PKA regulatory subunits from ODiNPred[30] including evolutionary information. The position of conserved structured domains according to Pfam[35] are highlighted in gray The predictions unambiguously identify the disordered linker joining the D/D domain and the first cNBD. The structured region adjacent to the linker, but not part of the cNBD, is the auto-inhibitory peptide. UNIPROT sequences used for prediction RIα (P10644), RIβ (P31321), RIIα (P13861), and RIIβ (P31323).

Table 1 Predicted fraction of disordered residues[a] and synonyms for AKAPs.

Name	UNIPROT	Synonyms	Fraction IDR[a]	Length[b]
AKAP1	Q92667-1	AKAP149, PRKA1, D-AKAP-1	59.8%	903
AKAP2	Q9Y2D5-3	KIAA0920, PRKA2, AKAP-KL	72.5%	859
AKAP3	O75969-1	AKAP110, SOB1, FSP95, CT82, PRKA3	24.5%	853
AKAP4	Q5JQC9-1	AKAP82, PRKA4	19.7%	854
AKAP5	P24588-1	AKAP79, H21 (Rodent: AKAP150)	82.2%	427
AKAP6	Q13023-1	AKAP100, KIAA0311, PRKA6	55.5%	2319
AKAP7α	O43687-1	AKAP18α, PRKA7α	65.4%	104
AKAP7γ	Q9P0M2-1	AKAP18γ, PRKA7γ	31.9%	348
AKAP8	O43823-1	AKAP95	67.5%	692
AKAP9	Q99996-2	AKAP350, AKAP450, KIAA0803, PRKA9, yotiao, CG-NAP	17.1%	3907
AKAP10	O43572-1	D-AKAP-2, PRKA10	29.3%	662
AKAP11	Q9UKA4-1	AKAP220, PRKA11	52.3%	1901
AKAP12	Q02952-1	AKAP250, Gravin	92.2%	1782
AKAP13	Q12802-1	BRX, HT31, LBC	65.7%	2813
AKAP14	Q86UN6-1	AKAP28, PRKA14	17.1%	197
AKAP17a	Q02040-1	CXYorf3, DXYS155E, SFRS17A, XE7, PRKA17A, 721P	33.2%	695

[a]Predictions using ODiNpred with evolutionary info enabled.
[b]Based on consensus sequence in UNIPROT.

IDRs take the form of flexible inter-domain linkers and tails. The lowest fraction is found in AKAP9 that does not appear to have any long IDRs, but rather a sprinkling of short IDRs through-out the sequence. These likely serve as short flexible turns and linkers joining folded segments. The high degree of predicted order, but absence of conserved domain, is likely due to a high propensity to form coiled-coils.[37]

In the absence of a conserved molecular architecture, AKAPs seem to be united by a high degree of intrinsic disorder. This mirrors a general abundance of IDRs in scaffolding proteins. In genome-wide surveys of the prevalence of disorder, proteins are typically screened for contiguous regions of

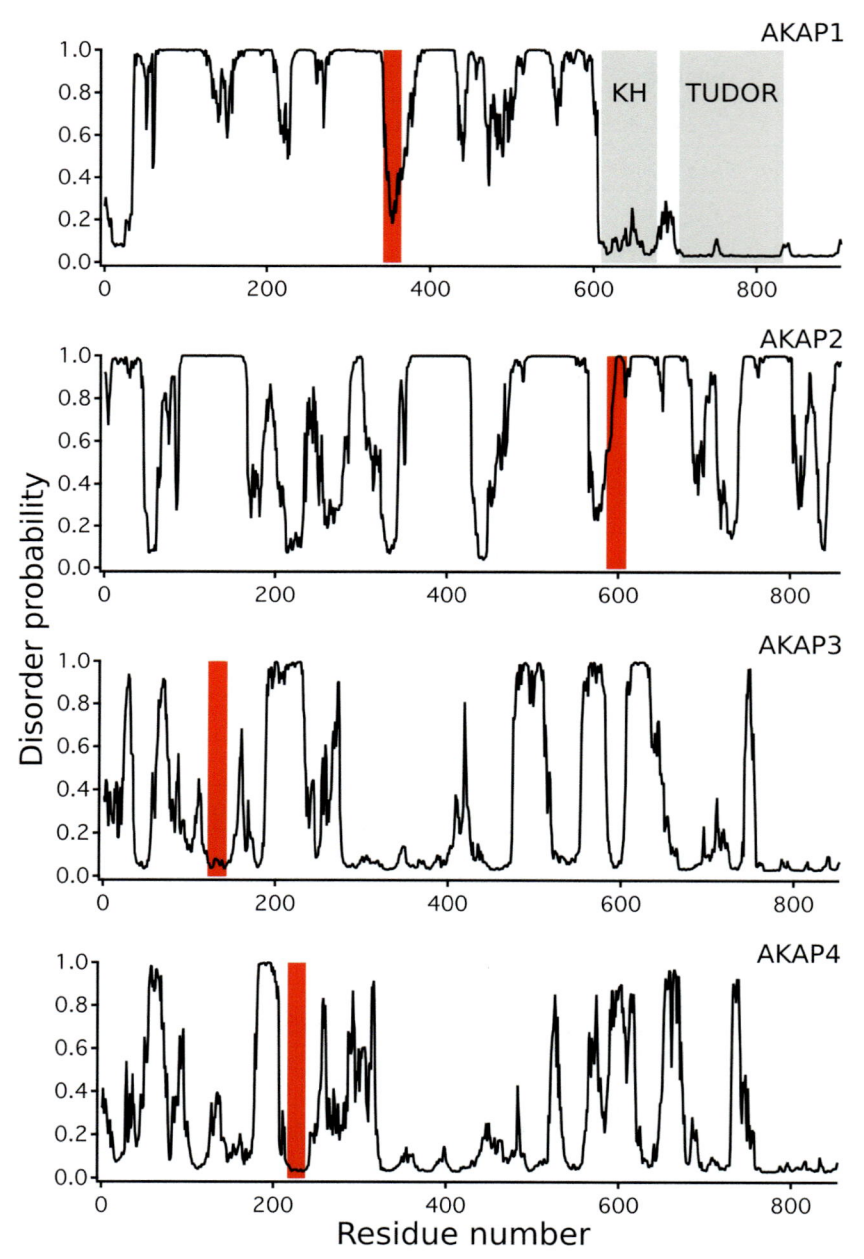

Fig. 4 Prediction of disorder and conserved domains in AKAP1-4. Prediction of the probability of intrinsic disorder by ODiNpred[30] based on the sequence of each AKAP. A score of 1 means near certainty that a regions is disordered, whereas a score of 0 means high propensity to be folded. Conserved domains as indicated in PFAM are shown in gray. The red bar shows the (predicted) location of the docking helix that binds the D/D domain based on either sequence similarity via PFAM or functional annotation in UNIPROT. The sequence used are provided in Table 1. KH, K homology domain.

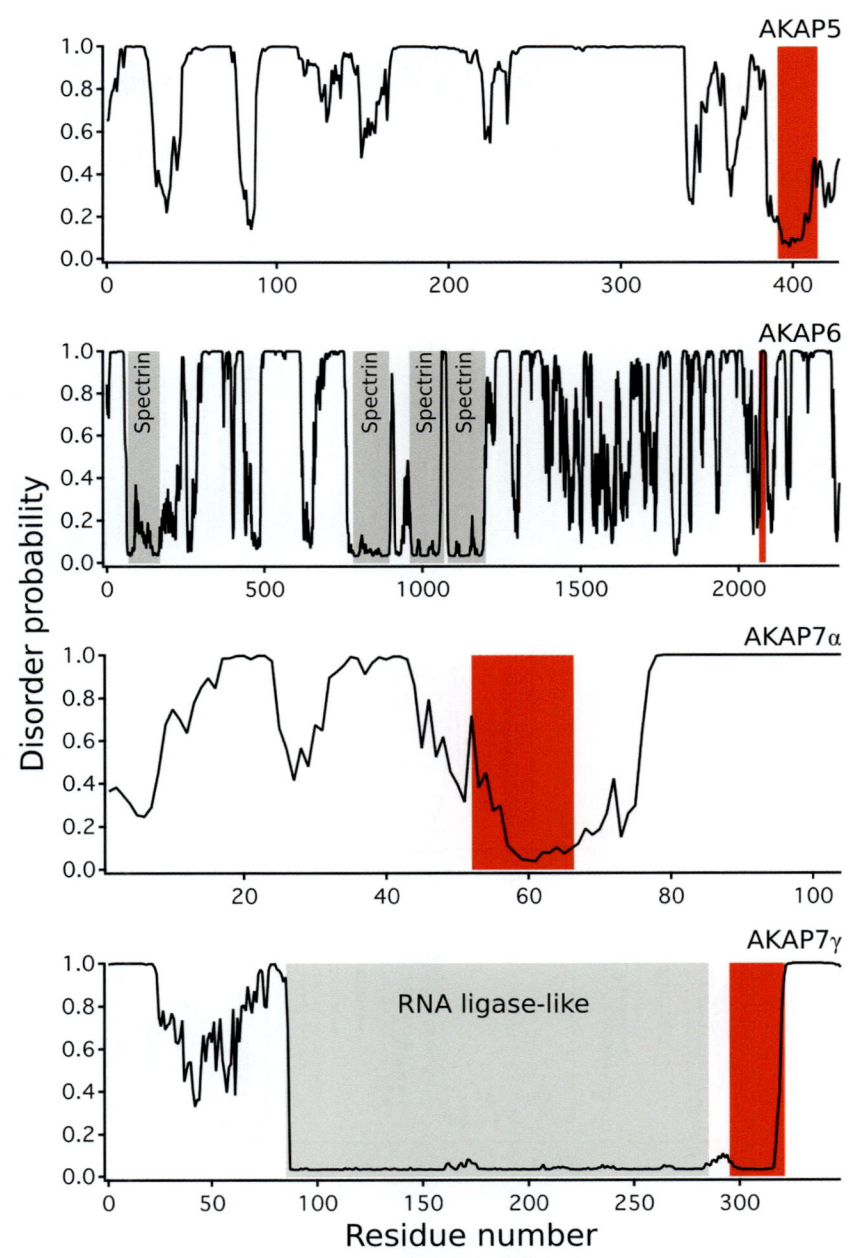

Fig. 5 Prediction of disorder and conserved domains in AKAP5-7γ. Details as in legend of Fig. 4. Note the widely different numbers of residues on the horizontal axes.

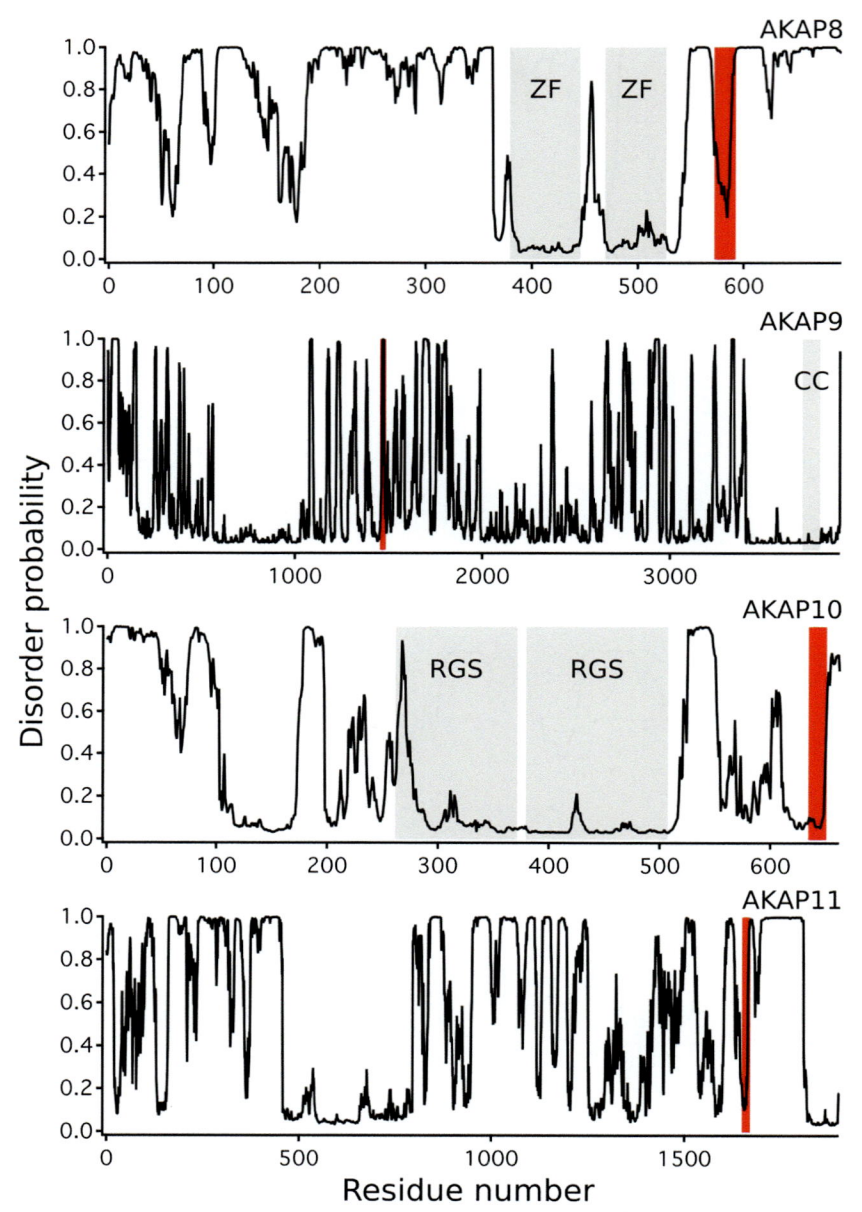

Fig. 6 Prediction of disorder and conserved domains in AKAP8-11. Details as in legend of Fig. 4. Note the widely different numbers of residues on the horizontal axes. CC, coiled-coil;. RGS, Regulator of G protein signaling domain; ZF, zinc finger.

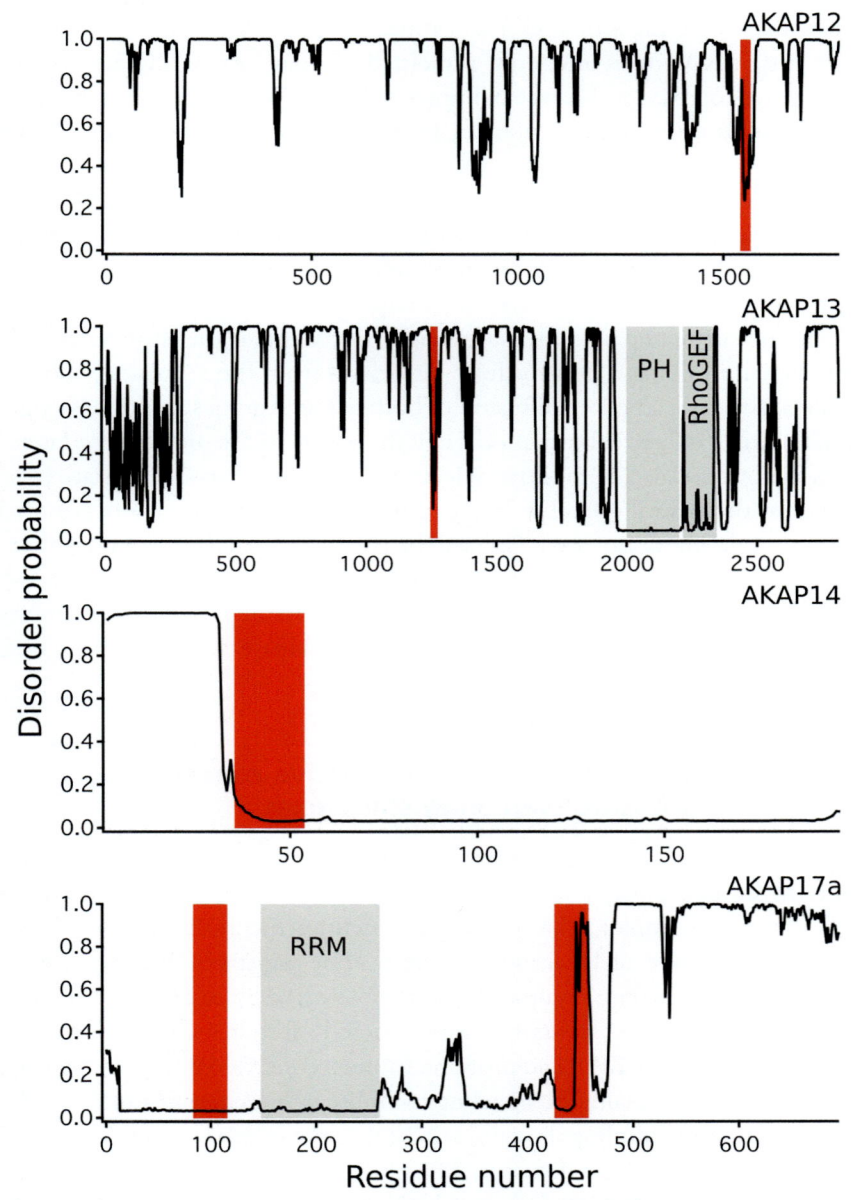

Fig. 7 Prediction of disorder and conserved domains in AKAP12-17a. Details as in legend of Fig. 4. Note the widely different numbers of residues on the horizontal axes. PH, Pleckstrin homology domain; RhoGEF, Rho guanine nucleotide exchange factor domain; RRM, RNA recognition motif.

more than 30 residues of a disorder probability above 0.5. Of the 16 AKAPs analyzed, AKAP7α is the only protein that does not fulfill this criteria. Compared to a 31% prevalence in human proteins,[38] this indicates a substantial over-abundance of intrinsic disorder in AKAPs that calls for an evolutionary explanation.

Two evolutionary hypotheses could explain the prevalence of IDRs in AKAPs: The neutral selection scenario suggests that IDRs do not harm AKAPs and are thus evolutionarily tolerated. Such a scenario can be envisioned if the role the AKAP is to target PKA to sub-cellular compartments and certain activators. Targeting to activators is the most important function if the kinase domain fully dissociates from the regulatory subunit upon activation. The free diffusion of both the activating second messenger and the free catalytic subunit means that the nature of the molecular connection is not critical. The positive selection scenario occurs if IDRs are beneficial to the AKAP, which is likely to happen in pathways where tethered phosphorylation dominates. This will be the case when the basal phosphorylation in the absence of cAMP is important, or when cAMP does not reach sufficiently high levels to lead to full dissociation. In such cases, the flexibility afforded by the IDRs in the AKAP allows the kinase domain to search its surroundings for substrates.

4. Modeling the reach and effective concentration of PKA-AKAP signaling complexes

The architecture of the signaling complex determines key parameters of tethered phosphorylation: For substrates that are not located directly in the signaling complex, the architecture defines the reach of the kinase, and thus the zone of local kinase activity. For substrates that are directly bound to the signaling complex, the architecture defines an effective concentration of tethered substrates. The tethered substrates include the regulatory subunit, the AKAP and proteins bound to the AKAP (Fig. 1). Starting from the kinase domain, the signaling complex architecture first includes the cyclic nucleotide binding domains (cNBD) of the regulatory subunit which is bound to the catalytic domains. The cNBD domains are tethered via the regulatory subunit flexible linker, which in the human regulatory subunits vary from ~35 to 90 flexible residues. The other end on the linker connects to the D/D domain, which binds the AKAP helix. This represents the shared component of the PKA-AKAP signaling complex, whereas the different AKAPs and the substrates bound to them are unique.

4.1 Estimating the reach of AKAP signaling complexes

The intrinsic disorder in AKAP signaling complexes make them a challenging target for structural characterization. The functional states of proteins of these complexes are unlikely to be represented by any single conformation, but should rather be represented as an ensemble of interconverting structures. NMR spectroscopy is usually the method of choice for structural investigations of disordered proteins, but the intact single-complexes are too large to be practical for NMR. Single-particle negative-stain electron microscopy has emerged as the most successful method for structural investigation of dynamic AKAP-PKA complexes.[11,13,14] While negative-stain electron microscopy cannot reach the same resolution as cryo-EM, the higher contrast in single micrographs allows low resolution structure determination based on a single particle by recording a tilt-series. This is advantageous for dynamic multi-domain proteins[39] as it is possible to reconstruct individual conformers without averaging different particles. Structural investigation of the complex between AKAP18γ and the PKA holoenzyme showed a highly flexible distribution of interdomain distances, and suggested that the kinase domain is restricted to a zone of ∼16 nm from the AKAP.[13] AKAP18γ only has a ∼10 residue linker between the central domain and the D/D binding helix,[34] so this radius represents the minimal reach of an AKAP tethered PKA. In contrast, the longer and more disordered AKAP5 (AKAP79) restrains the kinase to 25 nm from a tethered substrate.[11]

Single-particle negative-stain electron microscopy requires expensive, specialized equipment and is likely to remain a bottleneck for the foreseeable future. However, in many cases experimental characterization of the structural ensemble may not be necessary to understand intra-complex kinase reactions. Structures of conserved domains in AKAP-PKA complexes have either been solved or can be modeled based on homology. The main challenge is thus to describe how the IDR linkers join these modules into a functional signaling complex ensemble. Residue-specific properties of IDRs require characterization by NMR or molecular modeling,[40] but the overall properties such as end-to-end distances can usually be described by simple models developed for organic polymers.[41–44] The length of an IDR thus scales with the number of residues following a homopolymer scaling law:

$$R \propto N^{\nu} \tag{2}$$

where R is the end-to-end distance, N is the number of flexible residues, and ν is the polymer scaling exponent that depends on the compaction of

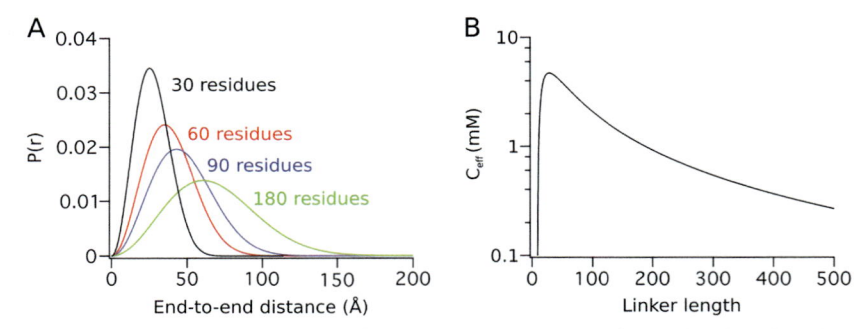

Fig. 8 Modeling of the reach and effective concentrations enforced by disordered proteins. (A) The reach of a disordered anchoring protein can be modeled from the end-to-end distance using a worm-like chain (WLC).[20] With increasing linker length, the reach and most probable end-to-end distance also increases. (B) The effective concentration is strongly dependent on chain length. When the linker is short compared to the distance it needs to span, a small increase in the linker length leads to a large increase in the effective concentration. Beyond the length needed to span the join the attachment sites without strain, the effective concentration decreases according to a polymer scaling law.[19,48] WLC simulations were performed using the "C_{eff} Calculator" app.[22] WLC parameters were chosen to represent an IDP of average sequence composition ($L_p = 4$ Å, monomer length = 3.8 Å). In (B), the distance between attachment sites was chosen to roughly match the dimensions of a kinase domain (30 Å).

the chain. For disordered segments ν may vary from about 0.4 to 0.7,[19,45] but most IDPs can be described using a value of about 0.5–0.6.[43] This value describes an ensemble average of a wide distribution of end-to-end distances. The distribution of end-to-end distances can be modeled using, e.g., a Gaussian chain or worm-like chain,[20,46,47] and the latter was recently made available as a convenient web applet.[22] In Fig. 8A, we have modeled the end-to-end distance distribution of linkers of different lengths using a worm-like chain. Each linker length samples a wide range of end-to-end distances, which suggests that the kinase domain effectively sweeps for potential substrates. A 30 residue linker (similar to RI subunits) has a predicted mean end-to-end distance of ∼30 Å and at most extends to 50 Å. A 90 residue linker (similar to RII subunits) has a predicted mean end-to-end distance of ∼50 Å and at most extends to 100 Å. The length of a disordered linker increases rapidly with the number of residues for short segments, but relatively modestly for longer linkers. With an ν of 0.5, a doubling of the a end-to-end distance requires four times as many flexible residues. Additional extension of the signaling complex has a modest effect on the mean and total dimension. The signaling complex also contains folded domains that add to the dimensions of the complex. Statistical mechanics suggest that a folded domain in disorder linkers can be approximated by a rigid rod with a length corresponding to the distance between its termini.[21] This thus represent the highest extension of the reach of the complex, although usually the dimensions will be lower than the sum of the disordered and folded parts.

4.2 Estimating the effective concentrations of tethered substrates

The architecture of the AKAP-PKA signaling complex also defines the effective concentrations of tethered substrates. As intra-complex reactions are concentration independent, the effective concentration governs phosphorylation rates (Fig. 2), and depends solely on the complex architecture.[15] As illustrated in Fig. 2, the phosphorylation rates reach a plateau at high effective concentrations. This suggests that depending on the architecture, intra-complex phosphorylation reactions may either exist in quite different regimes: At high effective concentrations, the reaction is relatively insensitive to changes in the connection, because the reaction rate is nearly saturated. Alternatively, if the effective concentration is below the saturation point of the substrates, a change in the connecting architecture that changes the effective concentration will also affect the phosphorylation rates. In the following, we will estimate effective concentrations for substrates in the AKAP-PKA complex. By necessity, any numbers quoted should be considered a ballpark estimate.

The potential substrate present at the highest effective concentration is the linker of the regulatory subunit. Regulatory subunit RIIα has a near consensus PKA site S99 (in P13861), whose phosphorylation has been reported to affect its affinity toward the catalytic subunit and the AKAP.[49] This serine is connected to the cNBD by $\sim$20 flexible residues (Fig. 3). The structure of the complex between the catalytic and regulatory subunits of PKA (PDB: 2qvs, 2qcs),[50,51] suggests that the attachment site of the linker is close in space to the active site with a spacing of at most 20 Å. When the linker is longer than needed to bridge the binding sites, effective concentrations can also be approximated by a polymer scaling law dependent on the length of the flexible linkers (Fig. 8B).[19] The effective concentration at this site is predicted to be 14 mM, well above the half-saturation C_{eff} of 0.14 mM observed for the similar Kemptide.[15] This site is thus effectively saturated and will be rapidly phosphorylated by even a tiny fraction of active kinase, which is likely important for its basal phosphorylation in the absence of cAMP.[12] At the end of the 90-residue disordered linker in RII, the effective concentration is predicted to drop to about 3 mM. It is thus likely that in the R-domain, effective concentrations are so high that even suboptimal sub-optimal substrates could be phosphorylated.

Next in line viewed from the tethered kinase domain is the AKAP. The complex between the D/D domain and AKAP helix impose a ~30 Å rigid link between the terminus of the R-subunit linker and the most proximal part of the AKAP. It is not clear how such a rigid body in a flexible linker will affect the effective concentration. Recently, it was suggested that the size of the folded domain should be added to linker length to calculate the new combined diffusion volume which determined the effective concentration. For a 90 residues linker with predicted mean end-to-end distance of 43 Å, a 30 Å increase in the distance would increase the diffusion volume (and thus decrease the effective concentration) by ~fivefold. This suggests that putative substrates in the AKAP are at most present at effective concentrations around a millimolar. The non-conserved architecture of AKAPs makes it difficult to put a lower bound on the effective concentrations experienced by AKAPs in general. We picked AKAP5 (AKAP79) as an example as it is highly disordered (Fig. 5), and is likely to be modeled well by simple polymer models. The N-terminus of AKAP79 is thus ~390 residues from the attachment site to the AKAP helix, and contains PKA phosphorylation sites that regulate its membrane interaction.[52] Combined with a 90-residue RII-subunit linker, this results in a total linker length of 480 residues and predicts an effective concentration of ~300 µM. As some AKAPs are larger, this suggests that substrate motifs in a tethered AKAP are present at effective concentrations in the range of 1 mM to 100 µM. Substrates tethered to AKAPs are likely to fall at the lower end of this range. This suggests that only optimal PKA substrate motifs proximal to the anchoring helix will experience saturation of their single-turnover rates. Most substrates, however, will be present at effective concentrations below their half-saturation points, and their phosphorylation rates will thus be sensitive to the architecture of the PKA-AKAP signaling complex.

5. Allosteric regulation of tethered reactions by AKAP structure

Allostery has classically been understood to propagate through the folded domains, but the rise to prominence of intrinsic disorder suggests that the concept should be broadened to include events mediated by disordered regions.[53,54] The definition of allostery involves regulatory interactions that are spatially separated, e.g., binding sites on opposite sides of a folded domain. In the context of tethered kinases, allostery may be defined as regulatory events that affect the phosphorylation rate without involving either the substrate or the enzyme domain. Changes in the dynamic architecture of the signaling complex are an excellent example of such allostery. As long as the kinase remains tethered to the regulatory subunit, any change in structure of PKA-AKAP complex will affect the contact probability between the kinase and tethered substrates and thus the phosphorylation rate.

Intrinsically disorder regions are much more malleable than folded domains, which makes them ideal for allosteric regulation (Fig. 9A). The simplest example may be that in the absence of a folded structure that needs to be conserved, the length of a disordered linker can readily by varied by, e.g., alternative splicing. For similar reasons, alternative splicing is much more common in intrinsically disordered regions.[55] Splicing in linker segments in AKAPs thus presents a straight-forward way to tune the effective concentration of tethered substrates. For disordered linkers of the same length the end-to-end distance distribution and the effective concentration they enforce varies depending on the compaction of the linker. The charge density and distribution of an IDR are the principal determinants of IDP compaction, as especially patches of concentrated charges can lead to collapse through interactions with patches of opposing charges.[41,44,45] In relation to allosteric regulation, especially multisite phosphorylation is of interest, as it can provide a condition-dependent change in the charge distribution and thus lead to linker compaction. Finally, binding of protein ligands in an IDR can change its end-to-end contact probability. Of special interest are multivalent interactions that bring together binding motifs that would otherwise be far apart in the sequence, and thus resulting in a large effective shortening of the linker. All these events lead to changes in the effective concentration, are independent of the kinase-substrate interactions, and should thus be regarded as allosteric regulation.

As estimated above, most PKA substrates are tethered at effective concentrations below the half-saturation point of all but the most optimal substrates. In general, it can thus be assumed that a tightening of the PKA-AKAP complexes will lead to increased phosphorylation rates. Furthermore, it may lead to a broadening of the substrate usage as very good substrates may approach saturation, and thus no longer increase in rate with increasing concentration. In Fig. 9B, we consider a change from lower to the higher end of the spectrum of effective concentrations of tethered substrates

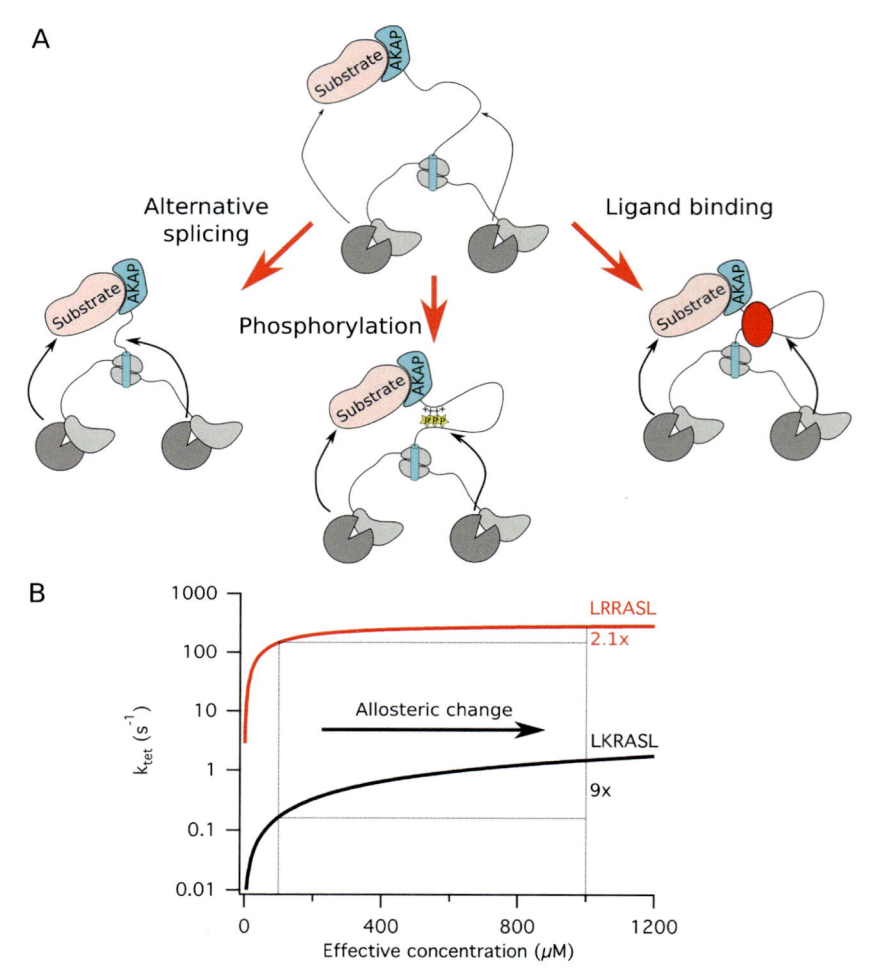

Fig. 9 Potential mechanism of allosteric regulation via the AKAP structural ensemble. (A) Tethered phosphorylation by PKA depends on the connection between the enzyme and substrate, and thus the structure of the AKAP. Phosphorylation rates will thus increase following changes in AKAP structure that shorten this connection. This can be achieved by shortening the length of a disordered regions by alternative splicing, by increasing the compaction of the linker region through post-translational modifications, or by ligand binding. (B) Modeling the effect of an allosteric change to the signaling complex architecture that increases the effective concentration from 100 μM to 1 mM. The phosphorylation rate of all tethered complexes will increase, but poor substrates will increase proportionally more [15].

for a good and a relatively bad substrate. The single-turnover rate increases for both substrates, but more for the worse substrate. As these are the extreme values, it likely also suggest that a factor of 10 might be the largest effect one could expect from an allosteric modification of the PKA-AKAP complex.

6. Conclusions

The discovery that PKA can be activated while remaining tethered to its regulatory subunit and anchoring protein, opens up new research avenues, where intra-complex phosphorylation is regulated by the architecture of PKA-AKAP signaling complexes. Intrinsically disordered regions are common in AKAPs, which suggests that descriptions of PKA-AKAP should aim for ensemble-function rather than structure-function relationships. Both experimental and theoretical investigations are needed to understand how these dynamic complexes allosterically regulate phosphorylation reactions.

References

1. Taskén K, Aandahl EM. Localized effects of cAMP mediated by distinct routes of protein kinase A. *Physiol Rev.* 2004;84(1):137–167. https://doi.org/10.1152/physrev.00021.2003.
2. Zaccolo M. cAMP signal transduction in the heart: understanding spatial control for development of novel therapeutic strategies. *Br J Pharmacol.* 2009;158:50–60.
3. Abel T, Nguyen PV, Barad M, Deuel TAS, Kandel ER, Bourtchouladze R. Genetic demonstration of a role for PKA in the late phase of LTP and in hippocampus-based long-term memory. *Cell.* 1997;88(5):615–626. https://doi.org/10.1016/S0092-8674(00)81904-2.
4. Kemp BE, Graves DJ, Benjamani E, Krebs EG. Role of multiple basic residues in determining the substrate specificity of cyclic AMP-dependent protein kinase. *J Biol Chem.* 1977;252(4):4888–4894.
5. Wong W, Scott JD. AKAP signalling complexes: focal points in space and time. *Nat Rev Mol Cell Biol.* 2004;5(12):959–970. https://doi.org/10.1038/nrm1527.
6. Newlon MG. A novel mechanism of PKA anchoring revealed by solution structures of anchoring complexes. *EMBO J.* 2001;20(7):1651–1662. https://doi.org/10.1093/emboj/20.7.1651.
7. Kinderman FS, Kim C, von Daake S, et al. A dynamic mechanism for AKAP binding to RII isoforms of cAMP-dependent protein kinase. *Mol Cell.* 2006;24(3):397–408. https://doi.org/10.1016/j.molcel.2006.09.015.
8. Mo GCH, Ross B, Hertel F, et al. Genetically encoded biosensors for visualizing live-cell biochemical activity at super-resolution. *Nat Methods.* 2017;14(4):427–434. https://doi.org/10.1038/nmeth.4221.
9. Bock A, Annibale P, Konrad C, et al. Optical mapping of cAMP signaling at the nanometer scale. *Cell.* 2020;182(6):1519–1530.e17. https://doi.org/10.1016/j.cell.2020.07.035.

10. Miller CJ, Turk BE. Homing in: mechanisms of substrate targeting by protein kinases. *Trends Biochem Sci*. 2018;43(5):380–394. https://doi.org/10.1016/j.tibs.2018.02.009.

11. Smith FD, Esseltine JL, Nygren PJ, et al. Local protein kinase A action proceeds through intact holoenzymes. *Science*. 2017;356(6344):1288–1293. https://doi.org/10.1126/science.aaj1669.

12. Isensee J, Kaufholz M, Knape MJ, et al. PKA-RII subunit phosphorylation precedes activation by cAMP and regulates activity termination. *J Cell Biol*. 2018;217(6): 2167–2184. https://doi.org/10.1083/jcb.201708053.

13. Smith FD, Reichow SL, Esseltine JL, et al. Intrinsic disorder within an AKAP-protein kinase A complex guides local substrate phosphorylation. *Elife*. 2013;2:e01319. https://doi.org/10.7554/eLife.01319.

14. Nygren PJ, Mehta S, Schweppe DK, et al. Intrinsic disorder within AKAP79 fine-tunes anchored phosphatase activity toward substrates and drug sensitivity. *Elife*. 2017;6: e30872. https://doi.org/10.7554/eLife.30872.

15. Dyla M, Kjaergaard M. Intrinsically disordered linkers control tethered kinases via effective concentration. *Proc Natl Acad Sci*. 2020;117(35):21413–21419. https://doi.org/10.1073/pnas.2006382117.

16. Speltz EB, Zalatan JG. The relationship between effective molarity and affinity governs rate enhancements in tethered kinase–substrate reactions. *Biochemistry*. 2020;59 (23):2182–2193. https://doi.org/10.1021/acs.biochem.0c00205.

17. Michaelis L, Menten ML. Die kinetik der invertinwirkung. *Biochem Z*. 1913;49:333–369.

18. Krishnamurthy VM, Semetey V, Bracher PJ, Shen N, Whitesides GM. Dependence of effective molarity on linker length for an intramolecular protein–ligand system. *J Am Chem Soc*. 2007;129(5):1312–1320. https://doi.org/10.1021/ja066780e.

19. Sørensen CS, Kjaergaard M. Effective concentrations enforced by intrinsically disordered linkers are governed by polymer physics. *PNAS*. 2019;116(46):23124–23131. https://doi.org/10.1101/577536.

20. Zhou H-X. The affinity-enhancing roles of flexible linkers in two-domain DNA-binding proteins. *Biochemistry*. 2001;40(50):15069–15073. https://doi.org/10.1021/bi015795g.

21. Van Valen D, Haataja M, Phillips R. Biochemistry on a leash: the roles of tether length and geometry in signal integration proteins. *Biophys J*. 2009;96(4):1275–1292. https://doi.org/10.1016/j.bpj.2008.10.052.

22. Kjaergaard M, Glavina J, Chemes LB. Predicting the effect of disordered linkers on effective concentrations and avidity with the "Ceff calculator" app. *Methods Enzymol*. 2021;647:145–171. https://doi.org/10.1016/bs.mie.2020.09.012.

23. Peng Z, Yan J, Fan X, et al. Exceptionally abundant exceptions: comprehensive characterization of intrinsic disorder in all domains of life. *Cell Mol Life Sci*. 2015;72 (1):137–151. https://doi.org/10.1007/s00018-014-1661-9.

24. Xie H, Vucetic S, Iakoucheva LM, et al. Functional anthology of intrinsic disorder. 1. Biological processes and functions of proteins with long disordered regions. *J Proteome Res*. 2007;6:1882–1898.

25. Hatos A, Hajdu-Soltész B, Monzon AM, et al. DisProt: intrinsic protein disorder annotation in 2020. *Nucleic Acids Res*. 2019;48(D1):D269–D276. https://doi.org/10.1093/nar/gkz975.

26. Iakoucheva LM, Radiovojac P, Brown CJ, et al. The importance of intrinsic disorder for protein phosphorylation. *Nucleic Acids Res*. 2004;32(3):1037–1049. https://doi.org/10.1093/nar/gkh253.

27. Uversky VN, Gillespie JR, Fink AL. Why are "natively unfolded" proteins unstructured under physiologic conditions? *Proteins Struct Funct Genet*. 2000;41:415–427.

28. Dunker AK, Lawson JD, Brown CJ, et al. Intrinsically disordered protein. *J Mol Graph Model*. 2001;19(1):26–59. https://doi.org/10.1016/S1093-3263(00)00138-8.

29. Nielsen JT, Mulder FAA. Quality and bias of protein disorder predictors. *Sci Rep*. 2019;9 (1):5137. https://doi.org/10.1038/s41598-019-41644-w.

30. Dass R, Mulder FAA, Nielsen JT. ODiNPred: comprehensive prediction of protein order and disorder. *Sci Rep*. 2020;10(1):14780. https://doi.org/10.1038/s41598-020-71716-1.

31. Vigil D, Blumenthal DK, Heller WT, et al. Conformational differences among solution structures of the type Iα, IIα and IIβ protein kinase a regulatory subunit homodimers: role of the linker regions. *J Mol Biol*. 2004;337(5):1183–1194. https://doi.org/10.1016/j.jmb.2004.02.028.

32. Hamuro Y, Anand GS, Kim JS, et al. Mapping intersubunit interactions of the regulatory subunit (RIα) in the type I holoenzyme of protein kinase a by amide hydrogen/deuterium exchange mass spectrometry (DXMS). *J Mol Biol*. 2004;340(5):1185–1196. https://doi.org/10.1016/j.jmb.2004.05.042.

33. Akimoto M, Selvaratnam R, McNicholl ET, Verma G, Taylor SS, Melacini G. Signaling through dynamic linkers as revealed by PKA. *Proc Natl Acad Sci U S A*. 2013;110 (35):14231–14236. https://doi.org/10.1073/pnas.1312644110.

34. Gold MG, Smith FD, Scott JD, Barford D. AKAP18 contains a phosphoesterase domain that binds AMP. *J Mol Biol*. 2008;375(5):1329–1343. https://doi.org/10.1016/j.jmb.2007.11.037.

35. El-Gebali S, Mistry J, Bateman A, et al. The Pfam protein families database in 2019. *Nucleic Acids Res*. 2019;47(D1):D427–D432. https://doi.org/10.1093/nar/gky995.

36. Davey NE, Van Roey K, Weatheritt RJ, et al. Attributes of short linear motifs. *Mol Biosyst*. 2012;8(1):268–281. https://doi.org/10.1039/C1MB05231D.

37. Lin JW, Wyszynski M, Madhavan R, Sealock R, Kim JU, Sheng M. Yotiao, a novel protein of neuromuscular junction and brain that interacts with specific splice variants of NMDA receptor subunit NR1. *J Neurosci*. 1998;18(6):2017–2027. https://doi.org/10.1523/JNEUROSCI.18-06-02017.1998.

38. Ward JJ, Sodhi JS, McGuffin LJ, Buxton BF, Jones DT. Prediction and functional analysis of native disorder in proteins from the three kingdoms of life. *J Mol Biol*. 2004;337 (3):635–645. https://doi.org/10.1016/j.jmb.2004.02.002.

39. Zhang X, Zhang L, Tong H, et al. 3D structural fluctuation of IgG1 antibody revealed by individual particle electron tomography. *Sci Rep*. 2015;5(1):9803. https://doi.org/10.1038/srep09803.

40. Jensen MR, Ruigrok RW, Blackledge M. Describing intrinsically disordered proteins at atomic resolution by NMR. *Curr Opin Struct Biol*. 2013;23(3):426–435. https://doi.org/10.1016/j.sbi.2013.02.007.

41. Das RK, Pappu RV. Conformations of intrinsically disordered proteins are influenced by linear sequence distributions of oppositely charged residues. *Proc Natl Acad Sci U S A*. 2013;110(33):13392–13397. https://doi.org/10.1073/pnas.1304749110.

42. Hofmann H, Soranno A, Borgia A, Gast K, Nettels D, Schuler B. Polymer scaling laws of unfolded and intrinsically disordered proteins quantified with single-molecule spectroscopy. *Proc Natl Acad Sci U S A*. 2012;109(40):16155–16160. https://doi.org/10.1073/pnas.1207719109.

43. Marsh JA, Forman-Kay JD. Sequence determinants of compaction in intrinsically disordered proteins. *Biophys J*. 2010;98(10):2383–2390. https://doi.org/10.1016/j.bpj.2010.02.006.

44. Mao AH, Crick SL, Vitalis A, Chicoine CL, Pappu RV. Net charge per residue modulates conformational ensembles of intrinsically disordered proteins. *Proc Natl Acad Sci U S A*. 2010;107(18):8183–8188. https://doi.org/10.1073/pnas.0911107107.

45. Müller-Späth S, Soranno A, Hirschfeld V, et al. Charge interactions can dominate the dimensions of intrinsically disordered proteins. *Proc Natl Acad Sci U S A*. 2010;107 (33):14609–14614.

46. Zhou H-X. Quantitative account of the enhanced affinity of two linked scFvs specific for different epitopes on the same antigen. *J Mol Biol.* 2003;329(1):1–8. https://doi.org/10.1016/S0022-2836(03)00372-3.

47. Zhou H-X. Polymer models of protein stability, folding, and interactions. *Biochemistry.* 2004;43(8):2141–2154. https://doi.org/10.1021/bi036269n.

48. Sørensen CS, Jendroszek A, Kjaergaard M. Linker dependence of avidity in multivalent interactions between disordered proteins. *J Mol Biol.* 2019;431(24):4784–4795. https://doi.org/10.1016/j.jmb.2019.09.001.

49. Manni S, Mauban JH, Ward CW, Bond M. Phosphorylation of the cAMP-dependent protein kinase (PKA) regulatory subunit modulates PKA-AKAP interaction, substrate phosphorylation, and calcium signaling in cardiac cells. *J Biol Chem.* 2008;283 (35):24145–24154. https://doi.org/10.1074/jbc.M802278200.

50. Wu J, Brown SHJ, von Daake S, Taylor SS. PKA type II holoenzyme reveals a combinatorial strategy for isoform diversity. *Science.* 2007;318(5848):274–279. https://doi.org/10.1126/science.1146447.

51. Kim C, Cheng CY, Saldanha SA, Taylor SS. PKA-I holoenzyme structure reveals a mechanism for cAMP-dependent activation. *Cell.* 2007;130(6):1032–1043. https://doi.org/10.1016/j.cell.2007.07.018.

52. Dell'Acqua ML, Faux MC, Thorburn J, Thorburn A, Scott JD. Membrane-targeting sequences on AKAP79 bind phosphatidylinositol-4,5-bisphosphate. *EMBO J.* 1998; 17(8):2246–2260. https://doi.org/10.1093/emboj/17.8.2246.

53. Tompa P. Multisteric regulation by structural disorder in modular signaling proteins: an extension of the concept of allostery. *Chem Rev.* 2014;114(13):6715–6732. https://doi.org/10.1021/cr4005082.

54. White JT, Li J, Grasso E, Wrabl JO, Hilser VJ. Ensemble allosteric model: energetic frustration within the intrinsically disordered glucocorticoid receptor. *Philos Trans R Soc B.* 2018;373(1749):20170175. https://doi.org/10.1098/rstb.2017.0175.

55. Buljan M, Chalancon G, Dunker AK, et al. Alternative splicing of intrinsically disordered regions and rewiring of protein interactions. *Curr Opin Struct Biol.* 2013;23 (3):443–450. https://doi.org/10.1016/j.sbi.2013.03.006.

56. Zhang P, Smith-Nguyen EV, Keshwani MM, Deal MS, Kornev AP, Taylor SS. Structure and allostery of the PKA RIIb tetrameric holoenzyme. *Science.* 2012;335 (6069):6.

CHAPTER NINE

Protein intrinsic disorder on a dynamic nucleosomal landscape

Sveinn Bjarnason[a,†], Sarah F. Ruidiaz[a,†], Jordan McIvor[b,†], Davide Mercadante[b,*], and Pétur O. Heidarsson[a,*]

[a]Department of Biochemistry, Science Institute, University of Iceland, Reykjavík, Iceland
[b]School of Chemical Science, University of Auckland, Auckland, New Zealand
*Corresponding authors: e-mail address: davide.mercadante@auckland.ac.nz; pheidarsson@hi.is

Contents

Abstract

The complex nucleoprotein landscape of the eukaryotic cell nucleus is rich in dynamic proteins that lack a stable three-dimensional structure. Many of these intrinsically disordered proteins operate directly on the first fundamental level of genome compaction: the nucleosome. Here we give an overview of how disordered interactions with and within nucleosomes shape the dynamics, architecture, and epigenetic regulation of the genetic material, controlling cellular transcription patterns. We highlight experimental and computational challenges in the study of protein disorder and illustrate how integrative approaches are increasingly unveiling the fine details of nuclear interaction

---

[†] These authors contributed equally.

Progress in Molecular Biology and Translational Science, Volume 183
ISSN 1877-1173
https://doi.org/10.1016/bs.pmbts.2021.06.006

networks. We finally dissect sequence properties encoded in disordered regions and assess common features of disordered nucleosome-binding proteins. As drivers of many critical biological processes, disordered proteins are integral to a comprehensive molecular view of the dynamic nuclear milieu.

Abbreviations

BD	Brownian dynamics
CG	coarse-grained
DBD	DNA binding domain
FRET	Förster resonance energy transfer
IDP	intrinsically disordered protein
IDR	intrinsically disordered region
MC	Monte Carlo
MD	molecular dynamics
NCP	nucleosome core particle
NMR	nuclear magnetic resonance
NRL	nucleosome repeat length
PTM	post-translational modification
TAD	transactivation domain
TF	transcription factor

1. Introduction

As organisms become increasingly complex, so too must they evolve a more sophisticated molecular alphabet. The recent discovery of proteins that can adopt multiple structural states is one way of addressing this complexity and it has dramatically changed our view of the protein structure-function paradigm.[1,2] Intrinsically disordered proteins (IDPs) either do not contain any well-defined secondary structure element or have long unstructured regions (IDRs), and they fluctuate between a multitude of isoenergetic structural states. These proteins, which comprise an estimated third of the human proteome,[3] are particularly prominent in the nucleus where as much as 70% have been shown or predicted to be IDPs.[4] The cell nucleus, which encompasses the genetic material, is a complex and moldable nucleoprotein landscape, shaped by frequent epigenetic changes that regulate the pattern of gene expression, and ultimately the organismal phenotype. It is thus unsurprising that a multivalent and dynamic nuclear proteome is needed to steer such a diverse environment. The conformational plasticity mediated by intrinsic disorder has been suggested to provide additional levels of functionality to complex cellular regulatory mechanisms.

In this chapter, we highlight protein disorder in the nucleus and emphasize the interplay between IDPs and the nucleosomal landscape that leads to a functional output. We first define the general components of the nuclear environment, before discussing the challenges and recent advances in understanding structural disorder within the context of transcription. We then compile and dissect a subset of important molecular systems in the nucleus that involve disordered interactions, including the effects of chemical modifications, and overview the resulting biological consequences. We exclusively review the interactions of structural disorder within nucleosomes and chromatin, but for reviews on IDP interactions with nucleic acids, we refer to excellent work on those topics.[5–7] Deciphering the complexity of molecular disorder in the chromatin landscape is an exceedingly challenging task. Yet, recent work has begun to map the functions of many constituent proteins of the nucleus by using innovative biophysical strategies, moving us ever closer to a comprehensive molecular view of the cell nucleus.

2. Protein intrinsic disorder on a nucleosomal landscape

2.1 Components of the nuclear environment

The importance of IDPs and IDRs in cell biology is now well established, and their prevalence in signaling and regulatory pathways has been clearly demonstrated.[6] It is their unique conformational properties that make them ideally suited for their roles. High structural heterogeneity, a consequence of their low complexity and biased amino–acid sequences,[8] imparts IDPs with multivalency in many cases, allowing them to interact with more than one biomolecular partner.[9] Even though the presence of disordered proteins in the nucleus has been recognized for decades, it is only relatively recently that their functions have surfaced. IDPs, which have sometimes been called constituents of the dark proteome,[10] are now increasingly being illuminated as key players in the nucleus of eukaryotes.

To appreciate the many roles played by IDPs and IDRs in the nucleus, we first need to clearly define the nuclear architecture that they operate within (Fig. 1). The genetic material for a typical human cell is composed of ~4.6 million basepairs of DNA, which contain the instructions for generating the cell's proteome. The DNA is substantially compacted to fit this enormous amount into the relatively tiny nucleus, and at all stages of DNA compaction we encounter dynamic protein disorder in one form or another. The first level of compaction is to wrap the DNA around an octamer of the core histones (H2A, H2B, H3, and H4) containing two copies of each, forming the nucleosome.[12] A chromatosome is then constructed by binding of linker histone H1 (H1), which attaches to the dyad of a nucleosome (Fig. 1).[13] Both the core and linker histones contain a large amount of

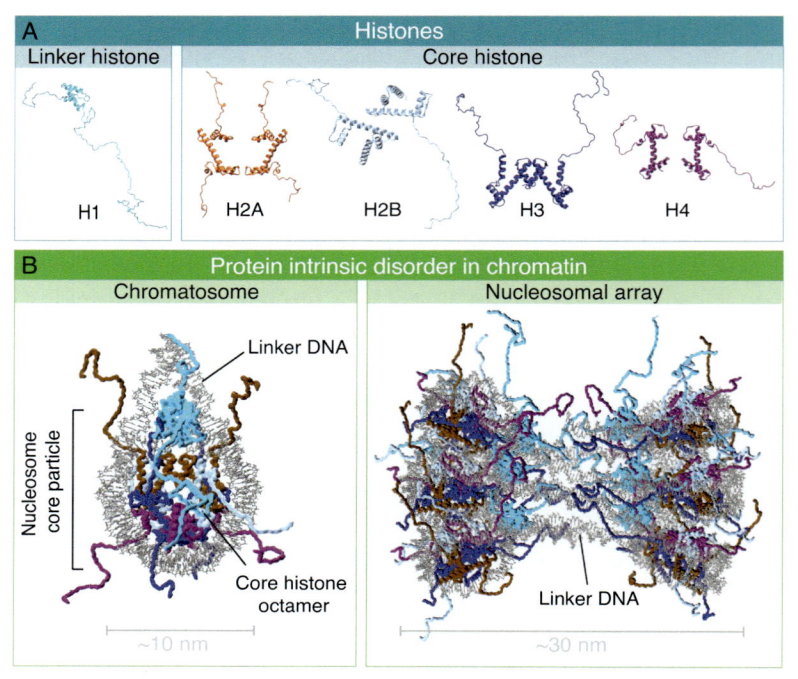

Fig. 1 Protein disorder on a nucleosomal landscape. Intrinsic disorder is a large component of the nucleosomal landscape, contributing to chromatin architecture, dynamics and overall function. (A and B) The nucleosome core particle (NCP) is composed of an octamer of core histones (H2A, H2B, H3, and H4), around which ~147 bp of DNA (gray) is wound in a left-handed super-helical manner.[11] Within the NCP, H2A (orange), H2B (light blue), H3 (dark blue) and H4 (magenta) homodimerize *via* interactions in the structured domains, while the intrinsically disordered N- and C-terminal regions extend into the local nucleosomal space. Linker histone H1 (cyan) binds on or close to the nucleosomal dyad, forming the chromatosome (B) and uses its long disordered and highly basic C-terminal domain to drive conformational changes in linker DNA, impacting the overall structure of poly-nucleosomal arrays and ultimately chromatin fibers.

disorder regulating nucleosome structure and dynamics and ultimately impacting global chromatin structure.[14] There are many histone variants, some cell- or tissue-specific, that can be dynamically exchanged to impart nucleosomes with distinct structural properties.[15] H1 rapidly exchanges between nucleosomes on the second to minute timescale *in vivo*,[16] using largely its positively charged and disordered C-terminal tail to drive orientational changes in linker DNA connecting adjacent nucleosomes.[17] Local interactions between nucleosomes, involving the disordered histone regions, and binding of various regulatory proteins modulate nucleosomal structure and dynamics (recently reviewed in[18]), and subsequently chromatin condensation into higher-order structures. Protein disorder thus plays an integral role in the formation, regulation, recognition, and modification of genome architecture.

2.2 Post-translational modifications fine-tune disordered interactions

To add yet another layer of complexity, most IDPs are chemically and reversibly modified after translation from the ribosome.[19] Histones and their variants have multiple post-translational modification (PTM) sites, mostly in their IDRs, where the pattern and number of modifications can fine-tune their interactions with nucleosomes and other biomolecules.[20] In general, PTMs render the proteome far more vast than the genome, with hundreds of thousands or even up to a million chemically distinct proteins at any given time in the cell.[21] Chemical modifications can change stability, concentration, localization, conformations, and interaction patterns of proteins, providing an important form of regulation and signaling. The most common modifications include (but are not limited by) covalent yet reversible chemical additions such as phosphorylations, acetylations, methylations, hydroxylations, and amidations, as well as attachments of sugar moieties or entire proteins involving sumoylation or ubiquitinylation.[19] In addition to protein modifications, DNA can be modified, most commonly involving cytosine methylation, and when located in CpG islands on promoters, this covalent modification is normally associated with gene repression.[22] Together, these modifications form an almost unfathomably complex and constantly evolving molecular surroundings that dictate the state of a cell.

Protein PTM sites are frequently located in IDRs, partly due to their accessibility to modifying enzymes such as kinases, acetylases, and methylases.[21] PTMs can induce or relieve secondary structure propensity or have a global effect on the structural ensemble sampled by the disordered region, potentially shifting the ensemble to a certain functional state, resembling conformational selection. They can also affect disorder-to-order transitions, which are a common interaction-mode for IDPs,[23] or affect the degree of disorder in fuzzy[24] or fully[25] disordered complexes. PTMs that affect charges will influence intrachain electrostatic interactions, which have an important role in determining the compactness of a disordered region.[26] In general, PTMs modulate the structural and dynamical properties of IDPs, fine-tuning their functional repertoire. We now explore the arsenal of experimental and computational approaches that can and have been used to engage with IDPs, ranging from simple gel-based binding experiments to sophisticated atomistic models.

2.3 Challenges in studying disordered protein interactions with nucleosomes

Quantitative measurements of structurally heterogeneous polypeptides binding to the dynamic nucleoprotein landscape is a daunting task. Nonetheless, technological advances that enable access to various levels of molecular detail are continuously emerging.[27] As an initial characterization of protein–DNA interactions, classical binding experiments have often involved using an electrophoretic mobility shift assay (EMSA). EMSA is a simple and rapid way to monitor the binding of proteins (structured or disordered) to DNA by observing the changed migration pattern of DNA as a result of protein binding.[28] The EMSA can provide information on binding affinity and specificity but may underestimate these parameters as during the electrophoresis the system is out of equilibrium. In addition, the EMSA does not give direct information on actual binding sites, *i.e.*, it does not detect the exact base pair sequence which is recognized. Exact sequence with base–pair resolution can be determined using footprinting assays (*e.g.*, hydroxyl radical footprinting[29]) or nuclease digestion (*e.g.*, Micrococcal nuclease or MNase). Isothermal titration calorimetry (ITC) and more recently microscale thermophoresis, enable quantitative determination of protein–DNA binding affinity and specificity.[30] Chromatin immunoprecipitation, which relies on chemical crosslinking of the target protein to DNA, combined with sequencing (ChIP-Seq[31]) is a powerful method to find protein binding sites *in vivo*. Similarly, ATAC-seq[32] (Assay for Transposase-Accessible Chromatin using sequencing) reveals genome-wide chromatin accessibility as a consequence of chromatin remodeling or other processes. In this elegant method a transposase is used to incorporate next-generation sequencing adapters into chromatin, which after sequencing provides a map of genome-wide chromatin accessibility. To understand local contributions from the polypeptide sequence, the before mentioned approaches can be combined with genetic and biochemical modifications of target proteins, such as introducing domain deletions/additions, charge reversal, domain swapping or local mutations. Still, without a view into microscopic molecular-level details, the underlying physical principles of protein function can be challenging to deconvolute. Cryogenic electron microscopy (cryo-EM) and X-ray crystallography enable determining atomic-resolution three-dimensional structures of macromolecules.[33] X-ray crystallography determines structures from diffraction patterns and it is the most widely used technique in structural biology.[34] Modern cryo-EM is rapidly catching up through recent advances in deep-frozen sample preparations, direct electron detection

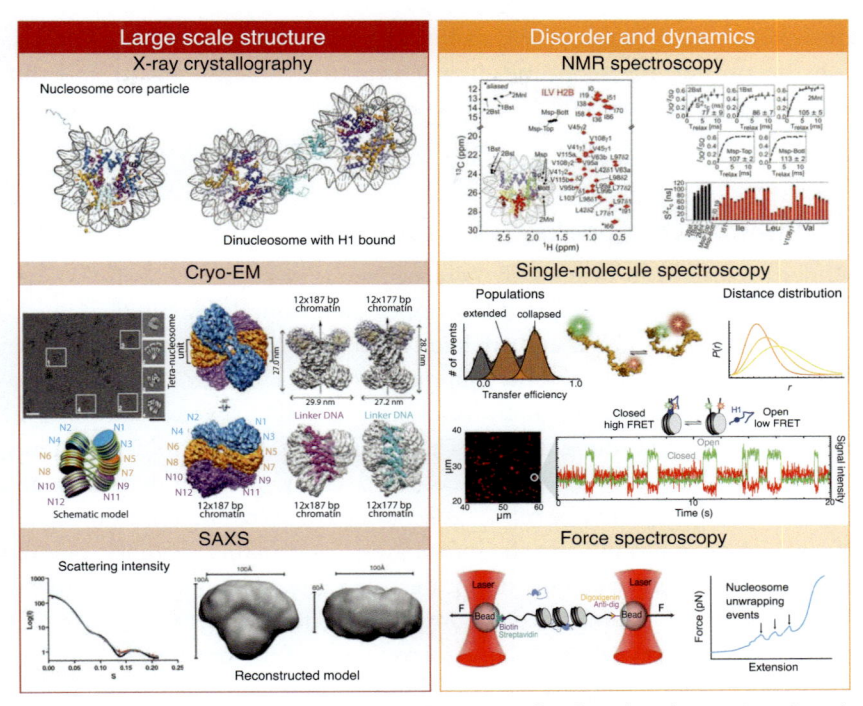

Fig. 2 Methods to study chromatin and intrinsically disordered proteins. Atomic-resolution structures can be determined from X-ray crystallography and cryo-EM while SAXS gives lower resolution information on the overall dimensions of molecules. NMR spectroscopy yields both atomic resolution three-dimensional models of biomolecules and their inter- and intramolecular dynamics, as well as providing residue-specific information on protein-protein or protein-DNA interactions. Single-molecule techniques can be used to study heterogeneous conformational ensembles, at equilibrium, intra- and intermolecular distance distributions, and reaction kinetics. The X-ray crystal structures show the nucleosome core particle[37] and a dinucleosome[38] with bound H1 (PDB codes 1AOI and 6LAB). *Cryo-EM, SAXS, and NMR data shown is reproduced from Song F, Chen P, Sun D, et al. Cryo-EM study of the chromatin fiber reveals a double helix twisted by tetra-nucleosomal units.* Science. *2014;344:376–380, Abramov G, Velyvis A, Rennella E, Wong LE, Kay LE. A methyl-TROSY approach for NMR studies of high-molecular-weight DNA with application to the nucleosome core particle.* Proc Natl Acad Sci U S A. *2020;117 (23):12836–12846, Yang C, Van Der Woerd MJ, Muthurajan UM, Hansen JC, Luger K. Biophysical analysis and small-angle X-ray scattering-derived structures of MeCP2-nucleosome complexes.* Nucleic Acids Res. *2011;39:4122–4135 with permission.*

cameras and sophisticated image analysis,[35] which take advantage of graphics processing units (GPU) acceleration. Recent studies using these methods have supplied us with an impressive view of large molecular assemblies, such as a translating ribosome[36] and entire chromatin fibers (Fig. 2).[38–40]

However, biomolecular processes involving extensive disordered interactions lie outside the scope of current structural biology efforts and thus require different approaches to understand their molecular underpinnings.

2.4 Integrative modeling of disordered protein interactions

Modern research on structurally heterogeneous systems such as IDPs often combines multiple techniques to decipher their underlying physical mechanisms (Figs. 2 and 3). To study dynamic and disordered systems, techniques that can resolve conformational subpopulations in bulk have proven particularly useful. Nuclear magnetic resonance (NMR) spectroscopy can be used to obtain three-dimensional structural models of well-folded proteins, and it has also been extensively used to study protein dynamics and disorder,[25,46,47] even in live cells.[48] After assignments of chemical shifts, protein NMR gives residue-specific information on structure, stability, binding sites, and dynamics on a wide timescale.[49] Despite still being limited to relatively small to medium-sized systems for structure determination, NMR spectroscopy has revealed dynamic movements of the disordered core histones and their interactions, even within entire nucleosomes (Fig. 2).[50–52] Small-angle X-ray scattering (SAXS), the solution-state counterpart to X-ray crystallography, gives information on the shapes of molecules, including IDPs and their dynamic populations, often aided by computer simulations.[53,54] Especially relevant to DNA binding proteins, fluorescence recovery after photobleaching (FRAP) probes the mobility of fluorescently labeled proteins, inside the cell nucleus, and has been used to study the dynamic exchange of histone H1 between nucleosomes.[55] These and other methods have over the years been extraordinarily influential in shaping our perception of IDPs.

Techniques that probe the behavior of individual molecules, and thus access molecular distributions, are an attractive approach to understanding disordered interactions and have been used to complement traditional ensemble methods. Single-molecule spectroscopy, usually in combination with Förster resonance energy transfer (smFRET), has emerged in recent years as an exceedingly powerful technique to study structured and unstructured proteins, *in vitro* and in living cells.[56–58] SmFRET enables sensitive site-specific probing of the distance and dynamics between two or more

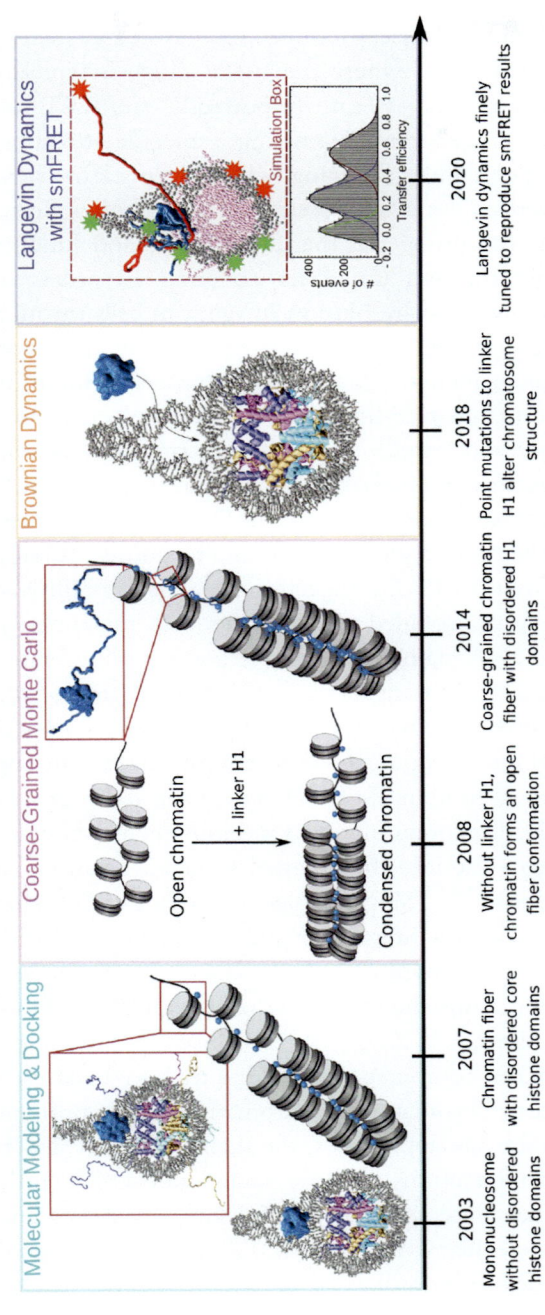

Fig. 3 Chronological overview of computational approaches adopted to study intrinsic disorder in chromatin topology and dynamics. Schematic illustration of the evolution of computational attempts used to investigate the interaction between proteins and DNA within the context of nucleosomes, chromatosomes and chromatin fibers. Early molecular modeling and docking investigated the binding of linker histone H1 to the nucleosome dyad and functioned as preliminary approaches to later attempts featuring molecular simulations on increasingly larger systems.[41,42] Coarse-grained Monte Carlo simulations were used to generate several possible topological arrangements of chromatin fibers with and without the linker histone H1.[43,44] Brownian dynamics,[45] which increases the amount of attainable sampling by scarifying internal motions, facilitated the understanding of how linker histone H1 diffuses toward and binds to the nucleosome dyad. Most recently, modeling and simulations featuring customized potentials, finely tuned to reproduce experimental findings, have provided a semi-quantitative overview of the disorder-mediated interactions between linker histone H1 and fully disordered chaperones involved in its displacement from the nucleosomal dyad.[17]

fluorescent dyes, *e.g.*, within a disordered region of a protein.[56] The rate of energy transfer between a donor and acceptor fluorophore is steeply dependent on the distance between them, where the useful range is typically on a convenient molecular scale of 1–10 nm. Importantly, smFRET can be applied to structurally heterogeneous systems one molecule at a time, avoiding the complication of ensemble-averaging, which can mask transient yet important molecular events. Confocal fluorescence microscopy offers a wide array of experiments that probe the thermodynamics and kinetics of biomolecular interactions through timescales covering 15 orders of magnitude,[59] as well as enabling high-resolution imaging in cells through fluorescence lifetime imaging (FLIM) and stimulated emission depletion (STED) microscopy.[60] Multi-parameter analysis of fluorescence intensity and photon timings allows quantitative investigation into molecular processes such as binding thermodynamics and kinetics, translational and intrachain diffusion, complex stoichiometries, misfolding and aggregation.[58] With total internal reflection fluorescence (TIRF), several surface-immobilized molecules can be excited and detected simultaneously, offering higher-throughput data analysis of FRET trajectories.[58] The versatility of the method has over the years provided new insights into fundamental biological processes such as DNA maintenance and repair, signaling, translation, transcription, and molecular transport.[57,61] On the flip side of the single-molecule coin are force spectroscopy techniques, such as optical tweezers or the atomic force microscope (AFM), that allow direct tethering and manipulation of individual proteins or DNA. Force spectroscopy can probe the microscopic molecular forces involved in biomolecular interactions and has enabled a fresh view into the energetics and mechanisms of protein-nucleosome interactions.[62,63] Single-molecule methods hold great promise for understanding chromatin interactions and when combined with technologies probing ensemble biophysics[64] and genome-wide approaches, these methods can provide a comprehensive view of dynamic and disordered protein-DNA interactions.[27]

In recent years, a plethora of computational techniques have been used alongside experiments to study chromatin and chromatin binding proteins (Fig. 3). By undertaking multiscale approaches, the finer molecular details of chromatin dynamics and interactions are now better understood. Early on, computational techniques were confined to molecular modeling and docking studies. All-atom molecular dynamics (MD) simulations are a gold standard technique for modeling biomolecular behavior, as they provide atomic-resolved information on the movements and interactions of molecules in their given environments.[65] By solving Newton's equations of motion for each atom, modeled as a van der Waals sphere, interatomic forces and their corresponding energies are calculated using molecular mechanics force fields describing both bonded and non-bonded interactions, either in implicit or explicit solvent conditions. In implicit solvent models, the solvent is treated as a structureless continuum, thereby reducing the number of interacting particles and degrees of freedom. In contrast to explicit descriptions, where the presence of each solvent molecule is explicitly accounted, implicit models do not include solute-solvent interactions. Although all-atom simulations provide an unparalleled level of detail, simulating nucleosomal arrays in this manner is unreasonable because of the high computational cost leading to insufficient sampling.

Due to the high computational costs associated with simulating these large and complex systems, an understanding of nucleosome and chromatin organization has started from simply creating models of single nucleosomes[66] or chromatosomes[41] that would fit experimental constraints. In particular, such studies focused on elucidating the binding mechanism of linker histone H1 to nucleosomes and were able to provide an idea, resembling that observed in electron microscopy studies,[41] of how H1 and other histones shape the conformational dynamics of single and di-nucleosomes[66] as well as nucleosomal arrays composed of up to 100 nucleosomes,[42] reporting on the polymorphic nature of chromatin.

A different approach involves the use of Monte Carlo (MC) simulations of coarse-grained (CG) representations. By generating conformational states according to Boltzmann probabilities, MC can be used to sample a Boltzmann distribution of configurations. Coarse-graining offers a computationally less expensive approach. In CG models, groups of atoms are embedded into beads, thereby reducing the degrees of freedom and allowing efficient generation of conformations, without the explicit time dependence of MD simulations. Consequently, larger models, such as those encompassing entire chromatin fibers, can be simulated yet at the expense of fine molecular details. MC simulations of CG chromatin fiber models have revealed many aspects of chromatin compaction. In particular, studies carried out using CG-MC found that H1 is required in the formation of higher order chromatin structures[43] and that, without H1, chromosomal arrays adopt an open fiber conformation.[44] Additionally, chromatin structures with highly variable nucleosome repeat lengths (NRL) produce more compact and uniform fibers,[67] while fibers with a longer NRL, corresponding to a more open chromatin structure, are likely to have a higher number of binding sites for chromatin binding proteins.[68]

Similar to MC, Brownian Dynamics (BD) simulations, which treat the simulated macromolecules as rigid bodies in implicit solvent, have been used to simplify the complexity of chromatosomes or chromatin. In BD simulations, the diffusion of solutes in a continuum solvent is simulated, and electrostatic interactions, which are particularly dominant in chromatin, are calculated by solving the Poisson-Boltzmann equation. BD approaches have successfully demonstrated the effects of sequence variation and PTMs on the H1-nucleosome ensemble. Due to lowering the complexity of the simulated system, docking simulations following the methodological paradigm of BD simulations, allowed a large number of PTMs to be considered, due to the increased computational efficiency associated with this technique.

Another approach to circumvent the computational constraints of all-atom MD is to coarse-grain (CG) the system. As discussed above, coarse-graining simplifies the system by reducing the degrees of freedom, making it an attractive technique for studies on chromatin fibers. In a recent study carried out by Watanabe et al., CG molecular dynamics simulations were used to create a model of an HP1α dimer bound to the histone H3 tail in a di-nucleosome complex.[69] In addition, reverse mapping of the CG structure was carried out upon the completion of the simulation to regain some of the lost atomic details. Within the context of CG models simulated by means of BD or MC, empirical potentials have added a certain level of integration with experimental data and specifically provided an experimentally-derived picture of the conformational ensembles of nucleosomes. Recently, smFRET and CG molecular simulations were tightly coupled in a complementary approach where the distance of multiple sites across protein-DNA within a nucleosome were mapped by smFRET and matched closely by the fine tuning of a single force field parameter describing van der Waals interactions between modeled beads. Integrative modeling has immense potential to deliver finer details of complex molecular systems and has already begun to uncover the physical principles governing IDP interactions.[64,70,71] We now move on to describe some recent work on disordered interactions with chromatin, highlighting new insights that biophysical methods have yielded on the role IDPs and IDRs in the nuclear environment.

3. Disordered interactions with nucleosomes
3.1 Nucleosome architectural proteins
3.1.1 Linker histone H1

Many proteins involved in generating and maintaining the overall nucleo-somal architecture contain long disordered regions. H1 is involved in chromatin condensation through stabilization of compact chromatin structures, and thus functions generally as a transcriptional repressor.[72,73] The polypeptide sequence is highly positively charged with two long disordered regions (N-terminal domain, NTD; C-terminal domain, CTD) flanking a small folded globular domain.[74] The globular domain of H1 is known to bind to the dyad axis of the nucleosome (Fig. 1), thus interacting with the nucleosomal core and both entry- and exit DNA linkers.[13,75] By binding to the nucleosome dyad, the H1 tails are free to form non-specific electro-static interactions with linker DNA to minimize charge repulsion, thereby facilitating chromatin condensation. Therefore, although the binding mode is facilitated through the structured domain of H1, function is largely con-ferred through the disordered tails and an on-dyad binding mode may pro-vide the freedom required for the H1 CTD to interact with one or both linker DNA arms. However, the conformational distributions of the disor-dered regions of H1 on the nucleosome have been more difficult to elucidate because of their pronounced dynamics and seeming lack of persistent struc-ture.[76] Importantly, single point mutations on linker H1 significantly affect chromatosome structure, indicating that small changes may alter the over-arching chromatin structure and, consequently, transcriptional regulation.[45] It has been suggested that H1 draws the two linker arms together, thereby reducing their mobility, and introducing a strong degree of asymmetry to the nucleosome.[13] Recent integrative studies of full-length H1 in complex with nucleosomes gave insight into the behavior of the long disordered tails.[17] The authors studied binding of human linker histone H1 to rec-onstituted nucleosomes using confocal single molecule spectroscopy and CG molecular simulations. Fluorescent labeling of the approximately 100 residue-long disordered CTD of H1 revealed that it becomes considerably more compact in complex with the nucleosome. This can be explained by screening of H1´s positive charges by the negatively charged nucleosomal DNA, which otherwise renders H1 highly expanded due to charge repul-sion. Labeling on the terminal end of nucleosomal linker DNA arms and addition of unlabeled H1, resulted in the expected closure of the linker DNA arms in the H1-bound nucleosome (in agreement with a crystal structure of the chromatosome). Comprehensive mapping of FRET efficiencies within the H1-nucleosome complex combined with nanosecond fluorescence correlation spectroscopy (nsFCS) showed that H1 lacks persistent structure and is extremely dynamic on the nucleosome, displaying sub-μs chain reconfiguration times. Using a simple CG model describing the system in terms of non-specific short-range and electrostatic interactions, combined with existing structural information on the nucleosome and H1's globular domain, the entire complex was simulated and the distances between the corresponding FRET pair locations back-calculated. After tuning the only free parameter in the model—the inter-bead interaction strength that was set globally for all beads—the simulation was able to capture the dynamic conformations of H1 on the nucleosome with high accuracy. Simulations were indeed in excellent agreement with the FRET-derived distances (a total of 57 FRET pairs), illustrating well how closely simulations can reconstruct experimentally determined parameters even in very complex systems.

The presence and conservation of multiple H1 variants within cells suggests that different variants may be linked to specific cellular functions. Within each variant, the structured globular domain shows the highest degree of conservation, while disordered tails are, expectedly, more variable. However, when H1 tail regions from different species are compared, a high degree of conservation is observed between orthologs. For example, human H1.4 and its mouse ortholog, H1e, share 93.5% sequence similarity,[77] indicating that H1 tail regions may confer a high degree of functional selectivity in cells.

In addition to the highly basic charge in the linker H1 CTD, recent studies suggest a direct link between CTD length and chromatin affinity. FRAP experiments on human linker histones found that H1 variants with shorter CTD tails, such as H1.1 and H1.2, have rapid recovery times compared to variants with longer CTD tails, such as H1.4 and H1.5. Moreover, longer histone H1 tails were found to have two or more cyclin-dependent kinase (CDK)-dependent S/T-P-X-K phosphorylation motifs. Therefore, recovery times may be dependent on the density of lysine residues, the CTD length and the distribution of DNA-binding S/T-P-X-K motifs.[78] Because the H1 CTD directly interacts with the linker DNA, each variant will have a different effect on NRL. A higher degree of chromatin folding will likely be achieved if neutralization occurs across the chromatin fiber. Although chromatin condensation is not a direct consequence of linker histone binding, it does stabilize higher order chromatin structures to an extent that depends on the corresponding H1 variant. In general, the affinity of H1 for chromatin increases with its compacting properties. In agreement with this observation, H1.0, H1.4 and H1.5 have a longer CTD and were found to stabilize higher order chromatin.

3.1.2 Post-translational modifications of linker histones

Linker histones are subject to a variety of PTMs, in both their IDRs, adding a large degree of compositional complexity to this protein family. The presence of multiple PTM sites in the H1 IDRs enables a number of regulatory mechanisms for H1 and finely regulates the affinity of each H1 variant for chromatin. The most prominent PTM for H1 is certainly phosphorylation, which occurs in a highly complex and dynamic fashion.[79] Phosphorylation mainly occurs in the CTD where S/T-P-X-K motifs (X is any amino-acid) are recognized by CDKs.[73] Although counterintuitive, H1 phosphorylation can trigger both chromatin expansion and contraction, based on the progression of the cell cycle. Such effects are likely to be a result of conformational rearrangements within H1, arising from site-specific modifications.[80] Phosphorylation levels are the lowest during the G1 phase, rise during the S phase and peak during mitosis, followed by a sharp decrease in the telophase.[11] *In vivo* studies showed that serine residues in H1.4 are generally modified during G1 and S phases, while threonine is phosphorylated in mitosis,[81] outlining the cell cycle dependence of phosphorylation. CDK1 and Cyclin B are primarily responsible for H1 phosphorylation during the mitotic phase. Additionally, recent studies suggested that several kinases phosphorylate the H1 NTD.[82–84]

The conversion of lysine into its methylated analogs (methyllysine, di-methyllysine or tri-methyllysine) is another important modification that can compete with or complement phosphorylation on a functional basis. For instance, the methylation of lysine 26 in the H1.4 NTD recruits heterochromatin protein-1 (HP1), resulting in heterochromatin formation,[85] and is controlled by a phospho-switch: when H1.4 is phosphorylated at serine 17, the interaction between HP1 and methyllysine 26 is inhibited, demonstrating the importance of crosstalk between PTMs.[85] In the cell, H1 phosphorylation is CDK2 dependent and is required for progression through the S-phase. Because CDK2 colocalizes with replication sites and H1 is crucial in the formation of higher order chromatin, CDK2 recruitment to replication foci by Cdc45 may result in H1 phosphorylation and drive fork progression,[86] linking H1 phosphorylation and active transcription.

Methylation of the H1 NTD and CTD may elicit variant-specific cellular responses *in vivo*. For instance, the two predominant human H1 variants, H1.2 and H1.4,[87] are methylated differently by the same methyltransferases.[88]

Interestingly, the methylation of H1.4 lysine 26, a highly conserved PTM in vertebrates, creates HP1 binding conditions, likely because H1.4 lysine 26 is part of an "ARKS" motif[85,88]: a conserved motif assumed to have a regulatory function in heterochromatin. Therefore, a link may be present between chromatin compaction and lysine 26 methylation.

Like phosphorylation, acetylation of the NTD leads to both heterochromatin formation and activation of transcription. Acetylation of H1.4 lysine 26 is related to the formation of facultative heterochromatin, which forms parts of the genome not shared across cell types and usually contains poorly expressed genes, which are task specific and mostly associated with cellular differentiation. Deacetylation by SIRT1, on the other hand, results in the formation of repressive heterochromatin.[89] The presence of an acetyl group on lysine 26 prevents methylation and subsequent recruitment of HP1, providing an additional level of regulation. Interestingly, an *in vivo* study using T47D cells expressing a lysine 26 to alanine H1.4 mutant, reported defects in gene regulation and cell proliferation, compared to wild-type H1.4.[90] Additionally, acetylation of H1.4 lysine 34 is also associated with transcription activation *in vivo*. In this position acetylation is, however, suggested to reduce H1-chromatin affinity and recruit TAF1; a subunit transcription factor TFIID.[91]

3.1.3 Core histones

The core histones, which are the main structural support of nucleosomes, contain relatively short yet crucial IDRs when compared to the linker histone. Core histones form an octamer around which DNA is wrapped in the initial stages of chromatin condensation.[73] Each core histone shares a common histone-fold domain of three helices connected by two loop regions. To complete the octamer, each core histone homodimerizes, followed by the formation of specific H2A-H2B and H3-H4 heterodimers to create the "handshake" shaped core[92,93] (Fig. 1). In addition to the structured domains, each core histone has an intrinsically disordered, solvent-exposed N-terminal tail; with only H2A having an additional C-terminal tail.[94] Like H1, these tails are enriched with highly basic residues that form electrostatic interactions with nucleosomal DNA, linker DNA and acidic patches on neighboring nucleosomes. Such interactions are believed to stabilize the histone-DNA and nucleosome-nucleosome associations.

Moreover, the dynamic nature of the tail regions makes them a target for PTMs and subsequent recruitment of histone chaperones, architectural binding proteins and chromatin remodelers.[95] The inter- and intramolecular contacts between the nucleosome and disordered core histone tails are crucial in the formation of the nucleosome core particle (NCP) and for the stabilization of higher order chromatin structures.

The intrinsically disordered tails of core histones have also been implicated in the formation of higher order chromatin structures and chromatin condensation through inter-nucleosomal interactions.[96] In the first nucleosome crystal structure published in 1997, an inter-nucleosome interaction between the N-terminal H4 tail and an acidic patch on the H2A/H2B dimer interface of a neighboring nucleosome was identified.[37] As for H1, the *in vivo* functions of core histones are complex and their ability to modulate transcription is largely dependent on their PTMs.[77] In the nucleosome, DNA accessibility is controlled by transient unwrapping from the NCP; a process that is modulated by each core histone to different degrees.[97,98] While the histone H3 tail suppresses nucleosome unwinding, the histone H4 tail enhances it.[99] Consequently, PTMs in the histone tails are especially important in their role of modulating protein–nucleosome interactions and regulating unwrapping.

Acetylation is an abundant modification in core histone tails, affecting chromatin compaction *via* the neutralization of positive charges.[20] Such effects have been demonstrated *in vitro*, where compaction of nucleosomal arrays required residues 14–19 in the NTD of the human histone H4[100] and the acetylation of lysine 16 prevents array compaction.[101] Moreover, acetylation is likely to reduce the electrostatic cross-talk between the DNA and the histone tails, decreasing the force required to unwrap nucleosomes[102] and increasing DNA accessibility to transcription factors and other modifying enzymes.[103] In line with this notion, histone acetylation has been shown to be strongly associated with transcription.[104] Additionally, many histone acetyltransferases interact with tri-methylated lysine 4 on H3; a modification associated with transcriptional activation.[20] Relevantly, *in vivo*, inhibition of transcription was shown to result in rapid histone deacetylation in mouse embryonic cells, indicating that much of histone acetylation occurs as a result of transcription.[105]

Phosphorylation of histones predominantly occurs in the intrinsically disordered NTD.[106] For instance, Aurora B kinase is known to phosphorylate H3 serine 10[107] and serine 28[108] during the mitotic phase *in vivo*, however, there is no evidence of both modifications being present on a single histone tail.[109] Nevertheless, phosphorylation of the histone H3 tails is likely required for cell cycle progression. Importantly, the Aurora B kinase is over-expressed in a number of human cancers,[110] suggesting that phosphorylation plays a significant role in nucleosome availability to transcription.

Methylation is another predominant modification in core histone tails and for the cross-talk between different PTMs. For example, H4 arginine 3 methylation is recognized by p300; the acetyltransferase responsible for acetylation in histone H4. Methylation in this position is indeed an impor-tant PTM in gene transcription, as it is required for the subsequent acetyla-tion that reduces electrostatic repulsion in the chromatin fibers.[111] In contrast, methylation of H3 arginine 8 is associated with gene repression, outlining the importance of site specificity in PTMs.[112]

3.1.4 HP1 proteins

Gene expression within the context of heterochromatin is facilitated by a series of important, conserved proteins called heterochromatin proteins (HP). These are fundamental units of chromatin packing that can be sub-divided into families, with HP1 being the dominant family composed of three isoforms in humans—HP1α, HP1β and HP1γ—, all of which have two highly conserved structured domains; the amino-terminal chromo domain (CD) and the carboxyl chromo shadow domain (CSD). The structured domains are separated by a disordered hinge region (HR), of varying length across paralogs. Additionally, shorter intrinsically disordered extensions are present at the N- and C-termini of HP1.[113] HP1 is a major component of heterochromatin and is involved in the regulation of DNA-mediated processes including heterochromatin formation, stabiliza-tion of telomeres and gene silencing in pericentric heterochromatin.[69,114]

In general terms, HPs are multivalent, structural chromatin effectors[115] that cause transcriptional repression by recognizing and binding di- or tri-methylated lysine 9 in histone H3 (H3K9me2/3) *via* the CD,[116] while remaining highly dynamic. Methylation of H3 provides an epigenetic mark, suitable for the hydrophobic binding pocket created by the CD. Despite the high degree of specificity between HP1 and H3K9me2/3, the binding affinity spans widely depending on the paralog.[117] Varying affinity is believed to provide a dynamic range in which HP1 paralogs are able to elicit different cellular functions. The dynamic nature of HP1α has recently been probed using *in vitro* techniques. By employing a chemically defined assay, well suited to cellular measurements, it was found that HP1α residence time increases with H3K9me3 density, due to rapid re-binding of dissociated factors on neighboring sites. Moreover, dimeric HP1α exhibited accelerated association rates; a key feature of effector multivalency, allowing fast and efficient binding in a competitive environment.[115]

PTMs of HPs also play a fundamental role in regulating affinity. For instance, HP1α phosphorylation further strengthens its multivalency, while simultaneously reducing DNA binding, ultimately increasing HP1α residence time.[118] Phosphorylation of serine residues in the disordered NTD in mouse HP1α was found to increase CD-H3K9me2/3 affinity such that the overall affinity close to that of mouse HP1β and HP1γ.[117] Moreover, in HP1β and HP1γ the serine residues are replaced by glutamate. Such findings suggest that CD-H3K9me2/3 affinity may be partially modulated by charge differences in distal regions. Therefore, regulating charge *via* phosphorylation of the serine residues within the NTD in HP1α may contribute to the protein binding to H3. In turn, this interaction may impact the activity of kinases or phosphatases, increasing or decreasing binding.

Between the HP1 paralogs, both the underlying amino acid sequence and length of the HR are variable and such differences may control localization and function.[113] For instance, there are 41 and 36 residues in the HP1α and HP1β hinge regions, respectively, and both variants localize to heterochromatic regions[119,120] mediating transcriptional gene silencing.[121] Comparatively, HP1γ, where the HR has only 31 residues, localizes to euchromatin[119,120] and plays a role in transcriptional elongation and RNA processing.[121] The molecular basis for functional divergence is suggested to arise from the non-conserved residues in the HR, since the positively charged domains (KRK and KKK) are conserved across all three variants.[121] These domains are crucial for the specificity of HP1-H3K9me2/3 binding *in vitro*[122] and for intranuclear localization *in vivo*.[123] PTMs in the hinge regions have been shown to affect HP1 functionality.[113,124] Phosphorylation of serine 83 in the hinge region of HP1γ increases its interactions with Ku70, a DNA repair protein, thereby increasing its localization to euchromatin.[113] Taken together, this may suggest that modifications in the disordered HR of HP1 paralogs are able to elicit specific cellular functions.

Recent studies show that HP1 proteins play an important role in heterochromatin by interacting with histones H3 and H4 and methyltransferase enzymes.[106,125] The binding of the HP1 CD to poly-methylated H3 lysine 9 (H3K9me2/3) and H1.4K26me[85] triggers a silencing mechanism, resulting in the formation of heterochromatin.[116] Moreover, this interaction may be influenced by PTMs,[126] especially those in the intrinsically disordered regions. In particular, phosphorylation of HP1α NTD poly-serine stretch 11–14, increases chromatin binding affinity by reducing tail flexibility in human and mouse cells.[117,127] Phosphorylation changes the conformation of the NTD, such that neighboring acidic residues (15E-DEE-E19) are able to interact with basic residues surrounding H3K9 (8R-Kme-STGGKAPR-K18).[128] Addition of the negatively charged residues formed upon phosphorylation results in repulsion, causing the HP1α NTD to behave as an extended intrinsically disordered region, in turn allowing the CD to dynamically bind H3K9me2/3.[128] Interestingly, in HP1β and HP1γ the residues corresponding to the poly-serine stretch of HP1α (12E-VL-E15 and 21K-VE-E24, respectively) are partially negatively charged.[117]

3.2 Intrinsically disordered proteins that interact or compete with linker histone H1

The state of chromatin compaction is tightly linked to the presence of H1. Therefore, the cell has evolved various regulatory mechanisms to actively remove H1 from nucleosomes. One such mechanism is proteins that compete with H1 for binding to the nucleosome or otherwise lead to its eviction. As mentioned above, H1 is highly disordered outside of the globular domain, a feature that is commonly shared among the diverse H1 competitors outlined here.

3.2.1 Protamines

Protamines are short (25–100 residues), highly basic, and disordered nuclear proteins[129,130] suggested to have evolved from histone H1.[131] Protamines replace core histones during the last stages of male germ terminal differentiation of spermiogenesis, where they are found to be the major packing units of DNA (Fig. 4B).[132] Most mammals have only one gene coding for protamine 1 (PMR1 or P1) which is expressed in spermatids as a mature protein[133] and is responsible for chromatin condensation in sperm. However, some mammals, including humans and mice, have a second protamine, PMR2 or P2. Protamines from the protamine 2 family are longer compared to P1 and are generated by proteolytic cleavage of a precursor. DNA packed by protamines in mature sperm cells is transcriptionally inactive and forms higher order structures vital for normal sperm function.[132,134]

Protamine packaging of DNA has been studied with chemical and physical studies of both natural sperm chromatin and synthetic DNA.[135] Earlier studies demonstrated that protamines can precipitate DNA from both

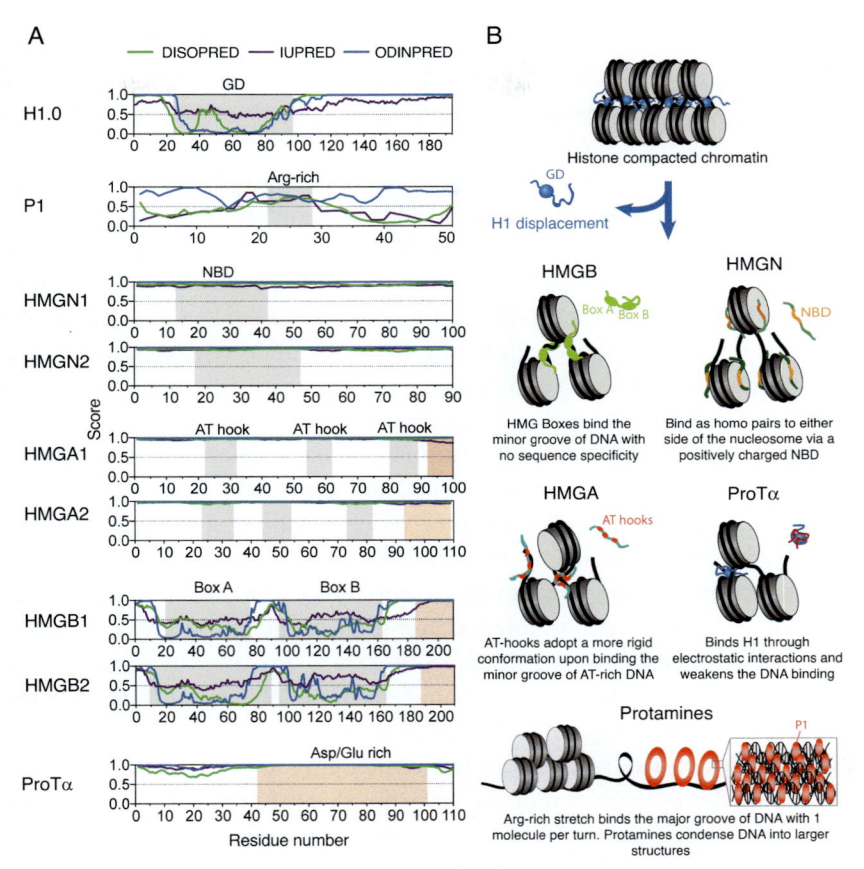

Fig. 4 Disordered H1 competitors and their nucleosome binding modes. (A) Disorder predictions of H1 and competitors are shown using three predictors. Domains are highlighted in gray: Linker histone H1.0 contains a folded globular domain, HMGN1 and 2 each have an NBD, HMGA1 and 2 each contain three AT-hooks, and HMGB1 and 2 are characterized by two folded Box domains. Acidic stretches are indicated in red. (B) Schematic illustration of the binding modes of H1 competitors in a nucleosomal context. The tightly packed structure of chromatin with bound H1 is remodeled by the disordered competitors, through eviction of H1. HMGB, HMGA, HMGN, ProTα and protamines all have distinct binding modes and their domains adopt different degrees of disorder essential to association with the nucleosomes.

assembled chromatin and native chromatin extracted from calf thymus tissue in a concentration dependent manner. By analyzing the supernatant composition using gel electrophoresis after addition of protamine to a chromatin solution, H1 was discovered to be the first histone to appear in solution. However, the release of H1 was slower in native chromatin compared to reconstituted nucleosomes although somewhat affected by sample preparation, suggesting that the mechanistic picture of H1 competition and histone-protamine transition is more intricate.[136] Later work unveiled that protamines replace histones though a complex and progressive transition mechanism.[134] After meiosis in spermiogenesis, the canonical histones are replaced with testis-specific histones, and subsequently replaced by transition proteins that causes alteration in DNA structure. Many protamine molecules then bind and reorganize DNA into tightly packed structures. Thus, the conversion from histone-packed to protamine-packed chromatin is a complex interplay between different DNA-binders, regulated by PTMs including hyperacetylation of histones, in steps that are crucial for the correct progression of chromatin maturation and spermiogenesis.

The primary structure of protamines is characterized by a conserved arginine-rich core functioning as a DNA anchor[135] and cysteine-rich N- and C-termini. The overall dimensions of different protamines have been predicted by simulations to be controlled by their net charge per residue, which shifts their conformational ensembles from collapsed globule to coil-like.[137] The central arginine-rich region provides a high net positive charge that facilitates strong binding to DNA. Protamines wrap around DNA in the major groove[138] and bind with one protamine molecule per turn of DNA helix (Fig. 4).[139] Even though protamines are known to be disordered when free in solution, it is challenging to probe conformational changes of the individual molecules within the context of the large chromatin structures because of the large number of protamines associating with chromatin. Potential disorder-to-order transitions upon DNA binding are yet to be clearly demonstrated experimentally for protamines. However, many cysteines partake in multiple intra- and intermolecular disulfide bridges that provide rigidity, which is essential to stabilizing the structure of sperm cell chromatin.[140,141] Single-molecule studies demonstrated that protamines can bend DNA into loops through multiple steps[142] leading to the formation of higher order structures similar to those induced by H1 DNA packing,[143] which suggests a common pathway for positively charged IDPs in chromatin condensation.

Protamines contain several conserved phosphorylation sites and have been shown to undergo various PTMs. These include phosphorylation, acetylation and methylation, which have been detected in protamines of mice sperm using peptide-based tandem mass spectrometry.[144] The data indicated that methylation, acetylation and phosphorylation do not occur at the same time on a single protamine, suggesting a complex network of PTMs affecting the epigenetic landscape of sperm cells.[144] However, the exact role of PTMs on the conformational ensembles of protamines and within the context of histone eviction in sperm remains unknown.

3.2.2 HMG proteins

High-mobility group (HMG) proteins belong to a family of disordered architectural transcription factors known to interact with nucleosomes and, together with H1, were the first nuclear proteins known to affect the structure of chromatin.[145] They were first discovered in isolated chromatin together with histones and named due to their unusually high electrophoretic mobility.[146] HMG proteins modulate local and global chromatin architecture by inducing formation of distorted DNA structures and promoting chromatin decompaction. The decompaction in turn enhances various DNA-dependent activities such as transcription, replication, and repair. HMG proteins are divided into three families—HMGA, HMGB and HMGN—, depending on their structural and functional properties, and we discuss here the role of their intrinsic disorder in nucleosome binding and H1 competition.[145,147]

The HMGN subgroup has five members; HMGN1 and HMGN2 were the first to be discovered, whereas HMGN3-5 were identified later. HMGNs are fully disordered (Fig. 4A) and bind the nucleosomal structure with high specificity in pairs to form complexes containing two molecules of either HMGN1 or HMGN2.[148–150] These proteins are characterized by a positively charged and conserved nucleosome binding domain (NBD), a nuclear localization signal (NLS), and an acidic C-terminal chromatin regulatory domain (CHUD) involved in modulating acetylation of histones.[147,151] HMGN proteins recognize generic nucleosome structures without specificity for DNA sequence or histones *via* the ~30 amino acid-long NBD.[152] The domain contains the canonical motif *RRSARLSA* which serves as an anchoring point on the nucleosome to a negatively charged patch formed by the H2A-H2B dimer surface.[153] The C-terminal domain of the HMGN protein interacts with the DNA in the two major grooves flanking the nucleosome dyad axis and is in close proximity to the N-terminal tail of histone H3.[149,154,155] HMGN also influences H3 phosphorylation and acetylation, by inducing local structural change that alters the accessibility of enzymes and thus the equilibrium of nucleosomal PTMs.[156,157] HMGNs are themselves modulated through phosphorylation,[157] which decreases the affinity to chromatin and allows kinases to access histone H3.[156] HMGNs also have many lysines placed in the NBD that are acetylated, which leads to less efficient binding to nucleosomes.[158]

NMR studies using paramagnetic relaxation enhancement (PRE) experiments, which probe long-range interactions, and methyl-labeled histones in assembled nucleosomes, showed that, upon addition of HMGN, an interaction occurs between an arginine-rich region of the NBD and the folded core of the H2A/H2B dimer.[155] In addition, it was demonstrated that the several lysine residues in the C-terminal end of the NBD have an affinity for DNA non-specifically. The experimental data was then used as restraints to model the binding orientation of the NBD on the nucleosome. In the calculated structural model, the N-terminal end of the two NBDs stacked on each side of the nucleosome are predicted to bind the H2A/H2B acidic patch, while the lys-containing ends associate with DNA near the exit/entry point.[155] Due to the orientation of the NBD on the nucleosome, the disordered C-terminal tail was expected to be located where H1 associates with nucleosomes. This suggests that the chromatin decompaction function of HMGN is a result of disrupting H1 binding to DNA and the core histone tails.

Surprisingly, native gel-shift assays and cross-linking studies showed that HMGN1 binds nucleosomes already bound to H1, without interfering with the specific contacts made between the H1 globular domain and the nucleosome.[159] This observation agreed with previous MNase digestion results on mammalian chromatin, which reported nucleosomes bound to two HMGN proteins and H1.[160] This implies that binding of HMGN proteins to nucleosomes engenders a dynamic rearrangement of H1 interactions leading to modulation of chromatin structure. However, the molecular mechanism is still unclear and whether HMGN folds upon binding to the nucleosomes or if the two bound HMGN remain fully unstructured in the complex, is yet to be determined.

HMGB proteins (HMGB1 and HMGB2) have two structurally conserved DNA binding domains (DBDs, Box A and B) that each fold into three helices when in complex with DNA.[145] A disorder-to-order transition observed by NMR spectroscopy on HMG domains from Sox-proteins,[161] demonstrated that HMG box domains retain a flexible structure in solution that folds upon DNA binding. In addition, HMGB proteins contain a disordered C-terminal region of ~30 amino acids (Fig. 4A) enriched with acidic residues. The Box DBDs of HMGB bind with low affinity to single-stranded, linear duplex, and supercoiled DNA,[162,163] but have a preference for bent and distorted DNA[164–166] which is increased upon acetylation.[167] Based on structures of closely related HMG-Box-DNA structures, the box domains distort DNA through intercalation of bulky hydrophobic amino acid residues into the DNA minor groove, resulting in bending the molecule toward its major groove (reviewed in[165]).

The acidic C-terminal region forms a flexible extended structure which was characterized by NMR spectroscopy and demonstrated narrow dispersion in ^{1}H-^{15}N-HSQC typical for IDPs.[168] The flexible tail is involved in dynamic intramolecular interactions, with the highest affinity for DBD Box B. SAXS and NMR studies, including PRE measurements, suggested that the C-terminal tail promotes a more compact conformation where the two basic boxes get closer to each other.[169] Apart from intramolecular interactions of the acidic tail with the HMG-boxes, the acidic C-terminal tail of HMGB1 is also engaged in intermolecular interactions with other proteins. Notably the HMGB tail associates with histone H3, an interaction suggested to modulate the biological functions of HMGB proteins.[170] MNase digestion data suggested that HMGB protects linker DNA on one side of the NCP at the entry/exit of nucleosomes opposite to the linker histone H1 binding site.[171] As observed for other HMG proteins, early studies using chromatin fractionation experiments reported strongly enriched HMGB1 and 2 in H1-depleted fractions of salt-soluble chromatin.[172] FRAP experiments showed that HMGB also enhances H1 mobility in cells, indicating its ability to displace H1 from chromatin.[173] Studies of HMGB and H1 interactions by chemical cross linking and gel filtration experiments showed that they form a 1:1 complex. The complex persists at physiological ionic strength, where it was reported by NMR spectroscopy that H1 binds through its basic C-terminal domain to the acidic tail of HMGB1, disrupting its interaction with HMG boxes. A consequence of this interaction is enhanced DNA binding and bending by HMGB1, followed by a lowered affinity of H1 for DNA.[174] This might facilitate H1 eviction in a chromatin context and supports the data showing increased H1 mobility in cells in presence of HMBG.[173]

As outlined above, several lines of evidence support that the disordered C-terminus of HMGB1 has a crucial role in its function, including orchestrating many different interaction partners (review in[175]) and modulating chromatin structure. HMGB function is also regulated by PTMs like acetylation, phosphorylation, methylation, as well as its oxidative state as formation of a disulfide bridge in Box A leads to reduced H1 displacement from hemicatenated DNA loops.[176] Most PTMs found or predicted in HMGBs are placed in the folded box domains, but several acetylation sites have been identified in disordered stretches within NLS regions. Although acetylation within the Box domain is known to affect DNA binding affinity, more studies point toward PTMs having a large impact on controlling the nuclear localization and export of HMGBs. However, how PTMs affect the conformational ensemble of the acidic disordered tail of HMGB and its interaction with H1 has yet to be fully defined.

The third class of highly disordered HMG chromatin binders are the *HMGA proteins (previously named HMG-I(Y))*. There are two genes coding for HMGA proteins, HMGA1 (and its splicing variants HMGA1a, b and c) and HMGA2, characterized by their very short DNA-binding AT-hook motifs. HMGA proteins are fully disordered in solution in absence of DNA as shown by biophysical techniques such as circular dichroism[177] and NMR spectroscopy.[178] An NMR study reported that the AT-hook DBD of HMGA transits from disordered to a well-defined crescent-shaped configuration upon binding to the minor groove of short AT-rich DNA stretches.[179] The specificity for AT-rich DNA regions was also demonstrated by a PCR-based systematic evolution of ligands by exponential enrichment (SELEX) approach, identifying nucleotide consensus sequences with two AT-rich stretches of 5–6 base pairs separated by four GC-rich base pairs.[180] EMSA studies show that HMGA binds isolated nucleosome particles with much higher affinity than to naked DNA.[181,182] It was previously suggested that HMGA1 also associates with core histones based on DNase footprinting and chemical cross-linking studies,[183] possibly explaining the preference for nucleosomal DNA.

HMGA1 proteins were found to co-localize with histone H1 at AT-rich DNA stretches called scaffold attachment regions (SARs) in mammalian cells.[184] T7 polymerase assays in combination with DNA binding assays showed that HMGA can compete with H1 on SAR, and even redistribute H1 onto non-SAR DNA.[183] Purification of HMGA and H1 from HeLa cell chromatin also demonstrated that HMGA is strongly enriched in H1-depleted fractions of active chromatin.[183] In a study of chromatin condensation in neural precursor cells of mice, it was found that HMGA proteins are essential for chromatin opening in the early developmental stage. Overexpression of either HMGA1a or HMGA2 in cells increase the sensitivity to MNase digestion of chromatin from extracted nuclei, whereas depletion of HMGA mRNA led to reduction in MNase digested DNA. This clear effect of HMGA proteins on MNase digestion of extracted DNA from nuclei suggests that HMGA induce chromatin opening and accessibility.[185] The chromatin was more resistant to digestion in absence of HMGA, which supports the notion that chromatin becomes more accessible in presence of HMGA, suggesting an inhibition of H1 driven compaction. Another work observed increased H1 mobility caused by HMGA by measuring FRAP in cells expressing GFP-H1 and microinjected with purified HMGA into the cytoplasm.[173] The apparent H1 displacement in the different studies was due to competition for chromatin binding sites, since HMGA mutants incapable of binding DNA did not increase H1 mobility in a similar manner nor compete with H1 for SAR binding.[173,183] This strongly indicates that HMGA competes with H1 on chromatin resulting in destabilization of higher order chromatin structure.

Like the other HMG proteins, HMGA undergoes various PTMs which have been extensively studied. In fact, HMGA1 proteins are among the most phosphorylated proteins in the nucleus by the action of various kinases like cdc2, protein kinase C (PKC), and casein kinase II (CK2).[157] Phosphorylation of HMGA leads to considerably lower affinity toward DNA, in part because two of the main phosphorylation sites are near the positively charged AT hooks and thus disrupt their binding to the negatively charged DNA. The acidic C-terminal tail of HMGA also undergoes phosphorylations *in vivo* which can lead to a conformational change.[186] NOE measurements and 1D proton spectra of C-terminally phosphorylated HMGA indicated a more rigid structure compared to the native and free HMGA which is fully disordered. Pull-down assays of truncated HMGA containing AT-hooks with the acidic C-terminal peptides showed a clear interaction to nucleosomes driven by electrostatics as the affinity increased with the number of phosphorylations in the C-terminal peptides. This supports the hypothesis that the phosphorylated acidic tail folds back onto positively charged clusters on the HMGA and through charge neutralization impairs binding mediated by these Arg/Lys-rich regions.

To summarize, all HMG proteins compete with H1 for chromatin binding sites (in a dose dependent fashion) although each HMG subfamily has distinct effects on the interaction of H1 with chromatin (Fig. 4).[152] HMG proteins all contain disordered regions—HMGA and HMGN being completely disordered in absence of DNA—but they have different structural features and folded domains. This results in distinct modes of action in their modulation of chromatin structures, although a common feature appears to be recognition of DNA conformation. Even though the regions where the HMG proteins bind nucleosomes is known, the sequence of events leading to H1 eviction from nucleosomes is still not fully clear. It has also been suggested that the different classes of HMG proteins can weaken H1 binding cooperatively without competing with each other, hinting that they distinctly affect H1 binding to the nucleosome.[173] Overall, the HMG proteins are part of a dynamic and elaborate interaction network that leads to H1 displacement, where disorder plays a fundamental role.

3.2.3 FoxA1

Forkhead box A (FoxA) transcription factor (previously called HNF-3) is part of a group of transcription factors evolutionary conserved in eukaryotes and is crucial in regulation of biological processes such as cell development, signal transduction, cell differentiation, and regeneration. FoxA1 is as a so-called pioneer transcription factor due to its ability to engage target sites on nucleosomal DNA.[187,188] FoxA1 is disordered outside of a highly conserved winged helix DBD which is structurally similar to that of histone H1.[189] The DBD contains a helix–turn–helix (HTH) motif that makes base-specific DNA contacts as well as two flanking loops (wings) that contact the phosphodiester backbone of DNA. FoxA1 is known to stably bind nucleosomes *in vitro* and *in vivo* near the nucleosome dyad[190,191] and decompact repressed chromatin compacted by H1 to make it accessible for other DNA binding factors.[192,193]

In vitro sequential binding experiments with purified proteins showed that FoxA1 displaces H1 prebound on assembled nucleosomes.[190] Further DNase footprinting of H1-compacted nucleosome arrays with and without FoxA1 demonstrated increased hypersensitivity in the digestion patterns and indicates that FoxA1 can open H1 compacted nucleosomes.[190,192] Truncation mutants of FoxA1 missing the 174 amino acid C-terminal domain failed to open the compacted arrays,[192] underlining the importance of the disordered regions of FoxA1. Early studies also indicated that the N- and C-terminal regions of FoxA1 are crucial for binding specificity to nucleosomes over free enhancer DNA.[190] More recent work identified a short region in the C-terminus of FoxA1, conserved among FoxA pioneer factors, that interacts with core histones and contributes to chromatin opening *in vitro* (see Fig. 7).[194] A single-locus study demonstrated that FoxA1 induction caused reduction of H1 occupancy at an enhancer site during retinoic acid–mediated differentiation of embryonic stem cells.[195] In later studies, an assessment of genome-wide occupancy of linker histone H1 in mouse hepatocytes showed FoxA occupancy on nucleosomes correlates with H1 displacement, whereas the FoxA deletion mutants had a striking increase in H1 disposition. All of these results indicate that FoxA binding displaces linker histones from the local chromatin, which could explain the subsequent increase in nucleosome accessibility and stimulation of transcription.

3.2.4 Prothymosin α

The nuclear protein prothymosin α (ProTα) is a linker histone chaperone that modulates H1 interaction with nucleosomes. Besides affecting chromatin condensation[196] and H1 mobility in the nucleus,[197] ProTα is involved in transcriptional regulation, cell proliferation, and apoptosis.[198] ProTα is fully disordered, with a highly negatively charged glutamate-rich (net charge −44) amino acid sequence and low hydrophobicity.[199,200] Borgia and co-workers used a combination of single-molecule FRET, NMR spectroscopy, and CG simulations to study the interaction between ProTα and histone H1, and showed that they form a tight complex with picomolar affinity yet remain highly disordered and dynamic in the bound state.[25] This novel interaction mode can be explained by the large opposite net charge of the two proteins which leads to complex formation through a mean-field type charge interaction without the need for defined binding sites or persistent interactions between specific individual residues. The CG simulations, which relied on a simple model involving non-specific short-range and electrostatic interactions, were able to reproduce the experimentally measured FRET efficiencies in the complex remarkably well. Later, Sottini et al. showed through an elegant set of kinetics experiments that ProTα and H1 can also form higher order but weakly interacting ternary complexes.[46] Again, integrating experiments and simulations, they showed that a second ProTα or H1 molecule can engage a preformed ProTα-H1 complex and lead to rapid exchange, keeping the system highly responsive despite the tight binding.

What is the purpose of forming such a disordered complex in the nuclear context? Heidarsson et al. addressed that question by studying the H1-ProTα interaction in the presence of reconstituted nucleosomes[17] (described also above, see Section "Linker histone H1"). Kinetic experiments using immobilized and fluorescently labeled nucleosomes showed that ProTα forms a ternary complex with H1 and the nucleosome, which accelerates the dissociation of H1 by almost two orders of magnitude through a competitive substitution mechanism. Further CG simulations confirmed the dramatic increase in dissociation rate as a function of ProTα binding and provided a molecular picture of how ProTα invades the complex by dynamically and gradually sequestering the H1 C-terminal IDR. The high negative charge in ProTα thus competes with the electrostatic interactions between the linker DNA and the disordered regions of H1, which reduces the interaction strength of H1 with the nucleosome and leads to an opening of the nucleosome linkers. These results provide clues toward resolving long-standing issues on histone H1 including the nature of the structural ensemble of H1 on the nucleosome and the discrepancy between *in vivo* (minutes) and *in vitro* (hours) residence times of H1 on the nucleosome.[55,201] Through integrative modeling of these challenging molecules, the authors suggested that it is precisely the high degree of dynamic disorder on the H1 IDRs that allows chaperones like ProTα to invade the complex and accelerate the dissociation of H1 from the nucleosomes. For such unspecific, charge dominated binding between dynamic and disordered proteins, the formation of higher order complexes may commonly occur, providing additional functionality and enabling a sensitive concentration-dependent response during signaling. Formation of higher order oligomers and the dynamic exchange within them may be particularly important to achieve dissociation of strongly interacting polyelectrolytes,[25] and to induce formation and regulation of phase-separated condensates.[5,202]

3.3 Chromatin remodelers and histone-modifying enzymes

Chromatin remodelers dynamically modify chromatin architecture to modulate access of the transcriptional machinery to DNA, and thus regulating gene expression.[203] Remodeling pathways are largely dependent on (i) various covalent modifications of histone tails driven by ATP-independent factors[203] such as deacetylase (HDAC), methyl transferase (HMT), acetyl Transferase (HAT), (ii) ATP-dependent chromatin remodeling complexes[204] which either slide, eject or restructure nucleosomes, and (iii) chaperones that bind to histones and stimulate their transfer onto DNA or other proteins.[205] On the basis of their functions, chromatin remodelers can be roughly divided into two families: ATP-dependent enzymes that include imitation switch (ISWI), chromodomain helicase DNA binding (CHD), switch/sucrose non-fermentable (SWI/SNF) and INO80,[204] and ATP-independent enzymes including the histone methyl/acetyl transferases, kinases, and isomerases. Despite differences in mechanisms and compositions, all ATP-dependent remodelers contain a structurally similar catalytic ATPase core which converts the chemical energy of ATP hydrolysis into conformational changes. Besides actively regulating gene expression, dynamic remodeling of chromatin imparts an epigenetic role in several key biological processes, *e.g.*, DNA replication and repair, apoptosis, and pluripotency.[206]

Chromatin remodelers have an extensive range of interacting partners. They can form multimeric complexes and interact with histones, transcription factors, nucleic acids, and various other machinery involved in the maintenance of chromatin structure.[207] Such a diverse range of interactions is difficult to explain with highly structured proteins. Predictions from amino acid sequence strongly suggest that chromatin remodelers contain substantial structural disorder,[208,209] involved in forming stable complexes and transient interactions with diverse interacting partners, potentially playing a more direct functional role than acting as simple linkers.[210,211]

3.3.1 ATP-dependent chromatin remodelers

Many ATP-dependent chromatin remodeling complexes are predicted to contain IDRs.[212] These IDRs range from relatively small regions, likely functioning as linkers, all the way to the BRG1/BRM-associated factor (BAF) complex which is made up of subunits that are predicted to contain long IDRs.[209,213] A recent study looked at the predicted disorder in BAF and found that 27 of the 30 subunits that were analyzed were predicted to be highly disordered.[209] The BAF complex is among the most frequently mutated complexes in many types of cancer, many of which are located in predicted disordered regions.[214] While the function of the predicted IDRs remains largely unknown, they are likely to assist with binding to histones, nucleic acids, and transcription factors.

ATRX (alpha thalassemia/mental retardation syndrome X-linked) belongs to the SWI/SNF family of chromatin remodeling proteins, and along with Death-associated protein 6 (DAXX), forms a complex that is necessary for H3.3 depositions into pericentric, telomeric, and ribosomal repeat sequences.[215,216] ATRX has multiple functions in the chromatin landscape, acting both as a chromatin remodeler and a histone chaperone.[217] ATRX is a large protein (2492 residues) and contains two structured domains; an N-terminal PHD-like domain and a conserved Snf2 domain.[218] The remaining ~1660 residues of ATRX sequence are predicted to be structurally disordered, with over 1300 residues in a single stretch separating the two domains.[219] The partner protein DAXX, a H3.3 histone chaperone, contains a long disordered C-terminal domain (residues 418–740).[213]

The involvement of the IDRs in ATRX and DAXX for catalyzing the deposition and remodeling of H3.3 nucleosomes, remains unclear.

Chromatin accessibility complex (CHRAC) is an evolutionarily conserved nucleosome remodeling complex that catalyzes histone octamer sliding on DNA.[220] Originally purified from *Drosophila melanogaster*, CHRAC consists of ISWI (ATPase), ACF1 and two histone fold subunits, CHRAC-14 and CHRAC-16.[221] A study looking into the function of CHRAC-14 and CHRAC-16 found unstructured N- and C-terminal domains on both proteins.[222] CHRAC-14 and CHRAC-16 form a heterodimer with a fold that resembles the geometry of histone dimer H2A-H2B,[222] which is predicted to create a surface for transient deposition of a segment of DNA as it is stripped from the core histone octamer. The C-terminal of both proteins is involved in DNA binding but with reciprocal effects; the C-terminal on CHRAC-14 increases DNA binding while the C-terminal on CHRAC-16 greatly decreases it but is still essential for sliding on DNA. It seems that the CHRAC-14/CHRAC-16 heterodimer enhances the catalysis of nucleosome sliding with weak and non-specific DNA binding. These findings were strikingly similar to the groups earlier work on the DNA chaperone HMGB1,[223] leading the authors to speculate that CHRAC-14/CHRAC-16 heterodimer serves as a built-In DNA chaperone.

3.3.2 ATP-independent chromatin remodelers

Post-translational modifications frequently occur in IDRs, as outlined above. Acetylation of the core histones enhances transcription by relaxing the condensed structure of the nucleosome, whereas deacetylation will promote chromatin condensation and transcriptional repression.[224,225] This effect is due to a charge neutralization of the acetylated lysine that weakens its interaction with the phosphate backbone of DNA. Both histone deacetylases and histone methylases are regulated by phosphorylations in predicted IDRs. Phosphorylations in HDACs 4,5,7 and 9 regulate shuttling between the nuclear and cytoplasmic compartments[226] and phosphorylations of sites flanking the nuclear localization sequence will promote chaperone protein binding and subsequent nuclear export.[227,228]

Histone methylation is a dynamic PTM central to eukaryotic transcription.[229] These modifications regulate gene expression by recruiting transcriptional cofactors that specifically recognize methylated lysine or arginine residues.[230,231] Dysregulation of histone methylation is associated with serious diseases such as cancers, developmental defects, and inflammatory bowel disease.[232,233] A recent study looked into PTMs of histone methylation enzymes in *Saccharomyces cerevisiae*, and found that phosphorylation was strongly enriched in predicted IDRs in methyltransferases while histone demethylases were phosphorylated within ordered regions.[234] Furthermore, the authors demonstrated that a phosphorylation cluster within an IDR of methyltransferase Set2p has a major effect on levels of H3K36 methylation *in vivo*. This decrease in H3K36 methylation leads to increased cryptic transcription, which can shorten the lifespan of cells.[235]

SIRT6 is an NAD^+-dependent histone deacetylase and is highly site-specific.[236] While early experiments, using H3 peptides,[237] demonstrated that SIRT6 has an ~1000 times slower catalytic activity then other related sirtuins, the low turnover rate did not match with recent studies using whole nucleosomes as substrates that found significantly higher catalytic rates.[238] This is likely due to interactions between the intrinsically disordered C-terminal region that has a high affinity to the nucleosome[239]; with SIRT6 tethered to the nucleosome the reaction can take place with greatly enhanced activity. Interestingly, while the SIRT6 interacts with nucleosomes in a 2:1 arrangement, only a single SIRT6 molecule can occupy the high affinity site. This arrangement may be due to the asymmetry of the two acidic patches, as observed with other chromatin remodelers that have a distinct response to each acidic patch.[240]

3.3.3 Chromatin remodelers with chaperone activity

Facilitates chromatin transcription (FACT) is a histone chaperone that has a dual-role as a nucleosome remodeler and chaperone.[225,241] In gene regulation, nucleosomes must temporarily unfold and then rapidly refold after the regulatory process. FACT increases accessibility of RNA polymerase II on chromatin by unfolding the nucleosome structure (Fig. 5).[242] FACT can then act as a histone chaperone that promotes nucleosome assembly by preventing some non-productive interactions between histones and DNA.[243] Both of FACT's two subunits, SSRP1 and SPT16, contain acidic and disordered regions that are implicated in histone binding.[241,244] Unlike most other histone chaperones, FACT can bind both H2A-H2B and H3-H4 dimers simultaneously,[245] with both subunits being involved in several interactions. Cryo-EM structures of FACT or SPT16 in complex with nucleosome constructs revealed that the CTD of SPT16, that includes an acidic IDR important for H2A/H2B binding, adopts a more ordered conformation when in complex with parts of the nucleosome.[246,247] Interestingly, the CTD appears to mimic DNA by compensating for the loss of histone DNA contacts (Fig. 5).[247] In a follow-up study using NMR

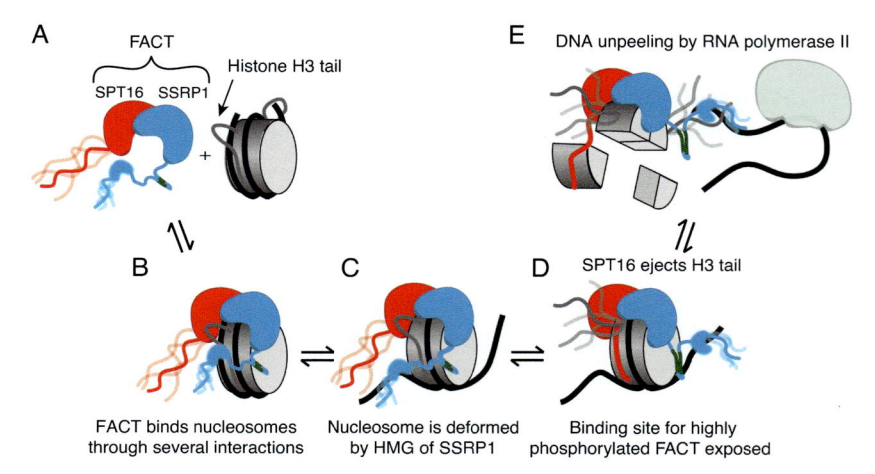

Fig. 5 Nucleosome assembly/disassembly by the histone chaperone FACT. (A) A canonical nucleosome with H3 tails (for clarity only the disordered tails of histone H3 is shown) buried in DNA gyres and the two subunits of FACT, SSRP1 and SPT16. (B) FACT binding to the nucleosome leads to deformation by the action of the HMG domain of the less phosphorylated (green area) state of SSRP1. (C) FACT with a highly phosphorylated SSRP1 has high affinity for deformed nucleosomes and replaces less phosphorylated FACT. (D) Deformation of the nucleosome exposes a binding site for the C-terminal domain of SPT16, causing increased solvent exposure of the histone H3 tail. (E) DNA is peeled off the nucleosome by RNA polymerase II (or other factors). Following transcription by RNA polymerase II, FACT can reassemble the nucleosome (not shown).

spectroscopy, it was revealed that one of the N-terminal tails of H3 adopts a different conformational ensemble when FACT is bound to the nucleosome.[248] NMR analysis of H3 tail chemical shifts indicated that it is buried in between two DNA gyres and that interaction is disrupted by the CTD of SPT16. This leads to increased solvent exposure of the tail, rendering it more susceptible to acetylation by HAT, indicating that FACT has a regulatory role in H3 acetylation. The intrinsically disordered domain (IDD) of SSRP1 has an acidic N-terminal part (AID) and a basic C-terminal part (BID). A recent study using NMR and CG molecular dynamics simulations, revealed how phosphorylation in the IDD change the intermolecular contacts between the AID and BID. These contact changes tune the affinity of SSRP1, with less phosphorylated states displaying high affinity to an intact nucleosome and highly phosphorylated states having high affinity to a deformed nucleosome, revealing an important mechanistic and regulatory role for the IDD.[210]

Another remodeler, decondensation factor 31 (Df31), is a fully disordered histone chaperone and an integral component of chromatin at all stages of *Drosophila melanogaster* lifecycle.[249,250] Df31 is suggested to have a role in the higher order structure of chromatin by promoting chromatin bridging *in vitro*.[251] Df31 binds to both histone H3 and H4 but has a higher affinity for H3.[252] Binding to H3 takes place through the intrinsically disordered H3 tail,[251] making PTMs to the H3 tail a likely modulator for binding. Recently, an RNA-dependent mechanism was discovered, where Df31 tethers chromatin-associated RNA (caRNA) to chromatin, resulting in an RNA-chromatin network which is more accessible and active.[252]

We have highlighted here how structural disorder is a prominent part of chromatin remodeling complexes but for most remodelers discussed here, detailed mechanistic insights remain hidden. FACT has, however, a well-established molecular mechanism, which was revealed with a close integration of NMR experiments and coarse-grained simulations, exemplifying the strength of such approaches.

3.4 Transcription through a nucleosomal barrier with disordered proteins

3.4.1 Transcription factors

The nucleosome represents a formidable barrier to transcription as the DNA sequence encoding a specific gene must become accessible to transcription factors in one way or another. The transcriptional machinery is rich with disorder and even the ribosomal assembly contains many disordered protein subunits.[253] The vast majority of transcription factors (TFs) (>85%) have long disordered linkers and transactivation domains (TADs) that flank their structured DBDs.[23,254] They bind cognate DNA sequences using predominantly their structured DBDs and may subsequently recruit other proteins to their binding site through their disordered TADs to initiate transcription. The TADs often contain hydrophobic residues (frequently aromatics) well interspersed with acidic residues, a feature that has been suggested to be important for keeping the region disordered and exposed in an active form allowing interactions with other proteins.[255] Nonetheless, the IDRs are not exclusively involved in protein-protein interactions: simulations have suggested that the affinity to DNA, cognate or non-specific, is tuned by disordered regions, especially those that have significant charges.[256] IDRs in TFs have also been linked to facilitating scanning for correct binding sites through non-specific interactions,[257] and to inter-strand exchange through a monkey-bar-like mechanism.[258] In fact, recent evidence points to TFs having multiple specificity determinants encoded in their IDR sequence, helping them to identify their specific binding sites by interacting with much broader DNA regions than are recognized with only their DBD cognate sites.[259] However, in the context of our nucleosomal landscape, traditional TFs require their binding sites to be accessible for binding, *i.e.*, within "open" chromatin states.

3.4.2 Pioneer transcription factors can alter cell fate

A unique class of TFs, called pioneer-TFs (pTFs), can bind to condensed, nucleosome-rich regions of the genome and open these previously inaccessible regions to transcription (Fig. 6).[187,260] This alters the transcriptional pattern of a cell—the main determinant of its fate[260]—and can initiate cell reprogramming. Despite the ultimate change in cell fate relying on subsequent recruitment of other factors, the initial binding ability to condensed chromatin is what distinguishes pTFs from other TFs. A remarkable example of pioneer activity is the so-called Yamanaka factors; a group of four pTFs (Oct4, Sox2, Klf4, and c-Myc) that can induce a fibroblast to revert to a pluripotent stem cell (iPSC)[261]—a process that earned the discoverers the Nobel prize in 2012. Other pTFs, such as FoxA1, Ascl1, and Pu.1, have since been shown to play key roles for inducing direct reprogramming from fibroblasts to hepatocytes, neurons, and macrophage-like cells, respectively.[187] Reprogramming cell fate has immense potential for human health, with recent reports showing extraordinary examples in regenerative medicine such as sight restoration in mice, *in vitro* disease modeling, and drug discovery.[262,263] However, to fully exploit the power of pTFs for cell reprogramming, a detailed and quantitative understanding of their molecular mechanism is critically needed.[187] For example, it is largely unknown whether pTFs bind to DNA that becomes spontaneously and transiently accessible on nucleosomes or whether they actively "open" nucleosomal DNA. In other words, how pTFs can dynamically invade compacted chromatin and initiate remodeling remains unclear. Some pTFs interact with enzymes that remodel chromatin besides recruiting other TFs, and in those cases, it can be challenging to separate the actions of the two classes of proteins: are the chromatin remodelers necessary for remodeling and do the pTFs just invade chromatin to initiate binding, or can those pTFs also remodel chromatin themselves? The answers to these questions remain hidden, in part due to the highly dynamic and heterogeneous conformations of pTFs and chromatin, which render these systems notoriously difficult to assay by classical structural biology methods.

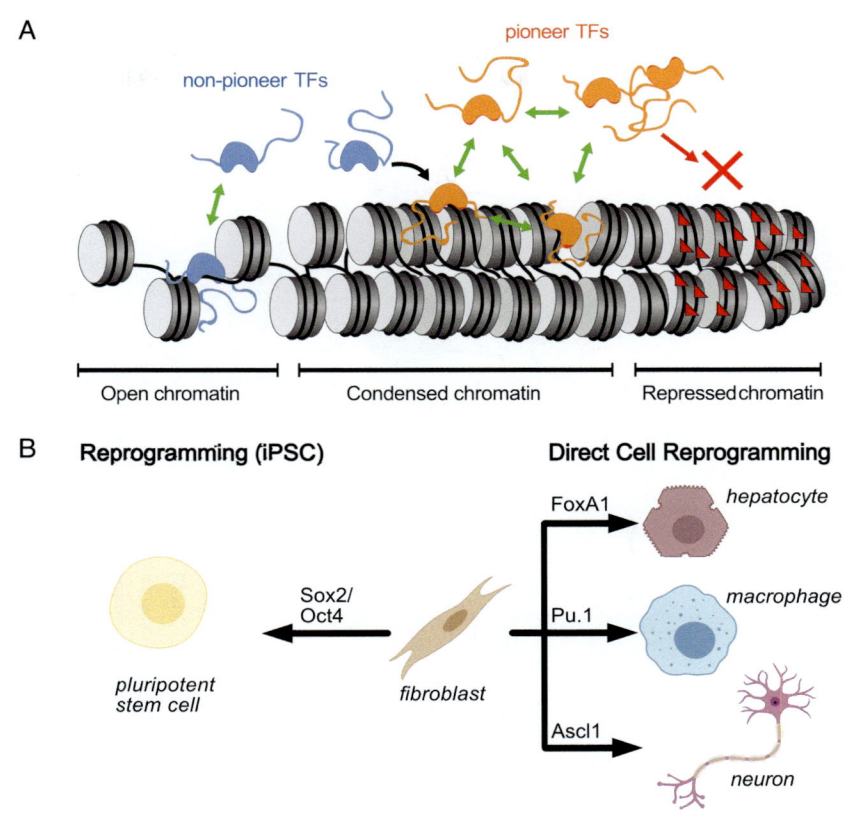

Fig. 6 Pioneer transcription factors can invade and open condensed chromatin and initiate cell-fate changes. (A) Pioneer-TFs (orange) can bind to condensed chromatin regions and render it accessible to traditional TFs (blue) or other components of the transcriptional machinery. (B) Pioneer-TFs can lead to cell-fate changes, either through reprogramming with formation of induced pluripotent stem cells (iPSCs), or through direct cell reprogramming. *Panel A based on Zaret KS, Mango SE. Pioneer transcription factors, chromatin dynamics, and cell fate control.* Curr Opin Genet Dev. *2016;37:76–81.*

Like the vast majority of TFs, pTFs are rich in disordered linkers and TADs (Fig. 7).[264] Despite their abundance in pTFs, IDRs have largely been overlooked thus far in studies of TFs, which is especially evident considering the vast number of TF DBDs in the Protein Data Bank and the total absence of 3D-structures containing entire eukaryotic TFs. Instead, intense focus has centered on the DBDs in attempts to explain pioneering activity, with impressive high–resolution structures revealing complexes between the pTF DBDs and nucleosomes.[265,266] The DBDs themselves are often disordered

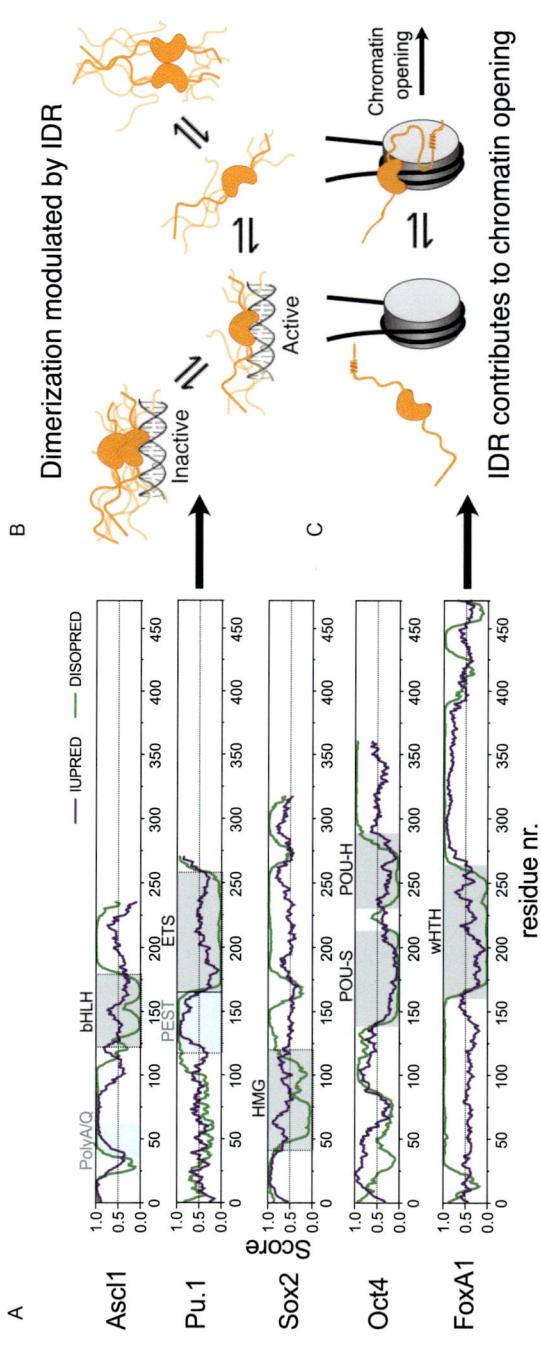

Fig. 7 Intrinsically disordered regions in pioneer transcription factors. (A) Disorder predictions for five pTFs based on two different predictors. Domains are highlighted. (B) The intrinsically disordered and acidic PEST domain in the pTF Pu.1 modulates the formation of a dimer on and off DNA. A 1:1 Pu.1-DNA complex activates transcription and dimerization negatively regulates the activity. The dimer in the absence of DNA is furthermore thermodynamically destabilized compared to the monomer. (C) FoxA1 interacts with the core histones in a nucleosome through a short motif in its C-terminal IDR. This interaction contributes to chromatin opening and thus its pioneering functions. The schematics of the conformations of structured domains and IDRs in panels (B) and (C) are purely for illustrative purposes. *Panel B reaction scheme is based on Xhani S, Lee S, Kim HM, et al. Intrinsic disorder controls two functionally distinct dimers of the master transcription factor PU.1. Sci Adv. 2020;6(8): eaay3178, Panel C based on results from Iwafuchi M, Cuesta I, Donahue G, et al. Gene network transitions in embryos depend upon interactions between a pioneer transcription factor and core histones. Nat Genet. 2020;52(4):418–427.*

before binding to their cognate DNA sequence, followed by a disorder-to-order transition upon complex formation.[267] The DBDs are also often the major contributors to DNA affinity and in some cases such as for Sox2, the IDRs seemingly weaken affinity for the cognate sequence.[268] A recent computational study also implicated rotational and sliding dynamics of the DNA on the nucleosome to be important for binding of pTFs. Using CG models and simulations, Tan and Takada showed that Sox2 recognizes a certain rotational phase of its binding site which induced sliding, affecting allosterically the binding of Oct4 or another Sox2 molecule.[269] Clearly, the DBDs are critical for pTF function but what possible role do IDRs play in chromatin opening and subsequent reprogramming pathways?

3.4.3 Roles of intrinsic disorder in pioneer transcription factors

Disordered regions are frequently involved in protein-protein interactions[270] and in TFs the TADs often recruit components necessary for transcription. Moreover, interplay between ordered and disordered regions is poorly understood but expected as IDRs have usually co-evolved with ordered regions and the conformational propensities of IDRs may therefore be modulated by folded domains and *vice versa*. It is possible that, after scanning and binding recognition sites, the DBDs act as anchors to allow the disordered regions to inflict interactions that disrupt internucleosome contacts in chromatin, leading to opening of the chromatin fiber. In that way, IDRs could be involved in actively opening chromatin, modulating oligomerization regulating pioneer activity (see below for Pu.1), or involved in recruiting chromatin remodeling enzymes, followed by opening of chromatin, and subsequent binding of other transcription factors to the exposed DNA. Ultimately, IDRs may have multiple, context-dependent roles regulated by cell-type, chromatin modifications, and local sequence determinants. Nevertheless, the role of IDRs has been glimpsed recently for many pTFs, suggesting a function in chromatin opening and a large-scale impact on gene expression networks.

Strong evidence of IDR involvement in chromatin opening comes from recent work from the Zaret lab, which revealed a role of IDRs in the prototypical pTFs FoxA1 and FoxA2 for modulating interactions with core histone proteins in a nucleosome.[194] Using a combination of sequence analysis, cross-linking, and mass spectrometry, the authors discovered a conserved 9 amino-acid sequence in the disordered C-terminal, which is critical for chromatin opening functions through an interaction with the core histones in a nucleosome. This short region likely forms a transient α-helix, as helix formation could be induced by addition of helix-promoting trifluoroethanol in a short peptide when monitored by circular dichroism spectroscopy. When this region was deleted, the chromatin opening ability, measured by DNase cleavage sensitivity, was severely reduced, as well as the ability to activate certain target genes. Using mouse embryos, the authors further went to show that deletion of the short a-helix led to a 60% reduction in target gene activation, severely impairing embryonic development by affecting gene expression and chromatin accessibility. Clearly, this disordered region plays a crucial role in the pioneering function of FoxA1. Beyond pioneering functions, the FoxA proteins also have heavy ties to cancer biology through their direct interaction with both the estrogen and androgen receptors,[271] and FoxA1 is currently hailed as a very promising therapeutic target. The interaction of FoxA1 with both receptors is influenced by PTMs in the disordered regions, including SUMOylation that has a negative effect on transcriptional activity and on association with the androgen receptor.[272]

The key Yamanaka factor Sox2 has a short N-terminal and a long, ∼200-residue C-terminal IDR flanking an HMG-box DBD.[273] The Sox2 HMG-box cooperates with the Oct4 POU-domain, and this interaction is critical for producing iPSC and maintaining pluripotency but the efficiency of reprogramming is conferred by the extreme C-terminal IDR[274] through a currently unclear mechanism. Recent studies have shown how Sox2 and Oct4 act in concerted fashion to invoke structural changes in the core nucleosome structure ranging from subtle local distortion to fully removing DNA from one side, depending on the cognate binding site location.[265] However, the dynamic events of scanning and binding that finally lead to chromatin opening are still mostly unknown. The IDR region immediately flanking the C-terminal side of the DBD (120–160) has recently been implicated in RNA binding, even concurrently with the DBD being DNA-bound.[273] The authors went on to show that deletion of the RNA binding domain severely reduced the efficiency of iPSC generation, demonstrating a clear link between the IDR and cell reprogramming.

Pu.1 is a hematopoietic master regulator pTF that contains an N-terminal TAD, a disordered anionic PEST domain (rich in prolines, glutamic acids, serines and threonines), and a structured DBD called ETS (Erythroblast transformation specific) domain. Xhani et al. showed that

Pu.1 dimerizes through its DBD and gene expression is regulated by two distinct dimeric states: a transcriptionally active 1:1 complex and an inactive ternary complex involving two Pu.1 molecules bound to a single DNA recognition site (Fig. 7),[275] forming a negative feedback mechanism that the authors confirmed *in vivo*. Using NMR spectroscopy and tryptophan fluorescence experiments, the authors showed that the intrinsically disordered PEST domain reduced the binding affinity of the second Pu.1 molecule to form a ternary complex. Interestingly, however, the PEST domain also promotes homodimerization in the absence of DNA. The two dimeric forms were found to be non-equivalent, with an asymmetric DNA-bound Pu.1 dimer and a symmetric homodimer in the DNA-free state. A legion of serines in the PEST domain is phosphorylated *in vivo*, which prompted the authors to introduce phosphomimetic substitutions in that region. Indeed, the degree of negative feedback was reduced with phosphomimetic substitutions which promoted the formation of a transcriptionally active 1:1 complex with DNA. It remains to be determined whether a similar regulatory dimerization mechanism would be observed on nucleosomes but the positively charged histone tails may provide an additional interaction interface for the negative charges in the PEST domain. There may furthermore be other complicating factors, as binding of Pu.1 to nucleosomes has been reported to be context-specific, suggesting a non-classical pioneering role for Pu.1.[276]

Yet another example of a disordered pTF is the achaete-scute homolog 1 (Ascl1), which drives the conversion of fibroblasts to neurons.[277] Ascl1 is a relatively small transcription factor that has a characteristic polyA/polyQ region in the N-terminal and a basic helix-loop-helix DBD in the C-terminal. In a clever, fragment-based approach, Baronti et al. were able to use NMR spectroscopy to dissect the highly aggregation-prone Ascl1[278] and found an extended and dynamic structure with transient helix formation yet no persistent tertiary interactions—a classical characteristic of an IDP. Little mechanistic information is available on the interactions between Ascl1 and DNA or nucleosomes but a genome-wide analysis showed that it is one of only a handful of TFs that binds strongly to both DNA and nucleosomes albeit likely as a heterodimer.[279]

We have highlighted a subset of pTFs that have been studied by biophysical approaches but many other established pTFs are predicted to contain long IDRs.[264] Molecular biology has over the years been extraordinarily powerful at identifying the key players in transcriptional regulation networks during cell development. Yet, the link between molecular properties of pTFs, especially the role of their IDRs, and cell reprogramming is still largely missing. Integrative modeling approaches, using available structural information in concert with biophysical studies and simulations, might be a potent strategy to understand the physical principles of cell-identity pathways, leading us closer to controlling cell fate.

4. Common sequence features of disordered nucleosome-binding proteins

In the disordered interactions and their regulation reviewed above, charge emerges as a recurring theme. Charge is a principal component of chromatin and is often utilized by IDPs to elicit a specific cellular response. While the DNA backbone is highly acidic, the linker and core histone tails are highly basic, creating an electrostatic balance in the NCP.[280] Opposite charges in the DNA and histone tails have been implicated in a number of inter- and intra-nucleosomal interactions, which act to either condense or decondense chromatin. Moreover, PTMs that alter charge in the disordered histone tails have been shown to affect nucleosome stability.[281] For instance, neutralization of positive charge by acetylation or introduction of negative charge by phosphorylation of basic residues in the histone H3/H4 tail regions, weakens the histone-DNA interactions by reducing electrostatic attraction.[281] Consequently, chromatin takes on an open structure, increasing nucleosome accessibility to modifying enzymes. Charge has an especially clear role for the highly disordered H1 competitors (protamines, HMG proteins, ProTα). A common feature among these proteins may be that the unspecific nature of charge interactions and the high fraction of charges allows these proteins to interact in complexes beyond a basic 1:1 stoichiometry, exchange rapidly in a concentration-dependent manner, and keep regulatory systems highly responsive despite high affinity binding. Those molecular parameters would in turn be finely regulated by PTMs that affect charge.

In the cell, several transcription factors, chromatin remodelers and architectural proteins function in a dynamic balance, ultimately controlling gene expression. Understanding the effects of charge in IDPs that interact with chromatin and chromatin-binding proteins may provide insight into their specific cellular mechanisms. To better understand charge properties, we calculated kappa (κ) values for the IDRs of proteins discussed in this review (Fig. 8). κ is a patterning parameter used to describe strong and weak

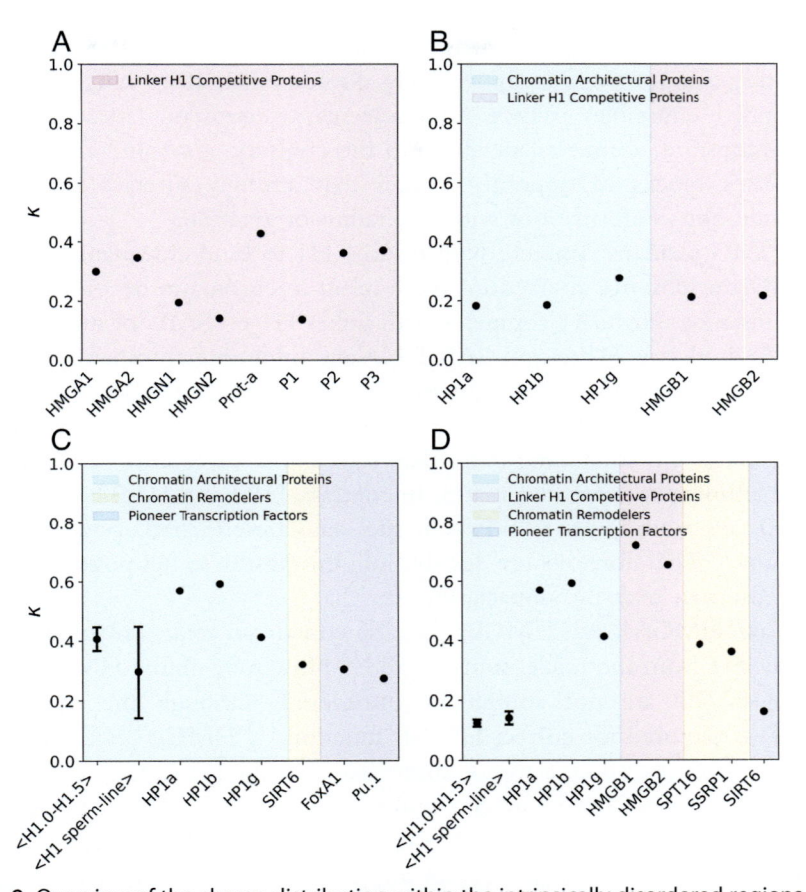

Fig. 8 Overview of the charge distribution within the intrinsically disordered regions of nucleosome-binding proteins. Understanding the role of charge in nucleosome-binding proteins may prove important to frame the functional space of different intrinsically disordered proteins within the context of transcriptional regulation. The parameter kappa (κ), which can take any value between 0 and 1 and has been formulated to link sequence properties to conformational behavior of intrinsically disordered regions,[26] describes the overall charge asymmetry in an amino-acid sequence. A low κ indicates more evenly distributed positive and negative charge, whereas κ increasingly close to 1 indicates blocks of positive and negative charge. The need to strongly coordinate DNA may render charge distribution an important factor to finely tune protein-DNA interactions. While for fully disordered proteins (shown in A) κ is found to vary considerably, for intrinsically disordered hinge regions linking structured domains (shown in B) κ is low and similar across all proteins. A clear difference can be noted between the κ values of the N-terminal (C) domains of linker histones that are involved in the physiological homeostasis of chromatin and those expressed by sperm-line cells, which are involved in extreme chromatin compaction and that show a lower κ. While no other trend can be clearly seen for the other proteins or for their C-terminal domains (shown in D), it is important to acknowledge that our classification is arbitrary and founded on the current understanding of the role that these proteins have within the nucleosomal landscape.

polyampholytes. A low κ value is indicative of well distributed negative and positive charges along an intrinsically disordered domain which generate extended ensembles, where intramolecular electrostatic attractions and repulsions are counterbalanced.[26] On the contrary, a κ value close to 1.0 indicates blocks of opposite charges that strongly interact leading to globule-like conformations with low radius of gyration.

HMG proteins compete with histone H1 to bind chromatin and thus HMG-nucleosome interactions often result in chromatin decondensation. For instance, HMGA1 competes with linker H1 on SARs of nucleosomal DNA, displacing H1 to non-SAR DNA and inhibiting chromatin compaction.[184] Interestingly, the κ value for HMGA1 is similar to that of the CTD of linker H1.1, H1.2 and H1.5 ($0.3 < \kappa < 0.4$). Linker histone H1.5 has a long CTD tail, containing more than two S/T-P-X-K sites, resulting in a high affinity for heterochromatin. In contrast, H1.1 and H1.2 have shorter CTD tails, with fewer S/T-P-X-K sites, and are enriched at euchromatic regions.[78] Therefore, charge distribution, in addition to net positive charge and disorder, may also impact affinity.

Like HMGA1/A2, HMGB1/B2 also contain an acidic tail and displace linker H1 from the nucleosomal dyad.[173] However, unlike HMGA1/A2, HMGB1/B2 are not completely disordered, although the disordered CTD is required for correct HMGB function.[175] HMGB1 is involved in the regulation of p53; a tumor suppressor that binds to DNA which acts by protect cells from malignant transformation.[282,283] HMGB1 has been shown to stimulate the linear DNA-p53 interaction *in vitro* and, *in vivo*, p53 activity is increased.[284] Additionally, HMGB1 and p53 have been shown to directly interact *via* the PXXPXP motif in the disordered NTD of p53 and HMG boxes in HMGB1. Moreover, the disordered acidic tail in HMGB1 is a direct determinant of this interaction, as it shields the positive charge in the HMG box decreasing p53-HMGB1 affinity and linking disorder to protein function.[285] In contrast to HMGB1, the interaction between H1.2 and p53 induces p53 repression in DNA damage response. Moreover, this interaction is negatively regulated by acetylation in the p53 CTD and phosphorylation in the H1.2 CTD. In both cases, PTM acts to disrupt the p53-H1.2 interactions, directly implicating charge and disorder in protein functionality.[286]

The connection between charge and disorder is prominent when considering the interplay of HMGN1/N2 and H1T2, H1oo and HILS1 variants. HMGN1/N2 promote chromatin decompaction by interacting with nucleosomal DNA at the major grooves flanking the dyad and competing with linker H1 for binding sites.[155] Furthermore, HMGN1/N2 has a low kappa value $(0.1 < \kappa < 0.2)$, that is similar to that of the H1T2, HILS1 and H1oo CTD and characteristic of disordered proteins. Therefore, HMGN1/N2 may use its disorder to compete with these H1 variants for binding sites. Interestingly, while most H1 variants have few arginine residues, H1T2 and HILS1 have an almost equal fraction of lysine and arginine residues in the CTD. Because arginine forms stronger interactions with the DNA phosphate backbone, the testis specific variants are likely to be harder to displace. For instance, during spermatogenesis, inactivation of the gene for H1T2 leads to defects in DNA condensation and chromatin packing; effects that are not favorable in cell development.[287]

5. Concluding remarks

The nucleus is enriched in proteins that are disordered and thus highly dynamic. These proteins play key roles in maintaining the genome and regulating its read-out. Despite decades of active research on IDPs and their well-recognized importance in ensuring the homeostasis of the nucleus, we still lack an exhaustive description of the interactions between chromatin components and IDPs, especially with respect to how they translate to biological function and regulation. New methodological paradigms are needed to tackle intrinsic disorder in the nucleus, because of both the intrinsic dynamic character and the physico-chemical properties of the interacting molecular partners, which frequently feature extremely strong electrostatics. Consequently, in recent years there has been a considerable upsurge in methodological development, especially for single-molecule techniques which can discriminate distinct conformational sub-populations and sequences of molecular events. Computational approaches that directly integrate single-molecule data and simulations, featuring customized potential energy functions tuned on the basis of experimental findings, have provided an unprecedented view of the ensemble of some key disordered interactions in the nucleus. Remarkably, simple potential energy functions that dominantly account for electrostatic contributions to binding, have been able to exhaustively reproduce experimental findings and provide a mechanistic understanding of protein-protein and protein-DNA interactions.[17,25,288] In the future, such simple customized potentials may evolve into more complex combinations of potential energy terms that might take into account, explicitly, the effects of post-translational modifications, such as methylation and acetylation, in specific sites along intrinsically disordered domains. Additionally, the modeling and parameterization of explicit ionic species, especially for coarse-grained simulations, would be a considerable advancement for a more accurate estimation of the energetics involved in the nucleosomal landscape, especially considering the primary role of ions in defining the association of strong disordered polyampholyte chains that interact with chromatin. Overall, access to integrative modeling approaches is still a challenge, as it requires strong collaborative efforts between different research groups. Nevertheless, creating synergy between experiments and simulations is key to refining our view of the disordered nuclear milieu.

Acknowledgments

This work was funded by the Icelandic Research Fund (Rannís grant nr. 206591-052 and 217392-051, to P.O.H.) and the Centre for eResearch (http://www.eresearch.auckland. ac.nz) at the University of Auckland for their help in facilitating this research. We apologize to the authors of other excellent work that we could not cite due to space limitations.

References

1. Wright PE, Dyson HJ. Intrinsically unstructured proteins: re-assessing the protein structure-function paradigm. *J Mol Biol.* 1999;293(2):321–331.
2. Schad E, Tompa P, Hegyi H. The relationship between proteome size, structural disorder and organism complexity. *Genome Biol.* 2011;12(12):R120.
3. van der Lee R, Buljan M, Lang B, et al. Classification of intrinsically disordered regions and proteins. *Chem Rev.* 2014;114(13):6589–6631.
4. Frege T, Uversky VN. Intrinsically disordered proteins in the nucleus of human cells. *Biochem Biophys Rep.* 2015;1:33–51.
5. Schuler B, Borgia A, Borgia MB, et al. Binding without folding—the biomolecular function of disordered polyelectrolyte complexes. *Curr Opin Struct Biol.* 2020;60: 66–76.
6. Wright PE, Dyson HJ. Intrinsically disordered proteins in cellular signalling and regulation. *Nat Rev Mol Cell Biol.* 2015;16(1):18–29.
7. Theillet FX, Binolfi A, Frembgen-Kesner T, et al. Physicochemical properties of cells and their effects on intrinsically disordered proteins (IDPs). *Chem Rev.* 2014;114(13): 6661–6714.
8. Habchi J, Tompa P, Longhi S, Uversky VN. Introducing protein intrinsic disorder. *Chem Rev.* 2014;114(13):6561–6588.
9. Hsu WL, Oldfield CJ, Xue B, et al. Exploring the binding diversity of intrinsically disordered proteins involved in one-to-many binding. *Protein Sci.* 2013;22(3):258–273.
10. Kulkarni P, Uversky VN. Intrinsically disordered proteins: the dark horse of the dark proteome. *Proteomics.* 2018;18(21−22):e1800061.

11. Fyodorov DV, B-r Z, Skoultchi AI, Bai Y. Emerging roles of linker histones in regulating chromatin structure and function. *Nat Rev Mol Cell Biol.* 2017;19:192–206. Nature Publishing Group.

12. McGinty R, Tan S. Nucleosome structure and function. *Chem Rev.* 2014;115(6): 2255–2273.

13. Bednar J, Garcia-Saez I, Boopathi R, et al. Structure and dynamics of a 197 bp nucleosome in complex with linker histone H1. *Mol Cell.* 2017;66(3):384–397.e8.

14. Peng Z, Mizianty MJ, Xue B, Kurgan L, Uversky VN. More than just tails: Intrinsic disorder in histone proteins. *Mol Biosyst.* 2012;8(7):1886–1901.

15. Martire S, Banaszynski LA. The roles of histone variants in fine-tuning chromatin organization and function. *Nat Rev Mol Cell Biol.* 2020;21(9):522–541.

16. Lever MA, Th'ng JP, Sun X, Hendzel MJ. Rapid exchange of histone H1.1 on chromatin in living human cells. *Nature.* 2000;408(6814):873–876.

17. Heidarsson PO, Mercadante D, Sottini A, et al. Disordered proteins enable histone chaperoning on the nucleosome. *bioRxiv.* 2020;2020.04.17.046243.

18. Kale S, Goncearenco A, Markov Y, Landsman D, Panchenko AR. Molecular recognition of nucleosomes by binding partners. *Curr Opin Struct Biol.* 2019;56: 164–170.

19. Walsh CT, Garneau-Tsodikova S, Gatto Jr GJ. Protein posttranslational modifications: the chemistry of proteome diversifications. *Angew Chem Int Ed Engl.* 2005;44(45): 7342–7372.

20. Bowman GD, Poirier MG. Post-translational modifications of histones that influence nucleosome dynamics. *Chem Rev.* 2015;115(6):2274–2295.

21. Darling AL, Uversky VN. Intrinsic disorder and posttranslational modifications: the darker side of the biological dark matter. *Front Genet.* 2018;9:158.

22. Luo C, Hajkova P, Ecker JR. Dynamic DNA methylation: in the right place at the right time. *Science.* 2018;361(6409):1336–1340.

23. Liu J, Perumal NB, Oldfield CJ, Su EW, Uversky VN, Dunker AK. Intrinsic disorder in transcription factors. *Biochemistry.* 2006;45(22):6873–6888.

24. Tompa P, Fuxreiter M. Fuzzy complexes: polymorphism and structural disorder in protein-protein interactions. *Trends Biochem Sci.* 2008;33:2–8.

25. Borgia A, Borgia MB, Bugge K, et al. Extreme disorder in an ultrahigh-affinity protein complex. *Nature.* 2018;555(7694):61–66.

26. Das RK, Pappu RV. Conformations of intrinsically disordered proteins are influenced by linear sequence distributions of oppositely charged residues. *Proc Natl Acad Sci U S A.* 2013;110(33):13392–13397.

27. Cuvier O, Fierz B. Dynamic chromatin technologies: from individual molecules to epigenomic regulation in cells. *Nat Rev Genet.* 2017;18(8):457–472.

28. Hellman LM, Fried MG. Electrophoretic mobility shift assay (EMSA) for detecting protein-nucleic acid interactions. *Nat Protoc.* 2007;2(8):1849–1861.

29. Jain SS, Tullius TD. Footprinting protein-DNA complexes using the hydroxyl radical. *Nat Protoc.* 2008;3(6):1092–1100.

30. Gontier A, Varela PF, Nemoz C, et al. Measurements of protein-DNA complexes interactions by isothermal titration calorimetry (ITC) and microscale thermophoresis (MST). *Methods Mol Biol.* 2021;2247:125–143.

31. Nakato R, Sakata T. Methods for ChIP-seq analysis: a practical workflow and advanced applications. *Methods.* 2021;187:44–53.

32. Buenrostro JD, Giresi PG, Zaba LC, Chang HY, Greenleaf WJ. Transposition of native chromatin for fast and sensitive epigenomic profiling of open chromatin, DNA-binding proteins and nucleosome position. *Nat Methods.* 2013;10(12): 1213–1218.

33. Nakane T, Kotecha A, Sente A, et al. Single-particle cryo-EM at atomic resolution. *Nature*. 2020;587(7832):152–156.
34. Shi Y. A glimpse of structural biology through X-ray crystallography. *Cell*. 2014;159(5):995–1014.
35. Wu M, Lander GC. Present and emerging methodologies in Cryo-EM single-particle analysis. *Biophys J*. 2020;119(7):1281–1289.
36. Loveland AB, Demo G, Korostelev AA. Cryo-EM of elongating ribosome with EF-Tu*GTP elucidates tRNA proofreading. *Nature*. 2020;584(7822):640–645.
37. Luger K, Mäder AW, Richmond RK, Sargent DF, Richmond TJ. Crystal structure of the nucleosome core particle at 2.8 Å resolution. *Nature*. 1997;389(6648):251–260.
38. Adhireksan Z, Sharma D, Lee PL, Davey CA. Near-atomic resolution structures of interdigitated nucleosome fibres. *Nat Commun*. 2020;11(1):4747.
39. Song F, Chen P, Sun D, et al. Cryo-EM study of the chromatin fiber reveals a double helix twisted by tetranucleosomal units. *Science*. 2014;344:376–380.
40. Garcia-Saez I, Menoni H, Boopathi R, et al. Structure of an H1-bound 6-nucleosome array reveals an untwisted two-start chromatin fiber conformation. *Mol Cell*. 2018; 72(5):902–915.e7.
41. Bharath MS, Chandra NR, Rao M. Molecular modeling of the chromatosome particle. *Nucleic Acids Res*. 2003;31(14):4264–4274.
42. Wong H, Victor J-M, Mozziconacci J. An all-atom model of the chromatin fiber containing linker histones reveals a versatile structure tuned by the nucleosomal repeat length. *PLoS One*. 2007;2(9):e877.
43. Stehr R, Kepper N, Rippe K, Wedemann G. The effect of internucleosomal interaction on folding of the chromatin fiber. *Biophys J*. 2008;95(8):3677–3691.
44. Kepper N, Foethke D, Stehr R, Wedemann G, Rippe K. Nucleosome geometry and internucleosomal interactions control the chromatin fiber conformation. *Biophys J*. 2008;95(8):3692–3705.
45. Öztürk MA, Cojocaru V, Wade RC. Dependence of chromatosome structure on linker histone sequence and posttranslational modification. *Biophys J*. 2018;114(10): 2363–2375.
46. Sottini A, Borgia A, Borgia MB, et al. Polyelectrolyte interactions enable rapid association and dissociation in high-affinity disordered protein complexes. *Nat Commun*. 2020;11(1):5736.
47. Milles S, Jensen MR, Lazert C, et al. An ultraweak interaction in the intrinsically disordered replication machinery is essential for measles virus function. *Sci Adv*. 2018;4(8): eaat7778.
48. Theillet FX, Binolfi A, Bekei B, et al. Structural disorder of monomeric alpha-synuclein persists in mammalian cells. *Nature*. 2016;530(7588):45–50.
49. Kay LE. NMR studies of protein structure and dynamics—a look backwards and forwards. *J Magn Reson*. 2011;213(2):492–494.
50. van Emmerik CL, van Ingen H. Unspinning chromatin: revealing the dynamic nucleosome landscape by NMR. *Prog Nucl Magn Reson Spectrosc*. 2019;110:1–19.
51. Zhou BR, Jiang J, Feng H, Ghirlando R, Xiao TS, Bai Y. Structural mechanisms of nucleosome recognition by linker histones. *Mol Cell*. 2015;59(4):628–638.
52. Abramov G, Velyvis A, Rennella E, Wong LE, Kay LE. A methyl-TROSY approach for NMR studies of high-molecular-weight DNA with application to the nucleosome core particle. *Proc Natl Acad Sci U S A*. 2020;117(23):12836–12846.
53. Bernado P, Svergun DI. Analysis of intrinsically disordered proteins by small-angle X-ray scattering. *Methods Mol Biol*. 2012;896:107–122.
54. Henriques J, Arleth L, Lindorff-Larsen K, Skepo M. On the calculation of SAXS profiles of folded and intrinsically disordered proteins from computer simulations. *J Mol Biol*. 2018;430(16):2521–2539.

55. Misteli T, Gunjan A, Hock R, Bustin M, Brown DT. Dynamic binding of histone H1 to chromatin in living cells. *Nature*. 2000;408(6814):877–881.

56. Brucale M, Schuler B, Samori B. Single-molecule studies of intrinsically disordered proteins. *Chem Rev*. 2014;114(6):3281–3317.

57. Asher WB, Geggier P, Holsey MD, et al. Single-molecule FRET imaging of GPCR dimers in living cells. *Nat Methods*. 2021;18(4):397–405.

58. Lerner E, Barth A, Hendrix J, et al. FRET-based dynamic structural biology: challenges, perspectives and an appeal for open-science practices. *Elife*. 2021;10.

59. Schuler B, Hofmann H. Single-molecule spectroscopy of protein folding dynamics—expanding scope and timescales. *Curr Opin Struct Biol*. 2013;23:36–47.

60. Gunther E, Klauss A, Toro-Nahuelpan M, Schuler D, Hille C, Faivre D. The in vivo mechanics of the magnetotactic backbone as revealed by correlative FLIM-FRET and STED microscopy. *Sci Rep*. 2019;9(1):19615.

61. Lerner E, Cordes T, Ingargiola A, et al. Toward dynamic structural biology: two decades of single-molecule Förster resonance energy transfer. *Science*. 2018;359: eaan1133.

62. Heidarsson PO, Cecconi C. From folding to function: complex macromolecular reactions unraveled one-by-one with optical tweezers. *Essays Biochem*. 2021;65(1): 129–142.

63. Rudnizky S, Khamis H, Malik O, Melamed P, Kaplan A. The base pair-scale diffusion of nucleosomes modulates binding of transcription factors. *Proc Natl Acad Sci U S A*. 2019;116(25):12161–12166.

64. Gomes GW, Krzeminski M, Namini A, et al. Conformational ensembles of an intrinsically disordered protein consistent with NMR, SAXS, and single-molecule FRET. *J Am Chem Soc*. 2020;142(37):15697–15710.

65. Vendruscolo M, Dobson CM. Protein dynamics: Moore's law in molecular biology. *Curr Biol*. 2011;21(2):R68–R70.

66. Fan L, Roberts VA. Complex of linker histone H5 with the nucleosome and its implications for chromatin packing. *Proc Natl Acad Sci U S A*. 2006;103(22):8384–8389.

67. Collepardo-Guevara R, Schlick T. Chromatin fiber polymorphism triggered by variations of DNA linker lengths. *Proc Natl Acad Sci U S A*. 2014;111(22):8061–8066.

68. Teif VB, Kepper N, Yserentant K, Wedemann G, Rippe K. Affinity, stoichiometry and cooperativity of heterochromatin protein 1 (HP1) binding to nucleosomal arrays. *J Phys Condens Matter*. 2015;27(6):064110.

69. Watanabe S, Mishima Y, Shimizu M, Suetake I, Takada S. Interactions of HP1 bound to H3K9me3 dinucleosome by molecular simulations and biochemical assays. *Biophys J*. 2018;114(10):2336–2351.

70. Alston JJ, Soranno A, Holehouse AS. Integrating single-molecule spectroscopy and simulations for the study of intrinsically disordered proteins. *Methods*. 2021.

71. Sanders JC, Holmstrom ED. Integrating single-molecule FRET and biomolecular simulations to study diverse interactions between nucleic acids and proteins. *Essays Biochem*. 2021;65(1):37–49.

72. Rhodes D, Robinson PJJ. Structure of the '30 nm' chromatin fibre: a key role for the linker histone. *Curr Opin Struct Biol*. 2006;16:336–343.

73. Hergeth SP, Schneider R. The H1 linker histones: multifunctional proteins beyond the nucleosomal core particle. *EMBO Rep*. 2015;16:1439–1453.

74. Hansen JC, Lu X, Ross ED, Woody RW. Intrinsic protein disorder, amino acid composition, and histone terminal domains*. *J Biol Chem*. 2006;281(4):1853–1856.

75. Goytisolo FA, Gerchman SE, Yu X, et al. Identification of two DNA-binding sites on the globular domain of histone H5. *EMBO J*. 1996;15:3421–3429.

76. Gibbs EB, Kriwacki RW. Linker histones as liquid-like glue for chromatin. *Proc Natl Acad Sci U S A*. 2018;115(47):11868–11870.

77. Izzo A, Kamieniarz K, Schneider R. The histone H1 family: specific members, specific functions? *Biol Chem*. 2008;389(4):333–343.

78. Th'ng JP, Sung R, Ye M, Hendzel MJ. H1 family histones in the nucleus: control of binding and localization by the C-terminal domain. *J Biol Chem*. 2005;280(30):27809–27814.

79. Lorch Y, LaPointe JW, Kornberg RD. Nucleosomes inhibit the initiation of transcription but allow chain elongation with the displacement of histones. *Cell*. 1987;49(2):203–210.

80. Roque A, Ponte I, Suau P. Post-translational modifications of the intrinsically disordered terminal domains of histone H1: effects on secondary structure and chromatin dynamics. *Chromosoma*. 2017;126(1):83–91.

81. Talasz H, Helliger W, Puschendorf B, Lindner H. In vivo phosphorylation of histone H1 variants during the cell cycle. *Biochemistry*. 1996;35(6):1761–1767.

82. Hergeth SP, Dundr M, Tropberger P, et al. Isoform-specific phosphorylation of human linker histone H1. 4 in mitosis by the kinase Aurora B. *J Cell Sci*. 2011;124(10):1623–1628.

83. Chu C-S, Hsu P-H, Lo P-W, et al. Protein kinase A-mediated serine 35 phosphorylation dissociates histone H1. 4 from mitotic chromosome. *J Biol Chem*. 2011;286(41):35843–35851.

84. Clausell J, Happel N, Hale TK, Doenecke D, Beato M. Histone H1 subtypes differentially modulate chromatin condensation without preventing ATP-dependent remodeling by SWI/SNF or NURF. *PLoS One*. 2009;4(10):e0007243.

85. Daujat S, Zeissler U, Waldmann T, Happel N, Schneider R. HP1 binds specifically to Lys26-methylated histone H1.4, whereas simultaneous Ser27 phosphorylation blocks HP1 binding. *J Biol Chem*. 2005;280(45):38090–38095.

86. Alexandrow MG, Hamlin JL. Chromatin decondensation in S-phase involves recruitment of Cdk2 by Cdc45 and histone H1 phosphorylation. *J Cell Biol*. 2005;168(6):875–886.

87. Meergans T, Albig W, Doenecke D. Varied expression patterns of human H1 histone genes in different cell lines. *DNA Cell Biol*. 1997;16(9):1041–1049.

88. Weiss T, Hergeth S, Zeissler U, et al. Histone H1 variant-specific lysine methylation by G9a/KMT1C and Glp1/KMT1D. *Epigenetics Chromatin*. 2010;3(1):1–13.

89. Vaquero A, Scher M, Lee D, Erdjument-Bromage H, Tempst P, Reinberg D. Human SirT1 interacts with histone H1 and promotes formation of facultative heterochromatin. *Mol Cell*. 2004;16(1):93–105.

90. Terme J-M, Millán-Ariño L, Mayor R, et al. Dynamics and dispensability of variant-specific histone H1 Lys-26/Ser-27 and Thr-165 post-translational modifications. *FEBS Lett*. 2014;588(14):2353–2362.

91. Kamieniarz K, Izzo A, Dundr M, et al. A dual role of linker histone H1. 4 Lys 34 acetylation in transcriptional activation. *Genes Dev*. 2012;26(8):797–802.

92. Arents G, Burlingame RW, Wang B-C, Love WE, Moudrianakis EN. The nucleosomal core histone octamer at 3.1 A resolution: a tripartite protein assembly and a left-handed superhelix. *Proc Natl Acad Sci U S A*. 1991;88(22):10148–10152.

93. Kurumizaka H, Kujirai T, Takizawa Y. Contributions of histone variants in nucleosome structure and function. *J Mol Biol*. 2021;433(6):166678.

94. Davey CA, Sargent DF, Luger K, Maeder AW, Richmond TJ. Solvent mediated interactions in the structure of the nucleosome core particle at 1.9 Å resolution. *J Mol Biol*. 2002;319(5):1097–1113.

95. Zentner GE, Henikoff S. Regulation of nucleosome dynamics by histone modifications. *Nat Struct Mol Biol*. 2013;20(3):259–266.

96. Allan J, Harborne N, Rau DC, Gould H. Participation of core histone" tails" in the stabilization of the chromatin solenoid. *J Cell Biol*. 1982;93(2):285–297.

97. Zheng C, Hayes JJ. Structures and interactions of the core histone tail domains. *Biopolymers*. 2003;68(4):539–546.

98. Li G, Levitus M, Bustamante C, Widom J. Rapid spontaneous accessibility of nucleosomal DNA. *Nat Struct Mol Biol*. 2005;12(1):46–53.

99. Andresen K, Jimenez-Useche I, Howell SC, Yuan C, Qiu X. Solution scattering and FRET studies on nucleosomes reveal DNA unwrapping effects of H3 and H4 tail removal. *PLoS One*. 2013;8(11):e78587.

100. Dorigo B, Schalch T, Bystricky K, Richmond TJ. Chromatin fiber folding: requirement for the histone H4 N-terminal tail. *J Mol Biol*. 2003;327(1):85–96.

101. Shogren-Knaak M, Ishii H, Sun J-M, Pazin MJ, Davie JR, Peterson CL. Histone H4-K16 acetylation controls chromatin structure and protein interactions. *Science*. 2006;311(5762):844–847.

102. Brower-Toland B, Wacker DA, Fulbright RM, Lis JT, Kraus WL, Wang MD. Specific contributions of histone tails and their acetylation to the mechanical stability of nucleosomes. *J Mol Biol*. 2005;346(1):135–146.

103. Anderson J, Lowary P, Widom J. Effects of histone acetylation on the equilibrium accessibility of nucleosomal DNA target sites. *J Mol Biol*. 2001;307(4):977–985.

104. Kurdistani SK, Tavazoie S, Grunstein M. Mapping global histone acetylation patterns to gene expression. *Cell*. 2004;117(6):721–733.

105. Martin BJE, Brind'Amour J, Kuzmin A, et al. Transcription shapes genome-wide histone acetylation patterns. *Nat Commun*. 2021;12(1):210.

106. Bannister AJ, Zegerman P, Partridge JF, et al. Selective recognition of methylated lysine 9 on histone H3 by the HP1 chromo domain. *Nature*. 2001;410(6824):120–124.

107. Sugiyama K, Sugiura K, Hara T, et al. Aurora-B associated protein phosphatases as negative regulators of kinase activation. *Oncogene*. 2002;21(20):3103–3111.

108. Goto H, Yasui Y, Nigg EA, Inagaki M. Aurora-B phosphorylates histone H3 at serine28 with regard to the mitotic chromosome condensation. *Genes Cells*. 2002; 7(1):11–17.

109. Rossetto D, Avvakumov N, Côté J. Histone phosphorylation: a chromatin modification involved in diverse nuclear events. *Epigenetics*. 2012;7(10):1098–1108.

110. Tatsuka M, Katayama H, Ota T, et al. Multinuclearity and increased ploidy caused by overexpression of the aurora-and Ipl1-like midbody-associated protein mitotic kinase in human cancer cells. *Cancer Res*. 1998;58(21):4811–4816.

111. Wang H, Huang Z-Q, Xia L, et al. Methylation of histone H4 at arginine 3 facilitating transcriptional activation by nuclear hormone receptor. *Science*. 2001;293(5531):853–857.

112. Wysocka J, Swigut T, Milne TA, et al. WDR5 associates with histone H3 methylated at K4 and is essential for H3 K4 methylation and vertebrate development. *Cell*. 2005;121(6):859–872.

113. Lomberk G, Bensi D, Fernandez-Zapico ME, Urrutia R. Evidence for the existence of an HP1-mediated subcode within the histone code. *Nat Cell Biol*. 2006;8(4):407–415.

114. Azzaz AM, Vitalini MW, Thomas AS, et al. Human heterochromatin protein 1α promotes nucleosome associations that drive chromatin condensation. *J Biol Chem*. 2014;289(10):6850–6861.

115. Kilic S, Bachmann AL, Bryan LC, Fierz B. Multivalency governs HP1alpha association dynamics with the silent chromatin state. *Nat Commun*. 2015;6:7313.

116. Jacobs SA, Khorasanizadeh S. Structure of HP1 chromodomain bound to a lysine 9-methylated histone H3 tail. *Science*. 2002;295(5562):2080–2083.

117. Hiragami-Hamada K, Shinmyozu K, Hamada D, et al. N-terminal phosphorylation of HP1α promotes its chromatin binding. *Mol Cell Biol*. 2011;31(6):1186–1200.

118. Bryan LC, Weilandt DR, Bachmann AL, et al. Single-molecule kinetic analysis of HP1-chromatin binding reveals a dynamic network of histone modification and DNA interactions. *Nucleic Acids Res*. 2017;45(18):10504–10517.

119. Hayakawa T, Haraguchi T, Masumoto H, Hiraoka Y. Cell cycle behavior of human HP1 subtypes: distinct molecular domains of HP1 are required for their centromeric localization during interphase and metaphase. *J Cell Sci.* 2003;116(16):3327–3338.

120. Bosch-Presegué L, Raurell-Vila H, Thackray JK, et al. Mammalian HP1 isoforms have specific roles in heterochromatin structure and organization. *Cell Rep.* 2017;21(8): 2048–2057.

121. Canzio D, Larson A, Narlikar GJ. Mechanisms of functional promiscuity by HP1 proteins. *Trends Cell Biol.* 2014;24(6):377–386.

122. Mishima Y, Watanabe M, Kawakami T, et al. Hinge and chromoshadow of HP1α participate in recognition of K9 methylated histone H3 in nucleosomes. *J Mol Biol.* 2013;425(1):54–70.

123. Muchardt C, Guillemé M, Seeler JS, Trouche D, Dejean A, Yaniv M. Coordinated methyl and RNA binding is required for heterochromatin localization of mammalian HP1α. *EMBO Rep.* 2002;3(10):975–981.

124. Badugu R, Yoo Y, Singh PB, Kellum R. Mutations in the heterochromatin protein 1 (HP1) hinge domain affect HP1 protein interactions and chromosomal distribution. *Chromosoma.* 2005;113(7):370–384.

125. Lachner M, O'Carroll D, Rea S, Mechtler K, Jenuwein T. Methylation of histone H3 lysine 9 creates a binding site for HP1 proteins. *Nature.* 2001;410(6824):116–120.

126. LeRoy G, Weston JT, Zee BM, Young NL, Plazas-Mayorca MD, Garcia BA. Heterochromatin protein 1 is extensively decorated with histone code-like post-translational modifications. *Mol Cell Proteomics.* 2009;8(11):2432–2442.

127. Nishibuchi G, Machida S, Osakabe A, et al. N-terminal phosphorylation of HP1α increases its nucleosome-binding specificity. *Nucleic Acids Res.* 2014;42(20): 12498–12511.

128. Shimojo H, Kawaguchi A, Oda T, et al. Extended string-like binding of the phosphorylated HP1alpha N-terminal tail to the lysine 9-methylated histone H3 tail. *Sci Rep.* 2016;6:22527.

129. Liang JF, Yang VC, Vaynshteyn Y. The minimal functional sequence of protamine. *Biochem Biophys Res Commun.* 2005;336:653–659.

130. Ward WS, Coffey DS. DNA packaging and organization in mammalian spermatozoa: comparison with somatic cells. *Biol Reprod.* 1991;44:569–574.

131. D'Ippolito RA, Minamino N, Rivera-Casas C, et al. Protamines from liverwort are produced by post-translational cleavage and C-terminal di-aminopropanelation of several male germ-specific H1 histones. *J Biol Chem.* 2019;294:16364–16373.

132. Rathke C, Baarends WM, Awe S, Renkawitz-Pohl R. Chromatin dynamics during spermiogenesis. *Biochim Biophys Acta.* 2014;1839(3):155–168.

133. Domenjoud L, Kremling H, Burfeind P, Maier WM, Engel W. On the expression of protamine genes in the testis of man and other mammals. *Andrologia.* 1991;23:333–337.

134. Carrell DT, Emery BR, Hammoud S. Altered protamine expression and diminished spermatogenesis: what is the link? *Hum Reprod Update.* 2007;13(3):313–327.

135. Balhorn R. The protamine family of sperm nuclear proteins. *Genome Biol.* 2007; 8(9):227.

136. Bode J, Willmitzer L, Opatz K. On the competition between protamines and histones: studies directed towards the understanding of spermiogenesis. *Eur J Biochem.* 1977;72(2):393–403.

137. Mao AH, Crick SL, Vitalis A, Chicoine CL, Pappu RV. Net charge per residue modulates conformational ensembles of intrinsically disordered proteins. *Proc Natl Acad Sci U S A.* 2010;107:8183–8188.

138. Balhorn R, Cosman M, Thornton K, et al. Protamine mediated condensation of DNA in mammalian sperm. In: *Male Gamete.* Cache River Press; 1999:55–70.

139. Bench GS, Friz AM, Corzett MH, Morse DH, Balhorn R. DNA and total protamine masses in individual sperm from fertile semen of selected mammals. *Cytometry.* 1996;23:263–271.

140. Balhorn R, Corzett M, Mazrimas J, Watkins B. Identification of bull protamine disulfides. *Biochemistry.* 1991;30:175–181.

141. Vilfan ID, Conwell CC, Hud NV. Formation of native-like mammalian sperm cell chromatin with folded bull protamine. *J Biol Chem.* 2004;279:20088–20095.

142. Ukogu OA, Smith AD, Devenica LM, et al. Protamine loops DNA in multiple steps. *Nucleic Acids Res.* 2020;48(11):6108–6119.

143. Hsiang MW, Cole RD. Structure of histone H1-DNA complex: effect of histone H1 on DNA condensation. *Proc Natl Acad Sci U S A.* 1977;74(11):4852–4856.

144. Brunner A, Nanni P, Mansuy I. Epigenetic marking of sperm by post-translational modification of histones and protamines. *Epigenetics Chromatin.* 2014;7:2.

145. Bustin M. High mobility group proteins. *Biochim Biophys Acta, Gene Regul Mech.* 1799;2010:1–2.

146. Nicolas RH, Goodwin GH. Isolation and analysis. In: Johns EW, ed. *The Chromosomal Proteins.* Academic Press; 1982:41–68.

147. Bustin M. Chromatin unfolding and activation by HMGN* chromosomal protiens. *Trends Biochem Sci.* 2001;26:431–437.

148. Phair RD, Misteli T. High mobility of proteins in the mammalian cell nucleus. *Nature.* 2000;404:604–609.

149. Alfonso PJ, Crippa MP, Hayes JJ, Bustin M. The footprint of chromosomal proteins HMG-14 and HMG-17 on chromatin subunits. *J Mol Biol.* 1994;236(1):189–198.

150. Postnikov YV, Trieschmann L, Rickers A, Bustin M. Homodimers of chromosomal proteins HMG-14 and HMG-17 in nucleosome cores. *J Mol Biol.* 1995;252:423–432.

151. Ueda T, Postnikov YV, Bustin M. Distinct domains in high mobility group N variants modulate specific chromatin modifications. *J Biol Chem.* 2006;281(15): 10182–10187.

152. Postnikov YV, Bustin M. Functional interplay between histone H1 and HMG proteins in chromatin. *Biochim Biophys Acta, Gene Regul Mech.* 1859;2016:462–467.

153. Ueda T, Catez F, Gerlitz G, Bustin M. Delineation of the protein module that anchors HMGN proteins to nucleosomes in the chromatin of living cells. *Mol Cell Biol.* 2008;28:2872–2883.

154. Trieschmann L, Martin B, Bustin M. The chromatin unfolding domain of chromosomal protein HMG-14 targets the N-terminal tail of histone H3 in nucleosomes. *Proc Natl Acad Sci U S A.* 1998;95:5468–5473.

155. Kato H, Van Ingen H, Zhou BR, et al. Architecture of the high mobility group nucleosomal protein 2-nucleosome complex as revealed by methyl-based NMR. *Proc Natl Acad Sci U S A.* 2011;108:12283–12288.

156. Lim JH, Catez F, Birger Y, et al. Chromosomal protein HMGN1 modulates histone H3 phosphorylation. *Mol Cell.* 2004;15:573–584.

157. Zhang Q, Wang Y. High mobility group proteins and their post-translational modifications. *Biochim Biophys Acta.* 2008;1784(9):1159–1166.

158. Bergel M, Herrera JE, Thatcher BJ, et al. Acetylation of novel sites in the nucleosomal binding domain of chromosomal protein HMG-14 by p300 alters its interaction with nucleosomes. *J Biol Chem.* 2000;275(15):11514–11520.

159. Murphy KJ, Cutter AR, Fang H, Postnikov YV, Bustin M, Hayes JJ. HMGN1 and 2 remodel core and linker histone tail domains within chromatin. *Nucleic Acids Res.* 2017;45:9917–9930.

160. Albright SC, Wiseman JM, Lange RA, Garrard WT. Subunit structures of different electrophoretic forms of nucleosomes. *J Biol Chem.* 1980;255(8):3673–3684.

161. van Houte LP, Chuprina VP, van der Wetering M, Boelens R, Kaptein R, Clevers H. Solution structure of the sequence-specific HMG box of the lymphocyte transcriptional activator Sox-4. *J Biol Chem.* 1995;270(51):30516–30524.
162. Sheflin LG, Spaulding SW. High mobility group protein 1 preferentially conserves torsion in negatively supercoiled DNA. *Biochemistry.* 1989;28:5658–5664.
163. Štros M, Štokrová J, Thomas JO. Dna looping by the HMG-box domains of HMG1 and modulation of DNA binding by the acidic C-terminal domain. *Nucleic Acids Res.* 1994;22:1044–1051.
164. Travers A. Recognition of distorted DNA structures by HMG domains. *Curr Opin Struct Biol.* 2000;10:102–109.
165. Thomas JO, Travers AA. HMG1 and 2, and related 'architectural' DNA-binding proteins. *Trends Biochem Sci.* 2001;26:167–174.
166. Jaouen S, De Koning L, Gaillard C, Muselíková-Polanská E, Štros M, Strauss F. Determinants of specific binding of HMGB1 protein to hemicatenated DNA loops. *J Mol Biol.* 2005;353:822–837.
167. Ugrinova I, Pasheva EA, Armengaud J, Pashev IG. In vivo acetylation of HMG1 protein enhances its binding affinity to distorted DNA structures. *Biochemistry.* 2001;40(48):14655–14660.
168. Knapp S, Müller S, Digilio G, Bonaldi T, Bianchi ME, Musco G. The long acidic tail of high mobility group box 1 (HMGB1) protein forms an extended and flexible structure that interacts with specific residues within and between the HMG boxes. *Biochemistry.* 2004;43:11992–11997.
169. Stott K, Watson M, Howe FS, Grossmann JG, Thomas JO. Tail-mediated collapse of HMGB1 is dynamic and occurs via differential binding of the acidic tail to the A and B domains. *J Mol Biol.* 2010;403:706–722.
170. Kawase T, Sato K, Ueda T, Yoshida M. Distinct domains in HMGB1 are involved in specific intramolecular and nucleosomal interactions. *Biochemistry.* 2008;47:13991–13996.
171. An W, Van Holde K, Zlatanova J. The non-histone chromatin protein HMG1 protects linker DNA on the side opposite to that protected by linker histones. *J Biol Chem.* 1998;273:26289–26291.
172. Jackson JB, Rill RL. Circular dichroism, thermal denaturation, and deoxyribonuclease I digestion studies of nucleosomes highly enriched in high mobility group proteins HMG 1 and HMG 2. *Biochemistry.* 1981;20:1042–1046.
173. Catez F, Yang H, Tracey KJ, Reeves R, Misteli T, Bustin M. Network of dynamic interactions between histone H1 and high-mobility-group proteins in chromatin. *Mol Cell Biol.* 2004;24:4321–4328.
174. Cato L, Stott K, Watson M, Thomas JO. The interaction of HMGB1 and linker histones occurs through their acidic and basic tails. *J Mol Biol.* 2008;384:1262–1272.
175. Štros M. HMGB proteins: interactions with DNA and chromatin. *Biochim Biophys Acta, Gene Regul Mech.* 1799;2010:101–113.
176. Polanska E, Pospisilova S, Stros M. Binding of histone H1 to DNA is differentially modulated by redox state of HMGB1. *PLoS One.* 2014;9(2):e89070.
177. Lehn DA, Elton TS, Johnson KR, Reeves R. A conformational study of the sequence specific binding of HMG-I (Y) with the bovine interleukin-2 cDNA. *Biochem Int.* 1988;16:963–971.
178. Evans JNS, Zajicek J, Nissen MS, Munske G, Smith V, Reeves R. 1H and 13C NMR assignments and molecular modelling of a minor groove DNA-binding peptide from the HMG-I protein. *Int J Pept Protein Res.* 1995;45:554–560.
179. Huth JR, Bewley CA, Nissen MS, et al. The solution structure of an HMG-I(Y)-DNA complex defines a new architectural minor groove binding motif. *Nat Struct Biol.* 1997;4:657–665.

180. Cui T, Leng F. Specific recognition of AT-rich DNA sequences by the mammalian high mobility group protein AT-hook 2: a SELEX study. *Biochemistry*. 2007;46: 13059–13066.

181. Reeves R, Nissen MS. Interaction of high mobility group-I(Y) nonhistone proteins with nucleosome core particles. *J Biol Chem*. 1993;268:21137–21146.

182. Li O, Vasudevan D, Davey CA, Dröge P. High-level expression of DNA architectural factor HMGA2 and its association with nucleosomes in human embryonic stem cells. *Genesis*. 2006;44:523–529.

183. Zhao K, Kas E, Gonzalez E, Laemmli UK. SAR-dependent mobilization of histone H1 by HMG-I/Y in vitro: HMG-I/Y is enriched in H1-depleted chromatin. *EMBO J*. 1993;12:3237–3247.

184. Saitoh Y, Laemmli UK. Metaphase chromosome structure: bands arise from a differential folding path of the highly AT-rich scaffold. *Cell*. 1994;76:609–622.

185. Kishi Y, Fujii Y, Hirabayashi Y, Gotoh Y. HMGA regulates the global chromatin state and neurogenic potential in neocortical precursor cells. *Nat Neurosci*. 2012;15: 1127–1133.

186. Wei T, Liu H, Chu B, et al. Phosphorylation-regulated HMGA1a-P53 interaction unveils the function of HMGA1a acidic tail phosphorylations via synthetic proteins. *Cell Chem Biol*. 2021;28:722–732.e8.

187. Iwafuchi-Doi M, Zaret KS. Pioneer transcription factors in cell reprogramming. *Genes Dev*. 2014;28(24):2679–2692.

188. Soufi A, Garcia MF, Jaroszewicz A, Osman N, Pellegrini M, Zaret KS. Pioneer transcription factors target partial DNA motifs on nucleosomes to initiate reprogramming. *Cell*. 2015;161:555–568.

189. Clark KL, Halay ED, Lai E, Burley SK. Co-crystal structure of the HNF-3/fork head DNA-recognition motif resembles histone H5. *Nature*. 1993;364:412–420.

190. Cirillo LA, McPherson CE, Bossard P, et al. Binding of the winged-helix transcription factor HNF3 to a linker histone site on the nucleosome. *EMBO J*. 1998;17:244–254.

191. Chaya D, Hayamizu T, Bustin M, Zaret KS. Transcription factor FoxA (HNF3) on a nucleosome at an enhancer complex in liver chromatin. *J Biol Chem*. 2001;276: 44385–44389.

192. Cirillo LA, Lin FR, Cuesta I, Friedman D, Jarnik M, Zaret KS. Opening of compacted chromatin by early developmental transcription factors HNF3 (FoxA) and GATA-4. *Mol Cell*. 2002;9:279–289.

193. Iwafuchi-Doi M, Donahue G, Kakumanu A, et al. The Pioneer transcription factor FoxA maintains an accessible nucleosome configuration at enhancers for tissue-specific gene activation. *Mol Cell*. 2016;62:79–91.

194. Iwafuchi M, Cuesta I, Donahue G, et al. Gene network transitions in embryos depend upon interactions between a pioneer transcription factor and core histones. *Nat Genet*. 2020;52(4):418–427.

195. Taube JH, Allton K, Duncan SA, Shen L, Barton MC. Foxa1 functions as a pioneer transcription factor at transposable elements to activate Afp during differentiation of embryonic stem cells. *J Biol Chem*. 2010;285:16135–16144.

196. Gómez-Márquez J, Rodríguez P. Prothymosin α is a chromatin-remodelling protein in mammalian cells. *Biochem J*. 1998;333:1–3.

197. George EM, Brown DT. Prothymosin α is a component of a linker histone chaperone. *FEBS Lett*. 2010;584:2833–2836.

198. Mosoian A. Intracellular and extracellular cytokine-like functions of prothymosin α: Implications for the development of immunotherapies. *Future Med Chem*. 2011;3: 1199–1208.

199. Cast K, Damaschun H, Eckert K, et al. Prothymosin α: a biologically active protein with random coil conformation. *Biochemistry*. 1995;34:13211–13218.

200. Uversky VN, Gillespie JR, Millett IS, et al. Natively unfolded human prothymosin α adopts partially folded collapsed conformation at acidic pH. *Biochemistry*. 1999;38: 15009–15016.

201. Bednar J, Hamiche A, Dimitrov S. H1-nucleosome interactions and their functional implications. *Biochim Biophys Acta*. 2016;1859(3):436–443.

202. Turner AL, Watson M, Wilkins OG, et al. Highly disordered histone H1-DNA model complexes and their condensates. *Proc Natl Acad Sci U S A*. 2018;115(47):11964–11969.

203. Kouzarides T. Chromatin modifications and their function. *Cell*. 2007;128(4): 693–705.

204. Gangaraju VK, Bartholomew B. Mechanisms of ATP dependent chromatin remodeling. *Mutat Res*. 2007;618(1–2):3–17.

205. Tagami H, Ray-Gallet D, Almouzni G, Nakatani Y. Histone H3.1 and H3.3 complexes mediate nucleosome assembly pathways dependent or independent of DNA synthesis. *Cell*. 2004;116(1):51–61.

206. Mossink B, Negwer M, Schubert D, Nadif KN. The emerging role of chromatin remodelers in neurodevelopmental disorders: a developmental perspective. *Cell Mol Life Sci*. 2021;78(6):2517–2563.

207. Längst G, Manelyte L. Chromatin remodelers: from function to dysfunction. *Gene*. 2015;6:299–324.

208. Sandhu KS. Intrinsic disorder explains diverse nuclear roles of chromatin remodeling proteins. *J Mol Recognit*. 2009;22(1):1–8.

209. El Hadidy N, Uversky VN. Intrinsic disorder of the BAF complex: roles in chromatin remodeling and disease development. *Int J Mol Sci*. 2019;20(21):5260.

210. Aoki D, Awazu A, Fujii M, et al. Ultrasensitive change in nucleosome binding by multiple phosphorylations to the intrinsically disordered region of the histone chaperone FACT. *J Mol Biol*. 2020;432(16):4637–4657.

211. Santofimia-Castano P, Rizzuti B, Pey AL, et al. Intrinsically disordered chromatin protein NUPR1 binds to the C-terminal region of polycomb RING1B. *Proc Natl Acad Sci U S A*. 2017;114(31):E6332–E6341.

212. Yan L, Wu H, Li X, Gao N, Chen Z. Structures of the ISWI–nucleosome complex reveal a conserved mechanism of chromatin remodeling. *Nat Struct Mol Biol*. 2019;26(4):258–266.

213. Warren C, Shechter D. Fly fishing for histones: catch and release by histone chaperone intrinsically disordered regions and acidic stretches. *J Mol Biol*. 2017;429:2401–2426.

214. Hodges C, Kirkland JG, Crabtree GR. The many roles of BAF (mSWI/SNF) and PBAF complexes in cancer. *Cold Spring Harb Perspect Med*. 2016;6(8):a026930.

215. Goldberg AD, Banaszynski LA, Noh K-M, et al. Distinct factors control histone variant H3.3 localization at specific genomic regions. *Cell*. 2010;140(5):678–691.

216. Drané P, Ouararhni K, Depaux A, Shuaib M, Hamiche A. The death-associated protein DAXX is a novel histone chaperone involved in the replication-independent deposition of H3.3. *Genes Dev*. 2010;24(12):1253–1265.

217. Dyer MA, Qadeer ZA, Valle-Garcia D, Bernstein E. ATRX and DAXX: mechanisms and mutations. *Cold Spring Harb Perspect Med*. 2017;7(3).

218. Gibbons RJ, Wada T, Fisher CA, et al. Mutations in the chromatin-associated protein ATRX. *Hum Mutat*. 2008;29(6):796–802.

219. Mitson M, Kelley LA, Sternberg MJE, Higgs DR, Gibbons RJ. Functional significance of mutations in the Snf2 domain of ATRX. *Hum Mol Genet*. 2011;20(13):2603–2610.

220. Varga-Weisz PD, Wilm M, Bonte E, Dumas K, Mann M, Becker PB. Chromatin-remodelling factor CHRAC contains the ATPases ISWI and topoisomerase II. *Nature*. 1997;388(6642):598–602.

221. Scacchetti A, Brueckner L, Jain D, et al. CHRAC/ACF contribute to the repressive ground state of chromatin. *Life Sci Alliance*. 2018;1(1):e201800024.

222. Hartlepp KF, Fernández-Tornero C, Eberharter A, Grüne T, Müller CW, Becker PB. The histone fold subunits of drosophila CHRAC facilitate nucleosome sliding through dynamic DNA interactions. *Mol Cell Biol.* 2005;25(22):9886–9896.

223. Bonaldi T, Längst G, Strohner R, Becker PB, Bianchi ME. The DNA chaperone HMGB1 facilitates ACF/CHRAC-dependent nucleosome sliding. *EMBO J.* 2002;21(24):6865–6873.

224. Asfaha Y, Schrenk C, Alves Avelar LA, et al. Recent advances in class IIa histone deacetylases research. *Bioorg Med Chem.* 2019;27(22):115087.

225. Gurova K, Chang H-W, Valieva ME, Sandlesh P, Studitsky VM. Structure and function of the histone chaperone FACT—resolving FACTual issues. *Biochim Biophys Acta Gene Regul Mech.* 2018;1861(9):892–904.

226. Di Giorgio E, Brancolini C. Regulation of class IIa HDAC activities: it is not only matter of subcellular localization. *Epigenomics.* 2016;8(2):251–269.

227. Zhang CL, McKinsey TA, Olson EN. The transcriptional corepressor MITR is a signal-responsive inhibitor of myogenesis. *Proc Natl Acad Sci U S A.* 2001;98(13):7354.

228. Wang AH, Yang XJ. Histone deacetylase 4 possesses intrinsic nuclear import and export signals. *Mol Cell Biol.* 2001;21(17):5992–6005.

229. Black Joshua C, Van Rechem C, Whetstine JR. Histone lysine methylation dynamics: establishment, regulation, and biological impact. *Mol Cell.* 2012;48(4):491–507.

230. Martin C, Zhang Y. The diverse functions of histone lysine methylation. *Nat Rev Mol Cell Biol.* 2005;6(11):838–849.

231. Bedford MT, Richard S. Arginine methylation: an emerging regulator of protein function. *Mol Cell.* 2005;18(3):263–272.

232. Sarmento OF, Svingen PA, Xiong Y, et al. The role of the histone methyltransferase enhancer of Zeste homolog 2 (EZH2) in the pathobiological mechanisms underlying inflammatory bowel disease (IBD)*. *J Biol Chem.* 2017;292(2):706–722.

233. Greer EL, Shi Y. Histone methylation: a dynamic mark in health, disease and inheritance. *Nat Rev Genet.* 2012;13(5):343–357.

234. Separovich RJ, Wong MWM, Chapman TR, Slavich E, Hamey JJ, Wilkins MR. Post-translational modification analysis of Saccharomyces cerevisiae histone methylation enzymes reveals phosphorylation sites of regulatory potential. *J Biol Chem.* 2021;296:100192.

235. Sen P, Dang W, Donahue G, et al. H3K36 methylation promotes longevity by enhancing transcriptional fidelity. *Genes Dev.* 2015;29(13):1362–1376.

236. Tennen RI, Berber E, Chua KF. Functional dissection of SIRT6: Identification of domains that regulate histone deacetylase activity and chromatin localization. *Mech Ageing Dev.* 2010;131(3):185–192.

237. Pan PW, Feldman JL, Devries MK, Dong A, Edwards AM, Denu JM. Structure and biochemical functions of SIRT6*. *J Biol Chem.* 2011;286(16):14575–14587.

238. Gil R, Barth S, Kanfi Y, Cohen HY. SIRT6 exhibits nucleosome-dependent deacetylase activity. *Nucleic Acids Res.* 2013;41(18):8537–8545.

239. Liu WH, Zheng J, Feldman JL, et al. Multivalent interactions drive nucleosome binding and efficient chromatin deacetylation by SIRT6. *Nat Commun.* 2020;11(1):5244.

240. Levendosky RF, Bowman GD. Asymmetry between the two acidic patches dictates the direction of nucleosome sliding by the ISWI chromatin remodeler. *Elife.* 2019;8:e45472.

241. Zhou K, Liu Y, Luger K. Histone chaperone FACT FAcilitates chromatin transcription: mechanistic and structural insights. *Curr Opin Struct Biol.* 2020;65:26–32.

242. Saunders A, Werner J, Andrulis ED, et al. Tracking FACT and the RNA polymerase II elongation complex through chromatin in vivo. *Science.* 2003;301(5636):1094–1096.

243. Gurard-Levin ZA, Quivy J-P, Almouzni G. Histone chaperones: assisting histone traffic and nucleosome dynamics. *Annu Rev Biochem.* 2014;83(1):487–517.

244. Winkler DD, Muthurajan UM, Hieb AR, Luger K. Histone chaperone FACT coordinates nucleosome interaction through multiple synergistic binding events*. *J Biol Chem.* 2011;286(48):41883–41892.
245. Wang T, Liu Y, Edwards G, Krzizike D, Scherman H, Luger K. The histone chaperone FACT modulates nucleosome structure by tethering its components. *Life Sci Alliance.* 2018;1(4):e201800107.
246. Liu Y, Zhou K, Zhang N, et al. FACT caught in the act of manipulating the nucleosome. *Nature.* 2020;577(7790):426–431.
247. Mayanagi K, Saikusa K, Miyazaki N, et al. Structural visualization of key steps in nucleosome reorganization by human FACT. *Sci Rep.* 2019;9(1):10183.
248. Tsunaka Y, Ohtomo H, Morikawa K, Nishimura Y. Partial replacement of nucleosomal DNA with human FACT induces dynamic exposure and acetylation of histone H3 N-terminal tails. *iScience.* 2020;23(10):101641.
249. Szőllősi E, Bokor M, Bodor A, et al. Intrinsic structural disorder of DF31, a drosophila protein of chromatin decondensation and Remodeling activities. *J Proteome Res.* 2008;7(6):2291–2299.
250. Crevel G, Huikeshoven H, Cotterill S. Df31 is a novel nuclear protein involved in chromatin structure in Drosophila melanogaster. *J Cell Sci.* 2001;114(1):37.
251. Guillebault D, Cotterill S. The drosophila Df31 protein interacts with histone H3 tails and promotes chromatin bridging in vitro. *J Mol Biol.* 2007;373(4):903–912.
252. Schubert T, Pusch MC, Diermeier S, et al. Df31 protein and snoRNAs maintain accessible higher-order structures of chromatin. *Mol Cell.* 2012;48(3):434–444.
253. Fuxreiter M, Tompa P, Simon I, Uversky VN, Hansen JC, Asturias FJ. Malleable machines take shape in eukaryotic transcriptional regulation. *Nat Chem Biol.* 2008;4:728–737.
254. Staby L, O'Shea C, Willemoes M, Theisen F, Kragelund BB, Skriver K. Eukaryotic transcription factors: paradigms of protein intrinsic disorder. *Biochem J.* 2017;474(15):2509–2532.
255. Staller MV, Holehouse AS, Swain-Lenz D, Das RK, Pappu RV, Cohen BA. A high-throughput mutational scan of an intrinsically disordered acidic transcriptional activation domain. *Cell Syst.* 2018;6(4):444–455. [e6].
256. Vuzman D, Levy Y. Intrinsically disordered regions as affinity tuners in protein–DNA interactions. *Mol Biosyst.* 2012;8:47.
257. Tafvizi A, Huang F, Fersht AR, Mirny LA, van Oijen AM. A single-molecule characterization of p53 search on DNA. *Proc Natl Acad Sci U S A.* 2011;108(2):563–568.
258. Vuzman D, Azia A, Levy Y. Searching DNA via a "monkey Bar" mechanism: the significance of disordered tails. *J Mol Biol.* 2010;396:674–684.
259. Brodsky S, Jana T, Mittelman K, et al. Intrinsically disordered regions direct transcription factor in vivo binding specificity. *Mol Cell.* 2020;79(3):459–471.e4.
260. Zaret KS, Mango SE. Pioneer transcription factors, chromatin dynamics, and cell fate control. *Curr Opin Genet Dev.* 2016;37:76–81.
261. Takahashi K, Yamanaka S. Induction of pluripotent stem cells from mouse embryonic and adult fibroblast cultures by defined factors. *Cell.* 2006;126(4):663–676.
262. Lu Y, Brommer B, Tian X, et al. Reprogramming to recover youthful epigenetic information and restore vision. *Nature.* 2020;588(7836):124–129.
263. Jang J, Yoo J-E, Lee J-A, et al. Disease-specific induced pluripotent stem cells: a platform for human disease modeling and drug discovery. *Exp Mol Med.* 2012;44(3):202–213.
264. Xue B, Oldfield CJ, Van YY, Dunker AK, Uversky VN. Protein intrinsic disorder and induced pluripotent stem cells. *Mol Biosyst.* 2012;8(1):134–150.

265. Michael AK, Grand RS, Isbel L, et al. Mechanisms of OCT4-SOX2 motif readout on nucleosomes. *Science*. 2020;368(6498):1460–1465.
266. Dodonova SO, Zhu F, Dienemann C, Taipale J, Cramer P. Nucleosome-bound SOX2 and SOX11 structures elucidate pioneer factor function. *Nature*. 2020;580(7805): 669–672.
267. Weiss MA. Floppy SOX: mutual induced fit in hmg (high-mobility group) box-DNA recognition. *Mol Endocrinol*. 2001;15(3):353–362.
268. Holmes ZE, Hamilton DJ, Hwang T, et al. The Sox2 transcription factor binds RNA. *Nat Commun*. 2020;11(1):1805.
269. Tan C, Takada S. Nucleosome allostery in pioneer transcription factor binding. *Proc Natl Acad Sci U S A*. 2020;117(34):20586–20596.
270. Shammas SL. Mechanistic roles of protein disorder within transcription. *Curr Opin Struct Biol*. 2017;42:155–161.
271. Robinson JL, Carroll JS. FoxA1 is a key mediator of hormonal response in breast and prostate cancer. *Front Endocrinol (Lausanne)*. 2012;3:68.
272. Sutinen P, Rahkama V, Rytinki M, Palvimo JJ. Nuclear mobility and activity of FOXA1 with androgen receptor are regulated by SUMOylation. *Mol Endocrinol*. 2014;28(10):1719–1728.
273. Hou L, Wei Y, Lin Y, et al. Concurrent binding to DNA and RNA facilitates the pluripotency reprogramming activity of Sox2. *Nucleic Acids Res*. 2020;48(7): 3869–3887.
274. Aksoy I, Jauch R, Eras V, et al. Sox transcription factors require selective interactions with Oct4 and specific transactivation functions to mediate reprogramming. *Stem Cells*. 2013;31(12):2632–2646.
275. Xhani S, Lee S, Kim HM, et al. Intrinsic disorder controls two functionally distinct dimers of the master transcription factor PU.1. *Sci Adv*. 2020;6(8):eaay3178.
276. Minderjahn J, Schmidt A, Fuchs A, et al. Mechanisms governing the pioneering and redistribution capabilities of the non-classical pioneer PU.1. *Nat Commun*. 2020; 11(1):402.
277. Chanda S, Ang CE, Davila J, et al. Generation of induced neuronal cells by the single reprogramming factor ASCL1. *Stem Cell Rep*. 2014;3(2):282–296.
278. Baronti L, Hosek T, Gil-Caballero S, et al. Fragment-based NMR study of the conformational dynamics in the bHLH transcription factor Ascl1. *Biophys J*. 2017; 112(7):1366–1373.
279. Fernandez Garcia M, Moore CD, Schulz KN, et al. Structural features of transcription factors associating with nucleosome binding. *Mol Cell*. 2019;75(5):921–932.e6.
280. Hiragami-Hamada K, Nakayama J-I. Do the charges matter?—Balancing the charges of the chromodomain proteins on the nucleosome. *J Biochem*. 2019;165(6):455–458.
281. Pepenella S, Murphy KJ, Hayes JJ. Intra-and inter-nucleosome interactions of the core histone tail domains in higher-order chromatin structure. *Chromosoma*. 2014; 123(1–2):3–13.
282. Rowell JP, Simpson KL, Stott K, Watson M, Thomas JO. HMGB1-facilitated p53 DNA binding occurs via HMG-box/p53 transactivation domain interaction, regulated by the acidic tail. *Structure*. 2012;20(12):2014–2024.
283. Štros M, Kučírek M, Sani SA, Polanská E. HMGB1-mediated DNA bending: Distinct roles in increasing p53 binding to DNA and the transactivation of p53-responsive gene promoters. *Biochim Biophys Acta, Gene Regul Mech*. 1861;2018:200–210.
284. Jayaraman L, Moorthy NC, Murthy KG, Manley JL, Bustin M, Prives C. High mobility group protein-1 (HMG-1) is a unique activator of p53. *Genes Dev*. 1998;12(4):462–472.

285. Thomas JO, Stott K. H1 and HMGB1: modulators of chromatin structure. *Biochem Soc Trans.* 2012;40:341–346.

286. Kim K, Jeong KW, Kim H, et al. Functional interplay between p53 acetylation and H1. 2 phosphorylation in p53-regulated transcription. *Oncogene.* 2012;31(39):4290–4301.

287. Tanaka H, Iguchi N, Isotani A, et al. HANP1/H1T2, a novel histone H1-like protein involved in nuclear formation and sperm fertility. *Mol Cell Biol.* 2005;25(16): 7107–7119.

288. Holmstrom ED, Liu Z, Nettels D, Best RB, Schuler B. Disordered RNA chaperones can enhance nucleic acid folding via local charge screening. *Nat Commun.* 2019; 10(1):2453.

CHAPTER TEN

Flexible spandrels of the global plant virome: Proteomic-wide evolutionary patterns of structural intrinsic protein disorder elucidate modulation at the functional virus–host interplay

Rachid Tahzima[a,*], Annelies Haegeman[b], Sébastien Massart[a], and Eugénie Hébrard[c]

[a]University of Liège (ULg), Department of Integrated and Urban Phytopathology, Gembloux, Belgium
[b]Flanders Research Institute for Agriculture, Fisheries and Food (ILVO), Plant Sciences Unit, Merelbeke, Belgium
[c]PHIM, Plant Health Institute, IRD, Cirad, Université de Montpellier, INRAE, Institut Agro, Montpellier, France
*Corresponding author: e-mail address: rachid.tahzima@uliege.be

Contents

Abstract

Intrinsically disordered proteins and regions (IDPs/IDRs) make up a large part of viral proteomes, but their real prevalence across the global plant virome is still murky, partly because of their massive diversity. Here, we propose an evolutionary quantitative proteomic approach to foray into genomic signatures that are preserved in the amino acid sequences of orthologous IDRs. Markedly, we found that relatively abundant IDP varies substantially in viral species among and within plant virus families, including according to genome size, partition or replication strategies. We also demonstrate that most encoded proteomic modules of the plant virome contain multiple disordered features that are phylogenomically preserved, and can be correlated to genomic, bio-physical and evolutionary strategies. Furthermore, our focused interactome-wide analysis highlights lines of evidence indicating that various IDPs with similar evolutionary signatures modulate viral multifunctionality. Moreover, estimated fractions of IDR in the vicinity of pivotal evolutionary structural domains embedded in interaction modules are strongly enriched with affinity binding functional annotations and relate to vector-borne virus transmission modes. Importantly, molecular recognition features (MoRFs) are abundantly widespread in IDRs of viral hallmark modules and their binding partners. Finally, we propose a coarse-grained conceptual framework in which evolutionary proteome-wide IDP/IDRs patterns can be, rather, reliably exploited to elucidate their foundational fine-tuning role in plant virus transmission mechanisms. While opening unexplored avenues for consistently predicting virus–host functions for many new or uncharacterized viruses based on their proteomic repertoire, other considerations advocating further structural IDP research in Plant Virology are thoroughly discussed in light of viral modular evolution.

1. The highly modular plant virus proteome

Viruses are, by orders of magnitude, the most abundant and dominant biological entities across all plant environmental niches, and constitute a major reservoir of genetic diversity affecting the different agroecosystems of the extant biosphere.[1] These intracellular obligate biotrophic pathogens can infect a broad spectrum of mainly angiosperms (flowering land plants) utilizing their host intracellular apparatuses for protein synthesis, replication and systemic movement in order to assure their multiplication. Most plant viral infections can be associated with disease symptoms inducing morphological and physiological alterations of the infected plant hosts, which often hampers plant performances and cause high yield losses for crops. Encompassing two viral realms (*Ribovaria* and *Monodnaviria*), the highly ubiquitous virome of land plants also underwent a notable evolutive expansion of very diverse and distinct virus repertoires.[2]

1.1 Plant virus diversity and evolution

Notwithstanding the fact that genetic diversity of plant viruses is hard to assess, plant viruses evolved a staggering diversity of genomic nature and size, mainly dominated by positive single stranded RNA ((+)ssRNA) viruses and, in a lesser extent by single stranded DNA (ssDNA) viruses, all determining distinct replication routes and protein expression strategies. Furthermore, genomes of plant viruses may also encompass other types of genomes, namely, from double-stranded DNA (dsDNA) which involve a Reverse-transcription step to negative single stranded RNA ((−)ssRNA), and double-stranded RNA (dsRNA).[3,4] Since most RNA plant viruses have genomes no longer than ~17 kb (e.g., *Closteroviridae*), genomic promiscuity may constrain virus evolution by reducing the number of sites that are not involved in complex epistatic and pleiotropic interactions, and limiting the possibility for gene duplication and lateral gene transfer. Paradoxically, RNA viruses are prone to high mutation rates associated with viral RNA polymerase mainly due to its low fidelity in the replication machinery and lack of proofreading activity. Hence, severe fitness costs are inflicted for those viruses that violate the error threshold, which in turn can place a cap on genome size.[5] This intricate balance between the variation generated by these high mutational rates and the biophysical constraints implicit with small viral genomes suggests that complex adaptive trade-offs are likely to shape the evolution of virus proteomes. Despite the similar structural and functional roles played by virions, they exhibit a remarkable apparent diversity of forms, including icosahedral, filamentous, and rod shapes. The icosahedral architecture represents another example of biophysical constraint preventing the capsid size expansion and resulting in the genome size stability. In addition, viral genomes are organized into single or multiple component nucleic acid molecules, the latter named segmented genomes. Some segmented viruses termed multipartite viruses encapsidate their genomic molecules in a set of separate virions which can harbor different sizes.[6] The multipartition represents another very challenging strategy endorsing trade-offs between packaging genomic information while maintaining viral genetic integrity and adaptive ability where viruses can complement the mutational load through replication with reassortment of genomic segments.[7,8] Nonetheless, sharing some common functionalities, plant viruses likewise have evolved specific and extensively various strategical mechanisms to also alleviate their lifecycles from the difficulties related to the immobility and escape most obstacles to their intra/intercellular spread.[9] On the functional basis, plant viruses have arguably become masters in fine-tuning viral replication, in highjacking host components and establishing specific interactions to modulate protein turnover machinery and quench of host's RNAi silencing defense pathways. Host range, local cell-to-cell movement and long-distanced systemic transport through the plant vasculature or vector-borne transmission are analogously functionally determined by both genomic, proteomic and extrinsic ecological factors as well and understanding the role of in host or vector range evolution remains challenging. Besides, plant viruses also present a remarkable diversity in their biology and can vary significantly in their hosts range (plants, plants and fungi, plant and arthropods), from a single plant species to more than 1000, and in their transmission mode (biotic/abiotic). Recent advances in metaviromics using detailed evolutionary RdRp-based phylogenomic helped organize the vast expansion of the global virome with enhanced coherence and revealed the extensive role of genetic exchange through horizontal gene transfer, gene loss and shuffling.[10] Evolution of *bona fide* plant viruses is paced by multiple and independent episodes of cellular structures recruitment or modular components. The origin and evolutionary forces accountable for such an abundant diversity of viruses spread sometimes between distantly related hosts remain still unclear. Henceforth, striving to deciphering their major or recurring tendencies through new approaches is unquestionably essential to outline their multiple evolutionary histories and the subsequent transition processes that molded this dazzling virus functional biodiversity.

1.2 Unique modular organization and functions of plant virus proteomes

Over the long-term, viruses encompass a wide pool of interchangeably conserved and highly evolvable functional modules contributing to their diversified emergence.[11] Irrespective of the selective constraints related to genome size, virion shape and biology, mainly three interchangeably functional types of virus hallmark proteins can delineate the interconnected plant virus lifestyle: proteins enabling semi-autonomous genome replication (replication modules), proteins responsible for virion formation and viral transmission (morphogenetic modules) and proteins involved in intra-host movement (MP-transport modules). Across the longstanding evolutionary timeline, optimized sets of individual or combined functionally compatible modules can be partially or entirely interchanged potentiating their independent evolution under a wide selective landscape and enduringly defying major genomic evolutionary limitations while contributing to their protracted viral diversification.[2] The genome replication module of plant viruses is mainly centered around two very distinct RNA or RT-dependent polymerases (RdRp, RT), the only gene that is conserved in all plant viruses within the realm of *Ribovaria*, to the exclusion of the rest of the global virome.[3] Replicases regularly involve several virus-encoded functional units, namely, the RNA-dependent RNA polymerase (RdRp), a helicase (HEL) that harbors NTP-motifs for displacing activity of complementary strands in duplex nucleic acids during genome replication and recombination, and a methyl-transferase (MET) leading the $5'$ capping of RNA. Depending on the translation strategy, viral proteinases can process polyproteins into functional elements, they are highly specific to their substrate and exhibit multifunctional catalytic activity. With regard to the morphogenetic modules, most plant viruses are encapsidated in helical, filamentous or icosahedral-shaped virions formed with the typical fCP and single jelly roll (SJR) folded capsid proteins (CPs), respectively, while some adopt more distinct structural architecture involving filamentous nucleocapsids, or less frequently, membrane envelopes ornamented with virus-encoded glycoproteins. Although well-conserved, the viral CPs are very often multifunctionally involved in genome protection, transmission and movement. By comparison, movement proteins (MPs) are evolutionarily and structurally very diverse.[12] Indeed, plant viruses need to escape from the cell wall, the pecto-cellulosic barrier surrounding plant cell, to reach the plant vasculature and to spread systematically in the host. Thus, MPs represent the most noticeable unique hallmark of plant viruses, although highly multifunctional, their main task at the interface with plant cell plasmodesma (PD) is to assure short-distance intracellular local transport of virions or viral nucleoprotein complexes and to widen or destroy the PD channel thus facilitating their passage to uninfected neighboring cells towards progressive systemic movement through the host's vascular system.[13,14] Diverse strategies have been selected alongside viral evolution involving single or multiple proteins such as MP from the 30 K superfamily, TGBp (triple gene block proteins) or MP-forming tubular elements. Besides fulfilling their dedicated function, many of these viral MP proteins evolved as RNA silencing suppressors (VSRs), to circumvent or inactivate a wide range of biological processes from the host plant involved in host antiviral defense.[15–17] Insofar, only few plant viruses encode no proteins directly involved in replication,[2] whereas others; like (−)ssRNA viruses (*Negarnaviricota*) have lost their morphogenetic module (e.g., capsidless viruses) or encode highly specific or unique proteins.[18] Nonetheless, most plant viruses encode these nonstructural and structural modules in their genomic make up, and untangling their evolutionary histories in light of proteomic features is essential to understanding plant virus origins and evolution from a new vantage.

1.3 Transmission mechanisms and virus–vector interactions of plant viruses

Plant viruses can be transmitted by very diverse ways.[19] Transmission can occur through abiotic means such as soil or water, or mechanically via grafting or contacts during the agricultural practices, or through seeds, pollen. However, the majority of plant viruses that cause disease in agricultural crops rely on an array of vectors for transmission and survival.[20] In a sense, vector-borne transmission represents a way to overcome the barriers to inter- and intra-host viral dispersion due to the motionless nature of plants and their pectocellulosic cell barrier.[21] The plant virus-transmitting vectors include mites, nematodes and fungi each relying on specific pathways but insects represent the largest class of vectors. The best-characterized plant viral insect vectors are found among Hemiptera and Thysanoptera like aphids, leafhoppers, planthoppers, whiteflies and thrips, respectively. Some of these biological characteristics of virus interaction are taxon-specific, yet without full congruency with the taxonomic classification or vector life style. Virus–vectors interactions are multiple, although some commonalities are noticeable. Generally, the infection cycle begins with the vector encountering the virus in the plant where virus acquisition by the vector takes place. The virus must then persist in or on the vector long enough for the virus to be transmitted and delivered into the cells of a new host plant. In most cases, specific viral determinants such as viral capsid, or to lesser extent membrane glycoproteins, are essential for vector-specific transmission. Specific virus–vector interaction can vary in time and occurs through identified to non–identified molecules or candidate receptors molecules in the cuticle of vectors.[22,23] Recently, integrative proteomics revealed, for example, in aphid and thrips, that specific receptors embedded in these proteins are directly accessible at the surface of the cuticle of specific organs, were shown to bind a plant virus and were consistently involved in viral transmission.[24–29] Additional viral nonstructural helper proteins can foster bridging the virion to the vector binding site. The different modes of viral transmission by vectors include nonpersistent, semi–persistent and persistent, whereby the transmission time window to disseminate the virus to a new host plant after feeding on an infected plant by the vector lasts from seconds to minutes, hours to days, or days to weeks, respectively.[30,31] Briefly, noncirculative nonpersistent (NC-NP) plant viruses are retained in the insect stylet, a specialized sucking mouth element. Semi-persistent (NC-SP) viruses are internalized in the insect by binding to chitin lining the foregut, but do not appear to enter tissues. Persistent viruses are taken up into and retained by insect tissues and are characterized by invading the salivary glands.[32] Persistent viruses can be further divided into circulative, nonpropagative (CP-NPr), and circulative, propagative (CP-Pr). Circulative viruses must escape the insect digestive tracts and extent to neighboring organs to reach the salivary glands for transmission implicating the interactions of viral-host proteins.[33] Circulative viruses generally follow a similar acquisition route by vectors and this begins in the gut of the insect. After entry into gut epithelial cells, the paths for virus movement between cells and dissemination to other tissues diverge, sometimes even for viruses within a family. The identification of cuticular protein receptors of plant viruses within their insect vectors constitute a crucial challenge to understanding the fundamental mechanisms of transmission and offers avenues for future alternative control strategies to limit viral spread.[34–38] Generally, the limited span of insect vectors exploited by viruses suggest relatively strong evolution pressure and bottlenecks by the necessities of vector transmission,[39] and often disclose large-scale coevolutionary patterns ruled by vector-type.[40] This coevolution process implies both a perennial arms race that fosters diversification of viral escape or cooperation strategies and elaboration of host defense pathways during the ubiquitous interaction with cellular hosts.

1.4 Evolutionary structural modules and proteomic make up

It is well established that most proteins often exhibit highly modular structure. Proteins domains, or modules, are compact and stable protein structural units which can occur either in the standalone form or as part of multidomain architectures. They are crucial macromolecular players. At this point, it is important to underline the fundamental notion of structural domain. Often, however, domains are characterized differently—as distinct regions of protein sequence that are highly conserved in evolution. These are essential to the functioning of the cell and fold into well-packed yet highly ordered structural units which are evolutionarily highly conserved. They fold in specific three-dimensional (3D) atomic spatial arrangements and function largely independently separated by flexible loops and relatively rigid regions in the form of turns and coils. They contribute to overall protein stability by establishing a multiplicity of intramolecular interactions mainly through interaction between hydrophobic residues in inter-domain interfaces. Interactions also augment the stability of individual domains, which constrains mutational substitution of interacting residues. These folds are generally defined by the composition, architecture and topology of their core "helix" and "sheet" secondary structure elements. This matches the common concept that surface residues are less conserved in proteins when compared to those that are buried in the structural core. In evolution, protein structures are evolutionarily highly conserved and capable of preserving an accurate record of genomic history over long evolutionary periods. They represent "relics" of molecular evolution and express the greatest levels of genomic redundancy and reuse. These compact, recurrent and independent folding units of protein structure sometimes combine with other domains to form multidomain proteins, a process driven mostly by the evolutionary forces of genomic rearrangement.[41] The salient features of structural domains (that is, independent folding and stability) conduce them to become distinct evolutionary units, which exist as stand-alone proteins or as parts of various domain architectures in multidomain proteins exhibiting various topologies, functions (here considered as modules). Often, the basic physicochemical interactions and the associated structural and sequence motifs are conserved throughout a fold or even across fold boundaries. Perhaps more important, on several instances, the same activity and/or function is performed by two or more unrelated folds in different viruses or in different viral systems in the same virus group. Comparisons of structural and functional modules repertoires reveals both substantial similarities between different viral species, particularly with respect to the relative abundance of modules, and major differences. The most notable manifestation of such differences is the viral lineage-specific expansion of these modules, which probably points to unique adaptations of viruses to their cellular context (host plant, vector insect, virion structure).

1.5 Chasing intrinsic disorder for plant virology: The promise of an emerging field

Plant viral proteins, and *Ribovaria* viral proteins specifically, display some level of disordered regions and loosely packed conformations. These features might endow viral proteins with increased structural malleability and newly exploited effective adaptability paths to interact with the components of the host.[42,43] They could also reflect high adaptability degrees and mutation rates observed in plant viruses, thus, denoting a unique strategy to buffer the mutational deleterious effects. While undeniably still in its early infancy, intrinsic disorder of land plant virus proteomes must be duly considered in order to provide fundamental generalizations on evolutionary patterns of transmission, on host range evolution, and on risk prediction of disease emergence. Nevertheless, our present fundamentally comprehensive proteome-wide analysis therefore highlights the extent to which evolution has fine-tuned viral proteomes, and not just the native structures, of complex biomolecules. Herein, using a coarse-grained approach, this chapter uncovers unique insight for the generality and importance of protein disorder in evolutionary proteomics of plant viruses and summarize the emerging global understanding on the IDP landscape across all evolutionary lineages of the global plant virome. We propose a complete, even though preliminary, framework of the forthcoming global plant virus disordome and invite the further inclusion of IDP into mainstream plant virology to advance our understanding of viral evolution.

2. Protein disorder composition in plant viruses: *Terra incognita*

In plant viruses, the vast majority of structural and nonstructural viral proteins of their proteome (e.g., replicase, capsid and movement proteins) are still poorly characterized. Although, well–conserved 3D structure (fold) and functions are vital for competent viral life cycle,[43–47] particularly at the interplay of host interactions, many proteins or regions of proteins never assume a specific fold and lack defined tertiary structure in their native functional state. Rather, they operate as ensembles of fast interconvertible conformers.[48–54] Natively unfolded proteins exist as a dynamic ensemble of conformations expressing several crucial biological functions under specific physiological conditions.[55,56] Proteins may be moderately or entirely intrinsically disordered, encompassing a wide array of malleable conformations depending on the amount of disorder. Intrinsically unstructured or disordered protein/regions (IDPs, IDRs) are pivotal to viral interactions networks including their flexibility and are often associated with diseases.[57–59] Contrastingly to cellular organisms, viruses rely much more extensively on disordered proteins, to optimally exploit the functional landscape of their proteome.[60–63]

2.1 Viral protein intrinsic disorder: Evolutionary functional modulator of adaptability

Particular properties of the viral lifestyle may predispose these obligate parasites toward use of IDR. Undeniably, some viruses are encoded within spatially restricted genomes, and the ability to execute diverse or even complementary functions, facilitated by structural reorganization, may carry advantageous features from the biophysical or coevolutionary stance.[64–67] The capacity to rapidly acquire motifs made accessible within disordered protein stretches that modulate interaction with host proteins may also constitute a valuable asset under particular biotic circumstances. While viral evolutionary switch often occurs between genetically unrelated hosts, viral evolvability afforded by genomically encoded flexibility may appear an important feature within certain families and foster coevolutionary coupling even with relatively well-folded partners.[68] Consequently, because they undergo faster evolutionary changes than folded domains, disordered regions may favor the rate of emergence of new adaptive phenotypes, in addition to the wide range of potential genotypes. Hence, while the functional protein interface involved in virus–host interaction may remain relatively conserved, short motifs can easily tolerate variability and evolve de novo more rapidly. Viruses can efficiently gain functional domains from their host, however, horizontally transferred genes are typically acquired only by larger viruses that can integrate a pinched genetic element from their host into their genome without functional loss of essential other proteins.[69]

Therefore, viruses with small genomes would better benefit from the structural and multifunctional flexibility and evolutionary optimal disordered regions to adapt to or explore different environmental niches.[43,46] Alternatively, these small-sized viral genomes may be so constrained to encode a limited and relatively more conserved set of structural proteins harboring largely ordered domains. Interestingly, some viruses can avoid such constraints by tolerating more genomic promiscuity with overlapping proteins over different reading frames. With respect to the global diversity of the land plant virome (Angiosperms), we expected that these evolutionary-driven constraints would result in significant variability in IDP prevalence across different distributions of IDRs among different plant viral groups. The use of compact protein interfaces allows increased redundancy by permitting functionality to be mediated by a set of short disordered regions (e.g., domains of HSP70h in the family *Closteroviridae* or retroviral Gag proteins with the family *Caulimoviridae*) rather than on a single globular interface, resulting in increased evolutionary robustness of viral proteomes.[70] Convergently, other studies have focused on human and animal viruses and the prevalence of IDP and its correlation with motifs involved in manipulating cell signaling, targeting host proteins for proteasomal degradation, directing viral proteins to the correct subcellular localizations, altering transcription of host proteins and deregulating cell cycle checkpoints.[71,72] While the functional importance of IDP interactions in viruses has been established, to date there is limited understanding of the nature and diversity of such IDRs across different viruses. Henceforth, a in depth understanding of the role of IDP in plant viruses is essential if we are to fully grasp the relationships between protein sequence and function, and furthermore to decipher the relationships between the overall diversity of plant proteome architectures and long-standing evolutionary strategies opted by plant viruses. Insofar, information about IDP for plant viral proteomes remain scarce. To date, only few attempts have been done in elucidating functional role of specific disordered regions of specific plant viral proteins such as the VPgs in the economically important families *Potyviridae* and *Solemoviridae*.[64,73–78] Furthermore, the structural analysis of IDPs may appear challenging because of their involvement in a plethora of pathogenicity pathways,[61,79,80] which makes IDPs attractive targets for disease management strategies. For this reason, herein, we set out to explore the striking variation in viral protein disorder in a representative group of plant viruses. Regardless of whether the proteomic evolution drives that of IDP, or vice versa, the exact relationship between these two fundamental biological features with virus biology has never been investigated. To explore the nature of these patterns across the whole plant proteome and its known biodiversity using comparative evolutionary proteomics in order to deepen our understanding of the still under-appreciated viral structural protein disorder, we summarize our findings and describe its extent and meaningful features in the plant virosphere.

3. The global plant virus proteome reveals overhauling IDP abundance and variability

To explore and grasp the importance of IDP across the global plant viral proteome, we retrieved predicted protein disorder of 1003 plant virus genomes from the RefSeq Database (NCBI)[38], looking systematically for trends within and between families of plant viruses. The emphasis was focused on the prevalence of IDP variation in whole proteomes or in part, together with variation in genome and proteome size, structural proteins and by other viral biological features such as the transmission mode, or nature of genome (DNA or RNA, single or double stranded, non-/ segmented, mono-/multipartite). Since various predictors outputs appeared relatively well correlated,[81] there were clear relationships between predicted disorder (%IDP) for viral genomes, we relied on IUPRED, PONDR VLXT and metapredictors D2P2 to survey the entire plant virome.[50,51,82,83]

We also focused on predictions within specific structural proteins, noting that the overall survey results were very strongly correlated, regardless which way the analysis was completed. The percent disorder (%IDP) for a plant viral genome represents the percentage of residues which the IUPred2A and PONDR softwares[84,85] predicts to be disordered, with a cut-off of greater than 0.5. The mean percent disorder for all surveyed viral genomes investigated was 4.7% (standard deviation 0.8%, data not shown). Plots of %IDP vs genome and proteome size for each studied plant virus within a family and replication type summarize the global tendencies (Fig. 1), and mean %IDP for a set of representative viral families was comprehensively tabulated (Table 1). We noted that plant viral genomes encode a strikingly high IDP amplitude, with minima-maxima values ranging between 0.01% (*Higrevirus, Kitaviridae*) and 26.83% (*Tymovirus, Tymoviridae*) (Fig. 1). Among viral categories based on genome nature, the mean %IDP was seen to significantly range between the major viral families, and within them, e.g., from 2.6% (*Reoviridae*) to 7.1% (*Amalgaviridae*) for the dsRNA, from 0.9% (*Nanoviridae*) to 8.2% (*Genomoviridae*) for the

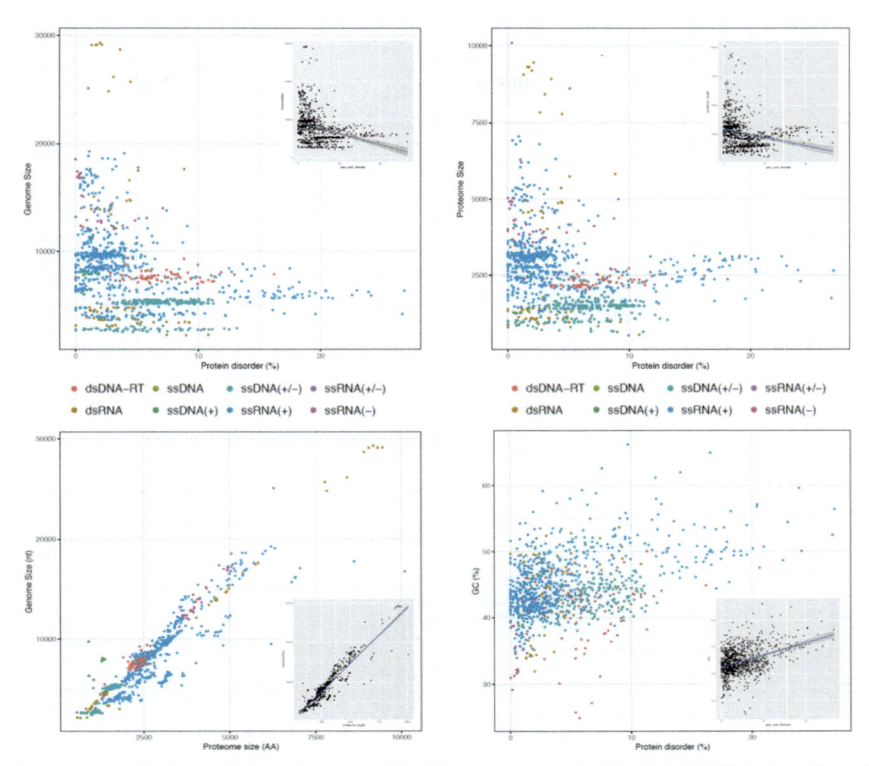

Fig. 1 Overall relationships of the predicted intrinsic protein disorder (IDP) in the Global Plant Virome. General IDP vs Genomic features Correlation between IDP with the genome and proteome features of 1003 plant viruses from the RefSeq database. Genomic replication types are single strand RNA positive strand ((+)ssRNA), single strand RNA negative strand ((−)ssRNA), double strand RNA (dsRNA), single strand DNA ((−)ssDNA and ambivalent (+/−)ssDNA), double strand DNA retrotranscribing (dsDNA-RT). Inside panel are the same plots with their respective R^2 slope in the middle matrix.

ssDNA, from 0.3% (*Fimoviridae*) to 2.73% (*Rhabdoviridae*) for the (−)ssRNA, and from 1.5% (*Secoviridae*) to 14.5% (*Tymoviridae*) and 17.3% (*Luteoviridae*) for the RNA viruses (Table 1). Furthermore, a considerable variability within families was also observed as summarized by the standard deviation for each family with the studied viral types (Table 1 and Fig. 1). Essentially, the highest standard deviation was observed in the family *Tymoviridae* and to a lesser extent in the family *Alphaflexiviridae* and *Tombusviridae*. We examined the relationships between viral disorder abundance and genomic size within every viral family at the genus level (data not shown). Amid a brief

Table 1 Prevalence of IDP across the diverse evolutionary lineages of the plant virome.

Virus family	Number of genomes (N)	Genome length (Kb)	Mean proteome size	Number of structural modules	Mean viral %IDP	Viral %GC	Correlation IDP-genome size	Correlation IDP-proteome size	Correlation IDP-%GC	Correlation IDP-number of proteins	Transmission mode[b]	Genome partition[c]
dsDNA-RT												
Caulimoviridae	65	7.6 (0.43)[a]	2.23	5	**7.3** (2.5)[a]	0.4 (0.05)	−0.010	0.370	0.332	0.039	NC-NP	Mo-NS
dsRNA												
Amalgaviridae	5	3.4 (0.01)	1.4	1	**7.1** (1.60)	0.48 (0.01)	0.138	**0.756**	0.033	NA	NA	Mo-NS.
Endornaviridae	14	14.7 (0.16)	4.8	4	**3.5** (2.04)	0.40 (0.04)	**0.704**	**0.723**	**0.790**	NA	Clonal	Mo-NS.
Partitiviridae	39	4 (0.36)	1.21	1	**3.1** (2.07)	0.45 (0.03)	−0.198	−0.211	0.040	0.068	Clonal	Mu-S.
Reoviridae	9	27.3 (1.8)	8.7	1	**2.6** (1.09)	0.37 (0.04)	**−0.644**	**−0.725**	0.534	−0.471	**CP-Pr**	Mo-S.
ssDNA												
Genomoviridae	3	2.2 (0.02)	0.74	1	**8.2** (1.10)	0.50 (0.01)	−0.418	**−0.970**	**−0.716**	**−0.845**	NA	Mo-S.
Geminiviridae	211	4.5 (1.21)	1.34	2	**6.6** (2.71)	0.40 (0.03)	0.280	0.249	−0.090	0.244	CP-NPr	Mo-NS/ Mu-S.
Nanoviridae	11	7.7 (1.0)	1.31		**0.9** (0.05)	0.40 (0.02)	−0.016	**0.762**	**−0.692**	**0.787**	CP-NPr	Mu-S.
ssRNA(−)												
Rhabdoviridae	16	12.8 (0.60)	3.8		**2.73** (1.96)	0.43 (0.02)	0.285	0.252	0.223	−0.034	**CP-Pr**	Mu-S.
Aspiviridae	2	11.8 (0.81)	3.6	1	**0.80** (0.06)	0.34 (0.01)	−0.297	−0.093	0.021	−0.408	Clonal	Segm.
Phenuiviridae	6	15.2 (7.00)	4.5		**0.56** (0.60)	0.37 (0.01)	0.379	**−0.698**	0.330	**−0.505**	**CP-Pr**	Mo-S.
Tospoviridae	3	17 (0.30)	5		**0.47** (0.07)	0.34 (0.12)	−0.010	**−0.832**	−0.825	**−0.832**	**CP-Pr**	Mo-S.
Fimoviridae	3	16.2 (1.71)	4.5		**0.33** (0.23)	0.31 (0.01)	**0.888**	**−0.825**	**0.722**	**−0.838**	**CP-Pr**	Mo-S.

ssRNA(+)

Luteoviridae	29	5.7 (0.18)	2.8	2	**17.3 (2.71)**	0.48 (0.01)	0.398	0.041	0.235	0.083	CP-NPr	Mo-NS.
Tymoviridae	27	6.3 (0.20)	2.5	3	**14.5 (5.97)**	0.55 (0.04)	−0.040	**0.678**	−0.094	0.473	CP-Pr/CP-NPr	Mo-NS.
Solemoviridae	19	4.2 (0.13)	2	3	**7.8 (2.81)**	0.50 (0.02)	0.519	0.290	0.434	0.156	CP-NPr	Mo-NS.
Alphaflexiviridae	50	6.9 (0.80)	2.3	3	**7.5 (4.51)**	0.50 (0.04)	**0.732**	**0.698**	0.218	0.528	NC-NP	Mo-NS.
Botourmiaviridae	3	4.8 (0.06)	1.3	1	**5.4 (1.81)**	0.52 (0.01)	**0.916**	0.573	**−0.615**	NA	NA	Mu-S.
Tombusviridae	61	4.2 (0.40)	1.7	3	**4.4 (4.81)**	0.50 (0.02)	−0.137	−0.116	0.491	−0.406	NC-NP	Mo-NS/Mu-S.
Benyviridae	4	12.4 (1.90)	4.1	3	**2.7 (2.02)**	0.43 (0.01)	−0.354	0.211	0.404	−0.103	CP-NPr	Mu-S.
Potyviridae	144	9.8 (0.90)	3.24	3	**2.6 (1.31)**	0.42 (0.02)	0.196	0.194	0.103	0.123	NC-NP	Mo-NS/Mu-S.
Bromoviridae	35	8.4 (0.20)	2.4	3	**2.5 (1.51)**	0.44 (0.01)	0.535	0.249	0.309	0.424	CP-NPr/NC-NP	Mu-S.
Betaflexiviridae	82	8.1 (0.60)	2.7	3	**2.1 (1.51)**	0.43 (0.03)	0.237	0.242	0.271	0.118	NC-NP	Mo-NS.
Closteroviridae	39	16.3 (1.70)	5.3	8	**1.8 (1.48)**	0.41 (0.04)	−0.236	−0.101	0.364	−0.367	NC-SP	Mo-NS/Mu-S.
Deltaflexiviridae	3	8.2 (0.10)	2.7	1	**1.7 (0.05)**	0.47 (0.14)	0.082	−0.532	**0.977**	**0.999**	NA	Mo-NS.
Virgaviridae	51	7.8 (2.10)	3.4	3	**1.5 (2.10)**	0.42 (0.01)	0.714	0.466	0.060	**0.706**	NC-SP	Mo-NS/Mu-S.
Secoviridae	58	11.3 (1.60)	3.3	3	**1.4 (1.08)**	0.44 (0.02)	**0.583**	**0.402**	0.426	−0.181	NC-NP/CP-NPr	Mo-NS/Mu-S.
Kitaviridae	3	10 (5.20)	2.8	2	**1.1 (1.61)**	0.43 (0.01)	**0.696**	0.684	**−0.726**	**−0.957**	**CP-Pr**	Mu-S.

[a]Standard deviation.
[b]Monopartite non-segmented.
[c]Transmission mode.
Relationship between IDP, Genomic-Proteomic Features and Transmission Mode. Mean intrinsic disorder (IUPRED Long) vs mean genome size, proteome size, and GC% content for 28 families and the 5 different types of genomic replication strategies. Relationship of disorder in viruses according to transmission mode in analyzed with pseudo–correlation (see plot). Data based on the average percentage from proteomes obtained from RefSeq (NCBI).
CP-Pr, Circulative persistent-propagative; CP-NPr, Circulative persistent-non propagative; Mo-S., Monopartite segmented; Mo-NS., Monopartite non-segmented; Mu-S., Multipartite segmented; Segm., Segmented; NC-NP, Non circulative-non persistent; NC-SP, Non circulative-semi-persistent

foray at higher taxonomic resolution within viral families, some genera show significantly high noticeable IDP values namely, *Alphaflexiviridae* (*Allexivirus*), Caulimoviridae (*Cavemovirus, Rosadnavirus*), Secoviridae (*Waikavirus*), Tombusviridae (*Tombusvirus, Umbravirus*), Tymoviridae (*Tymovirus*), Virgaviridae (*Pomovirus, Tobamovirus*) showed extensive variation, while for other families the abundance of protein disorder remains very comparable across all constitutive members.

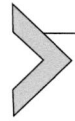

4. General affinity of phylogenomically unrelated small plant virus proteomes with increased IDP

The mean genome size of the investigated plant virus dataset varies from 2.2 kb (*Genomoviridae*, ssDNA) to 16.3–17 kb (*Closteroviridae*, ssRNA and *Tospoviridae*, respectively) with the exception of 27.3 kb (*Reoviridae*, dsRNA) (Table 1). When examining whether viral families with smaller genomes tend to have more disordered proteins, it appears that IDP prevalence within and between taxonomically and phylogenetically closely related viral groups can be highly variable. For example, within the viral order *Durnavirales*, both small genome-sized dsRNA families *Amalgaviridae* (7.1% ±1.6) and *Partitiviridae* (3.1% ±2.07), but also between the members of the family *Solemoviridae* (*Sobelivirales*) 7.83% ±2.8 with contrast to the related family *Secoviridae* (*Picornavirales*) 1.4% ±1.08. Within another group of genetically related ssRNA viruses, the order *Tymovirales*, additional striking variation can be uncovered within with the *Alphaflexiviridae* (7.5% ±4.5), *Betaflexiviridae* (2.1% ±1.5) and the *Tymoviridae* (14.5% ±5.9). Remarkably, among most investigated large-sized ((−) ssRNA), all orders belonging to the phylum *Negarnaviricota* namely, the *Bunyavirales* (*Fimoviridae, Phenuiviridae and Tospoviridae*), the *Serpentoivirales* (*Aspiviridae*) *and the Mononegavirales* (*Rhabdoviridae*), present relatively similar (with the exception of the *Rhabdoviridae*) and low %IDP, spanning 0.33% ±0.23 to 0.56% ±0.6, 0.8% ±0.06 and 2.73% ±1.96, respectively (Table 1). Using a nonparametric Spearman's correlation coefficient to quantify the degree of association between the IDP abundance and the rank of genomic features, substantial lines of evidence support this assumption. We examined the relationships between viral disorder abundance and genomic size within every viral family (Table 1). Within the dsRNA viral type, there is a significant strong negative correlation between mean %IDP among members of the family *Reoviridae* and mean genome or proteome size, considering only an average for each family as a single data-point in the analysis. Overall, except for the family *Reoviridae*, smaller viral proteomes tend to have more IDP (Table 1, Fig. 1). Contrastingly, in the same viral category, the family *Endornaviridae*, a moderate and strong positive correlation is observed between viral %IDP, and genome and proteome size, respectively. Within the ssDNA category, the family *Genomoviridae* exhibit the strongest negative correlation between %IDP and proteome size, whereas in the family *Nanoviridae*, a high positive correlation is observed. Likewise in several (+)ssRNA virus families with highly compacted proteomes and overlapping ORFs, such as *Luteoviridae*, *Tymoviridae* and *Solemoviridae*, high IDP correlated with lower genome size values together with proteome range, whereas the families *Bromoviridae* that harbor genome size similar to the large-sized *Tymoviridae* intriguingly exhibit very low IDP value. While some of the families are relatively small, or have little variability likely associated with relatively recent common genetic origins, for others there is a significantly extensive variation to admit further interpretation. In a number of families, genome size together with few other genomic features, such as genome and proteome size, was substantially positively correlated to protein disorder, and accounted for an appreciable proportion of the variation in disorder. These were the families *Amalgaviridae*, *Nanoviridae*, *Endornaviridae*, *Tymoviridae*, *Botourmiaviridae*, *Alphaflexiviridae*, *Virgaviridae*, *Secoviridae* and *Kitaviridae*. Out of all investigated viral taxa, the *Genomoviridae* and the *Phenuiviridae*, *Fimoviridae*, *Tospoviridae* viral families exhibited revelatory distinct negative correlations between genomic or proteomic size and %IDP, respectively, whereas others presented moderate extent of both correlations. Table 1 illustrates these trends for a representative set of families under study.

5. Emerging patterns in amino acid composition as a potential predictor of IDP and virus functional biology

Notwithstanding the relative universality of the genetic code and its conservation across species, synonymous codons are not equivalently used due to degeneracy, and codon biases can vary severely between across genetic sequences within the same genome.[86,87] As a result of the redundancy of the genetic code, adjacent pairs of amino acids can be encoded by many different pairs of synonymous codons raising the question of to what extent the actual encoding in plant viruses is related to IDP. Various factors can influence codon usage bias within and across genomes, including protein expression level and GC content.[88] Although each species clusters toward a specific setting of codons,[86,88] the origin and evolutionary drivers of these preferences remains largely unclear. Codon usage variation within genomes (intraspecies codon usage) is often attributed to selection, due to the significant positive correlation between protein expression levels and the presence of "preferred" or "optimal" codons.[89] Processes that drive codon usage variation across viral genomes (interspecies codon usage) are generally thought to be mutational,[90,91] although the magnitude to which these processes are ruled by mutation-selection principles, or biased genetic conversion remains largely debated.[92] Genomic GC content has been identified as the strongest determinant of codon usage variation across species.[93,94] Consequently, GC-rich organisms tend to favor GC-rich codons whereas AT-rich organisms are enriched in AT-rich codons.[88] However, codon usage bias is uneven across open reading frames (ORFs) of viral genomic organization, and many viral genes exhibit little or no codon usage bias. Most amino acids are encoded by two to six synonymous codons and it has been advocated that protein folding and functions can be influenced by synonymous codon substitutions.[95] Correlations between codon usage and propensities to form certain protein structural motifs have been previously investigated.[96–98] The viral fitness is also proposed to be affected by codon modifications of viral genomes.[99] Furthermore, amino acid biases in viral genome composition can also inform host–virus associations. In any case, genomic biases can coarsely discriminate viruses from different host taxa within several well-studied RNA viral taxa.[100–103] Virus–host biological interaction is a continuous coevolutionary process involving both host immune system and viral escape mechanisms. Moreover, the amino acids frequency in IDRs/IDPs significantly differs from that of ordered proteins[45,104,105]). Recently, regions predicted to be unstructured were found to be preferentially encoded by nonoptimal codons, whereas regions likely to be well structured are preferentially encoded by preferred codons.[106] These studies led to the hypothesis that codon usage and protein structures coevolved to allow proper folding of proteins. Underlying all hypotheses regarding the role of codon usage in plant virus IDP and although it is not clear why some codon pairs are under- or overrepresented, is the assumption that codon identity regulates protein folding through the influence of the rate of translation elongation. With respect to plant viruses, clear evidence supporting this assumption is lacking, however. Notwithstanding the expected role of codon usage in regulating protein folding, studies based on solved protein structures have not yielded clear correlations between codon usage and specific structural elements. Yet, it has long been recognized that systematic effects on amino acid usage are influenced by base composition (Novoa et al., 2019). Consequently, as disordered residues favor G and particularly C rich codons and has been reported to correlate host,[107] in plant virus investigation, it appears essential to tease apart potential confounding impact of disorder relationship with base compositional effects. In the present genome-wide investigation covering the plant virus proteome, we address such fundamental open questions and made two interrelated observations: (1) %IDP and %GC content are correlated (2) and %IDP, %GC and amino acid compositions can, in some viral taxon, be considered as potential predictor to host or vector. Hence, it appears crucial to characterize whether the base composition is higher because of selection on disorder, or vice versa, but most generally base composition is a strong biasing feature that occurs most prominently at synonymously variable sites, thus it is perhaps a more parsimonious explanation to consider base composition as a potential modulator in protein disorder variability, rather than contrariwise. At first glance, Table 2 suggests that both effects occur. First, using a comprehensive dataset, there is a substantial both positive (*Endornaviridae, Closteroviridae, Fimoviridae, Reoviridae*) and negative (*Genomoviridae, Nanoviridae, Botourmiaviridae, Kitaviridae and Tospoviridae*) correlation of %GC content in plant viruses with predicted protein disorder (Fig. 1), despite the fact that base composition preferences at these sites do not fundamentally modify the amino acid sequences. This suggests that viral %GC is indeed a substantial driver of the extent of predicted protein disorder in plant viruses. The stronger correlations observed for genomic %GC indicate that it is likely that selection pressures favoring or avoiding protein disorder also have an additional universal impact on viral %IDP within specific evolutionary viral lineages. Accordingly, we attempted to determine if IDP had a direct relationship with proteome size and number of proteins (Fig. 1). Among several families of the dsRNA viruses (*Amalgaviridae* and *Endornaviridae*), the ssDNA viruses (*Nanoviridae*) and (+)ssRNA viruses (*Alphaflexiviridae, Secoviridae, Tymoviridae*), strong positive correlation were observed, whereas a strong negative correlation were noticed for one family among the dsRNA viruses (*Reoviridae*), ssDNA viruses (*Genomoviridae*), (−) ssRNA viruses (*Fimoviridae, Phenuiviridae, Tospoviridae*). Intriguingly, a substantial number of virus taxa for which a negative IDP%–proteome size correlation has been reported are transmitted in a circulative persistent

Table 2 Overview to predicted intrinsic protein disorder (IDP) in a taxonomically representative selected subset of plant viruses the Global Plant Virome.

Realm	Kingdom	Phylum	Class	Order	Rep_type	Family	Genus	Viral Species	Abbrev.	GKb	GC%	#Prot	%IDP*	RdRp	MP	CP	CAI	NMoR	LMoR
Monodnaviria	Cressdnaviricota	Arfiviricetes	Muplavirales		ssDNA	Nanoviridae	Babuvirus	Banana bunchy top virus	BBTV	6.40	40	5	0.9	21.7	48.7	30.0	0.68	14.0	18.0
							Nanovirus	Subterranean clover stunt virus	SCSV	5.99	39	6	0.7	24.2	39.3	19.0	0.66	6.0	8.5
			Repensiviricetes	Geplafuvirales		Genomoviridae	Gemycircularvirus	Bromus associated gemycircularvirus 1	BaGV	2.24	48	2	10.8	15.4	NA	26.3	0.69	3.0	17.6
						Geminiviridae	Becurtovirus	Beet curly top Iran virus	BCTIV	2.85	41	5	0.7	23.2	38.8	35.7	0.69	10.0	14.5
							Begomovirus	Bean golden yellow mosaic virus	BGYMV	5.24	39	8	3.0	8.5	23.2	37.2	0.66	9.0	13.4
							Capulavirus	Euphorbia caput-medusae latent virus	EcmLV	2.68	40	7	10.2	14.6	NA	31.4	0.70	NA	NA
							Curtovirus	Beet curly top virus California	BCTV	2.93	39	6	2.7	7.5	34.9	34.2	0.67	NA	NA
							Eragrovirus	Eragrostis curvula streak virus	ECSV	2.75	49	3	2.8	18.3	31.4	29.0	0.69	5.0	13.2
							Grablovirus	Grapevine red-blotch associated virus	GRBV	3.21	41	5	5.3	16.3	NA	39.5	0.66	NA	NA
							Mastrevirus	Maize streak virus A	MSV	2.69	49	4	3.9	27.8	49.5	29.2	0.73	2.0	10.0
							Topocuvirus	Tomato pseudo-curly top virus	TPCTV	2.86	42	6	9.3	14.3	34.4	21.1	0.67	NA	NA
							Turncurtovirus	Turnip curly top virus	TCTV	2.98	42	6	4.8	17.5	40.3	22.5	0.66	NA	NA
Ribovaria	Lenarviricota	Alassoviricetes	Levivirales	ssRNA(+)		Botourmiaviridae	Ourmiavirus	Ourmia melon virus	OuMV	4.85	52	3	6.8	6.2	49.3	32.9	0.69	12.0	14.6
	Pisuviricota	Duploviricetes	Durnavirales	dsRNA		Amalgaviridae	Amalgavirus	Southern tomato virus	STV	3.44	48	2	5.4	27.1	NA	45.9	0.72	5.0	12.0
		Pisoniviricetes	Picornavirales	ssRNA(+)		Secoviridae	Comovirus	Cowpea mosaic virus	CPMV	9.37	42	5	0.6	33.3	11.7	12.4	0.68	NA	NA
		Sobelivirales		ssRNA(+)		Solemoviridae	Sobemovirus	Southern bean mosaic virus	SBMV	4.13	50	5	6.4	20.8	0.5	32.9	0.69	NA	NA
		Stelpaviricetes	Patatavirales	ssRNA(+)		Potyviridae	Potyvirus	Potato virus Y	PVY	9.70	42	2	18.7	9.7	9.4	29.9	0.67	2.0	20.0
	Kitrinoviricota	Alsuviricetes	Hepelivirales	ssRNA(+)		Benyviridae	Benyvirus	Beet necrotic yellow vein virus	BNYVV	15.91	40	10	2.2	9.7	11.4	48.8	0.64	27.0	13.3
				ssRNA(+)		Bromoviridae	Cucumovirus	Cucumber mosaic virus	CMV	8.62	46	5	5.4	21.4	20.8	22.0	0.66	12.0	15.3
			Martellivirales			Closteroviridae	Velarivirus	Little cherry virus 1	LChV1	16.93	35	8	1.8	11.3	2.3	6.3	0.65	28.0	11.1
							Ampelovirus	Little cherry virus 2	LChV2	15.05	41	10	3.4	15.3	5.9	18.6	0.64	18.0	12.7
						Virgaviridae	Tobamovirus	Tomato brown rugose fruit virus	ToBRFV	6.39	42	3	0.2	18.6	32.7	15.7	0.68	5.0	10.6
							Tobamovirus	Tobacco mosaic virus	TMV	6.40	43	6	0.3	11.1	17.7	16.7	0.65	NA	NA
			Tymovirales	ssRNA(+)		Alphaflexiviridae	Potexvirus	Potato virus X	PVX	6.44	47	5	0.8	23.1	31.9	27.9	0.70	5.0	10.0
						Tymoviridae	Tymovirus	Turnip yellow mosaic virus	TYMV	6.32	57	3	26.8	4.1	24.5	36.0	0.70	3.0	6.0
		Tolucaviricetes	Tolivirales			Luteoviridae	Luteovirus	Barley yellow dwarf virus	BYDV	5.68	46	7	14.0	22.5	76.5	28.5	0.69	6.0	12.2
							Polerovirus	Potato leafroll virus	PLRV	5.99	49	8	18.5	26.3	71.4	33.2	0.68	24.0	16.4
						Tombusviridae	Tombusvirus	Tomato bushy stunt virus	TBSV	4.78	48	7	0.3	18.6	35.8	13.2	0.67	NA	NA
	Duplornaviricota	Resentoviricetes	Reovirales	dsRNA		Reoviridae	Phytoreovirus	Rice dwarf virus	RDV	2.48	44	13	2.7	13.5	16.9	13.1	0.66	19.0	9.7
	Negarnaviricota	Ellioviricetes	Bunyavirales	ssRNA(+/-)		Tospoviridae	Orthotospovirus	Tomato spotted wilt virus	TSWV	16.63	35	5	0.3	7.7	28.5	15.5	0.66	NA	NA
		Monjiviricetes	Mononegavirales	ssRNA(-)		Rhabdoviridae	Cytorhabdovirus	Lettuce necrotic yellows virus	LNYV	12.81	43	6	1.8	19.6	2.0	20.8	0.68	NA	NA
			Mononegavirales	ssRNA(-)			Nucleorhabdovirus	Maize mosaic virus	MzMV	12.13	47	6	3.1	13.2	25.5	11.6	0.70	NA	NA
Pararnavirae	Artverviricota	Retraviricetes	Ortervirales	dsDNA-RT		Caulimoviridae	Caulimovirus	Cauliflower mosaic virus	CaMV	8.02	39	7	9.2	19.9	11.1	28.2	0.68	11.0	24.8

PONDR VSL2B

Taxonomic classification, general IDP and IDP per hallmark protein RdRp, MP, CP (IUPRED Long). Genomic features: Genome length, number of proteins per genome CAI (Codon Adaptive Index). NMoR (average number of predicted MoRFs), LMoR (average lengths of predicted MoRFs. Genomes and proteomes are part of the global dataset of 1003 plant viruses from the RefSeq database (NCBI).

and propagative mode, additionally suggesting that disordered regions are often found in viral genomes encoding overlapping regions of proteins. Since many plant viruses are transmitted by one or several vector in a specific mode, we also verified the possible relationship between all tabulated correlations and the transmission mode. Additionally, we examined whether these relationships were statistically significant. Using the important and genetically exceedingly variable family *Closteroviridae* as an example, we examined the genomic %IDP, %GC and profiles the amino acid composition of these CP-specific transmitted viruses. These semi-persistently transmitted viral species harbor remarkably long genomes encoding exceptionally large proteome size, and associated to different vector species (Aphid, Whitefly and Mealybugs) with one genus where the vector is still unknown (*Velarivirus*). Indeed, a complementary feature of putative disordered proteomic regions is a compositional bias toward polar and charged residues. Suggesting that IDP can depict enriched content of disorder-promoting residues (Lys, Glu, Pro, Ser, Gln, Gly, Arg) and a depleted content of order-promoting residues (Trp, Cys, Phe, Ile, Tyr, Val, Asn and Leu). The amino acids His, Met, Ala and Thr are not consistently enriched or depleted among intrinsically disordered proteins, consequently are considered disorder-order neutral residues.[108,109] The compositional preference in amino acids is detected by comparing the fractional difference in composition between a given set of proteins and a set of ordered proteins. To carry out our amino acid compositional analysis, we used the Composition Profiler tool (background sample PDB select 25) in a representative dataset. We investigated the propensity of intrinsic disorder derived from the whole proteome of each individual viral species and at genus level in the family *Closteroviridae* using well-established disorder predictors and computed the fractional compositional difference of the *Closteroviridae* proteome relative to a set of highly ordered proteins. Our comparative analysis on this data subset was directed by the critical observation that overall disordered proteins/regions have noticeably different amino acid compositions than do ordered proteins/regions. For the comparative analysis, the enrichment or depletion of individual amino acids was determined and plotted (from order-promoting to disorder-promoting residues with red for the most depleted to green for the most enriched) with annotation as to whether a residue was disorder-promoting (green in Fig. 2; W, I, H, M, A, G, Q, N, and P), order-promoting (red in Fig. 2; C, F, V, L, R, S), or disorder-neutral (R and T depending on genus). Overall, amino acid composition varied across *Closteroviridae* proteomes although with

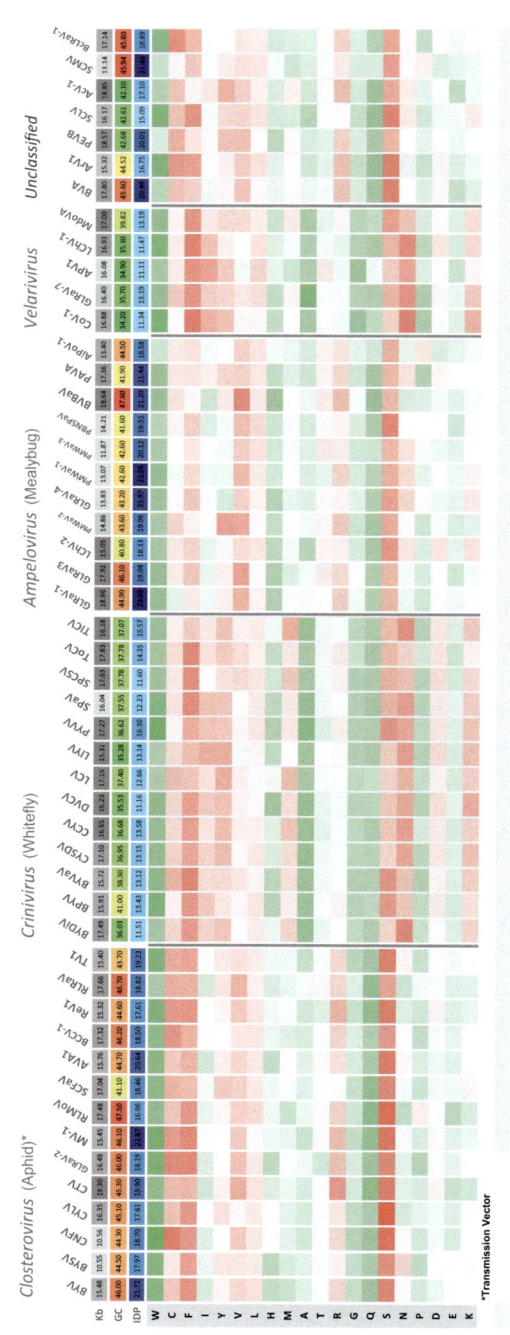

Fig. 2 Compositional profiling Heatmap of 50 proteomes within the Family *Closteroviridae* ((+)ssRNA) and their Vectors. Amino acid determinants defining structural and functional differences between the ordered and intrinsically disordered proteins. Positive and negative values, respectively, correspond to enrichment and depletion of given residues within query proteomes. Amino acids are represented as disorder-promoting, namely, residues found more abundantly in IDPs (red), whereas order-promoting correspond to residues in which IDPs are depleted (green), or disorder neutral (white). Members of the genus *Velarivirus* and recently sequenced or uncharacterized viruses (Unclassified) have no assigned vectors. (*) Information in parentheses indicate natural vectors.

strong genus-dependent congruence. Distinctively, two phylogenomically close genera *criniviruses* and *velariviruses* were depleted I, Y, N K amino acid, whereas both remaining genera, namely the the closteroviruses and ampeloviruses, where most enriched in V and R amino acids and weakly depleted or neutral in residue K. Moreover, these results showed significant genus/vector-dependent correlations between the proteomic amino acid composition profile and genomic %GC with %IDP among viruses and their vectors across different genera of this diverse family. Because the members of the family *Closteroviridae* tend to have large genomes of similar size, the overall trend noted initially of smaller viral families having more disorder was not allied with the effects of %GC on genome size (Fig. 2). Yet the overall variation in %GC and amino compositions could be clearly positively related to IDP and type of vector within this important family of plant viruses transmitted with different vector species in a noncirculative and semipersistent way.[31,110] Generally, it has been proposed that viral proteins of vector-borne plant viruses are subject to greater purifying selection on amino acid change than those viruses transmitted by other pathways and that virus–vector interactions impose greater selective pressure than those between virus and plant host.[137] Thus, an attractive proposition is that viral GC and IDP fluctuations can provide viruses with a variety of modulating leverages and may reflect evolutionary adaptations to their surrounding cellular environments. While this proposition is at face value plausible, more substantial evidence will be necessary to guarantee a robust conceptual scaffold to these assumptions.

Once the tracking effect of GC content on IDP is considered, further, until now unknown for plant viruses, universal amino acid specific rules governing the identity of selectively favored codons may be explored. To determine the relationship between codons, genomic features and predicted disordered residues in proteins within plant viral proteomes, we used the codon adaptation index (CAI,[91]) to score the codon usage bias for each codon for the protein-encoding genomes within same whole representative set of the global plant virome. CAI measures the deviation of a given gene sequence toward usage in highly expressed genomic regions and assigns a score between 0 and 1 for each codon. A higher CAI grade indicates a stronger preference to use optimal codons. For each genome of a taxonomically representative set of plant viruses, a simple Pearson correlation coefficient was calculated between CAIs, different genomic features and IDP scores. In our genome-wide analysis, negative correlation means that nonoptimal codons are preferentially used in disordered residues. Surprisingly, at

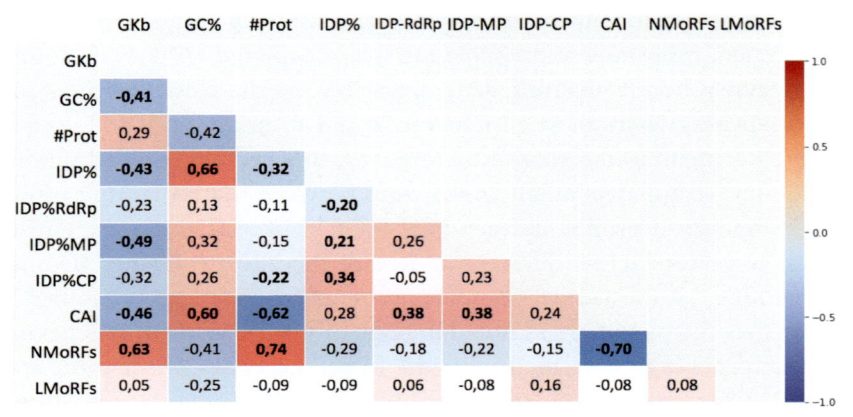

Fig. 3 Correlation matrix showing the relationship between CAI, IDP and proteomic features within a selected set of taxonomically representative plant virus species across all major plant virus evolutionary lineages. CAI measures the deviation of a given genomic sequence and assigns a score between 0 and 1 for each codon. A higher CAI score indicates a stronger preference to use preferred codons. NMoRFs : Avetage number of MoRFs. LMoRFs : Average length of MoRFs.

whole-genome level, most of plant viruses had a moderate to significant negative correlation with both genome and proteome size in contrast with both IDP and GC where a strong positive correlation was recorded (Fig. 3). Focusing at module-level (RdRp, MP, CP), CAI is positively correlated with their specific IDP. Remarkably, the average correlation is the highest and lowest in the RdRp and CP modules, respectively, which encode for most conserved domains. Yet they also exhibit lowest and highest percentage of disordered residues, respectively, meaning that the observed positive correlations between CAI and overall IDP scores are mostly due to disordered residues and are most likely biologically very relevant. Finally, and most staggering, inversely to both proteome and genome length, the strongest negative correlation was found between CAI and the abundance of MoRFs across all evolutionary lineages of the global plant virome. Together, this prime quantitative approach has proven productive and these results suggest consistently that nonoptimal codons are favorably exploited for encoding IDRs at the whole viral proteome-wide scale.

6. The global plant virus MoRFome: Gaining functional insights in viral IDP-mediated binding contextual landscape via harnessing interface molecular recognition features

Plant viruses are engaged, through their proteomic repertoires, in a complex array of functional and biophysical interactions often implying induced folding with other host cellular proteins in relatively hostile conditions. The structural malleability of IDRs enables particularly efficient and promiscuous interface connection, e.g., through ATP-, RNA- and protein receptor-binding, with a broad range of structurally diverse partners.[111–114] Congruently, functional and evolutionary high flexibility in protein–protein interaction networks, whether through permanent or transient binding, are most often significantly enhanced by IDR-enriched proteins.[115–117] Mediation by these IDPs/IDRs can be fine-tuned by either conformational selection (when the bound conformation is presampled in the disordered ensemble) or induced folding (disorder-to-order transition), as their structural flexibility enables them to ideally fit their target domain's binding surfaces.[118–120] Our understanding of the biological importance of IDP and its widespread occurrence in plant viruses is gradually expanding, thanks to the fine-tuning of computational genome-wide prospections. Henceforth, accurate IDP/IDR-based investigation can play a major role in researching the occurrence, the biological function, and the potential involvement of intrinsic disorder at different levels of complexity, including re-wiring plant virus–host protein–protein interaction networks linking pathogenesis-phenotypes and elucidating plant virus-associated disease.

Generally, IDR binding sites are classified under two overlapping categories: short linear motifs (SLiMs)[121] and molecular recognition features or elements (MoRFs).[122] In essence, MoRFs are defined as natively unstructured short binding motifs (5–25 residues long) strictly embedded in relatively longer IDRs that undergo different disorder-to-order conformational transitions upon partner-dependent binding, according to the induced-folding process.[111,113,122–124] MoRFs are mingled with both highly conserved and highly diverse residues and, on to a larger extent, can be highly locally conserved making them potentially highly praised features for coevolutionary studies.[125,126] Knowledge of these unique critical players enriched with amino acid with large hydrophobic side chains can essentially contribute better understand protein functions and facilitate comparative proteomic analysis of plant viruses. Therefore, accurate prediction and systematic annotation of MoRFs is an essential step in understanding functional properties of these proteins acting as molecular switches and in finding alternative applications for virus control. Moreover, accurate predictions of these abundant regions and their distinctive motif signatures from protein sequences can be generated amid machine learning-derived models where MoRFs are identified using putative disorder.[85,123,127–134] MoRF computational meta-predictors truthfully detect experimentally characterized MoRFs and, thus, can reliably contribute to the effective discovery of functional protein–protein interactions. Based on the structure they adopt upon binding, MoRFs are subcategorized into three basic groups: depending on the conformation they adopt. Hence, their careful scrutiny can deliver valuable insights into the opaquest parts of virus functional interactome.[121,135–137]

Viral MoRFs unique sequence signatures, abundance and how they advantageously modulate protein interactability have been extensively described across important viral processes including signaling and regulation,[61,138–141] yet MoRFs have never been substantially and systematically studied in plant viruses, which motivated us to extent our investigation by developing of a computational framework that predict MoRFs from unexplored plant viral proteins. Clearly, viral MoRFs do not appear to simply use these IDRs as "undefined linkers" among functional well-structured proteins, and instead are related with a variety of molecular biological functions, many of them intimately related to disease processes. To improve our fundamental understanding of these entities and highlight the increasingly paramount role that disorder plays in virus–host interactions, we predicted MoRFs putative conserved MoRF motifs potentially involved in protein–protein interaction. Here, we surveyed their presence using benchmarked metapredictors (MoRFpred, MoRFchibi and D2P2,[109,121,142]) to detect MoRFS and identify MoRFs residues based on local sequence physicochemical properties in a representative set of our plant virome dataset and their modular functional annotations. Toward this end, we exploit recent accurate well-performing SVM-based machine-learning sequence-based MoRFs predictors that have been specifically trained to detect them at reasonable computational costs and generate a propensity score of each residue to represent a MoRF.[109,121,142] The complementary advantages offered by a large spectrum of recently developed online platforms (MoRFpred, MoRFchibi and D_2P_2) lie in that they combine well-designed array of algorithmic architecture allowing feature-based SVM predictions with comprehensive alignment of sequence-derived markers of MoRF regions, including multiple disorder predictions, and evolutionary profiles.

Our complementary analysis reveals interesting patterns relating hallmarks MoRF regions located in plant virus proteomes and their interacting counterparts of their vector. We estimated the average abundance of putative MoRFs and IDRs across a representative set of plant viral proteomes (Table 2). Mapping the average abundance, with reasonable accuracy, is founded on content of residues located in the IDRs involving the MoRFs, that is the fraction of disordered MoRF residues among all residues in a given viral species within a specific taxonomic or evolutionary category. The content of average MoRF residues and their lengths is significantly variable between the investigated viral species from different families and replication strategies, albeit with some blanks or unresolved spots. Outstandingly, we discovered that the content of overall IDP and that of average MoRF abundance are negatively correlated together with the GC % (Table 2). The corresponding Pearson Correlation Coefficients (PCCs) equaled -0.29 and -0.41, respectively (Fig. 3). Moreover, the lowest value of MoRFs abundance in our studied dataset is observed within the *families Genomoviridae, Geminiviridae* (*Mastrevirus*), *Potyviridae* (*Potyvirus*). Whereas the highest values were recorded among the *Benyviridae* (*Benyvirus*) the *Closteroviridae* (*Velarivirus*) and the *Luteoviridae* (*Polerovirus*). A correlation where more protein numbers implies proportionally more MoRFs is also clearly evident for the largest viral proteomes (e.g., *Closteroviridae*). Lastly, faint positive correlation was observed between the average MoRF length. We also demonstrate that our comprehensive MoRFs-based framework can be used to dissect and predict putative new or uncharacterized plant virus–vector interactions. Their well-supported significance and potential impact in our understanding of viral transmission are thoroughly discussed using several examples of well-documented virosystems, including with respect to the displayed evolutive conservation across the plant proteome. Together, these innovative results are highlighted with a fairly detailed emphasis on the modular plasticity of the global plant virome with the hope that such enhanced knowledge may prove coherently informative regarding their associated biological functions in plant virus–host interactions.

7. Evolutionary proteomic comparative analysis of conserved viral modules make up reveals IDP landscape of viral–host interacting disordomes

Protein folding and binding rely on analogous processes, in which the protein explores favorable molecular interactions across a channeled energy landscape.[143] In viruses, many proteins are disordered under particular physiological circumstances, and fold into ordered structures only upon binding to their cognate host cellular partners.[113] In plant viruses, except for *Potyviruses* (*Potyviridae*), the underlying mechanisms by which folding is connected to binding at the host plant interface is poorly understood, but it has been hypothesized on speculative grounds that the binding to host plant proteins may be enhanced by a "fly-casting" effect,[75] where the IDP binds weakly and nonspecifically to its target and folds as it approaches the associated binding site. Intrinsically disordered protein (IDP) domains was predicted to occur in all analyzed proteomes. Despite the fact that these domains are not predicted to form stable three-dimensional conformation and functional structures, IDPs have been shown to endorse central functions in many biological processes.[44,60,144] Our whole-genome investigation has demonstrated that a fraction of proteins harboring substantial disorder regions are found in evolutionary strongly conserved domains of the hallmark movement, transmission and capsid proteins of plant viruses. The manifest structural and functional modularity of plant virus proteomes constitute, arguably, key features of virus evolution. Consequently, a productive approach to the investigate of the plant virosphere that complements evolutionary approaches toward modularity is the construction and analysis of networks of interconnected modules focused on their IDPs. Whether these modules preferentially modify their structures upon interaction with their host partners or whether they remain structured is a fundamental question. Extrapolation on the functional importance of codon usage these modules and the strong genome-wide positive correlation we observed between GC% and disorder tendency could suggest that nonoptimal codon usage allows IDPs to afford some conformational agility in certain domains. Furthermore, IDPs may serve as templates for explorative protein–protein interactions. Due to the uneven mutational pressures on these more conserved modules and because IDPs lack strong potential to fold into stable structural domains, these domains may readily misfold. The nonoptimal GC%-based codon usage in IDPs may either allow sufficient time for these IDPs domains to fold properly or may prevent interference by downstream well-structured domains in the IDP folding, suggesting a potential role of IDP in adaptation to ensure potential correct folding of proteins in new cellular hosts. As demonstrated in the previous section, the intrinsic disorder content in viral genomes can be highly divergent. In this section, we will illustrate in a systematic approach the molecular host–virus interplay with an emphasis on the IDP of the insofar module–module interactions landscape between the *Potyvirus*, the most prevalent family of plant RNA viruses, and the host plants. Succinctly, *Potyvirus* is the largest genus of plant viruses causing significant losses in a wide range of crops. Potyviruses are aphid transmitted in a nonpersistent manner and some of them are also seed transmitted.[145] As major economically important phytoviruses, potyvirids are extensively studied relatively to other plant virus genera and covering many facets, from molecular characterization to epidemiological evolution. The *Potyvirus* genome is highly modular and consist in a 10 kb (+)ssRNA.[146] The viral protein genome-linked (VPg) is covalently attached at its 5′-end and whereas the genomic 3′-end is terminated with a poly(A)-tail.[75] After post genomic translation, potyviruses express 11 mature proteins. The polyprotein is proteolytically processed by three *Potyvirus*-encoded serine protease P1, Helper component proteinase (HC-Pro), a protein with RNA silencing suppressor activity also involved in aphid-mediated transmission and Nuclear inclusion proteinase (NIa-Pro) into 10 mature proteins. CI, an RNA helicase with ATPase activity; the nuclear inclusion NIb the potyviral replicase act as a viral factory perform replication localized into intracellular specific membranes, the RdRp (RNA-dependent RNA polymerase); CP, the capsid protein; and two small peptides of unknown functions, 6K1 and 6K2. In addition to the 10 proteins ensuing from the polyprotein maturation, an eleventh protein, P3N-PIPO, embedded in the P3 sequence, a protein involved in involved in cell-to-cell movement is translated from a + 2 ribosomal frameshift.[147] After the translation and replication processes, viral genomes are processed toward different fates, comprising encapsidation into new virions, degradation, or cell-to-cell, long distance and vector-mediated movement. Network-based interactions among these viral proteins have been reported for several potyviruses.[148] Advances in the understanding of processes underlying the regulation of those pathways have been previously reviewed.[75] In this context, we propose here to examine the occurrence of intrinsic disorder in a well-described *Potyvirus* pathosystem and to assess its involvement in viral functions. Owing to the importance of the IDP in viral–host interaction, performing this analysis on the whole network of interaction is worth doing. In this section, we addressed for the first time the interactome of a representative well-studied *Potyvirus*, through in silico characterization of intrinsic disorder at the whole known interactome level. In addition, an attempt was made to identify biologically-relevant IDRs and establish a domain structural and functional annotation of these regions as part of the virus-host disordome. Inspired from multiple disperse sources, we incorporate experimentally information for protein–protein interactions of the viral proteome among them and also with plant proteins, gathered from, into an IDP-IDP interaction model. With these, we will draft out possible mechanisms by which viruses and plants interact. Finally, we will address these observations from an evolutionary perspective. Although it has been hypothesized that IDP could confer even to single domain essential plasticity to achieve multifunctionality, using well-established experimental data on the *Potyvirus* protein interactome,[75,148] lines of evidence support that the interactome landscape of the potyviral proteome is not correlated with the mean IDP of each interacting partner. IDP has been associated to RNA viruses evolvability toward escaping host immunity pathways to disrupt host resistances or broaden host range. Indeed, analysis of our domain-domain IDP-based interactome network indicate potyviral VPg, P1, 6K2, CP and P3N-PIPO widely associated with adaptive events, are highly disordered (Fig. 4 and Table 3). Finally, no correlation can be made so far between the whole-protein disorder content and the general molecular and biological functions. With the yet scarce available data describing impact of IPD in the virus–host interactome, we have been able to portray a unique first basis for the interactome network in this pathosystem from the IDP vantage. We have depicted generalizable structural features, such as viral IDP can target highly connected and disordered host proteins. The preliminary network here presented could, nevertheless, help in disentangling some of the properties associated to the infection by potyviruses. It is expectable that, the coming years, with the advent of fast and robust IDP scanning pipelines, will allow us to considerably enlarge the list of interactors, thus setting forward a more precise account of the different viral IDP strategies and their impact in shaping the plant virus–host interactome. Finally, a systems biology approach will be valuable framework to shed light on the intricate mechanisms operating during disordome coevolution, and it will help to identify emerging functional commonalities and specificities of the virus–host–vector tripartite including an emphasis on potential hub proteins. Future investigative work will provide new insights of the mechanisms by which intrinsically disordered proteins perform their diverse biological functions. Evidently, consideration of IDP as a key staple phenomenon stirring up the entire plant virosphere should be added to the future arsenal of comparative plant metaviromics targeting molecular-based plant viral disease management strategies.

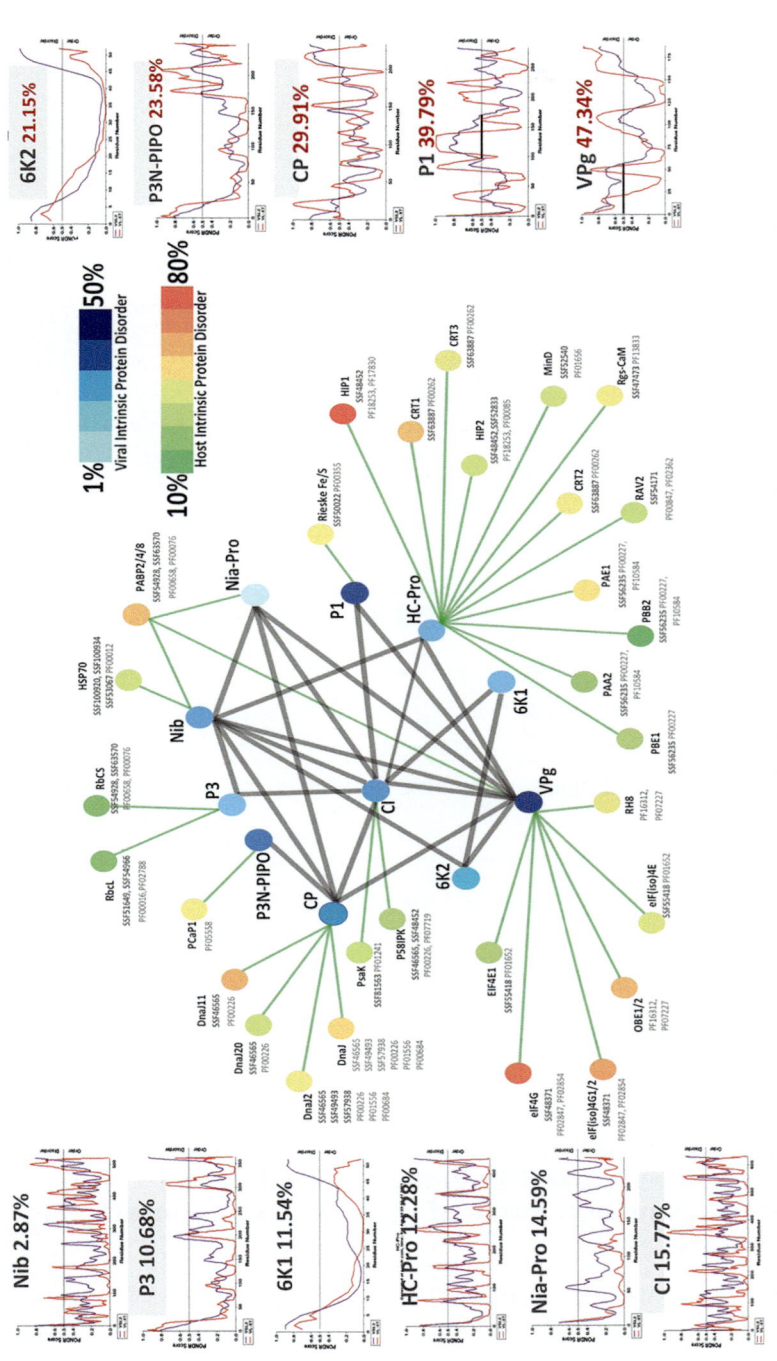

Fig. 4 IDP-based Interactome in Plant Viruses. Positional distribution of predicted intrinsic disorder for each protein in the *Potyvirus* proteome and their interaction with their respective experimental host-plant partner. Each line graph represents the distribution of mean disorder probability calculated for a given protein by averaging the per-residue disorder profiles generated by PONDRFIT (VLXT and VSL2B). Residues with scores above 0.5 are considered disordered.

Table 3 IDP-based interactome network in plant viruses.

Viral Module	Protein Funcion Descriptor	%IDP	Host Plant Interacting Partner (Gene Entry / Domain)		%IDP
P1	Serine protease	39.8	Rieske Fe/S	At4g03280 / SSF50022 (PF00355)	34.9
HC-Pro	Protease with RNA silencing suppressor	12.3	Rgs-CaM	At3g01830 / SSF47473 (PF13833)	34.3
	aphid-mediated transmission		RAV2	At1g68840 / SSF101936, SSF54171 (PF00847, PF02362)	28.1
			CRT2	At1g09210 / SSF49899, SSF63887 (PF00262)	35.5
			CRT1	At1g09210 / SSF49899, SSF63887 (PF00262)	47.2
			CRT3	At1g09210 / SSF49899, SSF63887 (PF00262)	32.6
			HIP1/FAM10	At4g22670 / SSF48452 (PF18253, PF17830)	75.5
			HIP2	At3g17880 / SSF48452, SSF52833 (PF18253, PF00085)	38.7
			MinD	At5g24020 / SSF52540 (PF01656)	30.4
			PAE1 20S-Alpha5	At1g53850 / SSF56235 (PF00227, PF10584)	38.4
			PAA2	At2g05840 / SSF56235 (PF00227, PF10584)	21.5
			PBB2	At5g40580 / SSF56235 (PF00227)	12.8
			PBE1	At1g13060 / SSF56235 (PF00227)	24.8
P3	Cell-to-cell movement	10.7	RubisCO RbcL	AtCg00490 / SSF51649, SSF54966 (PF00016, PF02788)	17.8
			RBCS-1A	At1g67090 / SSF55239 (PF12338, PF00101)	13.3
			RBCS-3B	At5g38410 / SSF55239 (PF12338, PF00101)	18.8
			RBCS-2B	At5g38420 / SSF55239 (PF12338, PF00101)	17.7
			RBCS-1B	At5g38430 / SSF55239 (PF12338, PF00101)	17.7
P3N-PIPO (P3)	Cell-to-cell movement	23.6	PCaP1	At4g20260 / PF05558	29.8
CI	RNA helicase with ATPase activity	15.8	PS8IPK	At5g03160 / SSF46565, SSF48452 (PF00226, PF07719)	28.6
			PsaK	At1g30380 / SSF81563 (PF01241, 1.10.286.40)	24.6
NIa-Pro	Protease	2.9	PABP2	At4g34110 / SSF54928, SSF63570 (PF00658, PF00076)	46.3
			PABP4	At2g23350 / SSF54928, SSF63570 (PF00658, PF00076)	46.4
			PABP8	At4g49760 / SSF54928, SSF63570 (PF00658, PF00076)	48.7
VPg		47.3	PABP2	At4g34110 / SSF54928, SSF63570 (PF00658, PF00076)	46.3
			PABP4	At2g23350 / SSF54928, SSF63570 (PF00658, PF00076)	46.4
			PABP8	At4g49760 / SSF54928, SSF63570 (PF00658, PF00076)	48.7
			EIF4E1	At4g18040 / SSF55418 (PF01652, 3.30.760.10)	27.7
			eIF(iso)4E	At5g35620 / SSF55418 (PF01652, 3.30.760.10)	31.8
			eIF4G	At3g60240 / SSF48371 (PF02847, PF02854)	68.9
			eIF(iso)4G1	At5g57870 / SSF48371 (PF02847, PF02854)	56.2
			eIF(iso)4G2	At2g24050 / SSF48371 (PF02847, PF02854)	52.6
			OBE1	At3g07780 / (PF16312, PF07227)	49.1
			OBE2	At5g48160 / (PF16312, PF07227)	48.1
			RH8	At4g00660 / SSF52540 (PF00270, PF00271)	32.9
Nib	RNA-dependent RNA polymerase	14.6	PABP2	At4g34110 / SSF54928, SSF63570 (PF00658, PF00076)	46.3
			PABP4	At2g23350 / SSF54928, SSF63570 (PF00658, PF00076)	46.4
			PABP8	At4g49760 / SSF54928, SSF63570 (PF00658, PF00076)	48.7
			HSP70	At3g09440 / SSF100920, SSF100934, SSF53067 (PF00012)	30.1
CP	Capsid protein	29.9	DnaJ	At3g44110 / SSF46565, SSF49493, SSF57938 (PF00226, PF01556, PF00684)	42.4
			DnaJ20 ATJ20	At4g13830 / SSF46565 (PF00226)	30.0
			DnaJ11 ATJ11	At4g36040 / SSF46565 (PF00226)	51.6
			DnaJ2	At5g22060 / SSF46565, SSF49493, SSF57938 (PF00226, PF01556, PF00684)	33.7
6K1	Small peptides of unknown functions	11.5			
6K2	Small peptides of unknown functions	21.2			

Predicted intrinsic disorder (PONDRFIT-VSL2B) for each protein in the *Potyvirus* proteome and their interaction with their respective experimentally validated host-plant partners.

8. Functional IDP and evolutionary conserved structures expand evidence for modulation of transmission mode of plant viruses by insect vectors

The now longstanding awareness that intrinsically disordered proteins (IDPs) are functional despite lack of well-structured protein[44,149,150] has been a paradigm-shifting remodelling of our understanding on the role of conformational flexibility in molecular interactions.[143,151,152] Until now most IDPs and their influent role in virus–host interaction have been studied animal and human viruses. In this context, given the unexplored nature and pivotal relevance of these aspects to plant viruses and their transmission, we wished to explore whether there is any correlation of viral disorder and vector-borne transmission to host. We were also interested to investigate whether the range and apparent substantial variation of disorder in viruses could be associated to functional IDP observed in the cognate interacting protein partners present in their vectors.

For this, we first systematically surveyed the prevalence of IDP according to modes of transmission. Table 1 suggest that this is indeed the case for vector-borne viruses transmitted in a circulative manner and showed a range of viral disorder that seemed to be specifically low (0.46–2.6%). While in noncirculative (NC-NP) transmitted virus, IDP spanned at mean of 3.12% (1.5% *Virgaviridae* to 11.25% *Alphaflexiviridae*), the noncirculative semipersistent viruses showed a much slightly higher mean IDP of 3.7% where *Caulimoviridae* and the *Tombusviridae* peak with the highest mean IDP values of 5.64% and 7.5%, respectively. Intriguingly, fungal-transmitted viruses (CP-NPr) show a similar IDP span, but a tendency toward lower disorder (mean IDP of 2.96%), so that the range for fungal viruses is 0.8–3.5%. The most striking range is seen for mite vectored viruses (*Tymoviridae* CP-NPr), with a mean IDP value of 14.5%. Obviously, plant viruses express a wide diversity of IDP profiles reflecting their adaptiveness to various cellular landscape.

Secondly, we explored the presence and location sites of potential MoRFs in major viral modules and discovered an abundant prevalence of MoRFs in the IDP-rich regions of different well-defined functional proteins of plant viruses (Fig. 5). Lastly, we surveyed the prevalence and abundance of IDP in know potential insect cuticular receptors recently described in confirmed vectors of plant viruses (Table 4). Intriguingly, several putative disordered–embedded MoRF of heterogeneous length and abundance were found in the highly disordered C-ter region of the stylin cuticular receptor of virus vectors (Stylin-001 to 005 in Aphids), further suggesting that these seemingly conserved ID regions could potentiate the simultaneous binding interactions of the protein with viral partners and underscore their paramount in plant virus transmission. These discovered MoRF interaction sites seem conserved within vector species and according to transmission mode, plausible supporting that induced folding events are associated upon cuticular proteins binding. While we currently have

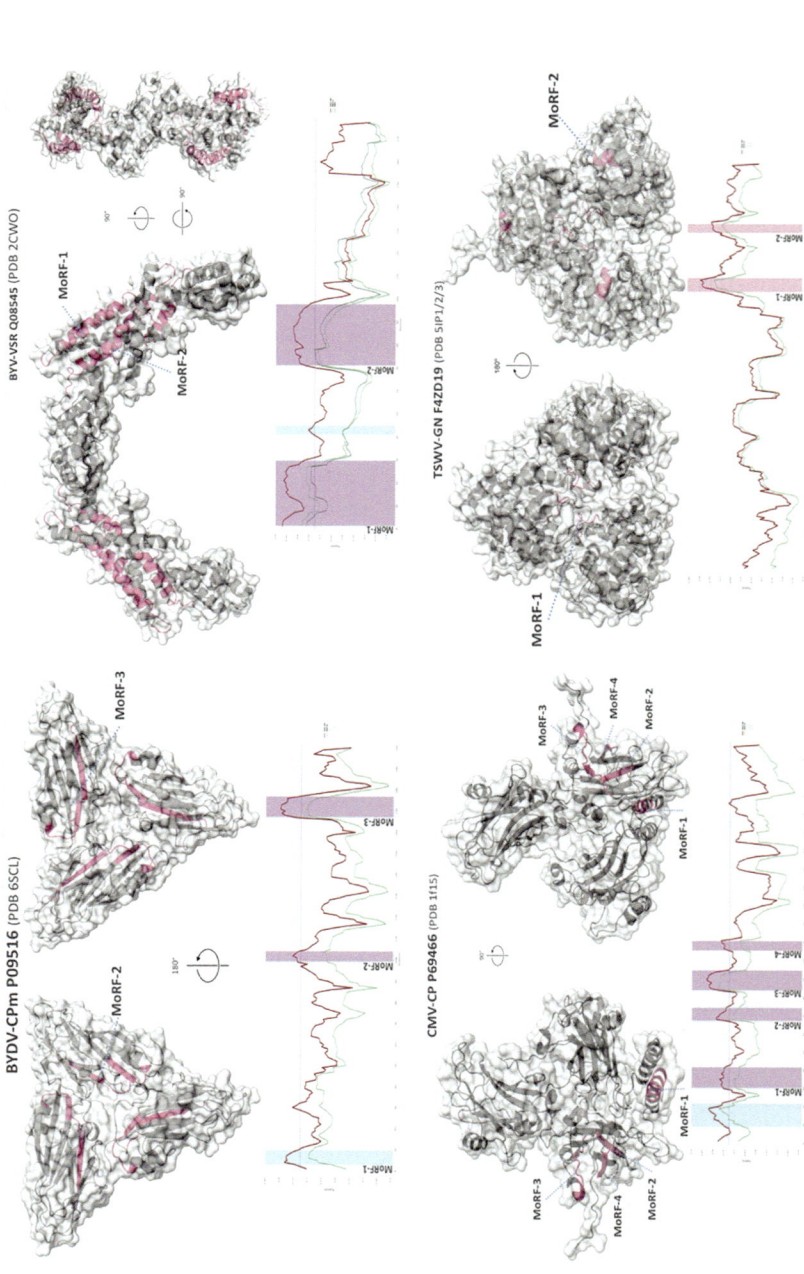

Fig. 5 Structure and prevalence of MoRFs enriched IDRs in viral proteins of plant viruses. Structural proteins resolved by X-crystallography showing their folded domains (ribbons colored in blue) and disordered regions. Protein represented with exposed and buried MoRFs regions highlighted (pink-magenta). (A) Minor Capsid (CPm) of *Barley yellow dwarf virus* (BYDV, *Luteoviridae*, PDBid 6SCL, 3.00Åres.), (B) capsid (CP) of *Cucumber mosaic virus* (CMV, *Bromoviridae*, PDBid 1f15, 3.30Åres.), (C) P21 Viral suppressor of RNA silencing (VSR) of *Beet yellow virus* (BYV, *Closteroviridae*, PDBid 2CWO, 3.30Åres.), and (D) the Glycoprotein (GN) of *Tomato spotted wilt virus* (TSWV, *Tospoviridae*, PDBid 5IP1/2/3, 2.70–3.00 Åres.). Graphs: MoRF prediction for each viral protein with predicted MoRFS residues highlighted (pink-magenta). The propensity scores provided by MoRFChibi (MC) and MoRFCHibi (MCW) and their IDP are shown in red, green, respectively. Others predictors are switched off.

Table 4 MoRFs occurrence and prevalence in insect vector CuP interacting with plant viruses.

Virus	Viral Protein	Transm.	Protein	Vector CuP Target (Domain)	%IDP	MoRFs	Reference
CaMV (Caulimoviridae)							
Acyrthosiphon pisum, M. persicae, B. brassicae	Helper Protein P2 (P03551, PDB3k4t)	NP-NC	Stylin-001 ACYPI009006 AWF70754 (MG188739)	Chitin_bind_4 / CPR RR-1 (PF00379)	43.7	89-95; 118-125	Webster et al., 2018*
Aphis craccivora		NP-NC	Stylin-001 AWF70758 (MG188743)	Chitin_bind_4 / CPR RR-1 / EndocuticleSgAbd-4-like (PF00379)	53.33	89-95; 118-125	Webster et al., 2018*
Aphis gossypii		NP-NC	Stylin-001 AWF70757 (MG188742)	Chitin_bind_4 / CPR RR-1 (PF00379)	54.07	89-95; 118-125	Webster et al., 2018*
Diuraphis noxia		NP-NC	SgAbd XP_015379180 (XM_015523694.1)	Chitin_bind_4 / CPR RR-1/ EndocuticleSgAbd-4-like (PF00379)	57.89	88-98; 116-123	Webster et al., 2018*
Acyrthosiphon pisum, M. persicae		NP-NC	Stylin-02 ACYPI003649 (BAH71241)	Chitin_bind_4 / CPR RR-1 (PF00379)	55.64	88-98; 116-123	Deshoux et al., 2020*
Acyrthosiphon pisum, M. persicae		NP-NC	Stylin-03 ACYPI001610 (BAH71700)	Chitin_bind_4 / CPR RR-1 (PF00379)	40.58	115-125	Deshoux et al., 2020*
Acyrthosiphon pisum, M. persicae		NP-NC	Stylin-04 ACYPI002877 (NP_001165731)	Chitin_bind_4 / CPR RR-1 (PF00379)	34.93	96-104; 111-116	Deshoux et al., 2020*
Acyrthosiphon pisum, M. persicae		NP-NC	Stylin-04bis ACYPI000308 (NP_001165739)	Chitin_bind_4 / CPR RR-1 (PF00379)	53.5	69-104; 111-116	Deshoux et al., 2020*
Acyrthosiphon pisum		NP-NC	Stylin ACYPI007911 (AK341247 BAH71681.1)	CBM_14 / CPAP3 (PF01607)	14.63		Deshoux et al., 2020*
Acyrthosiphon pisum		NP-NC	Stylin-05 ACYPI006031 (BAH71578)	CBM_14 / CPAP3 (PF01607)	14.41		Deshoux et al., 2020*
CTV (Closteroviridae)	CP	NP-SP	A0A3Q0J6I0-1 LOC103515660 (XP_026684079.1)	Chitin_bind_4 (PF00379)	39.69	167-173; 193-199	This Study
Diaphorina citri			A0A153DIB1-1 LOC103518937 (XP_008482240.1)	Chitin_bind_4 (PF00379)	36.36	83-89; 96-110; 124-135; 158-164	This Study
			A0A153DMB4-1 LOC103521394 (XP_008484727.1)	Chitin_bind_4 (PF00379)	30.56	223-246; 352-375; 387-398; 418-452; 497-510; 574-586; 593-614	This Study
BYDV-GPV (Luteoviridae)							
Rhopalosiphum padi, Schizaphis graminum	Readthrough protein (RTP)/CP	CP-NP	Cp62 ACYPI004113 (NP_001156154)	Chitin_bind_4 / CPR RR-2 (PF00379)	32.7	NA	Wang et al., 2015
							Wang et al., 2015
TYLCV (Geminiviridae)							
Bemisia tabaci	TYLCV-CP (P27265, PDB)	CP-NP	XP_018897334 (LOC109030698)	Chitin_bind_4 / CPR RR-1 (PF00379)	24.48	85-92; 147-163; 184-200; 211-219; 221-226;249-258; 273-293	This Study
	(P03561, PDB6EK5), (W5RUR4, PDB6F2S)	CP-NP	XP_022175931 (LOC111037581)	Chitin_bind_4 / CPR RR-1 (PF00379)	40.6	200-221; 242-280; 299-311; 329-345	This Study
		CP-NP	XP_018897335 (LOC109030698)	Chitin_bind_4 /CPR RR-1 (PF00379)	25.15	85-92; 147-163; 184-200; 211-219; 221-226; 249-258; 273-293	This Study
ScYLV (Luteoviridae)							
Melanaphis sacchari, Rhopalosiphum maidis		CP-NP	A0A2H8TGY1 (MBW13377, _9HEMI)	Chitin_bind_4 / CPR RR-1 / EndocuticleSgAbd-4-like (PF00379)	55.56	89-95; 118-125	This Study
RSV (Phenuiviridae)							
Laodelphax striatellus	Nucleocapsid NC3 (Q67897, PDB3AJF)	CP-Pr	CprBJ AGJ70277 (KC485263)	Chitin_bind_2 / CPR RR-1 (PF00379)	41.86	122-156, 172-181, 188-231	Liu et al., 2018*
		CP-Pr	(JAS02196) GEMB01000945	Tweedle motif /DUF243 / CPR RR-2 (PF03103)	56.4		Liu et al., 2018*
TSWV (Bunyaviridae)							
Frankliniella occidentalis	TSWV-GN (F4ZD19, PDB5IP1)	CP-Pr	CP-V A0A482JNJ7-1 QBP34368 (MH884758)	Chitin_bind_4 / CPR RR-2 (PF00379)	59.22	10-18; 39-44; 60-96; 99-121; 144-181; 211-238; 253-279; 303-313; 317-348; 388-399; 416-431	Badillo-Vargas et al., 2019*
		CP-Pr	EndoCP-GN A0A482JNN3-1 QBP34367 (MH884757)	Chitin_bind_4 / CPR RR-1 (PF00379) CP-GN	42.61	52-59; 79-110; 120-130; 163-174	Badillo-Vargas et al., 2019*
		CP-Pr	EndoCP-V A0A482JL26 QBP34366 (MH884756)	Chitin_bind_4 / CPR RR-1 (PF00379)	36.99	83-89; 96-110; 124-135; 158-164	Badillo-Vargas et al., 2019*

Cuticular proteins of different insect vectors transmitting plant viruses in different transmission modes. IDP profiles was predicted for a given CuP protein by averaging the per-residue disorder generated by PONDRFIT (VLXT, and VSL2B) and D_2P_2. Residues with scores above 0.5 are considered disordered.

few information about the position of the binding site, or whether these binding are specific, further exploring such the MoRFs and their exact mode of action would provide foundational insights on the mechanisms by which these critical IDP features could sequester the viral modules and colocalize the complex on the vector cuticular surface, thereby initiating replication of viral RNA and transmission of virions. Henceforth, a mechanistic rationalization of the role of viral modules and the entire plant virus disordome could emerge. Further structural IDP information and experimental trials will be indispensable in order to determine the subsequent sequence of processes that follow this initial recognition step, and ultimately lead to replication and transmission.

9. Outlook and new challenging prospects for plant virology

The principal and, in our opinion, striking conclusion of this study is that the global plant virome is vastly and abundantly populated by wide apparent yet underexploited IDP/IDR repertoire. This work was prompted by the recent metaviromics advances, which have considerably increased the insofar known diversity of plant viruses. Our reasoning vantage is that IDP can contribute decipher the functional diversity of the plant virosphere could improve understanding and fill glaring knowledge gaps of virus evolution. To our knowledge, the picture emerging from of the global plant virome IDP landscape viewed in light of modular virus evolution has not been presented previously. The large number and immense diversity of plant viruses included in our dataset, encompassing 1003 viral species distributed over 24 viral families, 13 orders, 10 classes across 5 viral phyla, created computational challenges for a systematic, proteome-wide IDP-based comparative analysis of each assigned evolutionary lineages of the global plant virome.

9.1 IDP viral diversity and evolution

Intrinsically disordered protein and regions are abundantly widespread across all investigated realms and families of the global plant virome. We have surveyed 1003 plant viral genomes to investigate how intrinsic protein disorder varies at different levels. Although this analysis has unveiled a very substantial intra- and intertaxon variation among plant viruses in their predicted disorder, overall the explored plant virus disordome seems taxon-specific and relatively conserved. This finding suggests that the underlying IDP mechanism is both evolutionary effective and robustly encoded in the global plant proteome. When mapping these results to what is known about the proteins structured and unstructured regions of the global plant virome, several original and untrivial insights can be emerged. Previous studies, rather focused on Potyviruses,[75] had highlighted the fact that many specific proteins from plant viruses are relatively highly disordered. One survey of disorder in viruses with a similar approach to our study[153] provided some consideration to this array of variation in viral disorder within few more plant virus families in which authors underlined that certain plant virus genomes have high levels of disorder, pointing out that plant viruses harboring smaller genomes have more IDP, and that for larger viruses IDP ranges typically between 2.9% and 23%. However, the authors also noted that certain viruses exhibit lower disorder. Though IDPs/IDRs of plant viruses can, likewise to any other organism, readily be identified based on their primary amino acid sequence[54,154] and given the aforementioned obvious evolutionary constraint governing the massive genetic and proteomic diversity, it is currently unclear how IDPs/IDRs persist at high frequency in the plant virus proteomes. Consequently, in stark contrast with for folded regions of modules where function can often be predicted and assigned with high specificity based on the presence of conserved protein domains, it remains fairly challenging to associate viral IDRs with specific biological functions based solely on their amino acid strings, limiting their systematic and accurate functional characterization.[155] One hypothesis for the preponderance of disordered regions despite high amino acid sequence divergence, is that the modular genomic features of disordered regions that are functionally essential, such as genome length,[156] amino acid composition,[157] complexity,[158] are under perpetual evolutionary constraints. Recovering clustering IDP patterns nested in or into the close vicinity of modular features and identifying their quantitative rates of evolutionary divergence would further valuably advance our understanding of the evolutionarily constrained genomic features operating at the level of the global plant virome. Based on robust correlations, our results indicate that our comparative proteomic analysis successfully captured additional interpretable viral features directly associated with IDP. Hence, we reported and hypothesize that multiple genomic tend to be of key importance to depict a wider and more accurate IDR functional landscape of the global plant virome.

9.2 Modular horizontal transfer and transport of intrinsic disordered proteins : The hidden gems of viral exaptations

One of the biggest challenges facing our understanding of the dynamics of disorder in plant viral proteomes is to grasp the causal origin and extent of strong disorder avoidance. It is well known that there is substantial variation among the main kingdoms of life in their distribution of disorder.[60,159] Recurrent, cross-boundary horizontal gene exchange (HGT) and shuffling of viral structural modules within viruses and among viruses and their hosts are important factors of plant virus evolution. Striking examples are presented by RdRp permutations such as in amalgaviruses, ancestral SJR-CP displacement by filamentous CP (e.g., potyviruses). Intrinsically disordered regions are more common among eukaryotic genomes than prokaryotic genomes, and since viruses frequently highjack their host genomes or exchange genetic material through extensive HGT, hence, we might expect that a substantial part of disorder proteins potential might have been transferred consequently (a phenomenon we term here Horizontal Intrinsically Disorder Regions Exchange "HIDRE"). Therefore, we have expected that, similarly to other viral entities, plant viruses would follow their hosts and vectors, a trend discussed in a previous survey on viruses.[56] Patterns observed among the Potyviruses interactome showed, although very faintly, some lines of evidence in this direction, yet this needs to be carefully investigated in more comprehensive research. On a more evolutionary general scale, this kind of broad HIDRE accompanied paramount life-style adaptive shifts of plant viruses to their new hosts, such as the emblematic movement proteins (MPs) which were, e.g., in the case of the chimeric ourmiaviruses, apparently acquired from genetically highly distant "picorna-like" viruses, and similarly exhibit high IDP-enriched profile. The complex IDP pervasiveness landscape seen through the HIDRE evolutionary prism indisputably emphasize its plausible key role as source of functional exaptations,[160,161] in light of modular plant virus evolution. Furthermore, plant virus vectors, whether arthropods, nematodes or fungi are particularly promiscuous reservoirs for viruses, often sharing the same virus categories with distantly interconnected organisms.[1,2] Invertebrates that endorse the essential role of vectors appear to be dominant agents of horizontal transfer within the global plant virome and therefore may be also putatively qualified as vessels of IDP and tightly associated to shaping the viral IDP landscape, albeit more investigative support is unequivocally needed. This phenomenon presupposes a critical role of IDP in viral diversification including host cellular preference and, while raising further complex question on the significance functional IDP conservation, its knowledge could help better enlighten the still misty routes of plant virus evolution.

9.3 The plant virome IDP as an akin direct consequence of nonoptimal codon usage

Synonymous codons are not used with equal frequencies in most plant virus genomes (Tahzima et al., 2021, under preparation). Codon usage has been suggested to act as modulator of kinetics and protein folding. The relationship between codon usage and protein folding and its impact on IDP of plant viruses remains to be clarified, however, genome-wide relationships and correlations between codon usage, GC% and protein disorder have never been established. Together, our investigation suggest that viruses preferentially encode various GC-content in intrinsically disordered regions across the plant virus proteome including in extremely conserved well-structured modules. Importantly, we found genome-wide correlations between GC-content choices and predicted proteomic IDP tendencies suggesting that plant virus strong codon choices and IDP coevolved to ensure adaptive proper protein folding according to the selective pressures enacted by their specific ecological niche, albeit the governing rules of this phenomenon remain elusive. Hypothetically, this may suggest that nonoptimal codons are preferentially used in intrinsically disordered regions, and more optimal codons are used in structured domains. For the first time in plant virology, we observed a significant negative correlation between CAI scores and protein disorder, meaning that there is a preferential use of nonoptimal codons in regions predicted to be intrinsically disordered. Furthermore, we highlighted that the correlation between codon usage and protein disorder tendencies is conserved across evolutionary plant viral lineages. Nevertheless, the functional significance of such correlations and their prevalence across the whole plant virome need to be robustly confirmed by experimental structure-based codon investigation. Additionally, the available solved protein structures are enriched for highly structured proteins and domains. Here, in our large-scale proteome-wide study, we discovered a strong proteome-wide correlation between GC content, transmission mode and protein disorder tendencies. As highly optimized adaptive entities, plant viruses have highly modular proteomic organization. Modular domains and multidomains often segregate binding sites for each ligand into the different structural domains. We proposed a global mapping of IDP enriched modules for Potyviruses, yet the general process, through which the different modules interact, however, has proven elusive. The observation that IDP/R reach domains appear in disproportionately higher amounts in interactions proteins denotes that intrinsic disorder is important for their functional role and allow us to put forward several hypotheses to attempt finding an explanation on the widespread functional significance of IDP. Without suggesting any exclusive role of IDP, the premises of such careful assumptions must consider the high-specificity/low-affinity of binding to the virus–vector interface, the ability to modulate interaction interfaces, and high specificity for multiple targets and primarily put the emphasis on the MoRF-promoted mechanisms of module-module interaction at the virus–vector interplay. The ability to understand the evolutionary determinants of IDP is the cornerstone to a quantitative description of these important processes and the discussion opened here is a critical step towards a unifying framework that connects intrinsic disorders with plant virus transmission modes. Predicted structural disorder is variably abundant in all investigated plant virus species, and due to its strong correlation with genome and proteome size, and %GC its level is significantly higher in ssDNA viruses than in RNA viruses.

By conservative estimations, most virus families contain significant disorder, that is, long disordered regions at least 30 residues in length. To the best of our knowledge, such explorative forays is the first of its kind to comprehensively embrace the whole RefSeq plant virus proteome. Undoubtedly, further large-scale proteome-wide disordome studies will constitute the cornerstone toward achieving in-depth robust deciphering of the proteomic IDP multifunctionality in the plant virome. Although it is considered almost conventional in the field that structural disorder increases with the complexity of the organism,[162,163] the highest IDP amplitudes are not witnessed in the most genomically complex host-changing plant viruses (e.g., in *Closteroviridae*), but in ssDNA (e.g., in *Geminiviridae*) viruses that lead a fairly specialist lifestyle.

9.4 IDP elucidates modulation of vector-borne transmission within the global plant virome: Underpinning evidence from MoRFs patterns through their evolutionary and biological relevance

The distinguishing and unifying feature of their extremely limited amount of proteins—if any—is their inability to fold into a unique and stable tertiary structure, while exhibiting a significantly higher array of structure-related functional organization. Clearly, from our comparative evolutionary proteomics of viral structure–function features, it becomes evident that structural disorder of plant viruses acts as an evolutionary amplifier of modular proteomic optimized functionality.[164] Of particular relevance is the growing beam of evidence that plant viral IDP is found in disproportionately higher amounts in RNA- or ATP-binding GO-annotated functions (UniProt) related to movement proteins, capsids, transmission factors and their targets partners, suggesting a prominent role of IDP in their interaction capacity.

In plant viruses, a major challenge to proteome-wide comparative analysis is the limited applicability of homology-based sequence analysis. Viral proteomes with a mixture of moderately disordered regions and structured domains can be assigned function and taxonomic category based on homology to their structured domains. Therefore, whereas highly divergent intrinsically disordered proteins and HGT make viral proteomes much more difficult to classify, we therefore asked, using a complete subset of well-characterized and unclassified plant viruses (*Closteroviridae*), whether hypotheses about amino acid compositional profiles and their propensity toward IDP could be generated using evolutionary proteome-wide IDP-signatures to cluster their genomes and predict viral biology. Based on heatmap profile analysis, though with some caution, lines of evidence are suggesting a generalizable predictive power of this approach and illustrate how proteome-wide IDP evolutionary signatures can be exploited to generate functional and biological hypotheses for unknown or uncharacterized viruses assigned to the expanding global plant virome.

Within the plant virus proteome IDPs are can be postulated to be involved in processes of receptor recognition because the common thread that connects most interacting viral proteins is that their function is modulated MoRFS-abundant IDRs. Without overstating, assumptions for the exact role of intrinsic disorder may involve both high-specificity/low-affinity binding and high specificity for multiple targets. However, despite the clear experimental substantiation for the dedicated involvement of IDRs in these functions, a mechanistic model that offers a quantitative rationale for their presence has yet to be established.[57,108] On a more hypothetical basis, IDP-related structural modules and domains of the plant virus proteome are predicted to contain short disordered regions (MoRFs) that are suggested to sample structured states within the conformational ensemble, and may become fully ordered upon binding to the partner. Binding by domains or motifs is considered usually weak, transient, and possibly of limited specificity,[113,122,165] which can be strengthen and/or made more specific by either cooperating with flanking residues, combining several modules through synergistic folding, with different dynamics, or affording mechanisms tolerating long-disordered domains.[46,150] MoRFS-associated structural disorder seems crucial at the functional virus–host interplay and may certainly impact further regulatory pathways beyond virus transmission. For example, once successfully transported and transmitted, and following entry, replication, and movement, plant viruses must escape the whole battery of signaling and defense pathways of the host cell, which is orchestrated through interactions of viral proteins with key host target proteins.

In few important plant viruses, several of their specific functional modalities related to virus transmission, such as adaptability in binding to specific cuticular insect proteins (CuPs) through weak but specific binding, and frequent regulation by post-translational modification, have been formally identified and experimentally validated.[24,28]

To demonstrate the association of evolutionary signatures with previously known functions, we associated IDRs with protein function and the occurrence and abundance of MoRFs, and indeed revealed that we could generate hypotheses about protein function for functionally unidentified IDRs. We present strong evidence that groups of IDRs predicted on specific insect vector CuPs share evolutionary signatures, and that these groups of IDRs can be associated with specific biological functions related to modes of virus transmission and species of insect vectors (Fig. 6). The inference of IDP patterns closely related to various vector-borne transmission routes is, arguably, the most unexpected outcome of the present analysis. Clearly, given that the primary evidence behind this derivation of the plausible role of IDP in modulation of active transmission of viruses, extreme caution is due in the interpretation of this observation. Moreover, the predominant abundance of predicted MoRFS in disordered region of viral binding partners (CuP) is conceptually compatible with our original findings in plant virus modules and reinforces these sheaves of evidence. The distinct occurrence and abundance of MoRFs in the direct vicinity of CuP modules associated to specific modes of transmission and vectors species also makes sense in terms of molecular logic of viral–host interaction and transmission strategies. Indeed, noncirculative transmitted plant viruses exploit the simplest interaction strategies to the vectors specialized cuticular tissues involving, in all likelihood, a relatively limited amount of vector's organs, namely, stylet and foregut, whereas circulative transmitted plant viruses conceivably evolved much diverse interacting partners while circulating across the vector internal tissues and metabolism. Hence, in a general

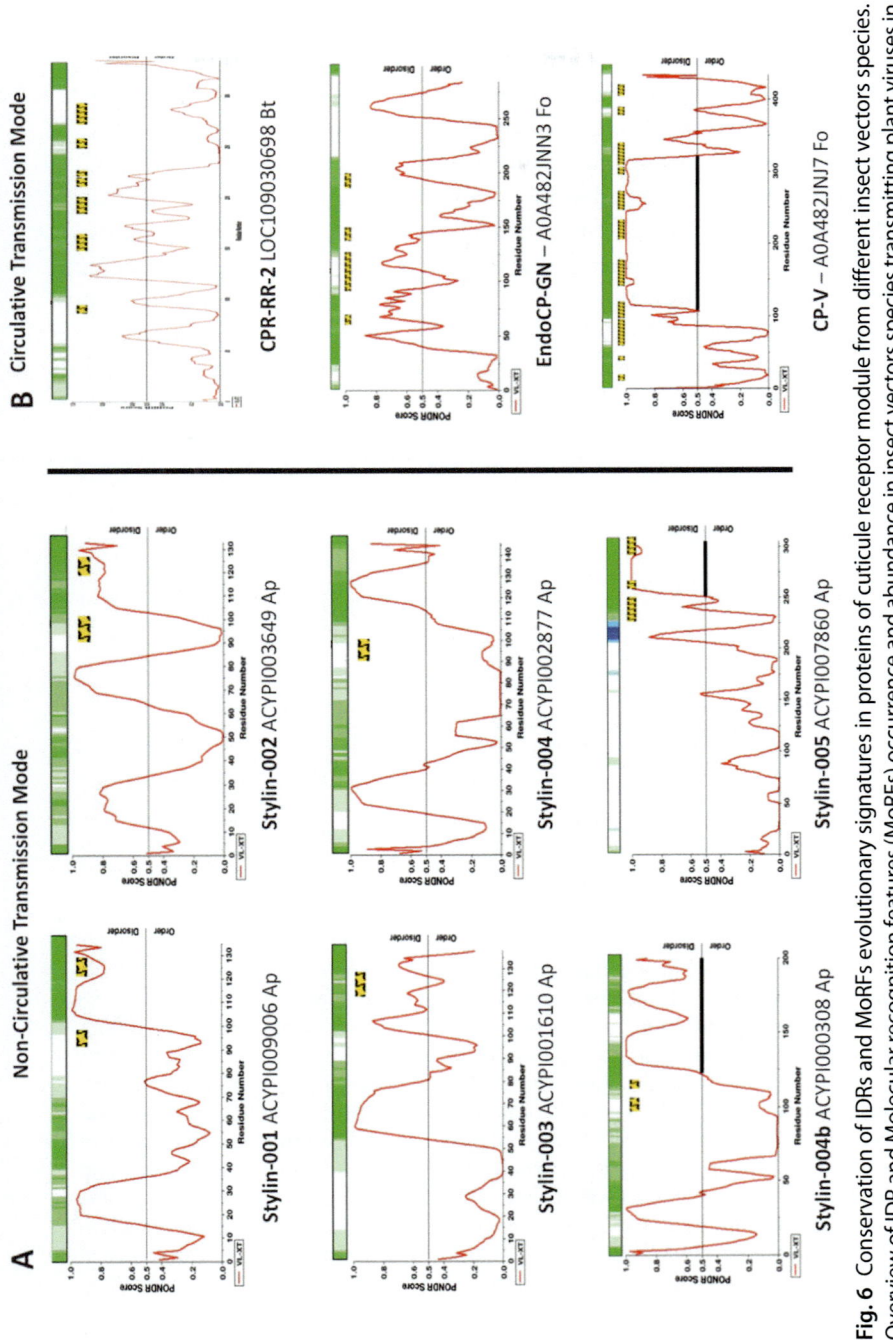

Fig. 6 Conservation of IDRs and MoRFs evolutionary signatures in proteins of cuticle receptor module from different insect vectors species. Overview of IDP and Molecular recognition features (MoRFs) occurrence and abundance in insect vectors species transmitting plant viruses in noncirculative (NC) mode by Aphid (*Acyrthosiphon pisum*) and in a circulative mode (CP) by Whitefly (Bt: *Bemisia tabaci*) and Thrips (Fo: *Frankliniella occidentalis*). Graph: PONDR® FIT-generated disorder profile of the corresponding proteins. The yellow tags indicate the occurence and abundance of predicted MoRFs. The green/white bar is the IDP profile generated by the D_2P_2 computational platform.

framework, we proposed that various kinds of domain–domain interactions involved in virus transmission are mediated by specific functional MoRFs signatures located near the chitin binding module (Chitin-bind-4, CPR-RR1/2, PF00379) and associated to evolutionary conserved IDRs harboring short motifs in their direct vicinity. Furthermore, although only few CuPs have been identified and experimentally characterized so far, the presence of IDRs is con served across orthologs in the vast majority of cases (Fig. 7). We also confirm that the vast majority of these CuP specific MoRFs–enriched IDRs are distinct from Pfam domains in mostly all transmission-associated CuPs. This can offer both extreme evolutionary agility and adaptive versatility and are usually bound by receptors domains

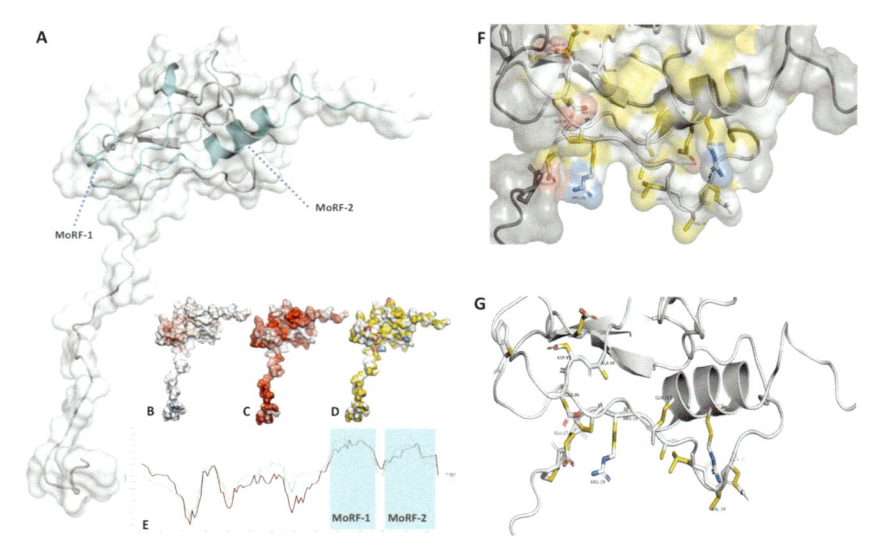

Fig. 7 MoRFs features indicate binding hotspots at the virus-insect vector cuticule receptor module surface. Overview Molecular recognition features (MoRFs) and both protein surface affinity and specificity of CuP-RR1 Stylin-001 (Pfam PF00379, ACYPI009006) displayed in (A) cartoon highlighting predicted MoRF residues (cyan), (B) according to electrostatic potential, (C) hydrophobic (red) to nonhydrophobic (white) gradient, or (D) highlighted according to hydrophobicity/affinity charges (red and blue). The color scheme reflects the relevant points for intuitive assessment of protein–protein interactions: yellow; negatively charged oxygens of glutamate and aspartate, red; nitrogen's of positively charged functional groups of lysine and arginine, blue; all remaining atoms including the polar backbone, white). (E) MoRF prediction in aphid cuticular proteins (*Acyrthosiphon pisum*) using MoRFchibi. Positions with scores of 0.752 or greater are considered MoRF residues. (F and G). Close up on the interacting residues in the chitin-binding domain (CBD) at the interface between the receptor and the virus are displayed in sticks.

of the partner proteins, keeping in mind that multiple IDRs may perform different function, thus further thwarting the mapping of functional IDRs of these modules. Most of these modules are adjacent to intrinsically disordered regions (IDRs). These functional advantages demonstrate the involvement of structural disorder in protein function amplification, as approached from a molecular and functional IDP perspective. While writing this chapter, we are presently developing a well-supported machine-learning approach to predict function for MoRFs-enriched IDRs based on evolutionary IDP signatures. On more challenging grounds, the systematic large-scale proteomic-wide survey at viral species level of these IDRs will greatly advance future functional and evolutionary proteomics in the field of plant virology.

9.5 IDP conservation and plant viral disease

Beside their evolutionary implications, widespread evidence and patterns of abundant functional and highly diverged portions of IDRs also bears several biological consequences for plant viruses and their hosts. The conservation of IDP features over long evolutionary time, despite accumulation of amino acid divergence, is coherent with the model of stabilizing selection[166,167] that postulates that individual amino acid sites are under relatively weak functional constraints. In this line, single point mutations are unlikely to dramatically impair IDR function, and hence large evolutionary divergence can accumulate, suggesting that disease-causing mutations in disordered regions are more likely to trigger gain of function, consistently with recent works.[168] Although the involvement of structural disorder was confirmed in many other important disease-associated proteins in humans,[47,108] its contribution in plant virus disease still remains unclear and needs to be further substantiated. Perhaps, large-scale transcriptomics focusing disordered proteins and domains involved in virus–plant pathogenic interaction, symptomatology and onset viral-associated disease syndromes can prominently lead to the unprecedented understanding in plant virology and reformulation of many rationale of plant pathology. Thus, the involvement of IDPs in disease points them as prime targets for future innovative design of control strategies to limit spread and transmission of plant viruses. Unquestionably, with further exhaustive structural–functional investigation of plant virus IDP we may foresee more comprehensive understanding underlying the evolutionary and structural foundations of intrinsic disorder in plant viral disease.

10. Conclusion

Intrinsic disorder makes up a large share of the global plant virus proteome. Through comparative and proteome-wide IDP analysis, many original aspects of the global plant virome involving its diverse viral linages have been addressed and clarified fostering the emergence of biologically IDP-centric plausible scenarios of virus–host evolution. Remarkably, we found it unexpected and noteworthy that, despite the colossal diversity of the rapidly expanding plant virome repertoire via the progresses of in-depths metaviromics, the contours of the evolutionary proteomic flexibility landscape across two major virus realms appear to be solidifying. Indubitably, retracing of the deepest events in the evolutionary history of plant virus proteomic IDP in relation to the intricate network of encoded functional modules involved in virus genome processing and expression, and proteins involved in virus–host interactions, such as MPs or RNAi suppressors, is bound to remain speculative, principally due to the highly adaptive nature of viruses itself and the complexity of their intimate relationship with their diverse hosts. Discovery of evolutionary signatures abundance and their peculiarities associated with specific functions in diverged IDR-rich modules suggests that underlying IDP-driven evolutionary mechanisms are shaping the global plant virus proteome on a much wider measure than currently assessed and advocates that both major structural domains as well as IDRs evolve under two different functional regimes. This original and much-anticipated comprehensive study of the global plant virome IDP enhance current knowledge and represents a traceable primer to better-reflecting foundational evolutionary processes of plant virus functional diversification and host interaction that are more sensical in light of intrinsic protein disorder.

References

1. Dolja VV, Koonin EV. Metagenomics reshapes the concepts of RNA virus evolution by revealing extensive horizontal virus transfer. *Virus Res.* 2018;244:36–52.
2. Dolja VV, Krupovic M, Koonin EV. Deep roots and splendid boughs of the global plant Virome. *Annu Rev Phytopathol.* 2020;58:23–53.
3. Wolf YI, Kazlauskas D, Iranzo J, et al. Origins and evolution of the global RNA virome. *MBio.* 2018;9. e02329-18.
4. Baltimore D. Expression of animal virus genomes. *Bacteriol Rev.* 1971;35:235–241.
5. Domingo E, Holland JJ. RNA virus mutations and fitness for survival. *Annu Rev Microbiol.* 1997;51:151–178.

6. Varsani A, Lefeuvre P, Roumagnac P, Martin D. Notes on recombination and reassortment in multipartite/segmented viruses. *Curr Opin Virol.* 2018;33:156–166.

7. Zwart MP, Elena SF. Modeling multipartite virus evolution: the genome formula facilitates rapid adaptation to heterogeneous environments. *Virus Evol.* 2020;6(1): veaa022.

8. Sicard A, Pirolles E, Gallet R, et al. A multicellular way of life for a multipartite virus. *Elife.* 2019;8:e43599.

9. Folimonova SY, Tilsner J. Hitchhikers, highway tolls and roadworks: The interactions of plant viruses with the phloem. *Curr Opin Plant Biol.* 2018;43:82–88.

10. Koonin EV, Dolja VV, Krupovic M, et al. Global organization and proposed megataxonomy of the virus world. *Microbiol Mol Biol Rev.* 2020;84(2):e00061–19.

11. McLeish MJ, Fraile A, García-Arenal F. Evolution of plant-virus interactions: Host range and virus emergence. *Curr Opin Virol.* 2019;34:50–55.

12. Wu X, Cheng X. Intercellular movement of plant RNA viruses: Targeting replication complexes to the plasmodesma for both accuracy and efficiency. *Traffic.* 2020;21:725–736.

13. Kumar G, Dasgupta I. Variability, functions and interactions of plant virus movement proteins: What do we know so far? *Microorganisms.* 2021;9:695.

14. Navarro JA, Sanchez-Navarro JA, Pallas V. Key checkpoints in the movement of plant viruses through the host. *Adv Virus Res.* 2019;104:1–64.

15. Csorba T, Kontra L, Burgyán J. Viral silencing suppressors: Tools forged to fine-tune host-pathogen coexistence. *Virology.* 2015;479-480:85–103.

16. Heinlein M. Plant virus replication and movement. *Virology.* 2015;479:657–671.

17. Schoelz JE, Angel CA, Nelson RS, Leisner SM. A model for intracellular movement of cauliflower mosaic virus: The concept of the mobile virion factory. *J Exp Bot.* 2016;67:2039–2048.

18. Koonin EV, Dolja VV. Virus world as an evolutionary network of viruses and capsidless selfish elements. *Microbiol Mol Biol Rev.* 2014;78(2):278–303.

19. Jones RAC, Naidu RA. Global dimensions of plant virus diseases: Current status and future perspectives. *Annu Rev Virol.* 2019;6(1):387–409.

20. Whitfield AE, Falk BW, Rotenberg D. Insect vector-mediated transmission of plant viruses. *Virology.* 2015;479-480:278–289.

21. Gutierrez S, Michalakis Y, Van Munster M, Blanc S. Plant feeding by insect vectors can affect life cycle, population genetics and evolution of plant viruses. *Funct Ecol.* 2013;27:610–622.

22. Uzest M, Gargani D, Drucker M, et al. A protein key to plant virus transmission at the tip of the insect vector stylet. *Proceeding of the National Academy of Science of the United States of America.* 2007;104:17959–17964.

23. Ziegler-Graff V, Brault V. Role of vector-transmission proteins. *Methods Mol Biol.* 2008;451:81–96.

24. Deshoux M, Masson V, Arafah K, et al. Cuticular structure proteomics in the pea aphid Acyrthosiphon pisum reveals new plant virus receptor candidates at the tip of maxillary stylets. *J Proteome Res.* 2020;19(3):1319–1337.

25. Mulot M, Monsion B, Boissinot S, et al. Transmission of turnip yellows virus by Myzus persicae is reduced by feeding aphids on double-stranded RNA targeting the ephrin receptor protein. *Front Microbiol.* 2018;9:457.

26. Schneweis DJ, Whitfield AE, Rotenberg D. Thrips developmental stage-specific transcriptome response to tomato spotted wilt virus during the virus infection cycle in *Frankliniella occidentalis*, the primary vector. *Virology.* 2017;500:226–237.

27. Wang H, Wu K, Liu Y, Wu Y, Wang X. Integrative proteomics to understand the transmission mechanism of barley yellow dwarf virus-GPV by its insect vector Rhopalosiphum padi. *Sci Rep.* 2015;5:10971. https://doi.org/10.1038/srep10971.

28. Webster CG, Pichon E, van Munster M, et al. Identification of plant virus receptor candidates in the stylets of their aphid vectors. *J Virol*. 2018;92. e00432-18.

29. Webster CG, Thillier M, Pirolles E, Cayrol B, Blanc S, Uzest M. Proteomic composition of the acrostyle: Novel approaches to identify cuticular proteins involved in virus-insect interactions. *Insect Sci*. 2017;24:990–1002.

30. Ammar e-D, Tsai CW, Whitfield AE, Redinbaugh MG, Hogenhout SA. Cellular and molecular aspects of rhabdovirus interactions with insect and plant hosts. *Annu Rev Entomol*. 2009;54:447–468.

31. Ng JC, Zhou JS. Insect vector-plant virus interactions associated with non-circulative, semi-persistent transmission: Current perspectives and future challenges. *Curr Opin Virol*. 2015;15:48–55.

32. Hogenhout SA, Ammar el D, Whitfield AE, Redinbaugh MG. Insect vector interactions with persistently transmitted viruses. *Annu Rev Phytopathol*. 2008;46:327–359.

33. Blanc S, Uzest M, Drucker M. New research horizons in vector-transmission of plant viruses. *Curr Opin Microbiol*. 2011;14(4):483–491. https://doi.org/10.1016/j.mib.2011.07.008.

34. Cilia M, Tamborindeguy C, Fish T, Howe K, Thannhauser TW, Gray S. Genetics coupled to quantitative intact proteomics links heritable aphid and endosymbiont protein expression to circulative polerovirus transmission. *J Virol*. 2011;85:2148–2166.

35. Dombrovsky A, Sobolev I, Chejanovsky N, Raccah B. Characterization of RR-1 and RR-2 cuticular proteins from Myzus persicae. *Comp Biochem Physiol B Biochem Mol Biol*. 2007;146:256–264.

36. Tamborindeguy C, Bereman MS, DeBlasio S, et al. Genomic and proteomic analysis of Schizaphis graminum reveals Cyclophilin proteins are involved in the transmission of cereal yellow dwarf virus. *PLoS ONE*. 2013;8(8):e71620.

37. Liu W, Gray S, Huo Y, Li L, Wei T, Wang X. Proteomic analysis of interaction between a plant virus and its vector insect reveals new functions of hemipteran cuticular protein. *Mol Cell Proteomics*. 2015;14:2229–2242.

38. Liang Y, Gao XW. The cuticle protein gene MPCP4 of *Myzus persicae* (Homoptera: Aphididae) plays a critical role in cucumber mosaic virus acquisition. *J Econ Entomol*. 2017;110:848–853.

39. Sacristan S, Malpica JM, Fraile A, Garcia-Arenal F. Estimation of population bottlenecks during systemic movement of tobacco mosaic virus in tobacco plants. *J Virol*. 2003;77:9906–9911.

40. Chare ER, Holmes EC. Selection pressures in the capsid genes of plant RNA viruses reflect mode of transmission. *J Gen Virol*. 2004;85:3149–3157.

41. Kim KM, Qin T, Jiang YY, et al. Protein domain structure uncovers the origin of aerobic metabolism and the rise of planetary oxygen. *Structure*. 2012;20(1):67–76.

42. Tokuriki N, Oldfield CJ, Uversky VN, Berezovsky IN, Tawfik DS. Do viral proteins possess unique biophysical features? *Trends Biochem Sci*. 2009;34:53–59.

43. Tompa P, Szász C, Buday L. Structural disorder throws new light on moonlighting. *Trends Biochem Sci*. 2005;30(9):484–489.

44. Dyson HJ, Wright PE. Intrinsically unstructured proteins and their functions. *Nat Rev Mol Cell Biol*. 2005;6:197–208.

45. Dunker AK, Babu MM, Barbar E, et al. What's in a name? Why these proteins are intrinsically disordered: Why these proteins are intrinsically disordered. *Intrinsically Disord Proteins*. 2013;1(1):e24157.

46. Tompa P, Fuxreiter M, Oldfield CJ, et al. Close encounters of the third kind: Disordered domains and the interactions of proteins. *Bioessays*. 2009;31:328–335.

47. Tompa P, Schad E, Tantos A, Kalmar L. Intrinsically disordered proteins: emerging interaction specialists. *Curr Opin Struct Biol*. 2015;35:49–59.

48. Goh GK, Dunker AK, Uversky VN. A comparative analysis of viral matrix proteins using disorder predictors. *Virol J.* 2008;5:126.
49. Oldfield CJ, Dunker AK. Intrinsically disordered proteins and intrinsically disordered protein regions. *Annu Rev Biochem.* 2014;83:553–584.
50. Xue B, Williams RW, Oldfield CJ, Goh GK, Dunker AK, Uversky VN. Viral disorder or disordered viruses: do viral proteins possess unique features? *Protein Pept Lett.* 2010;17(8):932–951.
51. Xue B, Dunbrack RL, Williams RW, Dunker AK, Uversky VN. PONDR-FIT: A meta-predictor of intrinsically disordered amino acids. *Biochim Biophys Acta.* 2010; 1804:996–1010.
52. Vogel C. Quantifying protein (dis)order. *Science.* 2017;355(6327):794–795.
53. He B, Wang K, Liu Y, Xue B, Uversky VN, Dunker AK. Predicting intrinsic disorder in proteins: an overview. *Cell Res.* 2009;19(8):929–949.
54. Uversky VN. New technologies to analyse protein function: an intrinsic disorder perspective. F1000Res. 2020; 9: F1000 Faculty Rev-101.
55. Uversky VN, Oldfield CJ, Dunker AK. Showing your ID: Intrinsic disorder as an ID for recognition, regulation and cell signaling. *J Mol Recognit.* 2005;18:343–384.
56. Xue B, Dunker AK, Uversky VN. Orderly order in protein intrinsic disorder distribution: Disorder in 3500 proteomes from viruses and the three domains of life. *J Biomol Struct Dyn.* 2012;30:137–149.
57. Dunker AK, Cortese MS, Romero P, Iakoucheva LM, Uversky VN. Flexible nets: The roles of intrinsic disorder in protein interaction networks. *FEBS J.* 2005;272: 5129–5148.
58. Kim PM, Sboner A, Xia Y, Gerstein M. The role of disorder in interaction networks: A structural analysis. *Mol Syst Biol.* 2008;4:179.
59. Uversky VN, Dave V, Iakoucheva LM, et al. Pathological unfoldomics of uncontrolled chaos: Intrinsically disordered proteins and human diseases. *Chem Rev.* 2014; 114:6844–6879.
60. Dunker AK, Silman I, Uversky VN, Sussman JL. Function and structure of inherently disordered proteins. *Curr Opin Struct Biol.* 2008;18:756–764.
61. Elrashdy F, Redwan EM, Uversky VN. Intrinsic disorder perspective of an interplay between the renin-angiotensin-aldosterone system and SARS-CoV-2. *Infect Genet Evol.* 2020;85:104510.
62. Uversky VN, Dunker AK. Understanding protein non-folding. *Biochim Biophys Acta.* 2010;1804:1231–1264.
63. Wright PE, Dyson HJ. Linking folding and binding. *Curr Opin Struct Biol.* 2009; 19:31238.
64. Xie H, Vucetic S, Iakoucheva LM, et al. Functional anthology of intrinsic disorder. 1. Biological processes and functions of proteins with long disordered regions. *J Proteome Res.* 2007;6:1882–1898.
65. Pancsa R, Zsolyomi F, Tompa P. Co-evolution of intrinsically disordered proteins with folded partners witnessed by evolutionary couplings. *Int J Mol Sci.* 2018;19 (11):3315.
66. Tokuriki N, Stricher F, Serrano L, Tawfik DS. How protein stability and new functions trade off. *PLoS Comput Biol.* 2008;4(2):e1000002.
67. Iserte JA, Lazar T, Tosatto S, Tompa P, Marino-Buslje C. Chasing coevolutionary signals in intrinsically disordered proteins complexes. *Sci Rep.* 2020;10(1):17962.
68. Elena SF, Agudelo-Romero P, Lalić J. The evolution of viruses in multi-host fitness landscapes. *The open virology journal.* 2009;3:1–6.
69. Murphy PM. Molecular mimicry and the generation of host defense protein diversity. *Cell.* 1993;72:8232826.

70. Pushker R, Mooney C, Davey NE, Jacque J-M, Shields DC. Marked variability in the extent of protein disorder within and between viral families. *PLoS ONE*. 2013;8(4): e60724.

71. Davey NE, Travé G, Gibson TJ. How viruses hijack cell regulation. *Trends Biochem Sci*. 2011;36(3):159–169.

72. Davey NE, Van Roey K, Weatheritt RJ, et al. Attributes of short linear motifs. *Mol BioSyst*. 2012;8(1):268–281.

73. Grzela R, Szolajska E, Madern D, et al. Virulence factor of potato virus Y, genome attached terminal protein VPg is a highly disordered protein. *J Biol Chem*. 2007;283:213–221.

74. Satheshkumar PS, Gayathri P, Prasad K, Savithri HS. Natively unfolded VPg is essential for sesbania mosaic virus serine protease activity. *J Biol Chem*. 2005;280:30291–30300.

75. Charon J, Theil S, Nicaise V, Michon T. Protein intrinsic disorder within the Potyvirus genus: From proteome-wide analysis to functional annotation. *Mol Biosyst*. 2016;12 (2):634–652. https://doi.org/10.1039/c5mb00677e.

76. Hébrard E, Pinel-Galzi A, Fargette D. Virulence domain of the RYMV genome-linked viral protein VPg towards rice rymv1-2-mediated resistance. *Arch Virol*. 2008;153:1161–1164.

77. Hebrard E, Bessin Y, Michon T, et al. Intrinsic disorder in viral proteins genome-linked: Experimental and predictive analyses. *Virol J*. 2009;6:23.

78. Rantalainen KI, Uversky VN, Permi P, Kalkkinen N, Dunker AK, Mäkinen K. Potato virus A genome-linked protein VPg is an intrinsically disordered molten globule-like protein with a hydrophobic core. *Virology*. 2008;377(2):280–288.

79. Deryusheva E, Nemashkalova E, Galloux M, et al. Does intrinsic disorder in proteins favor their interaction with lipids? *Proteomics*. 2019;6:e1800098.

80. Pauwels K, Lebrun P, Tompa P. To be disordered or not to be disordered: is that still a question for proteins in the cell? *CMLS*. 2017;74(17):3185–3204.

81. Nielsen JT, Mulder F. Quality and bias of protein disorder predictors. *Sci Rep*. 2019;9 (1):5137.

82. Dosztanyi Z, Csizmok V, Tompa P, Simon I. IUPred: Web server for the prediction of intrinsically unstructured regions of proteins based on estimated energy content. *Bioinformatics*. 2005;21:3433–3434.

83. Oates ME, Romero P, Ishida T, et al. D2P2: Database of disordered protein predictions. *Nucleic Acids Res*. 2013;41(D1):D508–D516.

84. Dosztanyi Z, Tompa P. Prediction of protein disorder. *Methods Mol Biol*. 2008;426:103–115.

85. Meszaros B, Erdos G, Dosztanyi Z. IUPred2A: Context-dependent prediction of protein disorder as a function of redox state and protein binding. *Nucleic Acids Res*. 2018;46: W329–W337.

86. Plotkin JB, Kudla G. Synonymous but not the same: The causes and consequences of codon bias. *Nat Rev Genet*. 2011;12:32–42.

87. Novoa EM, Jungreis I, Jaillon O, Kellis M. Elucidation of codon usage signatures across the domains of life. *Mol Biol Evol*. 2019;36(10):2328–2339.

88. Hershberg R, Petrov DA. General rules for optimal codon choice. *PLoS Genet*. 2009;5 (7):e1000556.

89. Sharp PM, Tuohy TM, Mosurski KR. Codon usage in yeast: Cluster analysis clearly differentiates highly and lowly expressed genes. *Nucleic Acids Res*. 1986;14 (13):5125–5143.

90. Chen SL, Lee W, Hottes AK, Shapiro L, McAdams HH. Codon usage between genomes is constrained by genome-wide mutational processes. *Proc Natl Acad Sci U S A*. 2004;101(10):3480–3485. https://doi.org/10.1073/pnas.0307827100 [Epub 2004 Feb 27. PMID: 14990797; PMCID: PMC373487].

91. Sharp PM, Emery LR, Zeng K. Forces that influence the evolution of codon bias. *Philos Trans R Soc Lond B Biol Sci*. 2010;365(1544):1203–1212.

92. Long H, Sung W, Kucukyildirim S, et al. Evolutionary determinants of genome-wide nucleotide composition. *Nat Ecol Evol*. 2018;2(2):237–240.

93. Knight RD, Freeland SJ, Landweber LF. A simple model based on mutation and selection explains trends in codon and amino-acid usage and GC composition within and across genomes. *Genome Biol*. 2001;2(4) [RESEARCH0010].

94. Palidwor GA, Perkins TJ, Xia X. A general model of codon bias due to GC mutational bias. *PLoS One*. 2010;5(10):e13431.

95. Yu CH, Dang Y, Zhou Z, et al. Codon usage influences the local rate of translation elongation to regulate co-translational protein folding. *Mol Cell*. 2015;59:744–754.

96. Pechmann S, Frydman J. Evolutionary conservation of codon optimality reveals hidden signatures of cotranslational folding. *Nat Struct Mol Biol*. 2013;20:237–243.

97. Zhou T, Weems M, Wilke CO. Translationally optimal codons associate with structurally sensitive sites in proteins. *Mol Biol Evol*. 2009;26:1571–1580.

98. Pechmann S, Chartron JW, Frydman J. Local slowdown of translation by nonoptimal codons promotes nascent-chain recognition by SRP in vivo. *Nat Struct Mol Biol*. 2014;21(12):1100–1105.

99. Bull JJ, Molineux IJ, Wilke CO. Slow fitness recovery in a codon-modified viral genome. *Mol Biol Evol*. 2012;29(10):2997–3004. https://doi.org/10.1093/molbev/mss119.

100. Greenbaum BD, Levine AJ, Bhanot G, Rabadan R. Patterns of evolution and host gene mimicry in influenza and other RNA viruses. *PLoS Pathog*. 2008;4(6):e1000079.

101. Greenbaum BD, Rabadan R, Levine AJ. Patterns of oligonucleotide sequences in viral and host cell RNA identify mediators of the host innate immune system. *PLoS One*. 2009;4:e5969.

102. Kapoor A, Simmonds P, Lipkin WI, Zaidi S, Delwart E. Use of nucleotide composition analysis to infer hosts for three novel picorna-like viruses. *J Virol*. 2010;84(19):10322–10328.

103. Lobo FP, Mota BE, Pena SD, et al. Virus-host coevolution: Common patterns of nucleotide motif usage in *Flaviviridae* and their hosts. *PLoS One*. 2009;4(7):e6282.

104. Uversky VN, Gillespie JR, Fink AL. Why are natively unfolded proteins unstructured under physiologic conditions? *Proteins*. 2000;41:415–427.

105. Tompa P. Intrinsically unstructured proteins. *Trends Biochem Sci*. 2002;27(10):527–533. https://doi.org/10.1016/s0968-0004(02)02169-2. PMID: 12368089.

106. Zhou M, Wang T, Fu J, Xiao G, Liu Y. Nonoptimal codon usage influences protein structure in intrinsically disordered regions. *Mol Microbiol*. 2015;97(5):974–987.

107. Mihara T, Nishimura Y, Shimizu Y, et al. Linking virus genomes with host taxonomy. *Viruses*. 2016;8(3):66.

108. Uversky VN. Intrinsically disordered proteins and their "mysterious" (meta)physics. *Front Phys*. 2019;7:10.

109. Vacic V, Oldfield CJ, Mohan A, et al. Characterization of molecular recognition features, MoRFs, and their binding partners. *J Proteome Res*. 2007;6:2351–2366.

110. Ng JC, Falk B. Virus-vector interactions mediating non-persistent and semipersistent transmission of plant viruses. *Ann Rev Phytopathology*. 2006;44:183–212.

111. Oldfield CJ, Cheng Y, Cortese MS, Romero P, Uversky VN, Dunker AK. Coupled folding and binding with alpha-helix-forming molecular recognition elements. *Biochemistry*. 2005;44:12454–12470.

112. Radivojac P, Iakoucheva LM, Oldfield CJ, et al. Intrinsic disorder and functional proteomics. *Biophys J*. 2007;92:1439–1456.

113. Yan J, Dunker AK, Uversky V, Kurgan L. Molecular recognition features (MoRFs) in three domains of life. *Mol Biosyst*. 2016;12:697–710.

114. Yang J, Gao M, Xiong J, Su Z, Huang Y. Features of molecular recognition of intrinsically disordered proteins via coupled folding and binding. *Protein Sci.* 2019 Nov; 28(11):1952–1965.

115. Cumberworth A, Lamour G, Babu MM, Gsponer J. Promiscuity as a functional trait: Intrinsically disordered regions as central players of interactomes. *Biochem J.* 2013;454:361–369.

116. Haynes C, Oldfield CJ, Ji F, et al. Intrinsic disorder is a common feature of hub proteins from four eukaryotic interactomes. *PLoS Comput Biol.* 2006;2:890–901.

117. Patil A, Kinoshita K, Nakamura H. Domain distribution and intrinsic disorder in hubs in the human protein-protein interaction network. *Protein Sci.* 2010;19:1461–1468.

118. Eliezer D. Biophysical characterization of intrinsically disordered proteins. *Curr Opin Struct Biol.* 2009;19(1):23–30.

119. Hsu WL, Oldfield CJ, Xue B, et al. Exploring the binding diversity of intrinsically disordered proteins involved in one-to-many binding. *Protein Sci.* 2013;22(3):258–273.

120. Hsu CC, Buehler MJ, Tarakanova A. The order-disorder continuum: linking predictions of protein structure and disorder through molecular simulation. *Sci Rep.* 2020;10(1):2068.

121. Malhis N, Jacobson M, Gsponer J. MoRFchibi SYSTEM: Software tools for the identification of MoRFs in protein sequences. *Nucleic Acids Res.* 2016 Jul 8;44(W1): W488–W493.

122. Mohan A, et al. Analysis of molecular recognition features (MoRFs). *J Mol Biol.* 2006;362:1043–1059.

123. Cheng Y, Oldfield CJ, Meng J, Romero P, Uversky VN, Dunker AK. Mining alpha-helix-forming molecular recognition features with cross species sequence alignments. *Biochemistry.* 2007;46(47):13468–13477. https://doi.org/10.1021/bi7012273.

124. Oldfield CJ, Cheng Y, Cortese MS, Brown CJ, Uversky VN, Dunker AK. Comparing and combining predictors of mostly disordered proteins. *Biochemistry.* 2005; 44:1989–2000.

125. Fang C, Noguchi T, Tominaga D, Yamana H. MFSPSSMpred: Identifying short disorder-to-order binding regions in disordered proteins based on contextual local evolutionary conservation. *BMC Bioinform.* 2013;14:300.

126. Fuxreiter M, Peter T, Istvan S. Local structural disorder imparts plasticity on linear motifs. *Bioinformatics.* 2007;23(8):950–956.

127. Ehrenberger T, Cantley LC, Yaffe MB. Computational prediction of protein-protein interactions. *Methods Mol Biol.* 2015;1278:57–75.

128. Meng F, Uversky VN, Kurgan L. Comprehensive review of methods for prediction of intrinsic disorder and its molecular functions. *Cell Mol Life Sci.* 2017;74:3069–3090.

129. Meng F, Kurgan L. High-throughput prediction of disordered moonlighting regions in protein sequences. *Proteins.* 2018;86:1097–1110.

130. Mizianty MJ, Uversky V, Kurgan L. Prediction of intrinsic disorder in proteins using MFDp2. *Methods Mol Biol.* 2014;1137:147–162.

131. Oldfield CJ, Uversky VN, Kurgan L. Predicting functions of disordered proteins with MoRFpred. *Methods Mol Biol.* 2019;1851.

132. Sharma R, Raicar G, Tsunoda T, Patil A, Sharma A. OPAL: Prediction of MoRF regions in intrinsically disordered protein sequences. *Bioinformatics.* 2018;34: 1850–1858.

133. Sharma R, Sharma A, Raicar G, Tsunoda T, Patil AOPAL. +: Length-specific MoRF prediction in intrinsically disordered protein sequences. *Proteomics.* 2018;19:e1800058.

134. Sharma R, Bayarjargal M, Tsunoda T, Patil A, Sharma A. MoRFPred-plus: Computational identification of MoRFs in protein sequences using physicochemical properties and HMM profiles. *J Theor Biol.* 2018;437:9–16.

135. Cozzetto D, Jones DT. The contribution of intrinsic disorder prediction to the elucidation of protein function. *Curr Opin Struct Biol.* 2013;23:467–472.
136. Katuwawala A, Peng Z, Yang J, Kurgan L. Computational prediction of MoRFs, short disorder-to-order transitioning protein binding regions. *Comput Struct Biotechnol J.* 2019;17:454–462.
137. Peng Z, Yan J, Fan X, et al. Exceptionally abundant exceptions: Comprehensive characterization of intrinsic disorder in all domains of life. *Cell Mol Life Sci.* 2015;72(1):137–151.
138. Alshehri MA, Manee MM, Alqahtani FH, Al-Shomrani BM, Uversky VN. On the prevalence and potential functionality of an intrinsic disorder in the MERS-CoV proteome. *Viruses.* 2021;13(2):339. Published 2021 Feb 22 https://doi.org/10.3390/v13020339.
139. Dolan PT, Roth AP, Xue B, et al. Intrinsic disorder mediates hepatitis C virus core-host cell protein interactions. *Protein Sci.* 2015;24(2):221–235.
140. Mishra PM, Uversky VN, Giri R. Molecular recognition features in Zika virus proteome. *J Mol Biol.* 2018;430:2372–2388.
141. Singh A, Kumar A, Uversky VN, Giri R. Understanding the interactability of chikungunya virus proteins via molecular recognition feature analysis. *RSC Adv.* 2018;8:27293–27303.
142. Disfani FM, Hsu WL, Mizianty MJ, et al. MoRFpred, a computational tool for sequence-based prediction and characterization of short disorder-to-order transitioning binding regions in proteins. *Bioinformatics.* 2012;28(12):i75–i83.
143. Sugase K, Dyson HJ, Wright PE. Mechanism of coupled folding and binding of an intrinsically disordered protein. *Nature.* 2007;447(7147):1021–1025.
144. Tompa P. Unstructural biology coming of age. *Curr Opin Struct Biol.* 2011;21(3):419–425.
145. Revers F, García JA. Molecular biology of potyviruses. *Adv Virus Res.* 2015;92:101–199.
146. Moury B, Desbiez C. Host range evolution of Potyviruses: A global phylogenetic analysis. *Viruses.* 2020;12(1):111.
147. Chung BY, Miller WA, Atkins JF, Firth AE. An overlapping essential gene in the Potyviridae. *Proc Natl Acad Sci U S A.* 2008;105(15):5897–5902. https://doi.org/10.1073/pnas.0800468105.
148. Elena SF, Rodrigo G. Towards an integrated molecular model of plant-virus interactions. *Curr Opin Virol.* 2012;2(6):719–724.
149. Dunker AK, Brown CJ, Lawson JD, Iakoucheva LM, Obradović Z. Intrinsic disorder and protein function. *Biochemistry.* 2002;41(21):6573–6582.
150. Tompa P. Intrinsically disordered proteins: a 10-year recap. *Trends Biochem Sci.* 2012;37(12):509–516.
151. Jensen MR, Communie G, Ribeiro Jr EA, et al. Intrinsic disorder in measles virus nucleocapsids. *Proc Natl Acad Sci U S A.* 2011;108(24):9839–9844.
152. Tompa P, Fuxreiter M. Fuzzy complexes: Polymorphism and structural disorder in protein-protein interactions. *Trends Biochem Sci.* 2008;33:2–8.
153. Pushker R, Mooney C, Davey NE, Jacque J-M, Shields DC. Marked variability in the extent of protein disorder within and between viral families. *PLoS ONE.* 2013;8(4):e60724.
154. Dosztanyi Z, Chen J, Dunker AK, Simon I, Tompa P. Disorder and sequence repeats in hub proteins and their implications for network evolution. *J Proteome Res.* 2006;5:2985–2995.
155. van der Lee R, Buljan M, Lang B, et al. Classification of intrinsically disordered regions and proteins. *Chem Rev.* 2014;114:6589–6631.

156. Schlessinger A, Schaefer C, Vicedo E, Schmidberger M, Punta M, Rost B. Protein disorder–a breakthrough invention of evolution? *Curr Opin Struct Biol.* 2011;21: 412–418.
157. Moesa HA, Wakabayashi S, Nakai K, Patil A. Chemical composition is maintained in poorly conserved intrinsically disordered regions and suggests a means for their classification. *Mol Biosyst.* 2012;8:3262.
158. Zarin T, Strome B, Nguyen Ba AN, Alberti S, Forman-Kay JD, Moses AM. Proteome-wide signatures of function in highly diverged intrinsically disordered regions. *Elife.* 2019;8:e46883.
159. Ward JJ, Sodhi JS, McGuffin LJ, Buxton BF, Jones DT. Prediction and functional analysis of native disorder in proteins from the three kingdoms of life. *J Mol Biol.* 2004;337:6352645.
160. Gould SJ, Vrba ES. Exaptation — a missing term in the science of form. *Paleobiology.* 1982;8:4–15.
161. Koonin, Eugene V, Mart Krupovic. The depths of virus exaptation. *Curr Opin Virol.* 2018;31:1–8.
162. Kumar, N., Kaushik, R., Tennakoon, C., Uversky, V. N., Longhi, S., Zhang, K., & Bhatia, S. Comprehensive intrinsic disorder analysis of 6108 viral proteomes: from the extent of intrinsic disorder penetrance to functional annotation of disordered viral proteins. J. Proteome Res. 2021a; 20(5), 2704–2713.
163. Kumar, N., Kaushik, R., Tennakoon, C., Uversky, V. N., Longhi, S., Zhang, K., & Bhatia, S. Insights into the evolutionary forces that shape the codon usage in the viral genome segments encoding intrinsically disordered protein regions. Brief. Bioinform. 2021b; bbab145.
164. Hilser VJ, Thompson EB. Intrinsic disorder as a mechanism to optimize allosteric coupling in proteins. *Proc Natl Acad Sci U S A.* 2007;104(20):8311–8315.
165. Tsvetkov P, et al. Operational definition of intrinsically unstructured protein sequences based on susceptibility to the 20S proteasome. *Proteins.* 2008;70:1357–1366.
166. Bedford T, Hartl DL. Optimization of gene expression by natural selection. *PNAS.* 2009;106:1133–1138.
167. Koch V, Otte M, Beye M. Evidence for stabilizing selection driving mutational turn-over of short motifs in the eukaryotic complementary sex determiner (Csd) protein. *G3: Genes|Genomes|Genetics.* 2018;2018:3803–3812.
168. Meyer K, Kirchner M, Uyar B, et al. Mutations in disordered regions can cause disease by creating dileucine motifs. *Cell.* 2018;175:239–253.

Further reading

169. Chothia C, Lesk AM. The relation between the divergence of sequence and structure in proteins. *EMBO J.* 1986;5(4):823–826.
170. Coleman JR, Papamichail D, Skiena S, Futcher B, Wimmer E, Mueller S. Virus attenuation by genome-scale changes in codon pair bias. *Science.* 2008;320(5884): 1784–1787. https://doi.org/10.1126/science.1155761.
171. Gunasekaran K, Tsai CJ, Kumar S, Zanuy D, Nussinov R. Extended disordered proteins: Targeting function with less scaffold. *Trends Biochem Sci.* 2003;28:81285.
172. Jones DT, Cozzetto D. DISOPRED3: Precise disordered region predictions with annotated protein-binding activity. *Bioinformatics.* 2015;31:857–863.
173. Leonard S, Viel C, Beauchemin C, Daigneault N, Fortin MG, Laliberte JF. Interaction of VPg-pro of turnip mosaic virus with the translation initiation factor 4E and the poly(a)-binding protein in planta. *J Gen Virol.* 2004;85:1055–1063.

174. O'Leary NA, Wright MW, Brister JR, et al. Reference sequence (RefSeq) database at NCBI: Current status, taxonomic expansion, and functional annotation. *Nucleic Acids Res*. 2016;44(D1):D733–D745.
175. Peng K, Radivojac P, Vucetic S, Dunker AK, Obradovic Z. Length-dependent prediction of protein intrinsic disorder. *BMC Bioinform*. 2006;7:208.
176. Uversky VN. Multitude of binding modes attainable by intrinsically disordered proteins: A portrait gallery of disorder-based complexes. *Chem Soc Rev*. 2011;40:1623–1634.
177. Uversky VN. Intrinsic disorder-based protein interactions and their modulators. *Curr Pharm Des*. 2013;19:4191–4213.
178. Uversky VN. Unusual biophysics of intrinsically disordered proteins. *Biochim Biophys Acta*. 2013;1834:932–951.
179. Uversky VN. The multifaceted roles of intrinsic disorder in protein complexes. *FEBS Lett*. 2015;589(19):2498–2506.

Index

Printed in the United States
by Baker & Taylor Publisher Services